sports health

The Complete Book of Athletic Injuries

sports health

The Complete Book of Athletic Injuries

William Southmayd, M.D. and Marshall Hoffman

Illustrations by Sarah Black
Design by Steven Black

New York • London

ISBN: 0-8256-3205-6

Library of Congress Catalog Card Number: 81-50613

In Canada: Gage Trade Publishing, P.O. Box 5000
164 Commander Blvd., Agincourt, Ontario M1S 3C7.

Printed in the United States of America.

Cover design by Tim Metevier.

To athletes and fitness enthusiasts;
to our wives, Sally Southmayd and Birgitta U. Hoffman;
and to our mentors, Arthur M. Pappas, M.D.,
Thomas B. Quigley, M.D., Ellis M. Haller and Lucien D. Agniel.

Table of Contents

	Acknowledgments	8
1.	Sports Mania	12
2.	Sports Myths	30
3.	Common Injuries	40
4.	Muscles, Joints, and Ligaments	70
5.	Fractures	86
6.	The Shoulder	100
7.	The Elbow	128
8.	The Wrist	154
9.	The Hand	168
10.	The Cervical Spine	186
11.	The Lumbar Spine	202
12.	The Hip and Pelvis	222
13.	The Thigh	234
14.	The Knee	244
15.	The Lower Leg	300
16.	The Ankle	310
17.	The Foot	336
18.	Other Injuries	382
19.	Nutrition	392
20.	Children and Sports	424
	Biographical Sketches	449
	Index	451

Acknowledgments

We are indebted to Arthur M. Pappas, the medical director of the Boston Red Sox, professor and dean of orthopedics at the University of Massachusetts Medical School, and nationally known authority on pediatric sports medicine and young adult orthopedics. He researched and wrote the chapters on children, muscles, and joints and he contributed many of his medical experiences to the book. He also reviewed the book and offered medical comments.

We are indebted to our colleagues at Sports Medicine Resource, Inc., who wrote the following chapters and sections:

Roy Roy McGregor, D.P.M., chief of podiatry, who wrote the sections on running shoes and orthotics.

Richard A. St. Onge, M.D., an orthopedic and hand surgeon, who wrote the chapter on the hand.

Nancy Clark, M.S., R.D. and author of the *Athlete's Kitchen: A Nutrition Guide and Cookbook,* who wrote the chapter on sports nutrition.

Debbi Hall, R.P.T., and Cynthia Rowe, R.P.T., who wrote the sections on physical therapy equipment.

Peter Stone, R.P.T., and Don LaBourr, R.P.T. and A.T.C., helped with the rehabilitation exercises.

We are also grateful to the following physicians and scientists for their advice and help with various parts of this book:

Chris Coughlin, exercise physiologist, The New England Heart Center; Earl F. Hoerner, M.D., M.P.H., clinical professor of rehabilitation medicine, Tufts University and coauthor of *Sports Injuries, The Unthwarted Epidemic;* Lewis Millender, M.D., a Boston-based surgeon and consultant to the Boston Red Sox; Alan J. Ryan, M.D., the editor of *The Physician and Sportsmedicine;* William D. Shea, M.D., Sports Medicine Resource; Louis Sternburg, a Boston-based sports psychologist; Bert Zarins, M.D., team physician of the Boston Bruins and codirector of the Sports Medicine Clinic at the Massachusetts General Hospital.

The following trainers were very helpful with advice, observations, and stories from their own experiences: Joe Bourdon, The Boston Teamen; Jack Fadden, Harvard University and Boston Red Sox (retired); Jim Kausek, Boston Bruins; John Lally, Washington Bullets; Ray Melchiorre, Boston Celtics; Charlie Moss, Boston Red Sox; Ralph Salvo, Baltimore Orioles; and Jim Warfield, Cleveland Indians.

We would like to acknowledge these athletes for their interviews: Carlton Fisk, Jim Rice, Jerry Remy, Butch Hobson, Carl Yastrzemski, Rudy May, Luis Tiant, Tommy John, Jim Palmer, and Ken Singleton.

We are grateful to athletes, coaches, and trainers of the following teams for interviews: Baltimore Orioles, Boston Bruins, Boston Celtics, Boston Red Sox, New York Yankees, and the Washington Bullets.

We are also grateful to Lucien D. Agniel, author and instructor of magazine journalism, who did the primary editing of the manuscript; and to editor-in-chief Jim Charlton of Quick Fox, whose support and help throughout is appreciated. A special thanks is due Ann Burch for coordinating and scheduling every stage of the manuscript, and to Vivi Mannuzza, whose production skills and editorial know-how made an especially difficult job a pleasure, and to Renée DuPerlé O'Brien for the patient and painstaking care that provided the finishing touches to this book.

Our beautiful illustrations and layouts were done by Sarah Black of New York City and the design was done by Steve Black, with the able assistance of Ed Bell on design and layout.

We would like to give special thanks to Martine Paim, my private secretary. Besides keeping my patients happy, which is a full-time job, she typed this manuscript, almost a second full-time job. Throughout this ordeal, she never complained. It is a tribute to her personality and good nature.

William Southmayd, M.D.
Marshall Hoffman

Sports Medicine Resource, Inc.
Chestnut Hill, Massachusetts 02167

sports health

The Complete Book of Athletic Injuries

Sports Mania

The drive to be fit has become a national mania. A recent Gallup Poll reveals that almost 50 percent of the adult population, or 57 million Americans, exercise daily.

According to a 1979 Nielsen survey, sports activities—swimming, bicycling, football, and jogging—draw about 571 million participants. Since there were only 214 million Americans in 1979, it's obvious that a lot of them take part in more than one sport.

Our streets and parks are filled with the almost 36 million people who jog. "It was impossible to run in the Falmouth (Massachusetts) Road Race because of the glut of runners," reports my coauthor, Marshall Hoffman. "For five miles, I was just boxed in, running with a moving mass."

Thirty-two million Americans play tennis, an increase of 60 percent in the last six years. The boom in tennis has set off a mini-construction blitz in home tennis courts and indoor tennis facilities. I am one of the 10 million racquetball enthusiasts. I golf and swim in the summer. Twenty-six million people swim at least three times each week. Richard St. Onge, a teammate on the Harvard football team in the early sixties and now a colleague in the practice of sports medicine, has

become a world-class mountain climber. About one-half million participate in that sport.

Rob Roy McGregor, the chief of podiatric medicine at Sports Medicine Resource, is a marathon runner. "Usually, I make at least one diagnosis while running a race," admits the nationally known foot doctor.

Fitness has become a national status symbol. Former President Jimmy Carter and Vice President George Bush are runners. During the 1980 presidential campaign, both liked to be photographed while exercising. Former President Gerald Ford fondly recalls his days as a University of Michigan football lineman. He swims and golfs to stay fit now. President Ronald Reagan, a former sports announcer, rides horses.

A popular television show is "The Battle of the Network Stars." In it, television stars, mostly from "sitcoms," compete in events like tug-o-war, relay cycling, and an obstacle course.

Four hundred and fifty marathons—26.2-mile-runs—were held in the U.S. in 1980. About 15,000 runners ran the New York Marathon. More than twice that number applied to enter the race.

Americans are eager to learn how their bodies work during competition and exercise. Because I am medical director of Sports Medicine Resource, Inc. (the largest sports medicine clinic in New England), a consulting physician for the Boston Red Sox, and a former competitive athlete, I am often asked the following questions:

What are the safest sports?
How can I prevent injuries?

THE MOST POPULAR SPORTS IN AMERICA (1979)

A.C. Nielsen interviewers asked heads of household: "Do you or members of your family participate in the following sports?" If the answer was yes, interviewers asked detailed questions on participation. The numbers in this chart represent at least five days a year of participation in any sport. For some sports, the criteria are much higher. For example, bicycling is nine days per month; football is four days a month; tennis is three days a month; ice skating is five days a year.

SOURCE: A.C. Nielsen Company.

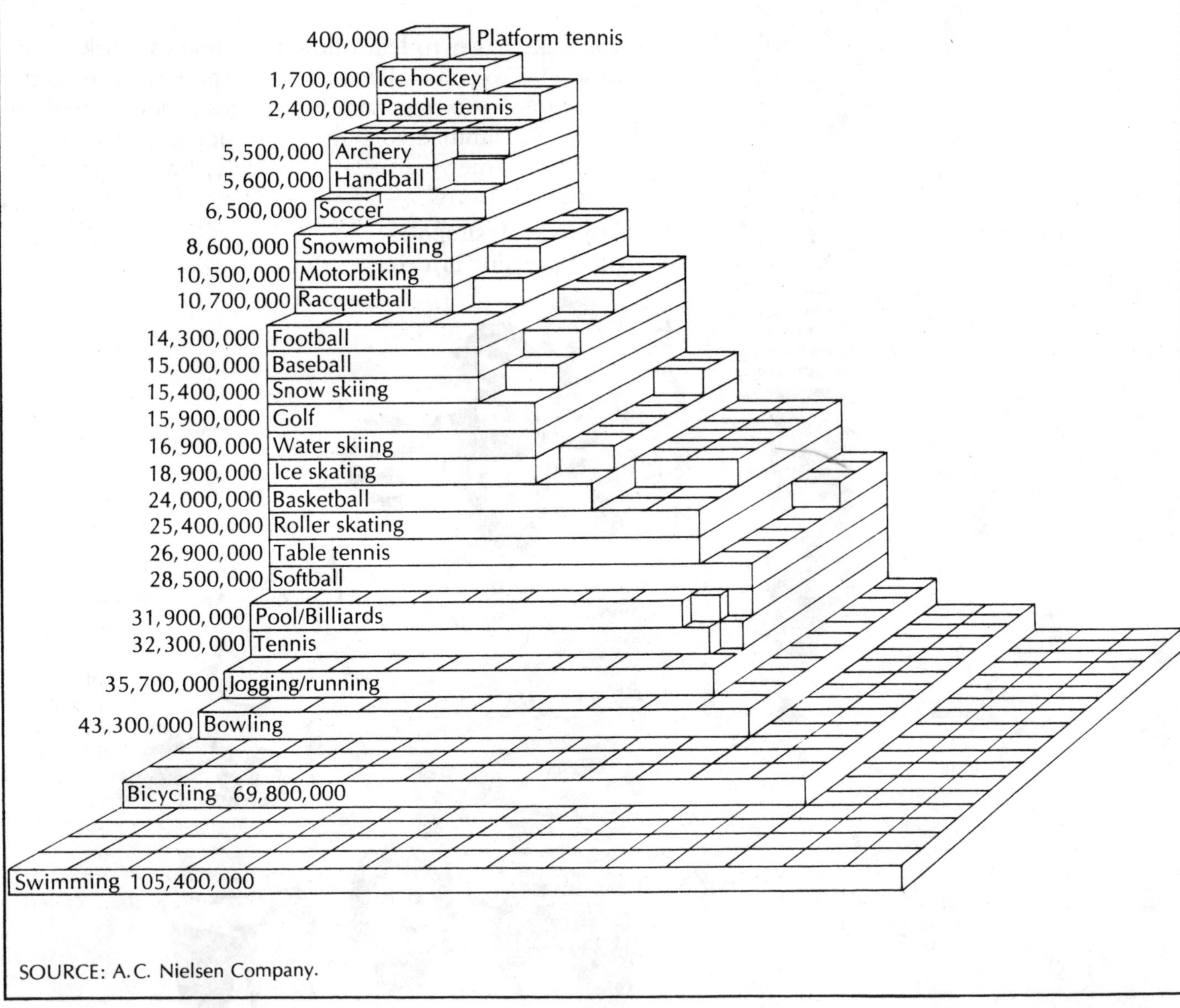

SOURCE: A.C. Nielsen Company.

SPORTS MEDICINE RESOURCE, INC.

In 1976, eight doctors and I organized Sports Medicine Resource (SMR), Inc., in an effort to offer, under one roof, all the medical services an athlete is likely to need. SMR has grown beyond our dreams. It is the largest sports medicine clinic in New England, perhaps in the U.S. We treat and rehabilitate about 18,000 athletes a year. We have treated professional and amateur athletes like Dave Cowens, Carl Yastrzemski, Dick Reuthven, Brad Park, Bill Rodgers, Wayne Grimditch (superstar champion) and Jerry Remy. But 75 percent of the athletes we see are recreational athletes—people who exercise for fun and health. We provide these services:

- Medical and surgical diagnosis and treatment of athletic injuries
- Physical therapy and rehabilitation of sports injuries
- Sports nutrition counseling
- Sports psychological counseling

The reason SMR is unique: All the professional staff make athletics an important part of their lives. Here are colleagues I practice with:

William D. Shea, M.D. Without Dr. Shea's hard work and determination, SMR would never have gotten off the ground. Shea was captain of the St. Francis Xavier (Nova Scotia, Canada) football team. He is a nationally known hip surgeon.

Arthur M. Pappas, M.D. Dr. Pappas, my mentor and teacher, played football at Harvard in the early fifties. He is the medical director of the Boston Red Sox and the president of the orthopedic section of the American Academy of Pediatrics. He is also the professor and chairman of the Department of Orthopedics at the University of Massachusetts Medical School. Dr. Pappas, a nationally recognized expert on children's orthopedics and young adult sports medicine, wrote the children's sections for this book.

Richard A. St. Onge, M.D. Dr. St. Onge, my teammate on the Harvard football team in the early sixties, is a talented orthopedic and hand surgeon. He is also a world-class mountain climber. In 1979, he was the team doctor for the U.S. Alpine Club expedition which climbed Mount Himal Chuli in Nepal. Dr. St. Onge wrote the chapter on the hand and fingers.

Tenley E. Albright, M.D. Dr. Albright, a general surgeon, is the 1956 Olympic Gold Medalist in figure skating. She was also a team physician for the 1976 Winter Olympic team. Dr. Albright, who still skates for exercise, holds skaters' clinics with Dr. Pappas at SMR each winter.

Rob Roy McGregor, DPM. Dr. McGregor, chief of podiatric medicine at SMR, is an exuberant and articulate enthusiast of sports medicine. He is known nationally for developing the Etonic Stabilizers, a running shoe. He has appeared with Cheryl Tiegs on "Good Morning, America." Jim Fixx, author of the *Complete Book of Running,* calls Rob "one of the nation's half-dozen best-known podiatrists."

Benjamin E. Bierbaum, M.D. Dr. Bierbaum was an excellent high school athlete. He was physician for the 1976 Winter Olympic team. He is a nationally known orthopedist. For exercise, he golfs and hunts.

Buddy LeRoux. Mr. LeRoux was an outstanding trainer for both the Boston Celtics and Boston Red Sox. He was one of the founding forces behind SMR.

Nancy Clark, M.S., R.D. Nancy, a registered dietician, is one of the few practicing sports nutritionists in the United States. In 1978, she led fifteen cyclists across the United States on a 4273-mile ride. She is the author of *The Athlete's Kitchen: A Nutrition Guide and Cookbook* and also wrote the chapter on sports nutrition.

Shana L. Bendix, ScD. Shana is a runner and a clinical psychologist. She counsels athletes on the emotional and interpersonal issues which inhibit performance. Shana is a pioneer in this new field.

We have six of the best physical therapists and trainers available to athletes: Charlie Moss, ATC; Peter Stone, RPT; Don LaBourr, ATC, RPT; Debbi Hall, RPT; Cynthia Rowe, RPT; and Dean Clement.

Jane White Venti manages our growing facility. Other staffers are Carol Abbott, Allan Nowick, Patty Gilligan, Ellen Herbert, Leslie Wood and Mariellen Porcaro. Martine Paim is my private secretary and keeps everyone smiling.

I am the medical director of SMR. In the sixties, I was captain of the Harvard football team and an All-Ivy guard. Besides working with the Red Sox, I write a column with Marshall Hoffman for the New York Times syndicate and am medical editor for *The Professional Athlete*. I also do a nationally syndicated television series on sports medicine, called "Sports and Fitness."

How fast will I heal?
Will I perform well after injury?
Why do I get hurt so often?
How do I prepare myself for an important game?
Can medication help me?

This book answers those questions and thousands more. It is the first book to explain sports injuries in a meaningful way. It describes symptoms, diagnosis, cures, prevention, and rehabilitation. It covers injuries ranging from bunions to fractured thighbones to the reconstruction of the knee joint.

Why are so many people exercising? The word is out: Vigorous exercise helps prevent heart attacks, aids in weight control, instills a feeling of well-being, and enhances creativity. Medically, it has been found to help patients with diabetes, ulcers, nervous tension, high blood pressure, back pain, recurrent headaches, and menstrual cramps. It cures hangovers, constipation, and insomnia.

In study after study, scientists have shown that exercise improves health. Here are the conclusions of some of those studies in the last fifteen years:

- Exercise improves your mood. The latest studies from Stanford University reveal that *endorphines,* a morphine-like hormone, are released into the brain by vigorous exercise. That is how scientists explain the "high" after exercise. Like morphine, exercise seems to be addicting, but it is a positive addiction.
- Exercise makes your heart work efficiently. The heart is like any other muscle. When it is exercised, it enlarges and becomes stronger. An athletic heart, because it is muscular, can pump the same amount of blood with fifty beats a minute as the average heart can with seventy-five beats. The difference adds up to 1 million beats per month.
- Exercise helps prevent heart attacks by enlarging the coronary arteries which feed the heart. It lowers the concentration of fat in the blood; fat obstructs the coronary arteries. Exercise teaches the heart to extract oxygen from the blood more efficiently. It also lowers blood pressure.
- Exercise increases deep sleep. Scientists cannot explain why exercise brings on deep sleep, but it does. Some explain it this way: Exercise alleviates stress, and people who are free of stress and tension normally sleep deeply.
- Vigorous exercise is an effective weight control method. Even though exercising in itself is not a great calorie burner, running burns about 900 calories an hour, and tennis, 420. It inhibits your appetite by controlling your blood sugar levels.After a hard racquetball game, I am not hungry. Marshall Hoffman eats approximately two hours after his long run.

Your body continues to burn calories for several hours after you stop exercising. Your body also absorbs fewer calories. When you exercise, food passes through your intestinal tract quickly. In medical terms, peristalsis is facilitated. In the nonathlete, the transit time for a meal is usually twelve to twenty-four hours. For a marathoner, it is as little as four to six hours.

Most people exercise because it is fun. I played high school and college football because I liked it. My coauthor, a high school trackman, runs because he enjoys it. Our sons ride bicycles for the same reason. My daughter water skis because she enjoys it. My wife loves golf and platform tennis. My coauthor's wife is an aerobic dancer and skier.

Patients ask me "What sport is best for me?" I answer "The sport you like best." The reason is simple. If you don't like a sport, you will not do it. For example, I can't get up any enthusiasm for running. Although my friends and patients praise the benefits of running, I find running boring. I like direct competition. That's why I play racquetball and golf.

"I recommend sports which are recreational and therapeutic," says Arthur Pappas, medical director of the Boston Red Sox and nationally known authority on pediatric sports medicine. "For leg injuries I recommend cycling. For shoulder injuries I recommend swimming, a form of hydrotherapy."

Pro golfer Dave Marr, who is now semi-retired, also likes direct competition. "I love to compete. When I was playing, I always liked finding out what I was made of," says Marr. "Some days I felt like Superman. Other days I found I was made of Jell-O, which was a little discouraging, but at least I learned something about myself."

I find that many of my patients exercise for relaxation and social camaraderie. That explains in part the phenomenal growth in the Road Runner Clubs, hunting and games clubs, and aerobic dancing groups.

For the 6600 professional athletes in America, sports may be the pot of gold at the end of the rainbow. For example, the average professional basketball player earns about $175,000 a year. For the pro hockey player, it is $100,000. In baseball, the average is $115,000. In football, $72,000.

Individual athletes can earn tremendous sums. Muhammad Ali earned $8 million for his losing effort against Larry Holmes in October 1980. Sugar Ray Leonard commanded $7 million for a few hours in the ring with Roberto Duran. In 1980, Martina Navratilova, a top tennis player, earned a reported $797,487 in prize money and bonuses, and eighteen-year-old Tracy Austin earned $655,378 at the same sport. Bjorn Borg, the best male tennis player in 1980, is said to have earned more than $3 million in prize money and endorsements.

Like victory and defeat, injury is part of sports and exercise. The number of sports injuries is overwhelming. Almost 11 million people each year are injured in the pursuit of fitness and competition, according to a special study prepared for this book by the Consumer Product Safety Commission. About

MONEY LEADERS IN TENNIS (1980)

Men	
1. John McEnroe	$602,383
2. Bjorn Borg	$523,212
3. Jimmy Connors	$409,641
4. Vitas Gerulaitis	$317,441
5. Guillermo Vilas	$302,126
6. Ivan Lendl	$292,286
7. Wojtek Fibak	$288,687
8. Gene Mayer	$254,119
9. Vijay Amritraj	$243,563
10. Brian Gottfried	$232,760
Women	
1. Martina Navratilova	$797,487
2. Tracy Austin	$655,378
3. Chris Evert Lloyd	$418,705
4. Hana Madlikova	$365,430
5. Billy Jean King	$321,309
6. Wendy Turnbull	$278,573
7. Andrea Jaeger	$218,431
8. Evonne Goolagong	$210,000
9. Virginia Ruzici	$187,113
10. Kathy Jordan	$178,593

one-third of these injuries are serious enough to require emergency room treatment.

"There is hardly a sport that is entirely safe," points out Joel Friedman of the Consumer Product Safety Commission. "Our figures show that more than 250,000 people get injured in volleyball, 84,000 in bowling, and 63,000 in golf."

Highlights from that study:

- More than 55 percent of all sports injuries occur in four activities: bicycling, baseball, football, and basketball. Sports like skeet shooting, mountain climbing and dart throwing—all of which seem to have a high degree of risk—report low numbers of injuries.
- About 70 percent of all sports injuries happen to athletes between the ages of ten and twenty-four. Those are the years in which most people are very active in competitive sports. Almost 40 percent of all injuries happen to children less than fifteen.
- Sixty percent of all sports injuries fall into these categories: sprains, strains, contusions, abrasions, and lacerations. Fractures account for less than 20 percent of all sports injuries, except in football (21.3 percent) and skiing (21.2 percent).
- Almost 90 percent of all sports injuries are treated by physicians. Fifty percent are treated in doctor's offices or training rooms. Almost 35 percent are treated in emergency rooms.
- Of the sports injuries seen in emergency rooms, 98 percent are treated and released.

In sports, the most injured body parts are the fingers and hands, but each sport takes a different toll. For example, ankles are injured most in basketball. The face, including the mouth, ears, and eyes, are the most injured in bicycling.

Government statistics show that the volume of sports injuries is increasing 8 percent a year. "The jump in sport injuries stems from the new hordes who are now competing and exercising," explains Joel Friedman. "There is no evidence that the injury rate is escalating."

In study after study, researchers find that sports are becoming safer. For example, football fatalities and serious injuries have been a major national concern for years. In 1905, President Theodore Roosevelt threatened to ban football if violence and serious injuries continued. But because of rule changes and improved equipment, football is becoming a safer sport. In 1979, four deaths were the direct result of playing football. That compares to an average of twenty-six deaths a year between 1964 and 1972. Football advocates point out that more people die from running than football.

The older you get, the higher the injury rate. Researchers at Southern Illinois University School of Medicine report 3 percent of all elementary school students (first to sixth grades) sustained sports injuries. The rate jumps to seven percent for

A PROFILE OF BICYCLE INJURIES (1979)

Number of Injuries 1,677,000

Males 66.2 Percent	Females 33.8 Percent
1,110,000	567,000

Scale: 0, 25, 50, 75, 100

Almost Three Out of Four Injuries Happen to Children Under 14

Age	Percent	Number
0-9	39.9	669,600
10-14	32.0	536,800
15-24	17.8	299,000
25-34	5.3	89,400
35+	4.9	82,200

Eighty-seven Percent Are Treated by a Physician

	Percent	Number
Treated in Emergency Room	34.0	570,600
Treated in Doctor's Office	53.3	893,400
Not treated	12.7	213,000

Types of Injuries

	Percent	Number
Contusions, Abrasions	37.5	213,700
Lacerations	31.4	179,100
Fractures	14.2	80,900
Strains, Sprains	9.2	52,700
Concussions	2.4	13,900
Others (Dislocations, Hematoma, Punctures, Avulsions)	5.3	30,300

Disposition of Injury

	Percent	Number
Treated and Released	94.8	541,100
Hospitalized	5.2	29,500

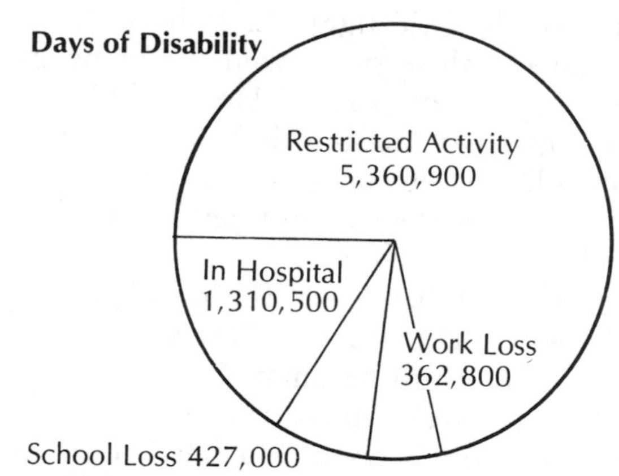

Injury to Body Part*

	Percent	Number
Face	20.2	115,300
Head	12.9	73,800
Hands, Fingers	8.7	49,400
Knee	8.6	49,300
Foot, Toes	7.2	41,200
Elbow	6.4	36,800
Shoulder	5.3	30,400
Lower Leg	5.2	29,900
Ankle	5.2	29,900
Wrist	4.7	27,100
Lower Arm	4.0	23,000
Upper Trunk	3.1	17,700
Others	8.2	46,800

More than 70 Percent of Bicycle Injuries Happen Between May and September*

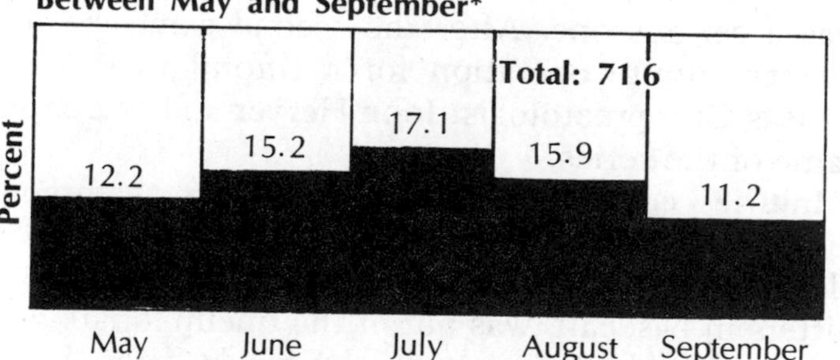

Note: Percentages don't add to 100 percent because of rounding.
*From the 570,600 bicycling injuries treated in emergency rooms.

SOURCE: Consumer Product Safety Commission.

junior high schoolers (grades 7 and 8). High schoolers (grades 9 to 12) have the highest rate—11 percent. The study documented all types of sports-related injuries received by school-age children in a midwest community of 100,000 for a one-year period.

In my twelve years of treating athletes, I've found the same result as the Southern Illinois University Medical School study. The more competitive the sport, the higher the injury rate. For example, I treat very few peewee football players. In the Boston area, about 30 percent of all high school football players get injured during the season. But the injury rate for the Boston Red Sox is almost 100 percent. Nearly every player gets injured at least once during the season. Professional hockey and football players usually sustain at least one injury per season.

The medical literature also cites another important trend: Unorganized sports—sandlot baseball and football, street hockey and skateboarding—produce twice as many injuries as their organized counterparts. The reason is a lack of protective equipment, coaching, and officiating.

Athletic injuries are now part of the American scene and language. Hamstring pulls, slipped discs, and tennis elbows are familiar to everyone. It was big news when former President Carter fractured his collarbone cross country skiing and collapsed in a 10,000-meter run.

In the first week in October 1980, the sports pages were dominated by the Muhammad Ali—Larry Holmes heavyweight title fight. After he lost the fight, Ali revealed that an overdose of thyroid medicine made him weak and "physically unfit" to fight Holmes. Ali admitted that he decided on his own double daily dosage from three grains, prescribed by his physician, to six grains. He said he later began feeling slow and weak, but never associated that with the drug overdose.

In mid-October 1980, the number one topic of the sports pages were hemorrhoids. That's the malady afflicting Kansas City Royals third baseman George Brett. He was removed in the fifth inning of the second game because of the hemorrhoidal pain. Brett's condition became a *cause célèbre.* It was discussed during nationally televised games and emerged as one of the top media stories of the series. Brett's condition was a serious one. After the second game, he underwent a twenty-minute operation for a thrombosed hemorrhoid by Kansas City proctologist John Heryer and he started the third game of the series.

Injuries can knock a team out of a championship. One reason the Boston Red Sox did so poorly in 1980 was the plethora of injuries to key players. Jim Rice, one of the top hitters in baseball, was out of the lineup for six weeks with a fractured wrist. But it was twelve weeks before he regained his all-star hitting form. Centerfielder Fred Lynn, now with the California Angels, fractured his toe. Second baseman Jerry Remy tore knee cartilage, which required surgery. Glenn Hoff-

A PROFILE OF BASEBALL INJURIES (1979)

Number of Injuries 1,471,000

Males 69.4 Percent	Females 30.6 Percent
1,021,000	450,000

0 25 50 75 100

Almost Four Out of Ten Baseball Injuries Happen to Young Adults

Age	Percent	Number
0-9	8.4	124,000
10-14	23.9	351,000
15-24	37.2	547,000
25-34	23.7	348,000
35+	6.8	101,000

Eighty-eight Percent Are Treated by a Physician

	Percent	Number
Treated in Emergency Room	31.3	460,000
Treated in Doctor's Office	57.1	840,000
Not treated	11.6	171,000

Types of Baseball Injuries*

	Percent	Number
Contusions, Abrasions	30.6	140,800
Strains, Sprains	30.5	140,300
Fractures	18.0	82,800
Lacerations	12.6	58,000
Dislocations	2.5	11,500
Hematoma	2.1	9,800
Others	3.7	16,800

Disposition of Injury*

	Percent	Number
Treated and Released	97.7	449,200
Hospitalized	2.3	10,800

Days of Disability*

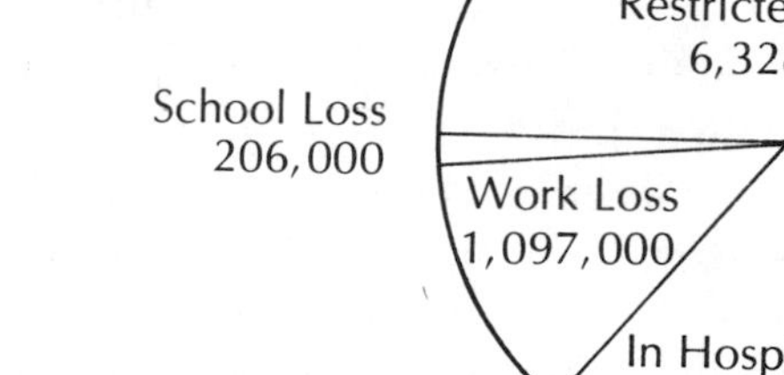

Baseball: Hands and Fingers Get Injured Most*

	Percent	Number
Hands, Fingers	22.7	104,600
Face (Eyes, Mouth, Ears)	18.2	83,500
Ankle	13.1	60,100
Knee	9.5	43,600
Head (Excluding Face)	6.1	28,200
Foot	4.7	21,700
Wrist	4.4	20,100
Shoulder	4.3	19,900
Lower Leg (Calf, Shins)	3.8	17,700
Elbow	2.8	12,700
Neck	0.6	2,600
Others	9.8	45,300

Three Out of Four Baseball Injuries Occur in the Spring or Summer*

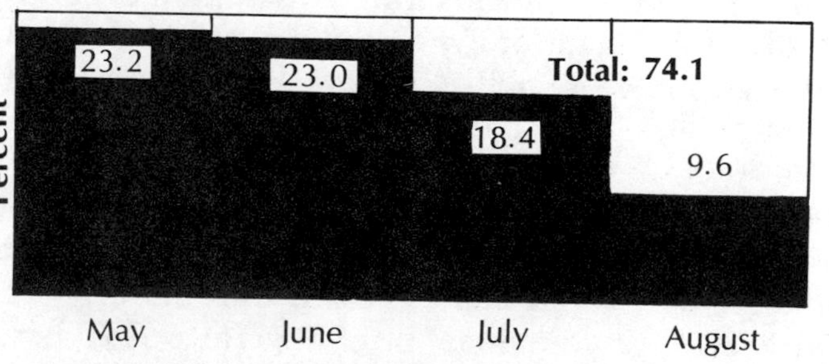

*From the 460,000 baseball injuries treated in emergency rooms.

SOURCE: Consumer Product Safety Commission.

man, a third baseman, had back spasms that sidelined him for two weeks. Butch Hobson, the starting third baseman, now also with the California Angels, was out the last half of the season with a hamstring pull. Ace starter Dennis Eckersly was out five weeks with back spasms caused by a structural abnormality of the back. Captain Carl Yastrzemski fractured his ribs. Outfielder Jim Dwyer was hit in the head with a pitched ball and hospitalized.

At one point, because of injuries, all-star catcher Carlton Fisk (now playing for the Chicago White Sox) was called into service to play both third base and left field. Marshall Hoffman, my coauthor, said half in jest, "I can play a little third base. Maybe Zimmer [the Red Sox manager in 1980] will give me a chance."

I can personally appreciate the mental and physical agony of an athletic injury. I recall only two out of twenty-seven football games in college where I played in complete health.

Athletic injury can threaten your career. Injuries ended the careers of Dave Cowens, Gayle Sayers, Dick Butkus, and Bobby Orr. An injury prevented Tai Bablonia and Randy Gardner, U.S. Olympic pair skaters, from winning a medal in the 1980 Winter Olympics.

Most important, an athletic injury can rob you of your main source of relaxation.

"When professional or competitive athletes are injured, they are concerned that their careers will be significantly altered from that point on," says Dr. Pappas. "In twenty years, working with athletes, I haven't found a way to dispel this fear until they are back in action."

The management of sports injuries has come a long way. When I was playing football at Harvard in the early sixties, there were no diagnostic tests used on my injured knee, such as the arthrogram or the arthroscope. Lucky for me that I only had sprained knee ligaments.

In the 1980s, many knee operations are done with the aid of a mini-television camera attached to the arthroscope. The incisions into the knee are very small, only about an inch long. In December 1980, I operated on a nineteen-year-old football player with a torn knee cartilage. At 7 A.M., he reported to the hospital. Using the arthroscope with the TV camera, I removed the cartilage through a small incision. Five hours later, on his way out of the hospital, he said to me, "Look, Doc, I can finally straighten out my knee." I sent him home with an Ace bandage and two Band-Aids to cover the knee incisions.

Five years earlier he would have been in the hospital for five days and on crutches for three weeks.

Chuck Rainey, a promising right-hander for the Boston Red Sox, had shoulder tendinitis and weakness of the shoulder muscles. Dr. Pappas and I tested him on an isokinetic dynamometer. We found his pitching arm was 20 percent weaker than his left arm. The same machine was used to strengthen the weak arm. Ten years ago, the Cybex machine didn't exist.

A PROFILE OF BASKETBALL INJURIES (1979)

Number of Injuries 1,292,000

Males 80.5 Percent	Females 19.5 Percent
1,040,000	252,000

0 25 50 75 100

About 58 Percent of Basketball Injuries Happen to Athletes Between 15-24

Age	Percent	Number
0-9	2.3	30,200
10-14	23.7	306,500
15-24	57.7	745,900
25-35	13.7	176,500
35+	2.6	32,900

Eighty-five Percent Are Treated by a Physician

	Percent	Number
Treated in Emergency Room	30.4	392,600
Treated in Doctor's Office	55.7	719,800
Not treated	13.9	179,600

Types of Basketball Injuries*

	Percent	Number
Strains, Sprains	54.4	231,500
Contusions, Abrasions	16.7	65,700
Fractures	13.9	54,700
Lacerations	8.4	33,100
Concussions	2.7	10,500
Others	3.9	15,100

Disposition of Injury*

	Percent	Number
Treated and Released	98.7	387,300
Hospitalized	1.3	5,300

Days of Disability

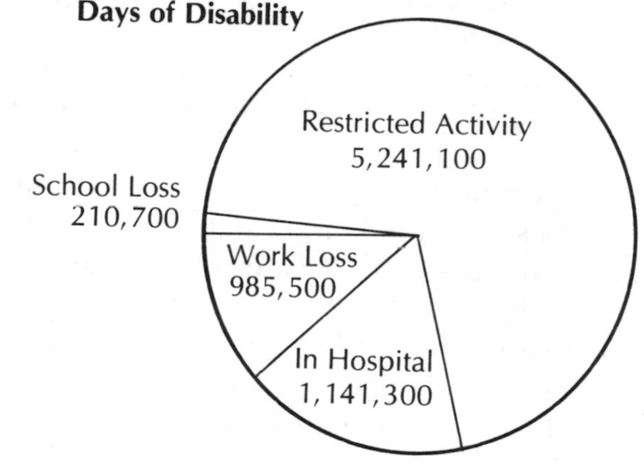

Ankles Are Basketball's Most Frequent Injury*

	Percent	Number
Ankle	33.1	130,000
Fingers, Hands	22.4	87,900
Face	9.2	36,100
Foot, Toes	8.4	33,100
Knee	7.6	29,900
Wrist	3.6	14,300
Elbow	2.4	9,500
Head	2.3	8,900
Shoulder	2.2	8,500
Others	8.8	34,400

Most Basketball Injuries Happen in Winter*

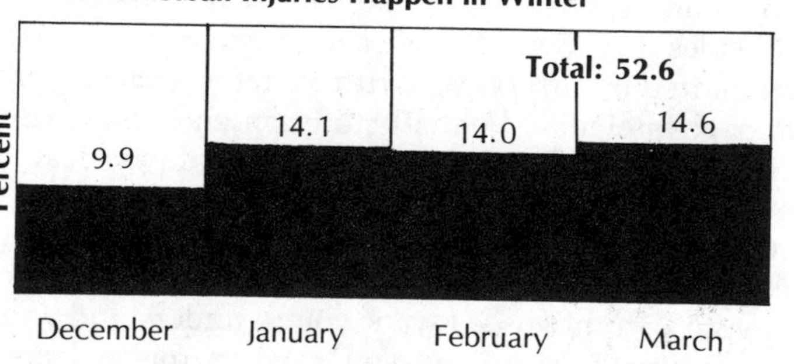

*From 392,6000 basketball injuries treated in emergency rooms.

SOURCE: Consumer Product Safety Commission.

athletes who will get injured. Using high-precision instruments like the isokinetic dynamometer, I can tell which muscle groups are strong or need strengthening.

Athletic injuries, even if they are small ones, are serious business. Think of the body as a chain. One link can weaken the entire chain. Dizzy Dean, the Hall of Fame pitcher of the St. Louis Cardinals, injured his big toe. Because of the injury, he changed his pitching motion. This damaged his throwing arm and eventually ended his career. Jim Palmer, a pitcher for the Baltimore Orioles, says that one of his legs is shorter than the other. The difference, Palmer says, almost ended his career, because it caused muscle spasms in his back.

In the 1980 National Football League's Wildcard playoff game, Billy Waddy, the split end for the Los Angeles Rams, had a difficult time reaching for the ball because of a hip pointer, an avulsion injury to the iliac crest.

Sports Health is a link between the injured athlete and the physician. My purpose in writing this book with Marshall Hoffman is to spread the word and share my experience as a sports medicine physician.

This book gives recommendations on when to see a doctor and guidelines on how long you can expect to be out of action. I explain the difference between a simple, a mild, and a severe injury. The book offers reasonable guidelines for how the injury should be treated. Is the best treatment rest, a brace, a cast, a cortisone injection or an operation, or maybe a combination?

The goal of every sports medicine physician, trainer, and coach is to minimize your "down-time" and safely return you to action.

The information in this book comes from twelve years of practicing sports medicine and thirty years of sports participation. I have personally treated more than 10,000 athletes and fitness enthusiasts. I perform more than 500 surgeries a year. I am medically responsible for athletes everyday of my life.

Besides my own practice, I attend medical symposia all over the U.S. In 1980, I attended medical meetings in Montana and Utah. I usually set aside at least four weeks a year for this type of learning. I teach at the University of Massachusetts Medical School and the New England Baptist hospital.

Besides my own knowledge, research for this book was augmented by interviews with athletes, coaches, trainers, and other physicians. Marshall Hoffman and I have tried to make the information in this book simple and practical. This is not a "how-to" book in the traditional sense. I don't advocate setting your own fractures or making your own diagnosis. But I can give you advice on taping your own ankles and puncturing your own blisters. I want you to understand your body and your injury. Ultimately, you have to make decisions about your body. This book will give you some of the knowledge needed to make intelligent decisions.

A PROFILE OF FOOTBALL INJURIES (1979)

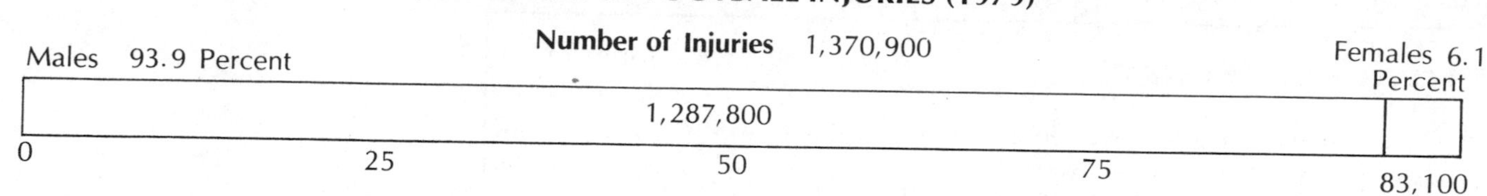

About 55 Percent of Football Injuries Happen to Athletes Between 15-24

Age	Percent	Number
0-9	5.9	81,200
10-14	32.9	451,200
15-24	54.2	742,400
25-35	5.6	77,300
35+	1.4	18,800

Eighty-seven Percent Are Treated by a Physician

	Percent	Number
Treated in Emergency Room	31.5	431,600
Treated in Doctor's Office	55.4	759,900
Not treated	13.1	179,400

Types of Football Injuries*

	Percent	Number
Strains, Sprains	34.2	147,600
Contusions, Abrasions	27.8	120,200
Fractures	21.3	92,100
Lacerations	7.9	33,900
Concussions	1.6	7,100
Others	7.2	30,700

Disposition of Injury*

	Percent	Number
Treated and Released	96.6	416,900
Hospitalized	3.4	14,700

Days of Disability*

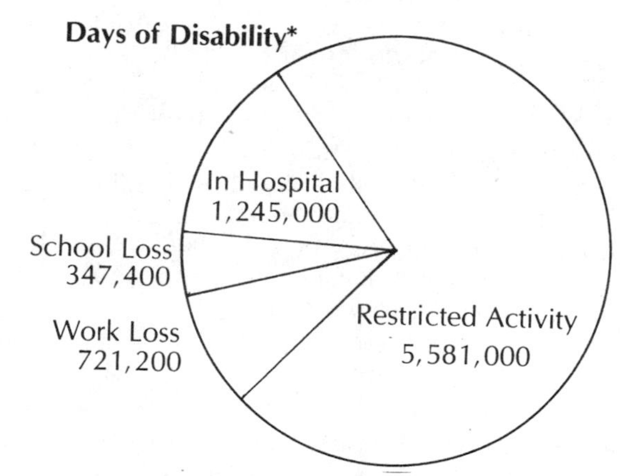

Hand and Finger Injuries Are the Most Frequent*

	Percent	Number
Fingers, Hands	22.0	95,000
Knee	11.4	49,200
Ankle	10.0	43,100
Shoulder	9.6	41,600
Face	8.0	34,700
Upper Trunk	6.7	28,800
Wrist	5.6	24,000
Foot, Toe	5.5	23,600
Head	4.1	17,900
Lower Area (Forearm)	3.3	14,300
Lower Trunk	3.3	14,100
Elbow	3.2	13,900
Lower Leg	2.8	12,000
Others	4.5	19,400

Most Football Injuries Happen in the Fall*

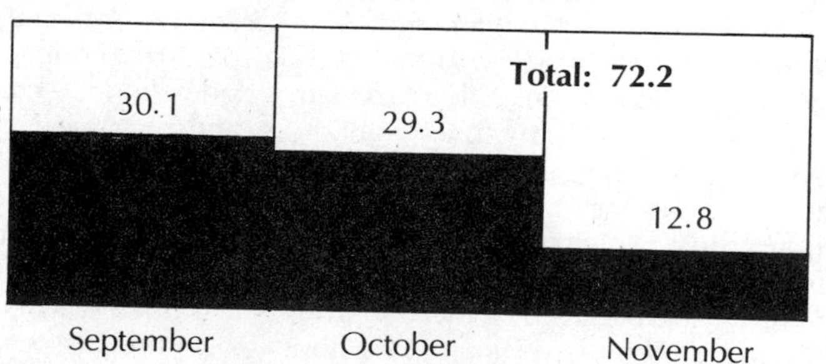

*From 431,600 football injuries treated in emergency rooms.

SOURCE: Consumer Product Safety Commission.

"I GAVE A LITTLE EXTRA"

Dave Cowens has always done things his own way, and his retirement was no exception. He wanted to transmit his thoughts directly to the public, without a writer serving as the middle man. Here is a partial text of his statement.

By Dave Cowens

EVANSVILLE, Ind.—I always took pride in my performance and thought that I gave a little extra something to the game. I never identified myself as a great player, only as one who set high standards regarding his performance. I worked diligently to live up to those standards so as to make my mark on the world of basketball. I wanted to offer something different. Perhaps it was just a certain flair or wildness or unexpected behavior on the court, but something critically different. I was given a fairly sound body to develop and train, and I worked it long and hard so that I would be asked to play a major role in the outcome of games.

However, I can no longer play that caliber of basketball, and it is unbelievably frustrating to remain in a rugged occupation with waning skills.

Enter here the fact that I have been playing basketball for sixteen years on two feet which can best be described by the observations that a team of foot and bone specialists made a couple of years ago. They were amazed I had been able to play up to that point in time without any radical, serious injuries, at which time I pointed out that I had sprained my ankles at least thirty times over the duration of my career, broken both legs, and fractured a foot. Now I am not explaining this for you to extend sympathy to me, or to use it as a crutch, but only to explain that there is something fundamentally wrong with my feet and ankles that would make me more susceptible to injury. Just last year I sustained a different type of foot injury when I severely stretched all the tendons around my big toe. This joint is now twice its normal size, and possibly always will be.

The primary reason that I will not remain on the active roster of the Celtics or any other professional ballclub is the fact that I have a highly weakened and worn out set of feet and ankles. Knowing this about my feet, and the fact that I have not been able to play a full season since 1976, I'd say my chances of getting hurt and not being able to contribute in all the games this year were highly probable.

So now I must assess my situation. I have given my feet a valid test to see if they were able to make the grade after three weeks of training sessions and exhibition games. The result is that I am basically playing on one leg. My right ankle is so weak that I can best describe it by saying that I have a sponge for an ankle. My left leg and ankle are therefore taking an extraordinary amount of abuse and they would no doubt give out before the year was over.

Accompanying all of these injuries is a notable amount of pain which I have been able to tolerate during each season, but which is progressively getting worse. I do not believe in taking medication which many others utilize to mask the pain and allow them to play more years and earn more income.

So I asked myself, "Why should I play?" To take up a slot that another man could fill? To jeopardize the success of the club? To risk injuring my ankles? To embarrass myself by not playing up to the expectations of the fans?

I'll tell you why it is such a difficult decision to make—because of the financial reward. So after all this deliberation and soul-searching over the past five months, and taking into consideration the combination of positive and negative factors such as the exceptional talent the Celtics have in their front line, the camaraderie among players, the rigors of traveling, etc., I have decided to retire.

I have enjoyed performing for you over the past ten years while hopefully engraving myself into the history of the Celtic organization.

An Emerging Science

Sports medicine and sports injuries are not new. They were well known in ancient Greece. Hippocrates, the father of modern medicine, studied under Heridocus, the physician for the ancient Olympic Games. Yet it has only been in the last twenty years that sports medicine has emerged as a medical specialty.

Today there are more than twenty full-time professorial posts in sports medicine in the U.S. and 18,000 officially designated team doctors in high schools, colleges, and universities. At least half of all residents in orthopedic surgery hope to make sports medicine a major part of their career.

"Sports medicine is a growth industry," says Allan J. Ryan, editor of *The Physician and Sportsmedicine.* "In September 1980, our magazine published a list of 230 sports medicine clinics and centers. Since that publication, we have received letters from about 100 other centers which were not mentioned.

"Secondly, the number of meetings and conferences grows annually. When we started *The Physician and Sportsmedicine,* we frequently didn't have enough meetings to fill a calendar page (two-thirds of a page). Now we are considering publishing three calendar pages per issue."

Sports medicine today is a broad, diverse area of medicine. At Sports Medicine Resource, we have a staff of six orthopedic surgeons, two podiatrists, three athletic trainers, three physical therapists, a sports psychologist, and a sports nutritionist. Soon we will add an exercise physiologist and a cardiologist. All of us practice sports medicine. It is impossible for one individual to know everything about the human body during exercise.

Will sports medicine become a full-dressed specialty of medicine, like orthopedics, pediatrics, or gynecology? "A group of seventy physicians has been working on a basic sports medicine curriculum for training physicians at both the undergraduate and graduate level," says Allan J. Ryan. "Our purpose is to keep sports medicine a broad field. We don't want it to become a narrow specialty. This would diminish the interest and participation of many physicians who treat athletes on a part-time basis in their own practices, like hand surgeons, podiatrists, orthopedists. If it becomes a narrow specialty, doctors might say 'Leave that problem to the sports medicine specialist.' We—the group of seventy physicians—don't want that to happen."

I believe that much more work has to be done on the prevention of athletic injuries. As Dr. James Nicholas, the team physician of many New York professional teams, puts it: "We do not have adequate information as to how to prevent injuries in sports. The prevention has not kept pace with the excellent early diagnostic abilities of many doctors for conditions such as painful shoulder, elbows, and knees."

Preventing athletic injuries will be the challenge of the eighties.

THE FATHER OF SPORTS MEDICINE

Since the beginning of time, physicians have treated injured athletes. Hippocrates, the father of medicine, studied under Heridocus, the physician for the ancient Olympic Games.

But the first physician to devote a major portion of his time to sports medicine was Galen, a doctor who was born and practiced in Pergamum in Asia Minor. After studying medicine for eleven years, he was appointed surgeon to the gladiatorial schools in 160 A.D.

The athletes of Galen's day participated in boxing, wrestling, gymnastics, track and field events, chariot racing, and gladiatorial contests.

Galen was a prolific writer. Historians estimate that he wrote about 2½ million words, equal to about eight copies of *War and Peace.*

Here's a sampling:

It is necessary to keep the wound continually moist where if the dressings dry out the ulcer becomes inflamed. This is true especially in the summer at which time I cured the most seriously injured by covering the wounds by cloth wet with astringent wine kept moist both day and night by a superimposed sponge.

His fame in treating athletes spread to Rome. He served as physician to three Roman Emperors—Marcus Aurelius, Commodus, and Septimus Severus.

George A. Snook, M.D.

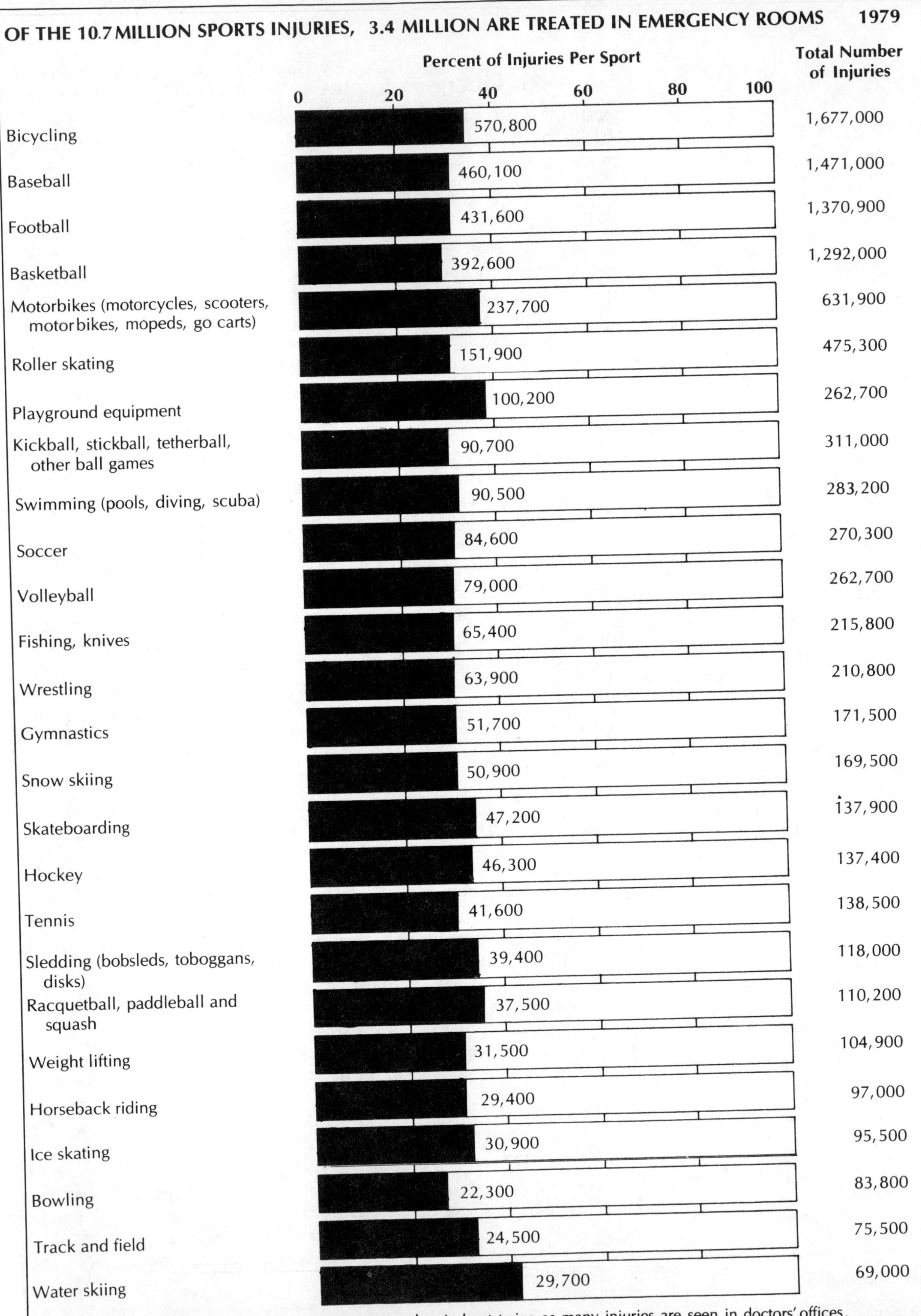
OF THE 10.7 MILLION SPORTS INJURIES, 3.4 MILLION ARE TREATED IN EMERGENCY ROOMS
1979
Percent of Injuries Per Sport
Total Number of Injuries
0
20
40
60
80
100
Bicycling
570,800
1,677,000
Baseball
460,100
1,471,000
Football
431,600
1,370,900
Basketball
392,600
1,292,000
Motorbikes (motorcycles, scooters, motorbikes, mopeds, go carts)
237,700
631,900
Roller skating
151,900
475,300
Playground equipment
100,200
262,700
Kickball, stickball, tetherball, other ball games
90,700
311,000
Swimming (pools, diving, scuba)
90,500
283,200
Soccer
84,600
270,300
Volleyball
79,000
262,700
Fishing, knives
65,400
215,800
Wrestling
63,900
210,800
Gymnastics
51,700
171,500
Snow skiing
50,900
169,500
Skateboarding
47,200
137,900
Hockey
46,300
137,400
Tennis
41,600
138,500
Sledding (bobsleds, toboggans, disks)
39,400
118,000
Racquetball, paddleball and squash
37,500
110,200
Weight lifting
31,500
104,900
Horseback riding
29,400
97,000
Ice skating
30,900
95,500
Bowling
22,300
83,800
Track and field
24,500
75,500
Water skiing
29,700
69,000
Note: These are injuries treated in emergency rooms only. At least twice as many injuries are seen in doctors' offices.

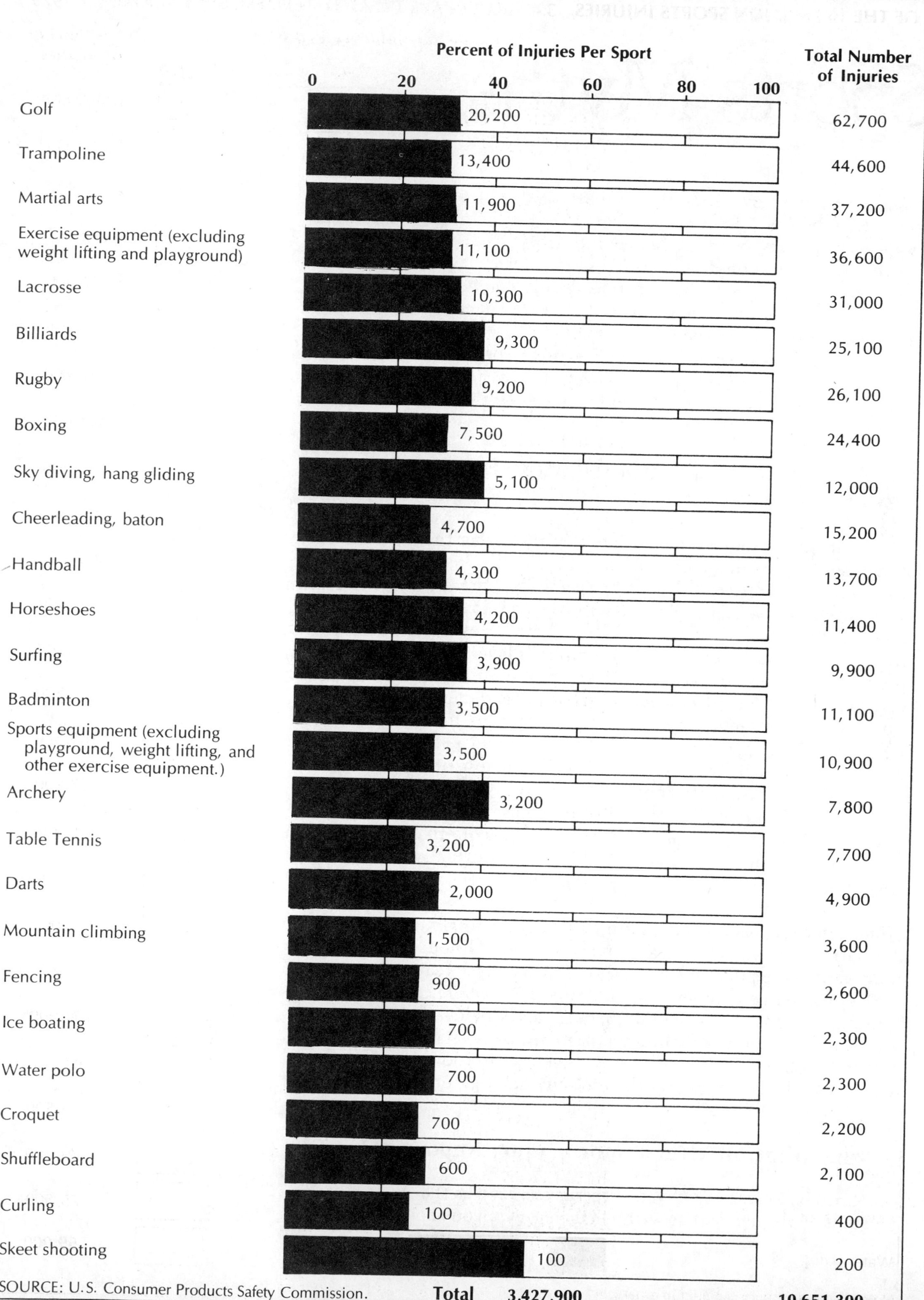

SOURCE: U.S. Consumer Products Safety Commission.

Sports Myths

It is amazing how little fitness buffs and athletes know about their bodies. I don't know where they get their information—from friends, enemies, or witch doctors.

I've treated million-dollar athletes who have a minimal concept of how their bodies operate during competition. If professionals are uninformed, how can recreational athletes be expected to understand the athletic body?

This chapter is designed to dispel many sports medicine myths. It is a sampling of questions I answer everyday in my practice—questions that are embedded in the myths.

Are Amateur Athletes Injured More Frequently Than Pros?

Many medical studies substantiate that the higher the level of competition, the higher the injury rate. On the Boston Red Sox, every player misses at least a few games because of injury. One reason is that they play 162 games and 35 preseason games with very few days off. There is no recovery time. Another reason is the intense competition. Players go at full speed. They sprint to first base. They run at full speed for batted balls. They give 100 percent physically and mentally. If you expose any athlete to injury risk often enough, inevitably something will happen.

The injury rate increases as the level of competition increases. About 10 percent of children under twelve years of age sustain sports injuries. Thirty percent of high school competitors miss games because of sports injuries. Seventy-five percent of college athletes sustain injuries, but virtually 100 percent of pro players miss some competition because of their physical ailments.

The difference between the injury rate and the total number of injuries nationally misleads many people. Although the injury rate in the pro ranks is nearly 100 percent per year, only 6600 pro athletes compete in the United States. Therefore, from the pro ranks the total number of injuries is low. On the other hand, millions of children participate in sports, and almost 4 million a year are injured. Even with a low injury rate, the totals are far greater than for any other group.

Is Rehabilitation After a Sports Injury Important?

Definitely. I prescribe physical therapy as soon as the swelling stops. This means that within a day or two, you should be

THE BILL OF RIGHTS
FOR THE SCHOOL AND COLLEGE ATHLETE

Participation in athletics is a privilege involving both responsibilities and rights. The athlete has the responsibility to play fair, to give his best, to keep in training, to conduct himself with credit to his sport and his school. In turn, he has the right to optimal protection against injury. Protection may be assured through good technical instruction, proper regulation and conditions of play, and adequate health supervision. Included are:

Good Coaching:

The importance of good coaching in protecting the health and safety of athletes cannot be minimized. Careful conditioning and technical instruction leading to skillful performance are significant factors in lowering the incidence and decreasing the severity of injuries. Also, good coaching includes the discouragement of tactics, outside either the rules or the spirit of the rules, which may increase the hazard and thus the incidence of injuries.

Good Officiating:

The rules and regulations governing athletic competition are made to protect players as well as promote enjoyment of the game. To serve these ends effectively, the rules of the game must be thoroughly understood by players as well as coaches and must be properly interpreted and enforced by impartial and technically qualified officials.

Good Equipment and Facilities:

There can be no question about the protection afforded by proper equipment and right facilities. Good equipment is now available and is being improved continually; the problem lies in the false economy of using cheap, worn out, outmoded, or ill-fitting gear. Provision of proper areas for play and their careful maintenance are equally important.

Good Health Supervision . . . Including:

FIRST . . . a thorough preseason history and medical examination. Many of the sports tragedies which occur each year are due to unrecognized health problems. Medical contraindications to participation in contact sports must be respected.

SECOND . . . a physician present at all contest and readily available during practice sessions. It is unfair to leave to a trainer or coach decisions as to whether an athlete should return to play or be removed from the game following injury. In serious injuries, the availability of a physician may make the difference in preventing disability or even death.

THIRD . . . medical control of the health aspects of athletes. In medical matters, the physician's authority should be absolute and unquestioned. Today's coaches and athletic trainers are happy to leave medical decisions to the medical profession. They also assist in interpreting this principal to students and the public.

working with a therapist or an athletic trainer.

My reasoning is simple. The therapist moves or mechanically stimulates the injured part, which increases the blood supply to the injured area. The blood brings the building blocks for healing—nutrients, oxygen, and inflammatory cells—to the injury.

Most injuries (even a simple ankle sprain) need at least five days to heal. Why let muscles and joints become weak or stiff from disuse? A therapist or an athletic trainer will prevent weakness and stiffness.

In my opinion, quick rehabilitation may cut disability time by at least 25 percent. Furthermore, some injuries like fractures, joint disruptions, torn ligaments, and cartilages are treated best with an expert physical therapy program.

I see about ten athletes a week who have permanently lost range of motion in their limbs because they have had no physical therapy program. At Sports Medicine Resource, we have six certified physical therapists and athletic trainers. Every team in professional sports has at least one full-time trainer. The New York Yankees have two top-notch trainers. So do other teams.

Are Casts Bad for Arms and Legs?

For broken bones to heal effectively, they need to be immobilized and rested. It is part of the ancient surgical principle to "rest the injured part."

The best way to make certain that a body part is rested is to use a cast. If a cast is not applied to a fracture, the bone ends wiggle, causing severe pain, inhibiting healing, and increasing the amount of healing bone that is produced. The result: large lumps at the site of the fracture.

The cast assures the broken bone will be rested. As soon as it is safe, however, the joints above and below a fracture should be removed from the cast. Long periods of joint immobilization can produce joint stiffness.

Are Women More Prone to Injury Than Men?

From a strictly medical and biological standpoint, there is no evidence that women have a higher incidence of injury. Male and female muscles look the same under a microscope. Bone samples also look identical. The musculoskeletal system, on a pound-for-pound basis, is the same.

Christine Haycock, M.D., who has studied this problem, says the number of injuries sustained by women athletes in some sports was initially higher because of poor coaching, improper training, and inadequate equipment. As these factors improved, the injury rate decreased to levels comparable with male injuries.

THE TED WILLIAMS STORY

"A Ted Williams is born once in a lifetime," said Jack Fadden, the former Boston Red Sox trainer and member of the Harvard Athletic Hall of Fame.

"Ty Cobb told me before his death: 'Jack, I'll be honest with you; Williams is the greatest hitter I have seen in my entire life. I've watched him behind the plate. If the ball isn't there, Williams won't go for it. He has tremendous willpower.'

"You mean to tell me that he is a better hitter than you?" asked Fadden.

"I have to admit it, Williams is a better hitter," Cobb replied.

"How can you explain that?"

"I never played any night games or Sunday baseball. I never missed five years of my baseball career because I was called into the military service," said Cobb.

Fadden continued: "One rainy day, while sitting in the Cadillac Hotel in Detroit, Rogers Hornsby told me the exact same thing about Williams. I said to Rogers that, well, tell you what I will do. I will make you the best right-handed batter of all times and Williams the best left-handed batter."

"I will settle for that," Rogers agreed.

"Bob Feller told me that Williams was the best batter he ever saw in his life—that you don't get anything from him; you have to earn every pitch."

In a study of female cadets at West Point, a similar conclusion was reached. Because the women had been less active physically in high school, their initial rate of stress fracture was 10 percent; in men it was 1 percent. When the physical training program at West Point was gradually increased, the female stress fracture rate fell to 3.3 percent. The study also found that the type of sports the female cadets had played did not promote muscular development of their upper extremities. Thus, 92 percent of the women failed the men's performance scale on upper body strength testing. Once they were given the appropriate training, their upper body performance improved greatly.

I have found this to be true in my own experiences. In 1976, I served as the physician for the female Olympic crew trials. The crews were elite female rowers from all over the country; I had never seen more finely trained athletes. It was clear to me that their upper body strength was comparable with any male's of the same body weight.

Size is critical in talking about female injuries in sports. You expect a 160-pound football player to be injured more frequently if he plays against 240-pound players. When comparing their performance and their injury rate with those of males, women must be judged according to their body weight.

Will Running Make Your Ankles and Knees Arthritic?

There is no evidence in the scientific literature to prove that sports will make your joints deteriorate quickly.

Many major league baseball players, including members of the Boston Red Sox, play full tilt with knees, ankles, shoulders, elbows, and wrists that are not perfect. I have studied many of these players firsthand and do not see their joints deteriorating because of sports.

Everyone's joints gradually wear out with time. When you are born, you have a surface on the ends of your bones that forms the joint surface cartilage. This is like the tread on a new tire. You lose tread as you age and use your joints, just as you lose tread from your car tire when you drive it.

If you have an injury to or surgery on a joint, it is clear that the tread will wear down faster. There is no scientific evidence, however, that hard use of that joint will lead to still faster wear and tear. I see as many women as men in their seventies with severe knee arthritis. Most of these women have not been involved in vigorous athletics because it wasn't acceptable to be in sports when they were young. Their knees have worn out just with daily activities. Therefore, even after ankle or knee surgery, I allow my patients to return to sports activities that they can do comfortably, with this one warning: Listen to your knee. If it swells and hurts with every workout, change sports or reduce the intensity of your sport.

Will a Back Corset Weaken Your Back Muscles?

Patients ask this question over and over. I often prescribe a back corset for people who have low back problems. It is a sound medical principle to use an external back support. It is the same as taping a sprained ankle or using a knee brace.

By returning to sports competition after back trouble, you will use the muscles of your back and abdomen more vigorously than if you do nothing. Even with the brace on, your muscles will become stronger. Realistically the back brace should be considered a way to build up the back muscles. It is not the enemy; it is a friend. It will not lead to weakening of your back muscles!

GUNG HO WILLIAMS

Ted Williams was really gung ho. I admired him for that. He never broke a curfew in his life. He never gave a manager trouble. He never belittled another ballplayer in his life. He never criticized anyone. If you asked him about a player, he would often say, "The kid has the tools. All he's got to do is work at it."

Jack Fadden, trainer for the Boston Red Sox

Is an Athlete Equally Prone to Injury All During the Season?

All studies show that most injuries occur early in the season. Take the study by Harlan C. Hunter, D.O., of 4393 football players in St. Louis. Dr. Hunter found that 51 percent, or 2277, were injured in one season. This is about the normal range for injuries in high school football.

About 50 percent of all the injuries occurred in practice during the third to fifth weeks. Injuries at the beginning of the season occurred more frequently in practice sessions than in games.

Dr. Hunter believes that lack of fitness and conditioning is the reason for the large number of injuries early in the season.

If You Work Hard Enough, Can You Be a Great Athlete?

No. The shape of your bones, your muscle type, your coordination, and your reflex speed are all inherited characteristics. Jim Rice, the Red Sox's all-star outfielder, has one of the best slugging percentages in the major leagues because he has the right physical ingredients. These talents are God-given. There are only a few Jim Rices in baseball.

The shape of your body (your morph type) determines the sports that best suit you. Muscular persons (mesomorphs) do best in contact sports and court sports. Tall, thin persons (ectomorphs) do best in endurance sport. Obese persons (endomorphs) are less likely to be professional or world-class athletes. Your body type is basically an inherited characteristic.

You can modify your basic body type. An endomorphic person can gain strength, shed weight, and become like a natural mesomorph. Tremendous psychological dedication to athletic activity can increase your skill level no matter what your basic body type.

"Ted Williams, the last player to bat .400, was the hardest-

YOU ARE NEVER TOO OLD

The most stirring moment of the America's Marathon in Chicago took place after the awards ceremony for the top ten finishers was over. As the clock approached 4:45, spectators shouted *"Here comes Ida!"* A moment later, the four happiest people in Chicago—twenty-one-year-old Ari Mintz, his fifty-year-old parents Alan and Gloria, and his seventy-four-year-old grandmother Ida—crossed the finish line of the 26-mile, 385-yard race side by side.

The applause was mainly for Ida, a great-grandmother who had just become the oldest woman ever to run a marathon. It was Ida's first attempt at the distance. The previous "oldest" had been Mavis Lindgren of Orleans, California, who ran her most recent marathon in 4:47:37 at the age of seventy-two.

Ida Mintz was trained by her son, Alan, a doctor who prescribed daily runs of eight, ten, twelve or eighteen miles during the months preceding the marathon. Ida supplemented her running with 150 sit-ups each day, increasing to 250 twice a week. The "galloping grandma" (she became a great-grandmother two weeks before the race) said she has been doing a daily routine of strenuous exercises for fifty years.

working athlete I ever knew," said Jack Fadden, the former trainer for the Boston Red Sox. "Sure, he had great talent, but he was dedicated to excellence. He would sit on the top step of the dugout studying every pitcher. To him, the pitcher-batter confrontation was serious business. 'The pitchers want to take bread out of my mouth,' Williams would say."

It is important to be realistic about your body build and your athletic interest. You would not expect Carlton Fisk, six-foot-three-inch catcher, at second base, or Jerry Remy, 165-pound second baseman, behind the plate. Sensitive guidance of children into appropriate sports helps them find happiness in athletic activities.

Can Doctors Heal You?

Nature is the great healer. Doctors simply assist nature. If I see a patient with a broken leg, I straighten it and hold it straight, but nature heals the fracture. I follow the basic medical axiom: "First, do no harm." I do not want to interfere with normal healing processes; I want to prevent swelling and to provide the necessary rest for the injured part so that a normal rate of healing can proceed.

Thomas B. Quigley, former Harvard football team physician and one of my mentors, used to say, "The person who discovers a way to increase the normal speed of healing of the body will win a Nobel Prize."

Many treatments, compounds, and medications claim to speed the rate of healing. I have not read a single scientific article that demonstrates that healing time can be increased artificially. You cannot beat nature. All doctors can do is to be very careful that they do not slow down nature.

Are More People Being Injured in Sports Than Ever Before?

If you read newspapers or watch TV sports, you get the impression that the number of sports injuries is increasing at an astronomical rate.

There is no doubt that the total number of sports injuries has increased. But, so has the total number of participants. Today about half the adult population, or 57 million Americans, exercise on a regular basis. That is about double the percentage of twenty years ago. The real question: What is the injury rate in any given sport and is it changing?

There is good evidence that the rate of injury is actually declining because of the diligent efforts of batallions of people who are making sports safer. In a study published in the *American Journal of Sports Medicine* in 1978, Edward M. Tapper, M.D., analyzed 4227 ski injuries that occurred in Sun

Valley, Idaho, from 1939 to 1976. He found that the rate of ski injuries had continued to decline steadily. For example, the rate in 1976 was one-third of what it was in 1960. The rate was fewer than three injuries per 1000 skier days. Dr. Tapper cites factors such as modern step-in bindings, better ski-boot design, and better ski-slope grooming techniques. Many eye-catching studies available today make you think that injuries are extremely common; I think just the reverse is true.

Does Cortisone Make You Heal Faster?

Your adrenal glands make cortisone daily. One of the purposes of body cortisone is to reduce irritation or inflammation. Excessive inflammation decreases the normal ability of the body to heal injuries. Cortisone cannot shorten the healing time of an injury, but it reduces the inflammatory response so that the normal healing rate can take place.

When I use cortisone to reduce inflammation, I inject it into the affected area with a small needle. This gives a high dose of cortisone where it is needed. Cortisone tablets increase the amount of cortisone throughout the whole body, not just in the injured area. Tablets have side effects—fluid retention, obesity, and hormonal imbalance. Thus, I prefer shots to tablets.

Too much cortisone can destroy a body part. Dick Butkus, the all-pro linebacker for the Chicago Bears, was allegedly given numerous cortisone shots for a knee problem. When he finally had surgery, his doctors discovered to their chagrin that the ligaments of his knee had disintegrated. Butkus sued for a million dollars and won. Sandy Koufax, all-star pitcher for the Los Angeles Dodgers, was given cortisone injections in the elbow numerous times. Many physicians believed that cortisone ended his career. Butch Buchholtz, tennis great and now the commissioner of World Team Tennis, received many cortisone injections for tennis elbow. He says the cortisone shots forced him into early retirement. Today he can barely carry a suitcase.

I give cortisone injections a maximum of three times in any one area of the body. If you haven't cured the problem by then, you should try some other form of therapy.

Do Children and Adults Heal at the Same Rate?

A child's body is always in a state of growth. It is like a young tree. Because children's bodies are building and rebuilding all the time, they heal much faster than adults. For example, a child's skin replaces itself every fifteen days. But it renews itself in only three days when damaged. That is one-third faster than adult skin renews. For certain fractures the rate of healing for children is twice that of adults. A fracture of the shaft of the thighbone heals in eight to ten weeks in children and in about eighteen weeks in adults.

Should a Sports Injury Be Treated Immediately By a Doctor?

The moment you feel some body part pull, sprain, strain, or break, start RICE, a four-part first-aid program for sports injuries. Do not wait for a doctor's okay.

R is for rest. If you continue exercise and other activities, you may extend the injury. I stands for ice, which is strong medication. Ice shrinks the torn blood vessels. The more blood that collects in a wound, the longer it takes to heal. C is for compression, which limits swelling. If uncontrolled, swelling retards healing. E is for elevation. Elevating the injured part above the level of the heart uses gravity to help drain excess fluid. Because swelling usually starts within seconds, start RICE as soon as possible.

Rushing to a doctor's office does not mean that you will be treated immediately. If you have been to a doctor's office recently, you know what I mean. Emergency rooms, where about one-third of sports injuries are treated, are even more crowded. Patients with sports injuries tell me that they have waited four hours to be examined by a doctor. Waiting can aggravate the problem.

Is Aspirin a Weak Medication?

Do not underestimate aspirin because it is cheap and can be purchased without a prescription. Aspirin is an effective painkiller because it depresses the pain centers in the brain. It is an effective anti-inflammatory agent. Joints love aspirin. In addition to pain, most doctors prescribe it for tendinitis, neuritis, and arthritis. Every parent knows that aspirin reduces fever quickly. Americans consume 27 million pounds of aspirin a year.

Is Heat the Best Immediate Treatment for Injury?

Never put heat on a new injury because heat opens up the broken blood vessels. The result is that more blood collects in the injury, resulting in more swelling. The more swelling, the longer it takes the injury to heal.

The best initial treatment is to ice the injury for forty-eight hours. Start heat treatments after the initial two-day period. You want to stimulate blood flow because it collects the nutrients, oxygen, and inflammatory cells that heal the injury. Heat at that stage is just the trick.

Can the Body Perform at High Levels Forever?

Athletic performance is a complex equation, of which age is a part. For a while, age is on your side because you are gaining experience. But eventually it turns against you. In my experience there are few major league baseball players who retire voluntarily.

The aging process robs athletes of their quickness. I remember how sad I felt in the Red Sox 1980 training camp when George Scott, a former slugger, could no longer hit a major league fastball. It ended his career. Why he could hit one year and not the next remains a medical mystery.

Do All Broken Bones Heal?

There are three bones in the body that are notorious nonhealers—the ulna in the forearm, the carpal navicular in the wrist, and the tibia, one of the two lower leg bones. They all have a poor blood supply. About one-third of the time the fracture ends do not unite.

Traditionally, the only answer was surgery. More recently, electrical stimulation has been used successfully. Electric currents passed across the fracture stimulate bone production. It works in 50 percent of the cases.

Is Inflammation Bad?

The inflammatory process, the way nature heals injuries, is the body's mustering system. It is a three-component process. The blood brings white blood cells to the injury to fight off infection and clean up dead tissue. Nutrient building blocks assemble to rebuild the weakened tissue. New blood vessels form to bring increased amounts of blood with oxygen.

Inflammation is a vital process, but sometimes chronic inflammation can be self-defeating. It overinflames. Chronic tendinitis is an example of overinflammation. This is when I prescribe anti-inflammatory medication or cortisone shots.

Are All Athletes Equally Prone to Injury?

Some players or fitness enthusiasts never get seriously injured. Mike Torrez of the Boston Red Sox is in his sixteenth year of professional baseball. At thirty-five, he has never had an arm injury. Younger pitchers on the staff are plagued by shoulder and elbow problems. Why is Torrez injury-free? The answers are not clear. He has the right mix of genes, experience, strength, conditioning, and timing.

Dick Kazmaier, an all-American and Heisman Trophy winner from Princeton in the early fifties, told me he never missed a game because of injury. He's another athlete with the right combination.

Most of the great athletes in American sports—Jim Brown, Ted Williams, Babe Ruth, Bill Russell, Dan Gable—remained mostly injury-free during their careers.

Does Surgery Restore You 100 Percent?

No. I find that the body never forgets. In most cases the purpose of surgery is to take a joint that is operating at 60 percent and try to get it to a 95 percent functional level. Such a gain makes most surgical operations well worth some risk. But a surgeon cannot restore you totally. Only the creator can provide a perfect human body. A surgeon's job, in my opinion, is to try to restore body function as close to normal as possible.

If you need surgery, do not despair. Many athletes who play in major league sports have had major surgery. I have removed torn knee cartilages from Jerry Remy and Dwight Evans, both top-flight baseball players. Both, however, play effectively with less than normal knee anatomy.

Can You Play into Shape after an Injury?

"Playing into shape" is one of the most dangerous myths circulating in sports circles. I do not know where this notion got started, but it is dead wrong.

I use the 95 percent rule. That means the injury must be 95 percent healed before you return to sports. I test the 95 percent rule by comparing body parts. For example, if you have a broken wrist, I compare it with the good one.

The best way to recover from an injury is to rehabilitate—to stretch and strengthen the injured part. With every injury there is some loss in strength and flexibility. Try practices or easy exercise. Playing back into shape is the best way I know to extend an injury.

However, the final stage of any rehabilitation program is to restart your sport. This is not what most people think of when you mention this phrase.

Common Injuries

What to Do About Common Injuries And How to Prevent Them

This chapter is about the aches, pains, strains, and sprains that you will encounter from time to time in athletics. Sports injury is inevitable. In my football career, I suffered a broken nose, three broken fingers, two front teeth knocked out, stretched knee ligaments, torn ankle ligaments, a mashed toe (the nail was surgically removed), cervical nerve burners, a broken hand, and a series of face lacerations that required thirty-two stitches. Despite my injuries, I missed only two games in eight years of football.

Probably, the least-injured athlete I know is my coauthor, Marshall Hoffman. During the writing of this book, he used to say, "I'm never injured. I am careful not to overdo it." But in the last week of working on the book, he came to my office with a limp.

"What's the matter?" I asked.

"I pulled a groin muscle," Marshall answered sheepishly. "I don't understand it. I was running slowly."

My point is that injury happens even in "safe" sports like shuffleboard, table tennis, and badminton. There is no absolutely foolproof form of injury protection.

Many injuries heal by themselves without medical supervision, but some, if left undiagnosed and untreated, can hamper or even end your participation in sports.

The purpose of this chapter is to give you guidelines and insight into injuries and how to prevent some of them from happening. I don't advocate self-care. I think that a physician can help your body to heal itself.

WHAT TO DO FIRST

Have you ever seen a football player on the sidelines with his foot propped up on a bench, with a bag of ice lashed to his ankle, wearing an Ace bandage? That player is practicing athletic injury first-aid. It is called RICE and is the immediate treatment for almost all athletic injuries, whether you've pulled a muscle, sprained a ligament, or broken a bone. The letters in the acronym RICE stand for:

Rest. Rest is necessary because continued exercise or other activity could extend the injury. Stop using the injured part the minute it is hurt. Use a sling or crutches.*

Ice. Ice decreases the bleeding from injured blood vessels because it causes them to contract. The more blood that collects in a wound, the longer it takes to heal.

*Excerpted from *The Sportsmedicine Book* by Gabe Mirkin, M.D., and Marshall Hoffman, Little, Brown & Co., 1978.

A SITTING TARGET

"I have been hit by foul and pitched balls all over my body," says Carlton Fisk, former catcher for the Boston Red Sox. "In 1980, I got hit twice on my left instep within three days. It rendered my left foot almost unworkable. One hit was so forceful that it tore part of the cover off the baseball.

"I was hit off my big toe on the left foot twice in the same game. As a result, I am losing my toenail. When your feet don't feel good, you don't feel good."

"NO REST"

Almost all trainers and team physicians who we talked to are faced with the same problem: how to get their players back to action quickly and safely after injury.

Charlie Moss, trainer for the Boston Red Sox, says, "Minor injuries are a particular problem in sports like baseball. We have one day off in every twenty days or so. It is tough to get a small injury rested. The problem is that minor injuries grow into major problems."

ICE THERAPY

"It is very difficult to get an athlete to believe that ice treatment is therapeutic," says Ray Melchiorre, the trainer for the Boston Celtics. "They think that ice is only important for its numbing and pain-relieving value. As soon as the injury becomes numb and pain-free, they'll stop using it."

Compression. Compression limits swelling which, if uncontrolled, could retard healing. Following trauma, blood and fluid from the surrounding tissues leak into the damaged area and distend the tissue. Swelling is sometimes useful since it brings antibodies to kill germs; but if the skin is not broken, antibodies are unnecessary and swelling only prolongs healing.

Elevation. Elevation of the injured part to above the level of the heart uses the force of gravity to help drain excess fluid.

Because swelling usually starts within seconds of an injury, start RICE as soon as possible. Don't wait for a doctor's orders. First place a towel over the injured area. Then apply an ice pack, ice chips, or cubes over the towel. Do not apply the ice directly to the skin as it can cause the skin to hurt.

For compression, wrap an elastic bandage firmly over the ice, around the injured part. Be careful not to wrap the area so tightly that you shut off the blood supply. The signs of a shut-off blood supply are numbness, cramping, and pain. If any of these occur, unwrap the area immediately. Otherwise, leave the ice pack and bandage in place for thirty minutes. Next, to allow the skin to rewarm and the blood to recirculate, unwrap the area for fifteen minutes. Then rewrap it. Repeat this procedure for three hours. If the area continues to swell or the pain increases, check immediately with a physician if you have not already done so.

If the injury is severe, you can follow the RICE program for up to twenty-four hours. If pain and swelling persist forty-eight hours after the injury, apply heat. Further treatment depends on the type of tissue that was injured.

"MY MAID SAID . . ."

"A doctor friend of mine recommended aspirin and moist, wet heat to help a patient who is suffering from back trouble.

"The patient returned a week later and told my doctor friend, 'The medicine and those heat treatments didn't do me one bit of good. My maid told me to put ice on my back and it feels great now.'

"My doctor friend retorted, 'That's funny, my maid always recommends moist, wet heat.' "

*From a story told to trainer John Lally of the Washington Bullets by team physician Stan Levine.

When Should You See a Doctor?*

In my opinion, you know your body best. If your intuition tells you that you are injured, see a doctor. I tell my patients that they have only one body. Don't gamble with it. Here are other guidelines.

1. **Pain.** Any injury that causes severe pain. Pain is nature saying that something is wrong. When it talks loudly, listen.
2. **All joint injuries.** All injuries to a joint or its ligaments should be examined by a physician. If they are not treated quickly, these injuries have a potential for permanence. A joint injury should be immobilized until it is seen by a physician.
3. **Loss of function.** If you cannot move a limb, an ankle or finger, for instance, then you have a loss of function.
4. **Pain in joint or bone that persists for more than two weeks.** These tissues are the ones in which the most serious injuries occur.

*Excerpted from *The Sportsmedicine Book.*

5. **Any injury that doesn't heal in three weeks.** All injuries that don't heal should be checked for a structural abnormality.
6. **Any infection in or under the skin manifested by pus, red streaks, swollen lymph nodes, or fever.** Infections, if uncontrolled, may lead to serious complications. Antibiotics generally bring relief quickly.

These are only guidelines. Every injury is an individual event. Use your common sense.

How Long Will It Take You to Recover?

The rule in orthopedics is three days, three weeks, or three months. That was almost doctrine in my medical training. But I have learned through hard experience that life and medicine are not so simple.

For example:

A fractured finger. Three weeks to heal on the average for children. Three to five weeks for adults.

A broken collarbone. Four weeks in children. Six to ten weeks in adults.

A broken toe. Three weeks on the average.

A simple sprained ankle. Five days.

A mild thigh contusion. Five days.

A hamstring pull. Three to five days.

A simple shoulder separation. Seven days.

Healing time depends mainly on blood supply. The blood brings the elements necessary for healing:

- **Nutrients.** Think of injury as a remodeling job. Nutrients are the building materials for healing.
- **Oxygen.** This is the energy source for the project. I tell my patients that you can't have fire without oxygen. The body can't heal without oxygen.
- **Inflammatory cells.** These are the workers. They carry away the old blood and dead tissue. They fight off infection. They assemble the building blocks. Unfortunately, the best they can do is a slick patch job. Most body tissue, except the liver, heals with scar tissue. The larger the injury, the bigger the scar. It is this biologic fact that explains why the body never totally "forgets" an injury.

Blood supply to certain parts is inherently better than to other parts. That is why injuries heal at different rates. Skin injuries to the face heal usually within a few days because the blood supply to the face is one of the best in the body. Skin

NOBODY EVER EXPECTED ME TO CATCH AGAIN

"My worst injury happened in a collision play at the plate," says former Red Sox catcher Carlton Fisk. "I was expecting a relay throw from the outfield. The runner who was trying to score ran into my leg and blew out my left knee. The collision caused an O'Donoghue triad. That means I tore the medial collateral ligament, the medial meniscus, and the anterior cruciate ligament. On July 1, 1974, I was operated on by John Malloy of Brighton, Massachusetts. Dr. Malloy said that my knee was the worst he had ever seen. He did not expect me to ever play baseball again.

"I worked my ever-loving tail off to rehabilitate my leg. I strengthened my hamstrings and quads. I did squats; I did sprints; I worked with weights.

"In 1975, I went to spring training with the naive concept that I could play. I had worked hard on my leg and knee. I did everything except catch. On March 16, 1975, I caught three innings. The next day was St. Patrick's Day, and I was going to catch my second game in spring training, but I got unlucky again. I was hit with a pitch which broke my arm. It wasn't until almost a year later that I was able to return to baseball."

injuries to the arms and legs heal in ten days because of a smaller blood supply.

Muscles. After skin, the next best blood supply is to the muscles. Muscles, like the calf, are filled with blood vessels. When you tear a muscle (a *muscle strain*), you rip apart these blood vessels. The blood rushes out and collects under the skin. To you, it is a black-and-blue mark. In medical parlance, it is called an *ecchymosis.* The average muscle strain will heal in three weeks.

Bones. They are full of blood vessels. Why? Bones are constantly rebuilding. In fact, every ten years you tear down and rebuild your complete skeleton. Without a rich blood supply, you could not do this. Also, the bones are an enormous storehouse of minerals. The bones store 98 percent of the body's calcium, 40 percent of its sodium, and 30 percent of its potassium.

Ligaments and tendons. These are composed of collagen material, which looks like chicken gristle. It is fibrous and doesn't have many living cells. Thus, it doesn't need a large blood supply. Collagen usually requires six weeks to heal solidly.

Spinal discs. The discs in your neck and low back area have a marginal blood supply. When I cut into these discs, I see only a few drops of blood. This is why disc problems take three months or more to improve. You can't rush nature. You can't change the blood supply.

Knee joint cartilage (*meniscus*). Knee joint cartilage has no blood supply. Therefore, it has no biologic ability to heal itself. Once injured, always injured. This is why so many athletes have to have knee cartilage surgery. There is no other way. In 1972, 62,000 knee cartilages were removed because of football injuries.

You will heal faster if you are in good condition. That seems obvious. But there is a technical reason. People in good shape have a better blood supply throughout their bodies. A marathon runner has to supply more blood to his muscles than a nonathlete, when running.

The degree of the injury influences the healing time. For instance, there are three grades of ankle sprains. A grade I sprain, with mild swelling and tenderness, takes you out of competition for four to five days. A grade II sprain, with severe swelling and tenderness, requires seven to ten days of inactivity. With a grade III sprain, the swelling is so bad you can't move your ankle up and down. This injury will cost you at least three weeks of inactivity.

One way to extend your healing time is to start back to sports too early. Injured tissue needs time to heal. Tennis elbow sufferers are a good case in point. A tear in the muscle origin which connects the forearm muscle to the outer elbow knob is the cause of tennis elbow in about 90 percent of tennis players. The tear heals with rest. Because the blood supply to

RUSHING BACK

For the last few years, Mitch Kupchak, the forward for the Washington Bullets, has been playing with back injuries. "In the 1979-80 season, Mitch tried to come back too quickly from his back surgery," notes John Lally, trainer of the Washington Bullets. "Our team was playing badly that year, and Mitch felt a personal responsibility to get back into the lineup." But, he struggled all year. "Mitch is a great ball player, but last year he was terrible, pathetic. He wasn't the same kind of player that you see out there today.

"I asked Mitch if he had any outside pressure on him to return to the lineup. He said, 'No.' He made the decision to play."

this particular muscle origin is minuscule, the healing time can be two to three months. Normally, after a week of rest, many tennis players return to the courts, only to extend the injury. Even daily activities—like lifting bundles or opening a car door—can cause reinjury. Each time this happens, you restart the biologic clock.

When Can You Resume Your Sports?

That is the number one question I am asked by athletes, their parents, their coaches, their agents, and the management of the Boston Red Sox.

There is never a pat answer. As I discussed earlier, different parts of the body heal at different rates. Also some people heal faster than others.

Here are my guidelines:

1. If the injured part hurts at rest, you should not exercise it.
2. As soon as the injured part does not hurt at rest, you may start exercising it minimally. That means *slowly*. If the pain starts up, stop exercising. Your body is telling you that something is wrong. Listen to your body signals.
3. As soon as you can exercise without pain, increase the intensity and the duration of your exercise program. Expect a little aching. But remember that the moment that sharp pain starts, stop.

When you are recovering from an athletic injury, it is important to maintain your cardiovascular fitness. Thus, if you have an injured ankle, perform a sport that doesn't require that you use the ankle strenuously. Try swimming. If you have a wrist injury, try bicycling. It takes only six weeks to lose your cardiovascular endurance. Any exercise will benefit you more than resting in bed or sitting in a warm bath.

I treat athletes according to their injuries, not the sport they play. I treat Fred Lynn's hamstring pull the same way I treat David Weinstein's hamstring pull. Weinstein is a top marathon runner in the Boston area. Rob Roy McGregor fits orthotics—devices which support the bottom of the foot—for skiers, runners, basketball and baseball players—in exactly the same way. A foot is a foot; a hamstring is a hamstring; a bone is a bone.

The big question is how to tell what has been injured. That is the job of medical professionals. You need a diagnosis. Becoming an expert at diagnosing any medical ailment, sports injuries included, takes training and experience. I was lucky in my athletic career. The Harvard football team physician was Dr. Thomas B. Quigley, one of the pioneers in sports medicine and sports injuries. He was assisted by trainer Jack Fadden, a

BOBBY LAYNE'S OVERTREATED ARM

"In the early fifties, I was sitting in the training room in Detroit with Ted Williams," recalls Jack Fadden, trainer of the Boston Red Sox at the time. Bobby Layne, the quarterback for the Detroit Tigers, stopped in for a visit.

"I had been a trainer for Layne early in his career," says Jack Fadden." He was one of the best quarterbacks I have ever seen."

"Jack, my arm, I can't throw the ball across the room," Layne said.

"What's happened?"

"Well, I don't know. I have seen a lot of doctors. One guy gave me a shot; another guy gave me heat treatments; another doctor gave me diathermy; another doctor gave me ice treatments."

"Well, that sounds like a lot of treatment. Rest your arm."

"For how long?"

"About two weeks."

"About a month later, I heard that Bobby was throwing just fine. A mutual friend asked me, 'What did you do for Bobby?'

"I didn't do anything for him. I just told him to rest his arm."

man with forty-five years of experience and practical knowledge. These two professionals never missed a diagnosis on my battered body. They taught me.

It is a physician's or a trainer's job to determine which one of the body's seven structures—muscles, bones, tendons, joints, ligaments, fasciae, or skin—is injured. The more experience that your medical professional has, the higher chance you have of getting a correct diagnosis. Without an accurate diagnosis, you are shooting in the dark. One out of every three patients who is treated at Sports Medicine Resource has been seen by at least two other doctors. The reason they come to us is that their problem isn't solved. Most of the time we have to dig harder to make the diagnosis.

No physician knows all the answers. I refer difficult hand injuries to Richard St. Onge, Lewis Millender, and Edward Nalebuff—all experienced in treating that body part. I consult with Arthur M. Pappas on children's problems because he is an expert in that field. Bill Shea consults on hip and pelvis problems, his specialty. Rob Roy McGregor knows more about the feet than anyone I've ever met. Our therapists and trainers know more about rehabilitating the athlete than I'll ever know. Making good diagnoses is a complex matter.

When you can return to competition is the hardest question to answer. After my injuries I've experienced an insecurity in the pit of my stomach. In one case, my knee was hurt. I had already missed two games. Even though I wasn't 100 percent healed, I wanted to play badly, but I had doubts whether I could play on the bad knee.

The only way to find out whether you've healed enough is to play on the injury. It is impossible to simulate game conditions in practice. There is a point where you must try out the part.

If you are not fully healed, you might overload your other body parts. In 1978, baseball player Carlton Fisk broke ribs on the right side of his body. Fisk felt that he had to change his throwing motion to compensate for the rib pain.

For the professionals, the question of when to resume play can affect their livelihood. On most teams, competition for a starting job is fierce. "When I injured myself at the end of the 1975 season, the year the Red Sox won the pennant, I found myself sitting on the bench," says Jim Rice, who batted .309 and knocked in 102 runs that season. "It was a shock to my system. I had to earn back my starting job."

Jerry Remy, who was disabled with knee problems in both 1979 and 1980, says, "The most difficult part of an injury is letting down your teammates. They want you to play and you want to play. It is an emotional strain. But if you return too early, you will just extend the injury and be disabled longer."

Whether you are a recreational, amateur, or professional athlete, the timing of when you return to sports after an injury is an important decision. Consider it carefully, not emotionally. If there is any doubt in your mind, don't make the decision by yourself.

When Should You Use Aspirin?

For centuries, people have chewed on willow bark to relieve pain and reduce fever. In 1827, a scientist whose name is lost in history identified salicylic acid as the active ingredient in willow bark. It was not until 1875, when a salt was added to it, that caustic salicylic acid became widely used, especially for rheumatic fever—a strep throat which spreads to joints. (Now penicillin and other antibiotics stamp out rheumatic fever before it spreads.) In 1899, modern aspirin, acetylsalicylic acid, was discovered by a German chemist named Dreser.

More than any other medication, I recommend aspirin. Just because aspirin is cheap and can be purchased without a prescription, do not discount it, because:

1. It is an excellent painkiller or analgesic.
2. It is an effective anti-inflammatory medication.
3. It reduces fever quickly.
4. It is a safe medication.

Even in large doses, aspirin has no harmful effects on your heart, lungs, blood vessels, or kidneys.

I use aspirin mainly for its pain-relieving and anti-inflammatory powers. I start patients with irritated tendons (tendinitis), irritated nerves (neuritis), or swollen joints on aspirin immediately. Aspirin thins the blood, which is an important plus for people who have injured legs or have been operated on. The development of blood clots in your legs is called *phlebitis*, and is a potentially serious condition in which the blood clots can break off and travel to your lungs. Occasionally, it is even fatal. If you are a bleeder, do not take aspirin. It will encourage bleeding.

Medical science knows all about aspirin—how it is absorbed into the blood, how it is distributed within the body, and how it is eliminated from the body. But, what is not understood is how it actually works.

An average dose is two aspirins with each meal—six a day. In some severe arthritis cases, the worst type of swollen joints, physicians prescribe eighteen aspirins a day.

If you are taking too much aspirin, your stomach will "speak up." You will develop heartburn. I recommend that you take aspirin with meals to avoid stomach upset. It does not matter if you take it before or after, as long as it goes into your stomach about the same time as the food.

Aspirin is readily and chiefly absorbed in the upper intestinal tract—the stomach itself. Within thirty minutes, you have appreciable blood concentrations of aspirin. Peak levels are obtained in about two hours, and then a slow decline occurs over the next six hours.

After absorption, aspirin is rapidly distributed throughout all body tissues. It is excreted from the body mainly by the kidney. About 50 percent of a given dose is eliminated within twenty-four hours. Traces of aspirin can still be found in the urine up to forty-eight hours after taking it. The blood level of aspirin can be measured by a plasma salicylate test.

As with all things in medicine, aspirin can be a two-edged sword. If it is taken in massive doses, it can cause death. This happens when small children ingest a whole bottle of candy-flavored aspirin. Two bottles of aspirin are a fatal dose for adults.

Salicylate poisoning—the medical name for aspirin overdose—is treated by giving patients large doses of intravenous fluid. Sodium bicarbonate also speeds up the excretion of aspirin from the blood stream. Plasma transfusions are used if the patient goes into shock.

Some people are allergic to aspirin. Medically it is called *idiosyncrasy*. It takes the form of a skin rash (red blotches), swelling of eyelids, tongue, lips, and face, or a combination. Occasionally, the voice box swells, which makes breathing difficult.

Aspirin allergy happens mainly to people who have a history of allergic disease, especially asthma. I always ask a patient, "Are you allergic to aspirin?" Unfortunately, skin tests of various sorts are of little value in anticipating aspirin allergy, but the overall incidence of sensitivity to aspirin is low, probably on the order of one person in every five hundred.

Throughout this book, I will recommend aspirin for various ailments.

Rehabilitation

At Sports Medicine Resource we use the team approach to the treatment, prevention, and rehabilitation of sports injuries. Rehabilitation is an art. There we have six certified therapists and trainers as part of our organization.

Rehabilitation should begin as soon as the swelling stops. That means that in forty-eight hours after an injury you should be working with a therapist or a trainer.

Why is it so important to start rehabilitation early? The sooner you start, the less atrophy and weakness you will have in your muscles, and the less ground you will have to make up to get back to athletics and competition.

If you do not rehabilitate after an injury, your muscles will remain in a weakened condition. Even if you go back to full sports, they will remain weak, and subconsciously you will favor that part. Your body part simply will not feel right until full strength has returned. Also, you will be very vulnerable to reinjury. In a weakened condition, an injured part simply cannot protect itself against the stress of sports participation.

I see ten patients every week who have residual complaints after an injury. All they need is rehabilitation. Once they regain full joint motion and full muscle strength, their performance comes back to normal.

Therapists cannot speed up the healing process; nature works at its own pace. But a therapist can get you back to your sport at least 25 percent faster than if you remain idle.

A therapist with modern machines and specific techniques can do the following:

1. By moving your limbs, muscles, tendons, ligaments, you can reduce joint stiffness and muscle weakness, which accompanies every sports injury.
2. Muscle imbalance is corrected by stretching and strengthening opposing muscle groups.
3. Stimulate the blood supply to the injured tissue. This can be done by either moving the injured part or stimulating it mechanically as with massage or heat therapy. By improving the blood supply to the tissue, you reduce swelling, which promotes healing.
4. After a serious injury and a therapy program, it is a therapist, with a physician, who okays you for competition. Our therapists and trainers use the "95 percent rule." They insist that at least 95 percent of muscle strength be regained before you go back to full sports participation.

THE PAPPAS VIEW

"In my opinion, you can't understand sports medicine by sitting in a clinic treating injured athletes," explains Arthur M. Pappas, medical director of the Boston Red Sox, who has been a physician for baseball, football, skating, soccer, and rugby teams. "You have to observe athletes in their own environment . . . in action. You have to be one of them.

"Because each sport uses muscles differently, the rehabilitation therapy must be variable. Also, the timing of your return will differ with each sport. For instance, a skater can return to action one week after a broken collarbone, whereas that same injury would sideline a hockey player for eight weeks."

A therapist can make you use your injured part far more than you can alone. Therapists know from experience how hard they can push the injured part in a given day. They know to back off if the joint becomes puffy. They understand your pain threshold. They know which pain is good and bad. They know when injuries have been rehabilitated. They are experts in muscles, nerves, tendons, and ligaments.

Rehabilitation is serious business. It is not all fun. A good therapist knows when to put an arm around you and say that you are improving, for in this sense every good therapist is an amateur psychologist. They set goals, measure improvement, and reinforce your sense of progress. Jack Fadden, my Harvard trainer, would place tape marks on the power leg machine during my rehabilitation. I still recall how hard Jack pushed me in rehabilitation. I admire him for his hard job. We have remained friends for twenty years.

Physicians should think about physical therapy as part of the entire injury process. With a physician referral, go to a therapist. Don't put your arms and legs up on a shelf. Why? Because muscle weakness and joint weakness will quickly set in.

The following section was written by therapists Debbi Hall and Cynthia Rowe. They explain some of the physical therapy modalities they use.

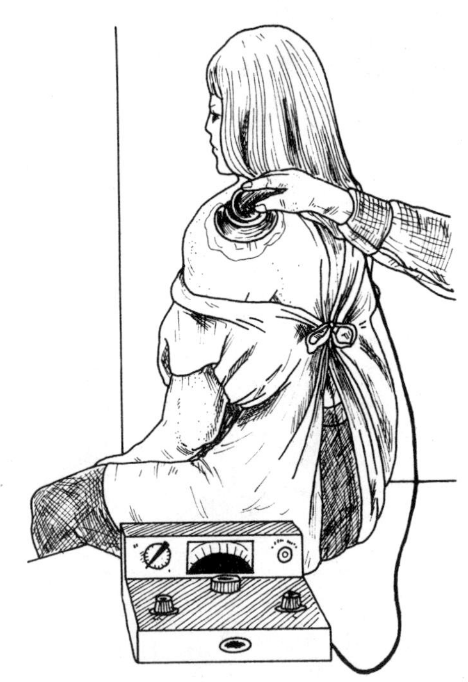

Ultrasound

Ultrasound

Ultrasound is a modality used to penetrate the skin and deep soft tissues with high-frequency sound waves. How does it work? The waves cause the tissue to vibrate, which in turn makes the tissues more accepting of the nutrients brought by the blood. The vibrations also aid in the elimination of cellular waste products. In therapy terms, it is called *membrane permeability*. A secondary effect of ultrasound is the production of deep heat. Any time cells move at a high speed, heat is produced.

Ultrasound can be used for many ailments, sprains, strains, contusions, tendinitis, bursitis, and scar tissue. Ultrasound should also be painless. We usually give seven to ten consecutive ultrasound treatments, with a gradual increase in dosage and duration. Length-of-treatment times range from four to ten minutes. If a patient feels an excessive sensation of heat, I reduce the dosage. This sensation indicates that the tissue is burning. Prolonged use of ultrasound, even at moderate intensities, should be avoided. Ultrasound is *very* safe under supervision of a therapist.

Ultrasound is not used in the areas of the eyes, ears, ovaries, testes, spinal cord, or pregnant uterus. It also should not be applied to an active infection.

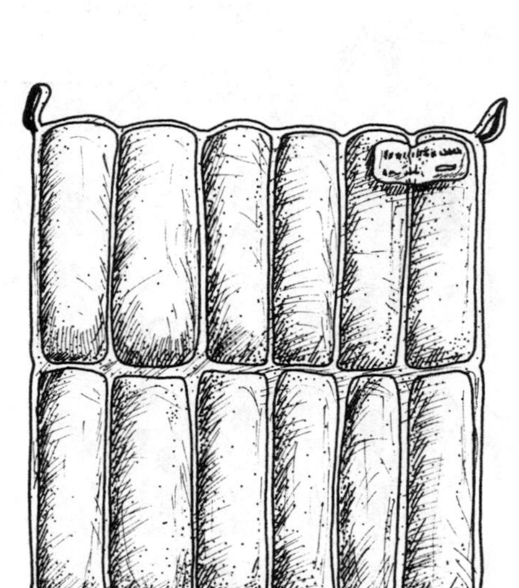

Hydrocollator

Hydrocollators

Hydrocollators produce moist heat and are used for their superficial heating effects. They are made of a combination of sand and silicone to enhance heat retention, and are stored in 126-degree water. To protect the skin, I place at least eight layers of towels between the skin surface and the hydrocollator. The patient uses the hydrocollator for twenty minutes per therapy session.

The heat penetrates into the subcutaneous tissues and superficial muscle layers. It has these physiological effects:

- It increases blood supply to the treated area.
- It increases temperature and thus relaxes the muscles.
- It decreases nerve excitability.

Hydrocollators are often used in conjunction with other modalities in physical therapy, such as ultrasound, range of motion, or exercise. Care should be taken when using hydrocollators over any desensitized area of skin or any metal implants. In these instances, extra layers of towels should be used.

Moist heat penetrates better than dry heat. Dry heat—like an infra-red lamp—penetrates only the top layer of skin, whereas moist heat goes deeper. Patients say that moist heat feels better. Having a patient relax before treatment aids in the progression of therapy.

Short-Wave Diathermy

Short-wave diathermy (SWD) is used as a physical therapy modality for its deep-heating effects. Short-wave diathermy involves high-frequency waves that penetrate up to two inches into the body's tissues.

Diathermy's heating effects are the same as any other heat modality. It is used primarily with inflammatory problems: osteoarthritis, rheumatoid arthritis, tendinitis, bursitis, tenosynovitis, and sprains. In sprains, I start the therapy after the danger of internal hemorrhaging has subsided. Diathermy is not used on desensitized skin, pregnant uterus, or over an area with a metal implant.

One of the disadvantages of diathermy over other forms of deep heat is the field size of waves emitted. The field size is so large that patients with pacemakers, even if *not* being treated with SWD, should be cautioned against being in the same room when the unit is in use.

Transcutaneous Electrical Nerve Stimulation

Transcutaneous electrical nerve stimulation (TENS) is a small, battery-operated device used to decrease or to eliminate pain. This pain can be from practically any problem; for example, migraine headache, chronic low back discomfort, after operations, or even during labor.

TENS

TENS is a low-voltage electrical stimulus that acts to bombard the sensory nerve endings in the affected area. The result is that those nerves send your brain messages describing the feeling of the TENS, rather than pain. It is thought that this nerve stimulation also acts to enhance the production of endorphines, the body's own "painkiller." When using a TENS unit, a patient wears two or four small carbon electrodes that are attached to a battery pack. This pack can be clipped to a belt or put in a pocket. On the pack are the dials that a patient uses to adjust the intensity of the stimulus to his or her pain tolerance.

Patients are advised to turn on the TENS intermittently, thirty to sixty minutes at a time, then shut it off and note the length of relief time. That way patients can record their progress. They wear a TENS unit only when they are in pain. Two advantages of using the unit are that (1) it can be used for any type of pain and (2) it enables patients to decrease the amount of pain medication they need to be comfortable.

Electric Stimulation

One way to get a muscle moving is to stimulate it electrically. With a sophisticated machine, electric current is passed into the muscles via electrodes. While electric stimulation sounds alarming, it is actually painless except for a prickling sensation directly under the electrodes as the current passes through to the muscles. *It doesn't damage the surface of the skin.* The intensity of electrical stimulation is easily adjusted to accommodate the pain tolerance of the patient.

Electric stimulation can be used as a diagnostic tool. I can determine if a muscle has lost its nerve supply. Without the nerve supplying the muscle, it atrophies. ES provides necessary stimulation to retard atrophy during nerve regeneration.

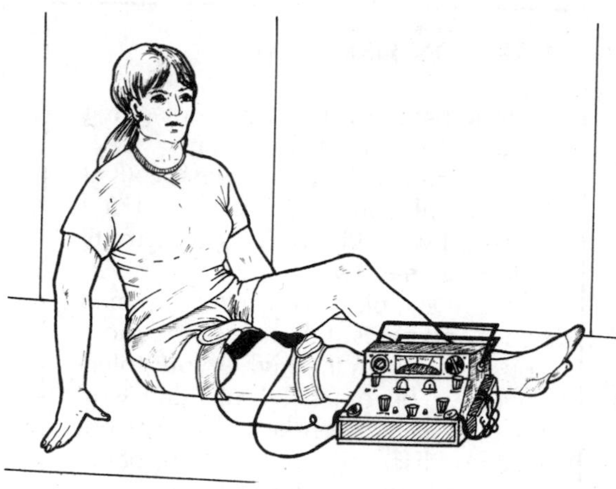

Electrical Stimulation

ES is also very useful in the treatment of muscles that have not lost their nerve supply. It can assist in strengthening a weakened muscle; "reeducate" a muscle involved in reconstructive surgery or one that has been immobilized for a long time; fatigue a painful muscle spasm; break up adhesions; and stimulate blood flow to the involved area, which promotes healing by reducing swelling and flushing away waste products.

ES is used to administer certain medicines called *ions*, by a technique named *iontophoresis*. The current produced by ES pushes the ions, such as hydrocortisone and acetic acid, through the skin to the injured area. Hydrocortisone is used as an anti-inflammatory agent and acetic acid is helpful in dissolving calcium in the muscle tendons or bursa sacs.

ES should be used only by a therapist. It is never used near the heart, on desensitized skin areas, or over an acute inflammation. ES should also never cause injury, exceed patient tolerance, stress an incision, or impede healing.

Cervical Traction

Cervical traction is a physical therapy modality used for whiplash, cervical arthritis, radiculitis, and cervical disc. Cervical traction does two things: It gives the small muscles of the neck region a slow, prolonged stretch, and it separates the spaces between cervical vertebrae to relieve pressure on nerve roots and/or discs in the neck region.

A patient can receive cervical traction either sitting or lying down. The traction device can be used two different ways, intermittently or static. The patient usually uses a unit for twenty minutes at a time and the poundage is gradually built up, depending upon patient tolerance, to thirty pounds. Cervical traction is usually used in conjunction with other physical therapeutic modalities, such as hydrocollators and ultrasound.

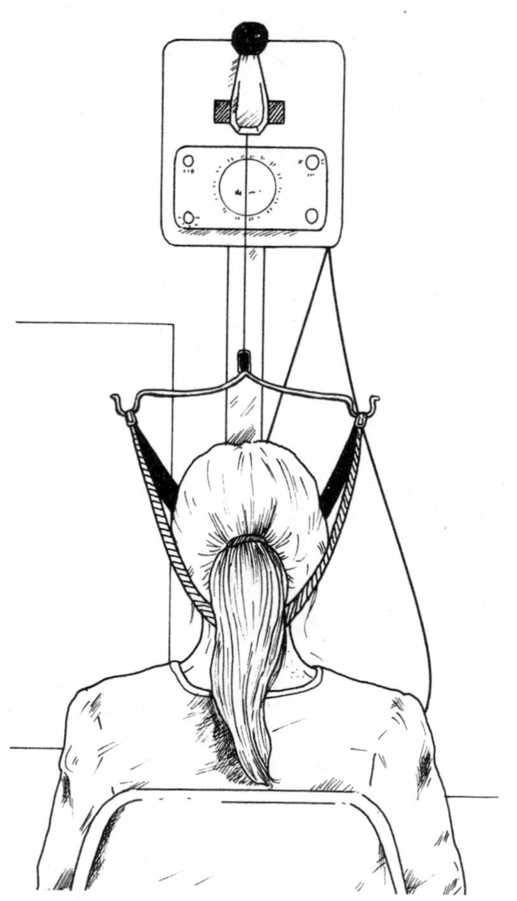

Cervical Traction

Pain

I am not an expert on pain. I am just a bone carpenter. But, I do deal with pain on a daily basis with my patients. Most people come to see me because they hurt. They want to know why they have pain and how to get rid of it. As a sports medicine physician, it is my job to answer these questions.

I understand pain personally. My high school and college football days were "painful" experiences. Like millions of others, I have had painful conditions, such as toothaches.

Because I have to interpret and analyze my patients' painful conditions every day of my life, I have changed some of my own pain perspectives. For instance, I do not take Novocain when I go to the dentist. I know that the pain is coming, and I have been able to get myself to relax and swing with it. The pain does not threaten me anymore.

When I was in college, a cortisone injection into my knee

was tremendously threatening. I would tense up and this would make the pain of the injection much worse. Having blood drawn from my arm for diagnostic tests no longer seems as painful as it was at an earlier age. Perhaps this comes as a result of my body aging and being less sensitive to pain in general, but I prefer to think that I have changed my mental attitude toward painful conditions.

Here is a simplistic view of the way pain works:

Your brain is the central switchboard, or monitoring station, of the body. It is consistently receiving information from your arms, legs, fingers, skin, eyes, ears, and mouth. The brain interprets the information and acts appropriately. If you touch a hot object, the temperature-sensitive devices in your skin send a quick message to your brain that the heat is on. This mechanism allows you to withdraw your hand quickly so that you will not be badly burnt. Your brain receives cold messages the same way.

Your brain is constantly picking up information from your joints. That is how your brain knows where your limbs are at every moment. We call this *proprioception*. Proprioception allows you to walk through a darkened living room in the middle of the night without falling. Your brain senses exactly where your hands and feet are in space.

You have special pain and pressure receptors in all the tissues of your body. They send their messages back to the brain by way of nerves which travel up the spinal cord to the brain. Pain receptors are activated when they are mechanically disrupted or are pressured by swelling.

Suppose you fracture your leg, which is full of pain receptors. The disruption of the bone sets off "scream" messages to your brain. In the first six hours, the swelling, which accompanies the fracture, adds to the pain.

Several scientific studies have shown that pain messages in humans vary from person to person. Some people have a high pain threshold; others have a low threshold. I do certain operations over and over again. I use the identical surgical technique every time. However, the amount of pain that people experience after that operation varies tremendously from person to person. Knee cartilage surgery is a good example. After I remove a torn knee cartilage, some people require morphine shots for five days; others need only aspirin to manage their discomfort.

Besides being sensitive to pain, if you are depressed or anxious, pain messages tend to be more intense. If you are a professional athlete and you earn your living with your body, any injury is going to be very threatening to you and your family.

One of the most frequent questions I am asked is "Should I play in pain?" It's a complex question with no easy answer. Pain is telling your brain that something is wrong. I do not believe in the "macho" image—playing through or exercising in severe pain. The main problem is identifying mild pain.

CARLTON FISK'S PAIN

"Bumps and bruises are not going to keep me out of the lineup," says all-star catcher Carlton Fisk. "If I played only on the days that I felt perfect, I would be playing only thirty to forty games a year.

"I always play with a lot of small aches and pains. I think that the more you play in pain the higher your tolerance level becomes. I think that's why I played with my broken rib. I was not able to discriminate between being hurt and being injured."

A SIMPLE CURE

"In 1979, I had chronic low back pain problems on my left side," says Jim Palmer, the outstanding pitcher for the Baltimore Orioles. "The pain continued for three or four months until it was discovered that one of my legs was a quarter inch shorter than the other. The cure was simple. I had a podiatrist put a lift in my shoe. I have never been bothered by any low back problem since."

There are some guidelines: You should *not* play if the following occurs:

- Pain comes from a swollen joint
- Pain intensifies when you use your body part
- Pain is associated with an infection
- Pain continues unabated for three days
- Pain suddenly increases the morning after the injury

You know your body better than any doctor, coach, or trainer. I find that most of the athletes I treat are very savvy about their bodies. If you are involved in regular competition, you will almost always have some part of your body producing discomfort. Most athletes play through these ailments.

Painkilling medicine, *analgesics*, work mainly in the central switchboard of your brain. They diminish the brain's ability to pick up the painful messages from the injured area. When you have intense pain and cannot bear the discomfort, analgesics are merciful. I do not favor people taking strong painkillers so that they can return to sports activities. It gives them a false sense of security and causes them to overuse or to abuse the injured part, thus leading to extension of the injury.

Avoiding and Preventing Injury

Injuries can devastate a professional team. One reason that the Red Sox finished poorly in 1980 was the huge number of injuries to their key players.

After winning the National Basketball Association championship in 1978, the Washington Bullets slumped in 1979 mainly because of injuries to three key players: Mitch Kupchak, Phil Chenier, and Wes Unseld.

Teams which remain injury-free win championships. That is one reason the Pittsburgh Steelers won four National Football League championships.

Luis Tiant, who pitched for the New York Yankees in 1980, claims that his longevity and injury-free performance in baseball is due to a conditioning program which he follows year round. "I keep my legs strong," says Tiant who says that he is forty years old. (Others estimate he is forty-eight). "Pitching involves the entire body. About 50 percent of the power and finesse in pitching comes from the legs. That is why I stretch and strengthen my legs," Luis adds.

Injury is the main cause of recreational athletes quitting sports. "I'm never going to play football again," a seventeen-year-old football player said to me after I took out his torn knee cartilage.

"Why?" I asked with surprise. "You are physically able to play."

"I'm afraid of getting hurt. I don't want to be a cripple."

In my opinion, there is no guaranteed way to prevent injury. I do not have a magic strengthening exercise, better equipment, or super training methods. I cannot examine a group of athletes and point out who will get injured because of overpronated feet or tight hamstring muscles. Yes, tight hamstrings strain more easily and overpronated feet do contribute to knee and hip pain; but certainly, not everyone who has these structural abnormalities sustains injury.

I could not predict Fred Lynn's hamstring pull. Lynn, one of the best conditioned athletes I have ever treated, stretches his hamstrings all the time. In fact, he is a compulsive stretcher. On the other hand, I examined the overpronated feet of Marshall Hoffman, my coauthor. Medically, he should be a candidate for knee pain. But in his six years of running, he says that his knees have never bothered him.

The risks of injury can be dramatically decreased by stretching, conditioning, and using good equipment. For example:

- Don Zimmer, the 1980 manager of the Boston Red Sox and now manager of the Texas Rangers, has a metal plate in his head. While playing with the Dodgers, he was struck on the head by a pitched ball. The blow caused internal bleeding in the brain. This is called *subdural hematoma*. His skull had to be drilled to release the pooled blood. The hole was about three centimeters across. It was covered by a steel plate. Such an injury might be avoided today because all batters wear helmets.
- I broke my nose and lost two teeth in football. If I had worn a mouthpiece or a football mask, I could have avoided those injuries.
- "Most running shoes are so well designed that they head off injuries before they happen," explains Rob Roy McGregor, chief of podiatric medicine at Sports Medicine Resource and designer of the Etonic Stabilizers. "They support the foot extremely well and limit the excessive rolling inward of the foot, which causes many of the running injuries."

If you are concerned about injury, there are a host of things you can do for yourself. For example:

- Pick a safe sport.
- Train, but do not overtrain.
- Stretch.
- Use strength training.

PICK A SAFE SPORT

Some sports are inherently more dangerous than others. About 20 percent of the one million youngsters who play football sustain an injury to the knee, shoulder, or ankle each year. Fifty thousand nonprofessional football players require knee surgery. Football is a relatively high-injury sport. In a study prepared by Arnold. E. Reif, for *Sports Injuries, The Unthwarted Epidemic*, by Paul F. Vingar and Earl Hoerner, football was rated as a very high risk sport for long-term serious injury. It was in the same risk category as mountain climbing, downhill skiing, snowmobiling, and hang gliding. Surfing and mountain hiking were medium risk. Other contact sports were not measured.

A similar study by Pennsylvania State University on the injury rate of college athletes showed that football was the most injury-prone sport, followed by ice hockey, soccer, basketball, wrestling, and field hockey. Sports which registered low injury rates were softball, volleyball, and baseball.

The data showed that the less contact there is in sports, the less injury—something you probably anticipated. Of course, I love the game of football and consider it well worth the risk. Every athlete must make his or her own decision.

TRAIN, BUT DO NOT OVERTRAIN

There are many things you can do to reduce the chances of injury. One is not to overtrain. The body is built to work and to play. Remember, it has limits. About 60 percent of the injuries that we see at Sports Medicine Resource are due to overtraining.

Why is overtraining so common? Athletic improvement happens when you push your body. The reason so many records have been slashed in sports is that athletes are performing prodigious workloads. These athletes are literally extending their bodies to the outer limits. We have professional athletes who work out on the Nautilus equipment and can bench press twice their body weight. I treat teenage swimmers who swim six hours a day. I see gymnasts who work out five hours daily.

There is an upper limit to the amount of work even the most highly conditioned body can perform. There is a fine line between work and overwork. Our office is full of athletes, recreational and professional, who have gone over that line.

The average athlete can sustain body stress only twice a week. That means going all out. The point at which this is reached varies, depending on conditioning, age, sex, and weight. Marshall Hoffman told me that when he started running, he would become breathless doing twelve-minute miles. In the beginning, that was his stress point. As he improved, he got down to eight-minute miles. In college, I could bench press 240 pounds, ten times. I never improved beyond that point.

Great athletes like Bill Rodgers, Patti Lyons Catalano, Dave Cowens, relief pitcher Tom Burgmeier, and wrestler Dan Gable can stress their bodies three times a week.

The problem is identifying stress and overstress, work and overwork. Here is a list which may help you recognize it.

- Localizing pain, usually in a joint
- Stiffness in a joint with a slight loss in motion
- Swelling
- No athletic improvement even with rest
- Change of the contour of a bone, muscle, or joint
- A burning sensation, tingling, and numbness in a limb
- Unexplained redness of the skin
- Unexplained black-and-blue marks

These symptoms should alert you that an injury is underway. Don't extend it. Rest! Back off. Start RICE.

STRETCHING

John Lally, the trainer of the Washington Bullets, is a scientific trainer. "I know that stretching cuts down on injuries," Lally says. Lally is more than talk. Before the Bullets practice or play, they stretch—not as individuals, but as a team. Lally is the leader. Only injured players are excused from the fifteen-minute program.

Joe Bourdon, trainer for the Boston Teamen, the professional soccer team, has devised a two-man strengthening routine for his charges. "The players like working together," says Bourdon. "They can stretch each other better than working alone."

Almost every professional team understands that stretching is an important way to prevent strains and sprains. It also improves speed, balance, agility, strength, and endurance. But many individual athletes haven't got the message. From personal observation, players of tennis, golf, and volleyball don't stretch before exercising. I don't know why. They just don't.

Why is stretching so important? Everytime you compete or exercise, your muscles are slightly injured. I call them "microinjuries." When the muscles heal, they heal slightly shorter. It is the same as the healing of a scar. The scar tissue draws the wound together.

Stretching lengthens the muscles and tendon units. It also fills the muscles with blood and makes them pliable. The more pliable the muscles and tendons are, the less likely they are to sprain or to strain.

Before you stretch, your muscles are like a tight string which can snap easily. After you stretch, the muscle-tendon units are like rubber bands.

There is a difference between the feeling of a "good stretch" and pain. Stretching should feel comfortable. When you stretch to a point where you are feeling an easy stretch, then hold and relax by "thinking" about the feeling of the stretch. As you relax and the feeling of the stretch decreases, then stretch a little further until you get the feeling of a "good stretch" again. This is stretching. If you go too far it will hurt and you won't be able to relax. It is necessary to learn how to

WHY STRETCH

Stretching, because it relaxes your mind and tunes up your body, should be part of your daily life. You will find that regular stretching will do the following things:

- Reduce muscle tension and make the body feel more relaxed.
- Help coordination by allowing for freer and easier movement.
- Increase range of motion.
- Prevent injuries such as muscle strains. (A strong, pre-stretched muscle resists stress better than a strong, unstretched muscle.)
- Make strenuous activities like running, skiing, tennis, swimming, cycling easier because it prepares you for activity; it's a way of signaling the muscles that they are about to be used.
- Develop body awareness. As you stretch various parts of your body, you focus on them and get in touch with them. You get to know yourself.
- Help loosen the mind's control of the body so that the body moves for "its own sake" rather than for competition or ego.
- Promote circulation.

SOURCE: *Stretching* by Bob and Jean Anderson, Shelter Publications, Bolinas, CA, 1981. Reprinted by permission.

RICE IS STRETCHING

"I don't stretch," says all-star outfielder Jim Rice of the Boston Red Sox. "All my muscles need are some greens."

Just after he made that statement, I noticed him down on the floor in the Red Sox clubhouse being stretched by third baseman Butch Hobson, one of the strongest players on the team. He just smiled.

relax and you can't relax if you are straining.

Here are my stretching rules:

- Stretching should be done slowly. Don't bounce or do ballistic stretches. That is one way to create strains. To be effective each stretch should last twenty to sixty seconds. Stretching should not be painful.
- You should stretch for at least fifteen minutes prior to exercise or competition. It takes that long to fill the muscles with blood and stretch out the fibers.
- You should stretch those muscles which you use during the exercise or competition. For example, tennis players should do wrist curls with small weights for their forearms and runners should stretch their legs. Baseball players should do runners' stretches and swing a bat, gradually increasing the weight of the bat. In the on-deck circle, some baseball players use a weighted bat or swing two bats at a time.
- The older you get, the more you need to stretch. With aging, your muscles become shorter. Also, some people are naturally tighter than others. It is even more important for these people to stretch.
- Marshall Hoffman, my coauthor, is one who stretches all the time. "No matter how much I stretch, my hamstring muscles are still tight. Because they are so tight, I can't touch my toes."

Each sport has its own set of stretches. Here are the four stretches that I recommend for running sports.

Stretching can be fun. Asta O'Donnell, of Brookline Physical Fitness, Inc., has stretching and exercise classes. Her students find the classes relaxing. They also have flat stomachs, good posture, and wonderful flexibility.

FOUR IMPORTANT STRETCHING EXERCISES

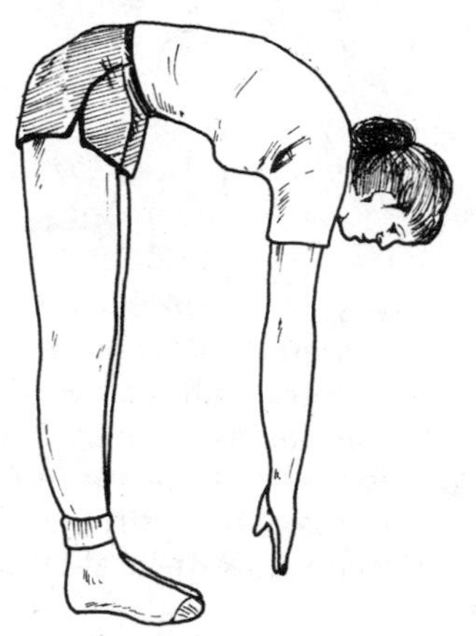

Toe-touching to stretch the hamstrings: With your heels together and knees straight, try to touch the ground or floor with your fingers. Do not bounce. Hold the position to the count of ten. Release, then repeat at least five times.

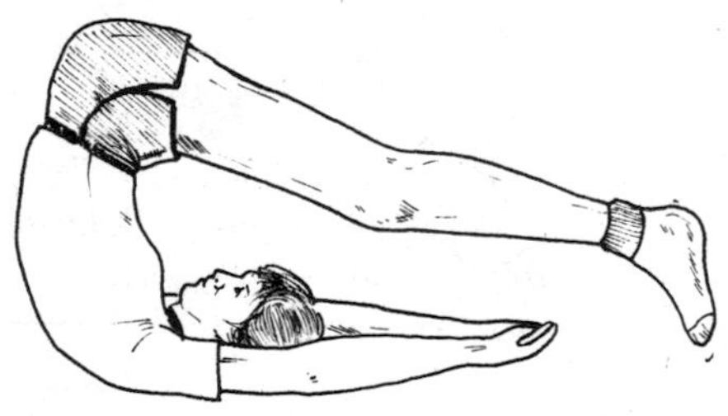

The plow to stretch the lower back and hamstrings: Lie on your back. Without bending your knees, raise your legs over your head and try to touch the ground with your toes. Do not force. Hold to the count of ten. Lower your legs and repeat at least five times.

WARMING UP

After stretching, you should warm up. Warming up increases the blood supply to the muscles and raises their temperature. Like stretching, warming up makes muscles and tendons more pliable and resistant to injury.

How should you warm up? The same way as the professional athletes do it. The Boston Celtics do a lay-up drill. The Boston Bruins skate slowly around the rink. The Boston Patriots have a light contact drill. Some marathon runners jog a mile or so before a race. My point: Use your muscles in the same way as when you participate in your sport. Start off slowly and gradually increase the intensity of your exercise. The key to a good warmup is to increase the pace of your workout so gradually that your muscles can adjust to the increased pace and remain free from injury.

STRENGTH TRAINING

Strength training has always been important in football. But now athletes in other sports—long-distance running, baseball, hockey, soccer, and even tennis—are working on the weights. Patti Lyons Catalano, the woman runner of the year in *Running Times Magazine*, works out on the Nautilus equipment at Sports Medicine Resource. So does Red Sox second baseman Jerry Remy.

Why are so many athletes using strength training? One reason: Strong athletes get injured less. Many studies on animals show that strength training strengthens ligaments and tendons. Both become heavier and thicker.

Animal studies also show that bones from weight-trained animals have a greater resistance to breaking. Scientists explain that the bones, which are always growing, deposit calcium salts along the lines of stress, which strengthens the bones. But the detailed physiological mechanism which leads to stronger bones is not known for certain.

Studies on humans also reveal that the stronger you are the less likely you are to be injured. Bernard A. Cahill and Edward Griffith of Peoria, Illinois, studied football knee injuries for eight years. Before the introduction of a vigorous preseason conditioning program, including weight training, one athlete in fifteen sustained a knee injury. Following the introduction of a total body conditioning program, the number of knee injuries decreased 61 percent in linemen and 20 percent in backs. Besides the reduction in numbers, the severity of the knee injuries declined.

As far as injury prevention is concerned, there is no data to say that one form of strength training is superior to another. The choices are:

1. Dead weights—these are the standard weight-lifting plates with bars to put the plates onto in order to increase the load.
2 Universal gym equipment.
3. Nautilus strengthening machines.
4. Cybex isokinetic dynamometer muscle-strengthening machine.

I am asked if strength training slows you down. I believe in the old adage "with strength comes speed." The athletes that can run the 100-yard dash fastest tend to be the strongest participants in the event. Their degree of leg strength can be demonstrated on the Cybex isokinetic dynamometer

Medical scientists have studied weight lifters and found they are not "muscle bound." That is, they do not have restricted motion of their joints. They are not "tighter" than athletes with less muscle bulk. Body builders, however, are a different class of athlete. By the sheer bulk and physical size of their muscles, they have difficulty participating in sports such as golf or baseball.

Sometimes, upper body strength can hinder some athletes.

Japanese split to stretch the inner thighs: Stand erect. Without bending your knees, gradually spread your legs apart as far as you can. Place your palms on the floor for balance. Hold your maximum stretch for at least ten seconds. Repeat at least five times.

Wall push-up to stretch the calves: Face a wall, standing at least four feet away. Place your palms on the wall, keeping your back straight. Bend your elbows so your upper body will move closer to the wall. If you keep your heels on the ground, the calves and Achilles tendons will be stretched. Hold this position to the count of ten. Straighten your elbows. Then repeat at least five times.

Butch Hobson, former third baseman of the Boston Red Sox, who played football at the University of Alabama, told me that he had to lose about ten pounds of upper body muscle bulk before he could hit effectively in the majors. His average got steadily better at the plate as his upper body bulk decreased somewhat. In a sense, Butch was trying to play baseball with a water skiing life vest wrapped around his upper body.

I am a firm believer in strength training in all sports. The training regimen has to be tailored to the individual needs and the individual sport. That means that the best use of weights are specific. You should use weights to approximate the sport you are exercising or competing in; for example, football players should use sleds to build up shoulder and leg muscles, and runners should use pulley weights to copy the arm movements in running.

STRENGTHENING YOUR HEART

I am asked frequently "Should I train my heart?" Yes! Besides contributing to better health and longer life span, cardiovascular exercise can improve the quality of your life. It can decrease anxiety and alleviate depression, help you sleep, add to your ability to concentrate on work, increase your stamina, and help you lose weight.

Sports and exercise can prevent the ultimate injury—death from heart disease. Almost one million Americans die every year because of heart disease. It happens because of negligence of the body's most important muscle, the heart.

The rule of heart (cardiovascular) fitness is simple. Pick a sport which will increase your heart rate to a level where you are breathing hard. I do not want you to be completely breathless, just enough to huff and puff a little. This is the point where your heart is beating about 120 beats per minute. It is at that point where your heart is benefiting from exercise. You must exercise for thirty minutes continuously. No stopping! You need to do this three times a week. So, you will have to pick a sport like jogging, bicycling, rowing, ice or roller skating, cross country skiing, swimming, racquetball, or aerobic dancing.

Sports like bowling, golf, sailing, and softball do little to increase your heartbeat. Thus, they do not lead to cardiovascular fitness.

Even if your sport does not require cardiovascular fitness, it can be important. "The more fit you are, the relatively lower the stress level on your sport, and the longer it takes you to fatigue," says Chris Coughlin, exercise physiologist at the New England Heart Center. "In many sports, fatigue can ruin your concentration, and thus, your sport.

"Secondly, I've noticed that cardiovascular-trained athletes eat differently and perhaps more healthily. They need and eat more complex carbohydrates. The result is that their level of blood sugar is more constant. Fluctuations in blood sugar create hunger. Thus, they are less hungry."

Sports are becoming safer because of the work of others, too. For example, there have been improvements in coaching, rules, equipment, and environment.

Coaching

"In high school football, I had been taught to fake away from a defensive halfback who had an angle on me, and then run over him," says Dr. Richard St. Onge, now a hand and orthopedic surgeon at Sports Medicine Resource. "I was never more embarrassed when the Harvard freshman coach came screaming out on the field after I had run sixty yards, but failed to score a touchdown because I tried to run down a defensive back on the one-yard line.

" 'For crying out loud, for a Harvard kid, you've got to be the stupidest player I ever saw,' the coach screamed in front of all my teammates. 'Your head is not a battering ram—it's to think with. You're at Harvard now. We don't run over people when we don't have to.' "

Good coaching can be important in injury prevention. The most effective treatment for tennis elbow is to see a tennis coach. A coach will teach you to stroke the ball with your entire arm. Thus, the force of hitting the ball is passed up to the large shoulder muscles. This relieves stress on the outer part of your elbow. Hitting the ball with your elbow bent can lead to a ripping of the elbow tendon, which constitutes tennis elbow. I advise all my patients with tennis elbow to see a tennis coach. In the long run, a coach can do more good than I can.

Coaches, from experience, have learned the safe method of training and competing in sports. Tommy Harper, a Red Sox coach in 1980, explained to the team in spring training the correct way to slide into second base. I was surprised that there was so much to it.

"If you slide directly into the base with your foot flat, there is a good chance that you will jam your ankle," Harper instructed. "You will end up in Doctors Pappas's or Southmayd's offices. You are forcing your foot to take the full weight of your body. It is not a car bumper. A hook slide is safer. With a hook slide, you are using the top of your foot to touch the base. Because your knee is bent and your ankle is flexed, it should be easier to touch the bag. Your other leg is also providing tremendous help in decelerating your body."

The coach is the number one safety official in sports today. In many school systems and recreational leagues, the coach functions as trainer and doctor. After a simple injury, it is the coach who has to make the decision as to whether an athlete continues to play. Unfortunately, many coaches do not have the benefit of medical instruction. This puts the coach in a compromised position—morally, ethically, medically, and legally. Many coaches are very concerned with their legal responsibility to their athletes as well as their liability. I recommend the conservative approach. You can never be criticized for holding athletes out of competition, but you can be severely penalized for allowing them to continue once they have been injured.

Besides improving technique, coaches are important in putting the right body in the right position. I was a lineman in high school and college, because I was short and squat. I could not run as fast as a back. If I had played baseball, I would have been the catcher. My body type determined the sport and my position in the sport. My son David is a wing in hockey because he is a quick, strong skater and has a left-handed shot which goalies have difficulty stopping. If you play a position where your body is ill-suited, you are more likely to get injured.

At least thirty times a year, I tell injured high school athletes, "Change your sport or your position." Recently, I told an outstanding high school pitcher and his parents that pitching was too much for the boy's arm. "The tendinitis in his shoulder will never heal if he continues to pitch," I warned.

"He loves baseball," the parents argued.

"Fine, wonderful. Why can't he play first base for a few months?"

They looked at each other somewhat puzzled.

The mother said, "We hadn't even considered the possibility. Sure, let's try it."

For a runner with repeated stress fractures of the foot, I recommend bicycling. For a swimmer with shoulder bursitis, I recommend jogging.

Any experienced coach could offer the same advice to a player.

Changing Rules

Because of a record number of serious injuries, President Theodore Roosevelt in 1905 threatened to ban football in the United States. A series of rule changes in 1906 improved the safety of the game. Today, football has almost a million male players in secondary schools alone, plus 70,000 in two- and four-year colleges.

Two recent rule changes, involving spearing and crackback blocking, are designed to protect football players from being hit from the blind side. These types of blocks were responsible for a high number of knee injuries.

Teammate Ernie Zissus, who played football at West Point and later at Harvard, taught me how to use my head as an offensive weapon. This technique, called *spearing*, contributed to Ernie's four broken noses and numerous facial lacerations. It did not do my face any good either. After my football days had ended, spearing became an automatic penalty. It was found to be a dangerous technique for the player giving the punishment, as well as the player receiving it.

Penalties for "high sticking" in hockey, lifting the stick above the shoulders, were instituted to prevent the stick from being used as a weapon. In amateur hockey, fighting gets a player expelled from a game. In my opinion, that is one reason that there are few fights in the amateur ranks.

"When I first started running long-distance races in the summer, there might only be one or two water points in a ten-kilometer race," points out Marshall Hoffman. "Today, race directors are much more concerned about dehydration, heat exhaustion, and heatstroke. In the Falmouth Road Race, held in August ever year, I was hosed by friendly spectators six times. Nothing feels better to a runner on a hot day than a cold shower. Showers were important in keeping my body temperature down."

Now, most race directors are careful to provide water and other medical services. In the Schlitz ten-kilometer race held in Washington, D.C., in the spring of 1980, Marshall reports that three ambulances were stationed at the finish line. Wayne Leadbetter, a sports medicine physician and orthopedic surgeon, treated runners at the end of the race mainly for heat exhaustion.

Hundreds of other rule changes have been made in other sports. In a high percentage of cases, these changes improve safety and prevent injury. Governing bodies like the NCAA, the Amateur Athletic Union, and the Little League should be congratulated for the work they have done to improve safety and prevent injury.

Athletic Equipment

One reason that I was so battered in football—a broken nose, lost teeth, and thirty-two facial stitches—was that good protective equipment hadn't been developed. When I first started playing football, I used a leather helmet with no face mask. Later on, I wore a helmet with a single bar face mask. It wasn't until my junior year at Harvard that I finally got smart and started wearing a helmet with a full face mask.

Marshall Hoffman describes the sneaker-type shoes that he competed with in indoor high school track. "They were a black canvas shoe which came only in D widths. If you had a thin or wide foot, you were out of luck. It had a flimsy counter, a double piece of canvas, to support the Achilles tendon. It didn't have an Achilles collar, a flared heel, a saddle, or a flexible sole. About one-third of the team suffered from *plantar fasciitis*, pain on the bottom of the foot, at the beginning of the track season. The problem was caused by the stiffness of the track shoes. About halfway through the season, the plantar fasciitis would clear up mysteriously. It wasn't until I started to write about sports medicine that I figured it out. It took about half the track season to wear down the soles."

HOCKEY WRIST

"About 80 percent of the Boston Bruins sustain wrist injuries ever year," says Jim Kausek, sports therapist for the team. "I protect wrists with tape or a Velcro elastic band. The players who wear the wrist bands and the tape seem to have fewer sprains, but many of the players find that the tape or the band interferes with their shots."

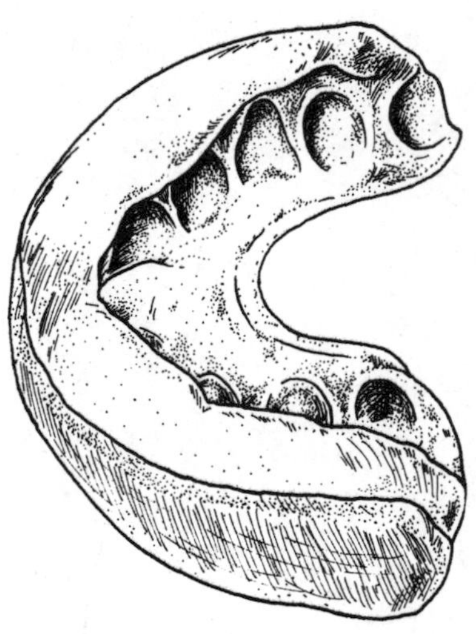

Protective Mouthpiece

The improvement of the football head gear has cut down many head injuries. The most important single element of the helmet is the system of webbed suspension straps. These straps fit the head and keep it from contact with the plastic shell. The helmet protects the back of the head and ears.

Wearing protective equipment is becoming a mandatory part of sports. Both Little League and major league baseball players wear batting helmets. Every professional player wears a cup-type jock strap to prevent injury to the genital area from line drives or inside fast balls. College hockey players now must wear helmets. At the high school level, goalies must wear face masks. In children's soccer, shin pads are essential. They prevent broken bones and "barked shins."

Various forms of protective equipment have been developed to protect female athletes who are playing softball, field hockey, and ice hockey. A specialized bra is used by runners and other female athletes to protect the breasts during sports activities.

Developing and selling protective sports equipment is a major part of the sporting goods industry. One estimate puts the dollar value of the equipment at $200 million a year.

One of the most dramatic examples of the value of protective equipment is the mouthpiece. Injuries to the teeth had statistically occupied a top rate of occurrence in high school football, as well as other contact sports. The injury is disfiguring and permanent. Insurance companies report a 50 to 90 percent decrease in dental injuries with a mouthpiece. In addition to preventing dental injuries, medical scientists have found that wearing a mouthpiece decreases incidence of concussion as much as 75 percent.

Some athletes resist wearing a mouthpiece. "I can't be understood by other players," one quarterback told me. In response, I put on a mouthpiece and started to talk normally. Sometimes when I address high school football players, I talk with a mouthpiece in my mouth. About halfway through my little talk I spit it out and give a few words about why wearing a mouthpiece is important. I got this idea from Dr. Victor Della Giustina, a dentist from Augusta, Georgia. Dr. Giustina, who did an outstanding study on the effectiveness of mouthpieces in the early sixties, is partially credited with the drive that made it mandatory to wear protective mouthpieces in high school football.

As high school budgets become tighter and tighter, it is the contact sports with their expensive protective equipment that come under greatest fire. One of the appeals of soccer over football is that the protective equipment necessary to outfit a soccer team is cheap compared to football. However, if you are going to run a contact sports program, you must have adequate protective equipment. To do anything less would be dangerous to the competing athletes. Here is a list of the necessary protective equipment in various sports:

Ice Hockey: Skates with Achilles tendon protector, shin pads, protective pants with thigh pads and hip pads, cup jock, shoulder pads, elbow pads, gloves, and helmets with face masks.

Football: Shoes, knee pads, thigh pads, hip pads, rib pads, shoulder pads, forearm pads, hand protectors, and helmets with face masks; flack jackets for quarterbacks.

Lacrosse: Shoes, knee pads, shoulder pads, elbow pads, gloves, and helmets with masks.

Soccer: Shoes, shin pads, cup jock, and elbow pads (optional).

Wrestling: Knee pads, jock supporter, cups, elbow pads, and ear pads.

Track and Field: Shoes and jock supporters.

Fencing: Shoes, chest protector, face mask, and gloves.

Baseball: Shoes, cup jock, and batting helmet.

Catchers: Chest protector, chin pad, face mask, and shin pads.

Basketball: Sneakers, knee pads, and jock supporter.

It is not enough just to give an athlete good equipment. It must fit properly and be kept in good repair. Equipment managers, coaches, and team physicians must constantly evaluate the nature and condition of the protective equipment offered their athletes.

Tape is important athletic equipment. A good tape job can make almost any joint more stable and less likely to sprain. Most college and professional teams understand the value of taping.

"I tape everyone's ankles before a game," says Ray Melchiorre, the trainer for the Boston Celtics. "It is a team rule. Taping cuts down on ankle sprains, a major injury for our players."

For many high school and recreational athletes, trainers are not always available. It is almost impossible to tape yourself effectively. That is why I sometimes prescribe a Lenox-Hill brace for patients with lax knee ligaments. The reason I use tape and the brace is to prevent injuries and minimize those you already have.

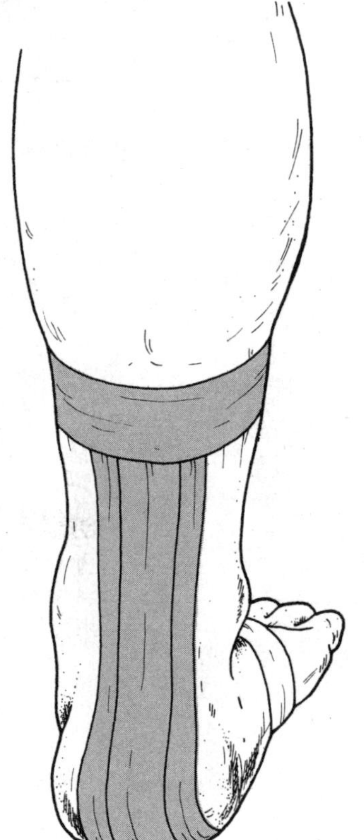

Taping for the Achilles Tendon

Environment

Exercising or competing on poor surfaces can cause injuries. Many endurance runners have leg complaints when running on hard pavement; these complaints completely disappear when they run on grass or soft composition track. The soft surface cushions the foot strike.

Many football and soccer stadiums are now covered with artificial turf instead of natural grass. While complete data is not in yet, it seems reasonable that there may be an injury-saving quality to the artificial turf, because it gives a predictable, level surface.

Potholes and uneven surfaces are the main causes of ski

accidents. Skiing on icy terrain obviously poses more of a risk than skiing in deep powder snow. Many basketball floors in a professional league are laid down on top of ice hockey surfaces. This can cause condensation of moisture on the floor surface and make it treacherous. This list of "surface" problems goes on and on in sports. It is up to league officials, game officials, and coaches to decide in team sports whether a particular contest should or should not be played on a given athletic field. I treat many ankle and knee sprains on rainy days when playing fields are sloppy. Many contusions happen while players play on frozen fields. Take environmental and atmospheric conditions into consideration before exercise or competition. Use good judgment and a reasonable degree of caution in judging the conditions of the turf under foot.

Medical Coverage

In 1978, at a Boston hockey rink, a high school player was cut across one of the main arteries in the neck which lead to the brain. He bled profusely. Although the hospital was only five minutes away from the hockey rink, he died before arriving at the emergency room. If a doctor or trainer had been on hand, the boy might have been saved.

In the best of all worlds, you should have a doctor attend every contact practice and game played in the country. The fact of the matter is that there are not enough physicians to do the job. Although 18,000 physicians are assigned to athletic teams, only a small percentage actually cover the games.

Being a team physician is expensive in time. Dr. Pappas, the medical director of the Boston Red Sox, spends about six hours at Fenway Park on game days. He usually arrives at the park at about 4:30, just before the Red Sox pregame practice. He checks on every player who is injured. It is a baseball sick call.

While the Red Sox are out on the field, he has sick call for the visiting team. Because of his national reputation and his good-natured demeanor, he has built up a following among many of baseball's leading players. Though they have their own team doctors, they like to consult with Dr. Pappas. Dr. Pappas and I rarely leave Fenway Park before 11 P.M.

The preseason physical is a standard fare in professional sports. I recommend it for all competitive athletes. I do them for both high school and professional players. In five instances, I uncovered scars on high school players. "What's this scar?" I asked.

"Oh! About two years ago, I was in a car accident and lost one of my kidneys."

"Have you had any problems with the other kidney since that accident?"

"No, Doc. Not a bit."

"Did it ever occur to you that you might hurt the other kidney in football? That would leave you with no kidneys."

MAKING FOOTBALL SAFE

- First-down markers and end-zone markers should be flexible and soft. Doctors call it "non-traumatic."
- "The chain gang" should be instructed to throw their sticks flat on the ground if the play moves toward the sideline.
- There should be ample room well beyond the end zone so that pass receivers do not charge into walls or stumble over curbing of the running track.
- Goal posts should be padded.
- The field itself should be level and firmly sodded.

A SAFE BALL COURT

- The basketball floor should be smooth and free from splinters. It should not be too slick.
- Since basketball players often dive for balls going out of bounds, the sidelines should be wide, even at the cost of making the court narrower. Spills are a serious business in basketball since the players do not wear much protective equipment.
- In many high school gyms, basketball nets are hung near obstacles. This should be avoided.

MEDICAL COVERAGE

My responsibilities as team physician:

Do preseason physicals.

Cover contact practices and games.

Diagnose, treat, and rehabilitate injuries.

Decide playing status of players.

Treat noninjuries; e.g., colds, fever, and splinters.

Arrange emergency room facilities, including ambulance and medical equipment.

Modify and design athletic equipment.

Recommend off-season training and rehabilitation programs.

"What does that mean?"

"The only way you could stay alive is to be hooked up to a kidney machine. Do you want to risk that?"

"I see what you mean, Doc."

"Have you ever thought of golf or tennis?"

Once or twice every year, I see this problem. It is my policy and the recommendation of the American Medical Association not to certify a player for contact sports with a loss of one of the paired organs—the eyes, kidneys, testicles, inner ear, or ovaries.

Besides the loss of a paired organ, I look for:

1. Blindness in one eye
2. Hernias
3. Pilonidal sinuses
4. Joint instability, deformity, or pain.

Not everyone believes in preseason physicals. Allan J. Ryan recently reported that only 15 percent of athletes have physicals by their team physicians. The National Collegiate Athletic Association suggests that a physical be done only once in four years. However, most schools still reexamine their athletes on an annual basis.

Besides covering the game and doing preseason physicals, I decide the playing status of players. I have told amateur and professionals players "You aren't ready to return to play yet."

I am responsible for treating injuries during the game. I answer questions such as "Is it a sprained ankle, a hamstring pull, or a torn cartilage?" "Can the player continue in today's competition or practice?"

Even though I am a specialist in sports medicine and orthopedic surgery, I am expected to treat common colds, give advice on baldness and insomnia. I am the source of second medical opinions in fields where I have no expertise. Players ask if I would look at their children or pregnant wives. That is all part of being a team physician.

Dr. Pappas and I are designers of equipment. In 1979, we had a special shoe designed for Carl Yastrzemski to help his Achilles tendinitis. I have devised numerous taping methods to support injured joints and limbs.

At the end of the season, Dr. Pappas, trainer Charlie Moss, and I give each player an off-season training and workout schedule. Over the winter months, players can put on weight and lose muscle. No professional athlete can get into playing shape in spring training. Even high school players should train year round. In the off season, Carl Yastrzemski, at forty-two, runs five miles a day. Carlton Fisk works out on weights.

I am responsible for having all emergency medical supplies on hand; that includes an ambulance with emergency medical technicians. At the beginning of the season, I make arrangements at a local hospital to handle x-rays and medical emergencies. We can put an injured player in a hospital in three minutes. (Fenway Park is in the middle of the hospital district in Boston.) A doctor can never go wrong by preparing for the worst.

EYE GUARDS

1. Champion Eye Guard
   Champion Glove Co.
   2200 E. Ovid
   Des Moines, IA 50313
   (515) 265-2551

2. Protec Eye Guard
   Protec Inc.
   11108 Northrop Way
   Bellevue, WA 98004
   (206) 828-6595

3. Protec Wire Guard
   Protec Inc.
   11108 Northrop Way
   Bellevue, WA 98004
   (206) 828-6595

4. Sports Spec
   Odyssey Optics
   Division of Plastic Safety Co.
   Box 47
   Eastmeadow, NY 11554
   (516) 481-4458

5. American All Sport Eye Guard
   Criss Optical Manufacturing Co.
   Box 220
   Augusta, KS 67010
   (316) 775-6346

6. Action Eyes
   Bausch and Lomb Inc.
   1400 N. Goodman St.
   Rochester, NY 14602
   (716) 338-6209

7. Safe-T-Gard
   Imperial Optical Co.
   Dundas Sq.
   Toronto, Ontario
   (416) 595-1010

8. Ektelon Goggle
   Ektelon
   7079 Mission Gorge Rd.
   San Diego, CAL 92120
   (800) 540-2662

9. Gargoyle Eye Guard
   Protec Inc.
   11108 Northrop Way
   Bellevue, WA 98004
   (206) 828-6595

SOURCE: *The Physician and Sportsmedicine.*

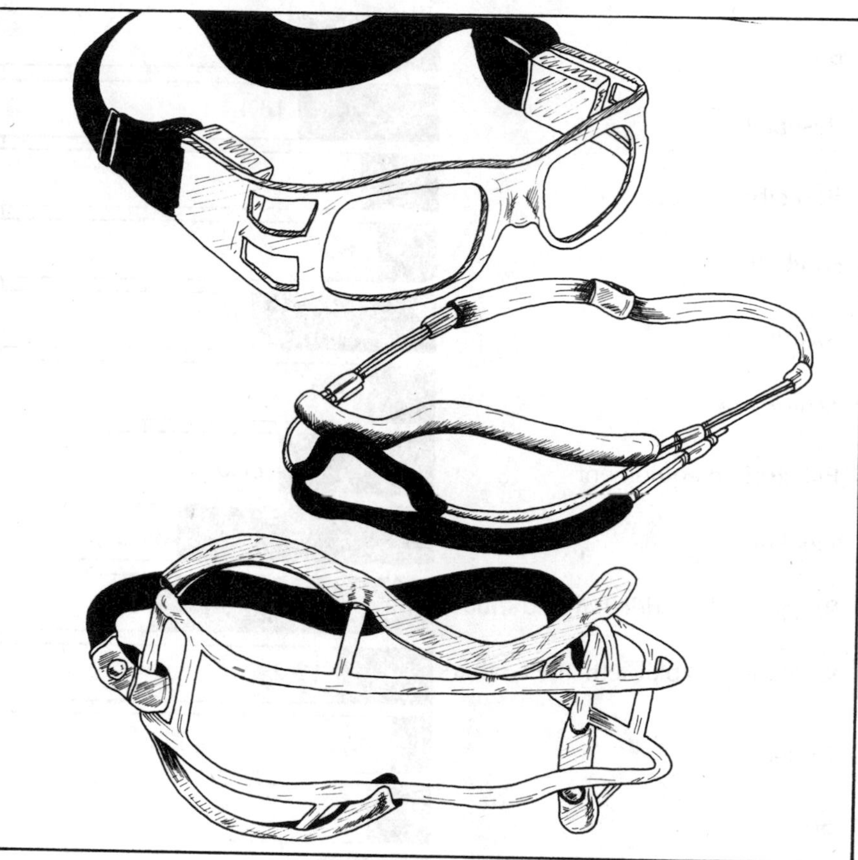

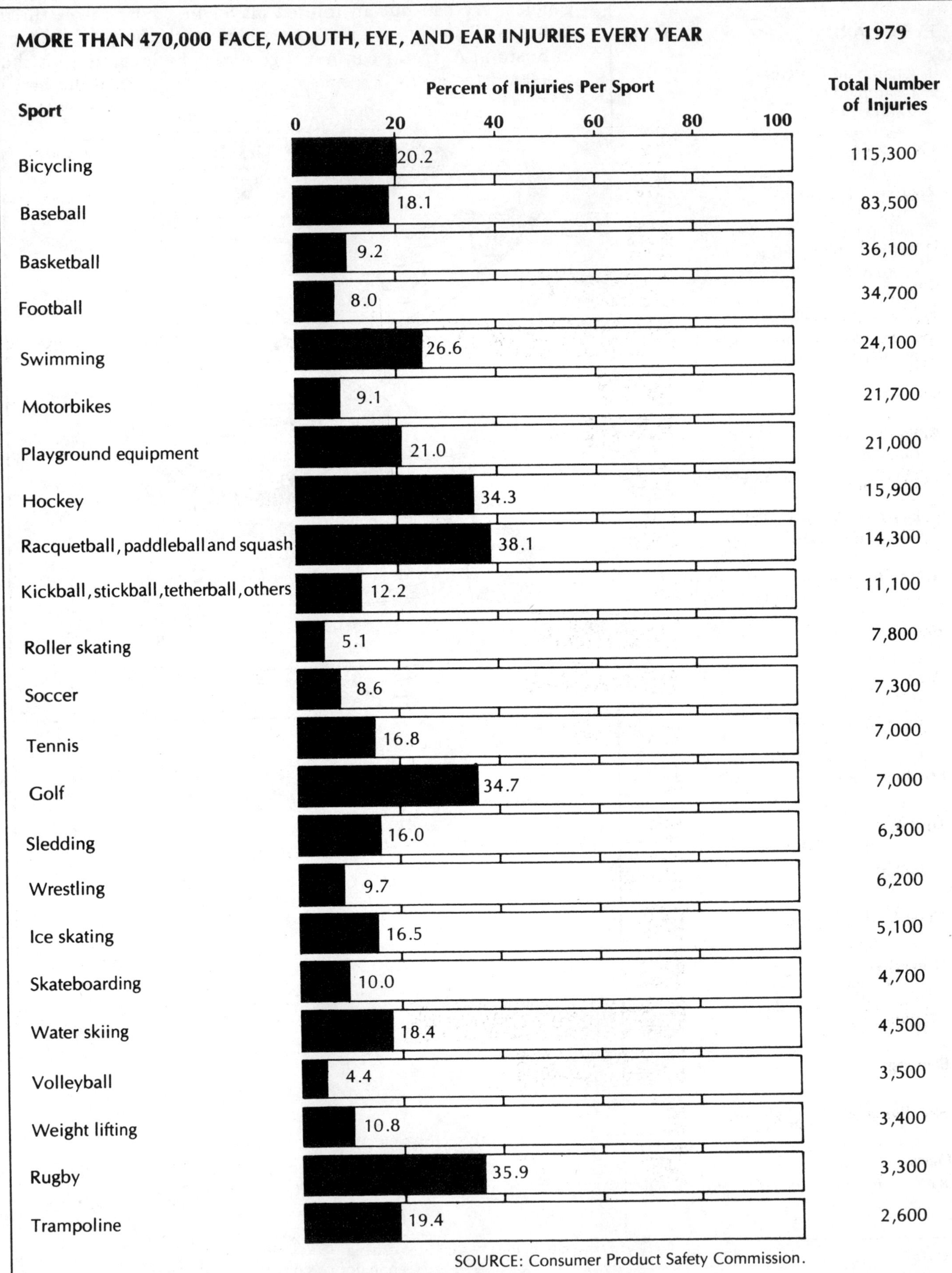

MORE THAN 470,000 FACE, MOUTH, EYE, AND EAR INJURIES EVERY YEAR
1979
Sport
Percent of Injuries Per Sport
Total Number of Injuries
0
20
40
60
80
100
Bicycling 20.2 115,300
Baseball 18.1 83,500
Basketball 9.2 36,100
Football 8.0 34,700
Swimming 26.6 24,100
Motorbikes 9.1 21,700
Playground equipment 21.0 21,000
Hockey 34.3 15,900
Racquetball, paddleball and squash 38.1 14,300
Kickball, stickball, tetherball, others 12.2 11,100
Roller skating 5.1 7,800
Soccer 8.6 7,300
Tennis 16.8 7,000
Golf 34.7 7,000
Sledding 16.0 6,300
Wrestling 9.7 6,200
Ice skating 16.5 5,100
Skateboarding 10.0 4,700
Water skiing 18.4 4,500
Volleyball 4.4 3,500
Weight lifting 10.8 3,400
Rugby 35.9 3,300
Trampoline 19.4 2,600
SOURCE: Consumer Product Safety Commission.

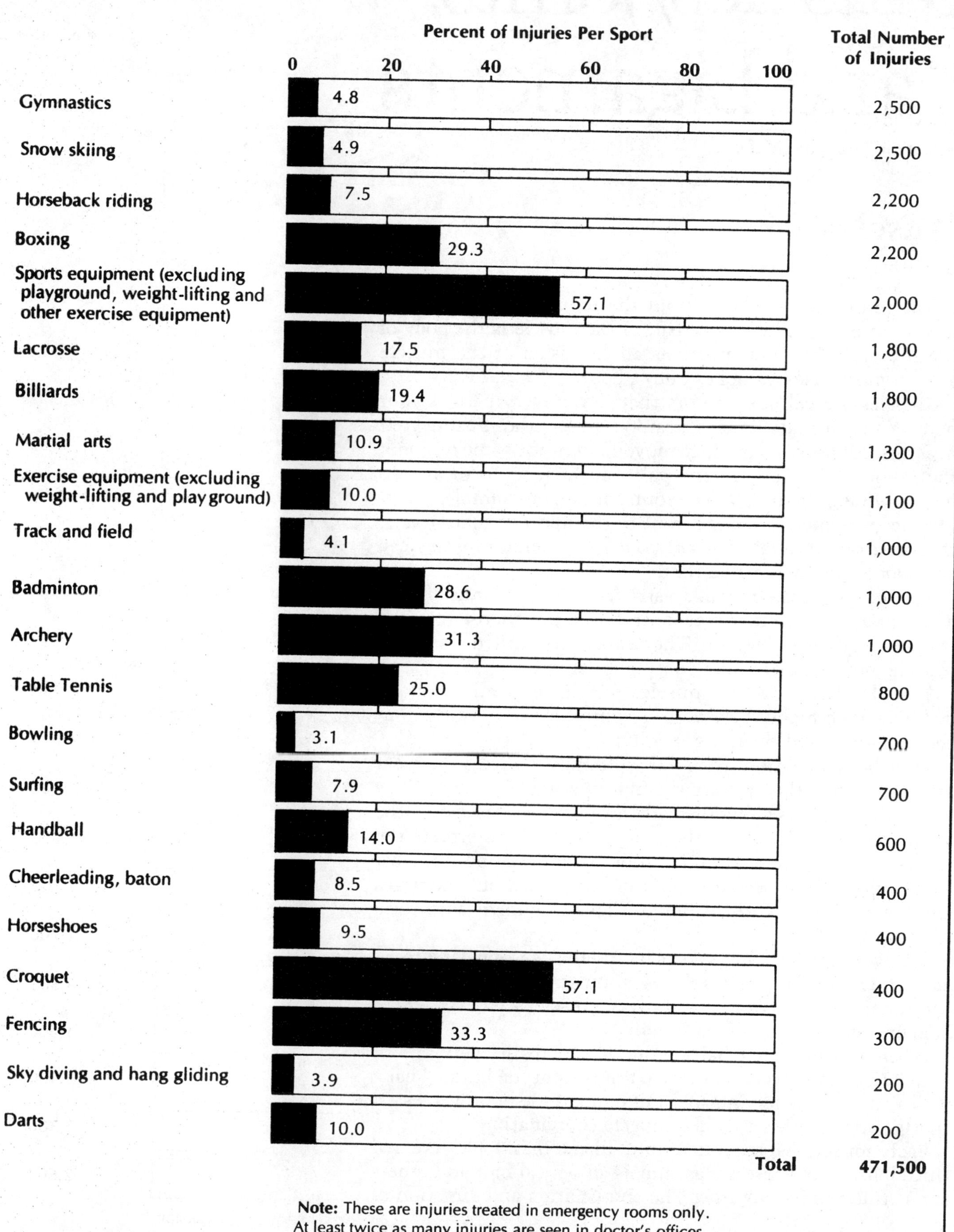

Note: These are injuries treated in emergency rooms only. At least twice as many injuries are seen in doctor's offices.

Muscles, Joints, and Ligaments

Arthur M. Pappas, M.D.

Muscles

The term *muscle* comes from the Latin word *mus* which means "mouse." The mouse part of the muscle is the body of the muscle. The tendon acts as an extension of the muscle and, in most cases, attaches onto a bone.

Muscles are cellular motors that move every part of your body. You can't talk, breathe, eat, or blink without using your muscles. All muscles produce movement by the same method: they shorten. The shortening pulls on the tendons or attachments which, in turn, move your bones. For example, when the biceps, which is on the front of the upper arm, contracts, the forearm is brought toward your body. Some muscles can contract as much as 50 percent.

The biceps muscle is called a striated or skeletal muscle. It is a voluntary muscle that you control. You can choose to make it operate at a specific rate. The heart is also called a striated muscle, but it is not under conscious control.

You also have smooth muscles within the walls of blood vessels which make these vessels contract. You have them in the intestines and these are involuntary.

Your body has more than 425 voluntary muscles. In most athletic events, the eyes are the first link in the system. They send visual images along the pathway of the nerves, called the visual neural pathway, to the brain. The brain interprets the signal, decides which muscles should contract and how fast and hard the contraction should be, and sends out messages along another network of nerves. This is known as a time reaction variable.

For every muscle that contracts, there is an opposing muscle, or antagonist, that relaxes simultaneously. The brain coordinates the movements of every muscle, a process which happens in a fraction of a second.

When you hit a baseball or throw a football, every one of your 425 muscles receives instructions from the brain. That's why you may spend your life improving these brain waves. In sports parlance, this is called muscle coordination.

Each muscle is separated into fibers that look like red thread. Many of these are as thin as an optical hair and others are as thick as fishing line. The composition and direction of

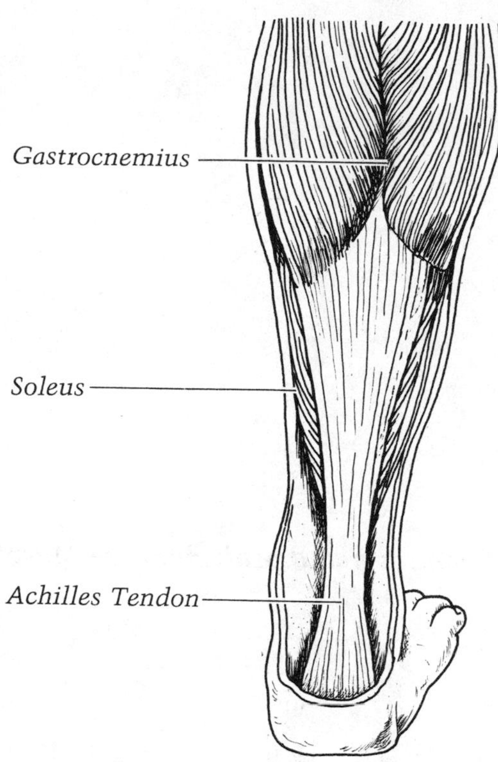

Calf Muscle and Achilles Tendon

the fibers determine the strength and appearance of the muscles. For instance, the finger flexor muscles are made up of fibers on a parallel axis which terminate in a flat tendon. These fibers are known as *penniform* and *unipennate* muscles and give the muscles ability to contact or to shorten. Examples of these muscles are the small muscles of the hand and foot.

A second muscle fiber configuration is the multiple herringbone pattern which is called *multipennate.* An example of this muscle type is the deltoid muscle of the shoulder. The herringbone pattern gives these muscles a smaller range to shorten, but they are more powerful.

There are two types of muscle fibers: fast twitch and slow twitch. The fast twitch muscle fibers have a high level of internal energy which make the muscle fibers contract at about twice the speed of the slow twitch fibers. Fast twitch fibers are activated for short bursts of speed—a 100-yard dash, a net volley in tennis, a 40-yard sprint in football, a fast break in basketball, a finishing kick in a long-distance race.

The slow twitch fibers move your muscles at a slower pace. Distance runners, cross country skiers, rowers, bicyclists, and skaters usually have more slow twitch fibers.

You are born with a certain number of fast and slow twitch muscle fibers. I am a heavily muscled man and therefore have more slow twitch fibers in my muscles than fast twitch ones. I could never have been a great sprinter. I played guard on Harvard's football team. On the other hand, Dick St. Onge of Sports Medicine Resource (who wrote the chapter on the hand and fingers and was a former resident on my service) is taller and leaner. St. Onge played fullback on Harvard's football team. He has a greater proportion of fast twitch fibers.

How can you find scientifically if you have more fast or slow twitch fibers?

A muscle biopsy can reveal the proportion of fibers. Under local anesthesia, a small piece of your muscle is removed with a special needle. The muscle is stained with dye and its differential composition is observed under a microscope.

Muscle size is due to the width of the fibers in your muscle. When you exercise, especially against resistance, the diameter of the muscle fibers thicken. This process is called *hypertrophy.*

Muscle size can also increase by splitting, which is called *hyperplasia.*

"The majority of scientific opinion is that hypertrophy is the main factor in muscle size," says exercise physiologist Chris Coughlin of the New England Heart Center. "These studies show that the number of fibers appear to be established at birth and only increase under extreme, repetitive muscular exertion."

Testosterone, the male sex hormone, contributes to the degree of muscular hypertrophy: the higher the level of testosterone, the greater the propensity for large muscles. Generally,

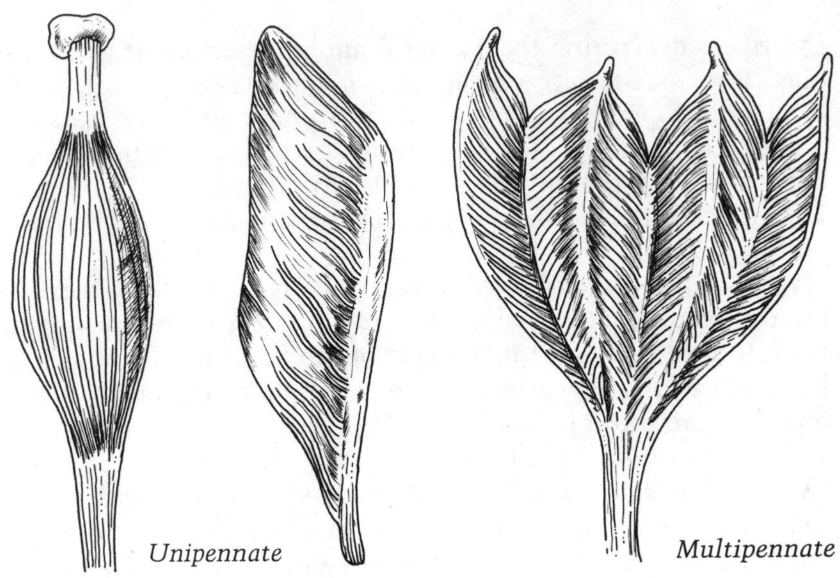

Penniform and Multipennate Muscles

females have lower levels of testosterone and smaller muscles than men.

Muscles get food and oxygen from the blood—the same way the rest of the body is fed. The blood, propelled by the heart, arrives at the muscles via arteries, circulates through the muscles via capillaries, and exits the muscles via veins. The blood flow is enhanced by the muscle's pumping action. As the exercised muscle relaxes, the blood flow expands. As the muscle contracts, the blood flow decreases. This pulsating action is responsible for boosting the blood supply in the muscle.

To further understand the functioning of a muscle, we must comment on the micro-energy generator system. Within each muscle fiber, there are a number of muscle fibrils that lie in parallel patterns. Each fibril contains units of two separate proteins: *actin* and *myosin.* These proteins account for at least 75 percent of muscle energy. It is the physical relationship of actin and myosin that creates a gliding mechanism and energy source as a muscle shortens and lengthens. The biochemical activator of this sequence is *adenosine triphosphase* (ATP).

The food that the muscle burns is called muscle glycogen or muscle sugar. The muscle stores the sugar for use during exercise. The muscle sugar that you have in your muscle at the time of exercise is the amount of fuel that you will have during the exercise. It comes from the food you have eaten about twelve hours before exercising. The amount of food that the blood stream can deliver during the exercise is minuscule. Only athletes in endurance sports—marathon running, long-distance bicycling, rowers, boxers, and cross country skiing—require large stores of muscle sugar before exercising. The way that these athletes increase the amount of sugar affixed to their muscles is to deplete—exercise all the sugar out of their muscles. The more an athlete depletes, the more sugar affixes to the muscle. Muscles are depleted when they become heavy.

Boxers call the condition "arm weary." Runners call it

"hitting the wall." Cyclists call it "bonking." (See page 401 on carbohydrate loading, a dietary manipulation to load the muscle with sugar.)

When muscle sugar burns, it breaks down into a chemical called pyruvate. If there is enough oxygen available, pyruvate converts to carbon dioxide and water, which are expelled from the lungs. However, if there's not enough oxygen in the muscles, the pyruvate converts to lactic acid, which builds up in the muscle and then overflows into the bloodstream. The lactic acid impedes muscle contractions and makes it increasingly difficult for muscles to move. A high level of lactic acid will eventually stop the muscle from contracting altogether.

A shortage of oxygen is signaled when you begin to breathe hard—huffing and puffing. Your body is trying to take in more oxygen to compensate for your oxygen debt.

As the body takes in enough oxygen to fire the muscles, the lactic acid converts to pyruvate again. The better condition your muscles are in, the more lactic acid they can hold without breaking down.

The harder you work out or exercise, the quicker you build up an oxygen debt and create lactic acid. You will quickly limit the amount of work that you can perform. Exercising without oxygen is called anaerobic exercise. Sprinters are anaerobic athletes. When you exercise with enough oxygen, it is called aerobic exercise.

Muscle Endurance

Muscle endurance is part of any sport. Endurance, in a medical sense, is the ability of muscle groups to respond to prolonged activity. This response is dramatically improved by training, especially aerobic training. This training strengthens your heart muscle so that it can pump a greater volume of blood with each beat.

- It enlarges your arteries so more blood can flow through them.
- It increases the number of red blood cells so your blood's oxygen-carrying capacity is increased.
- Training increases the rate at which enzymes in your muscles pick up oxygen from the blood.

The maximum ability of your body to deliver oxygen to your muscles is limited by your genetic makeup and varies substantially from individual to individual. The peak rate at which the body can take in and use oxygen is called the VO_2 max (max = maximum, V = volume, O_2 = oxygen).

Even though you have a genetic limit, with training you can get close to that limit and sustain it for a long period of time. I have one patient, a nationally known marathoner, who has a maximum heartbeat of 160 beats a minute. His heart can't beat faster. He can run at 80 percent of his VO_2 max for two hours and thirty minutes, the time it takes him to run a marathon.

Other athletes don't need this amount of training. I know of very few baseball players who have the same muscle endurance as a marathoner. There is no reason that they should. On the other hand, I don't know any marathoner who can throw a baseball at ninety miles an hour for nine innings. Every sport is sport energy specific.

As a muscle tires, it loses some of its ability to respond to the nerve impulses and is less efficient in eliminating waste products. In precision sports, like baseball, tennis, competitive ice skating, bowling, and some track and field events, slightly fatigued muscles can mean the difference between winning and losing.

What are the limiting factors in muscular endurance?

The loss of muscle sugar is number one. When you run out of sugar, your muscle becomes uncoordinated. It is the same as a car running out of gasoline.

A second factor is the lack of oxygen to burn the muscle sugar. This is called *hypoxia*. I have already talked about how muscles get food and oxygen.

Another factor is the buildup of lactic acid, which diminishes the ability of the muscle to respond. This is called *acidosis*.

You can also run out of fat reserves, which are secondary fuel for the muscle.

Still another limiting factor is heat buildup in the muscles, called *hyperthermia*.

Athletes often ask me if participating in different sports will improve their performance in their own sport. The answer is maybe. The best training for a sport is to practice that sport. However, participation in other sports tends to increase phyical fitness and provide physical and emotional diversion and prevent boredom and burnout—long-term problems for all competitive athletes.

Muscle training in any sport requires work on a combination of the following factors:

Coordination: Use your muscles in the same manner as you would use them in competition. If you are a tennis player, play as much tennis as you can. If you are a bowler, bowl. If you are a boxer, box.

Speed: Use your muscles in the same manner as you would use them in competition. I followed that rule when I played college football. That is how I sharpened my reflexes. Trainers and coaches all over the country are finally getting the message that speed exercise develops the fast twitch fibers of the muscle. Slow endurance exercises develop the slow twitch fibers.

Strength: Use your muscles against resistance in the same manner as you would use them in competition. Baseball players should swing with a weighted bat; golfers, weighted clubs.

THE CYBEX ISOKINETIC MACHINE

The Cybex is a muscle-testing and strengthening device that can give a detailed profile of the condition of various muscle groups, including the measurement of strength through the full range of motion of a particular joint.

This instrument now plays an integral part of evaluation, screening, and rehabilitation of athletes.

The Cybex is an orthotron machine with a computer. When an arm or leg is strapped to the machine and you flex your muscle, the muscle contraction is recorded by the computer.

The speed at which you produce this motion is preset on the machine itself. For example, suppose the Cybex is attached to your leg. You start the exercise with your knee bent ninety degrees. The purpose of the exercise is to straighten out your knee and to use the quad muscle of your thigh to achieve this motion. If halfway through your exercise, you develop minor pain in your knee and slow down, the machine lightens the resistance against your leg. This allows you to complete the motion at the same speed.

The Cybex machine uses a dual channel recorder which provides a continuous printout curve on graph paper of the range of joint angles and peak torque across the entire range of motion of the muscles being tested.

Isotonic testing, commonly performed on weight machines like the Universal gym and the Nautilus, measures strength by assessing the greatest weight that can be moved one time through the full range of motion.

Strengthening Muscles

Every athlete wants to be strong and a main objective of many training routines is to improve strength. Strong is synonymous with athletic excellence. "How do I get strong, Doc?" is one of the most common questions that I am asked and is a difficult one to answer, too. In general, an effective muscle-building program must be individualized for the athlete and the sport. For example, I give Priscilla Hill, a world-class figure skater, different muscle-strengthening exercises than Dwight Evans, the right fielder for the Boston Red Sox.

The key to strength training is to work the muscle to capacity. As your strength increases, the workload should be increased. The best strengthening exercises are those which duplicate the athletic skill exactly.

One of the most effective football exercises is to push a football sled with your shoulders. This duplicates the motion of the line charge. It strengthens the shoulder, neck and upper torso muscles and ligaments.

My coaches used the football sled all the time. As the season approached, we used the sled more and more. Eventually, a coach would stand on the sled so I would push the sled and the coach. This technique builds upper body and leg strength quickly.

Kyle Rote, Jr., twice winner of the superstar competition in the late seventies and all-star soccer player for the Dallas Tornados, practices hitting headers (hitting the ball with his head) 100 times. This exercise strengthens his neck muscles.

Most of the Red Sox use a weighted bat to practice swings. It builds the wrist, arm, and shoulder muscles in the same way as the batter swings.

Some track coaches recommend running in the sand, which is a resistance exercise for distance runners. Track coaches in New Zealand and Australia have used this technique to produce world-class runners and world champions.

When you strengthen your muscles, you must stretch them, too. Every time you use your muscles vigorously, they tear slightly. These are called *micro-tears.* When they heal—usually in twenty-four to forty-eight hours—they heal slightly shorter. If you don't include some flexibility training with your weight training, you will end up with a decrease in joint mobility. Some call this being *muscle bound.*

I've had that happen to one Red Sox pitcher. He strengthened the muscles in the front of his shoulder, but failed to stretch the muscles in the back of his shoulder. He did this without consulting me. The muscle imbalance dramatically changed his pitching motion for the worse. The ideal program is to increase strength and flexibility at the same time.

Muscle and joint flexibility is so important in professional baseball players. (That is not to say that flexibility is not important in other sports.) We test every player at spring training for many factors of physical conditioning. This way I

The limitation: the strength that is measured relates to the weakest point in the range of motion.

With the Cybex, muscles must contract at the speed of the machine. No matter how hard you push, the testing lever will not move faster than the predetermined angular velocity.

I use the Cybex mainly to diagnose the extent of injuries and to determine when an athlete can return to action.

For example, I test four motions in assessing leg injuries: knee flexion, knee extension, hip abduction and adduction, and I added a factor for strength.

I always compare the injured muscle to the unaffected muscle. The key measurement is the peak torque force that can be generated by each limb.

Peak muscle torque is measured on the machine at both 30°/S, a slow speed, and at 240°/S, a high speed. If the strength difference between the two limbs is greater than 10 percent, then the muscle is injured.

The only major drawback to isokinetic testing is that it measures strength generated by specific muscular contractions, not the coordinated effect of muscle balance or imbalance. Also, it is difficult to duplicate athletic functions that require complex muscle interactions.

Arthur M. Pappas, M.D.

know before the season starts who needs to stretch and which muscle groups need strengthening.

Some players are naturally strong. Jim Rice, one of the most muscular men in baseball, never works out with weights. "Weights make me too muscle bound," Rice explains. "It changes my swing." In the off season, Jim plays a lot of golf.

There are three forms of strength training.

- Isometric Exercises. In this exercise you tighten the muscles without actually changing the length of the muscle or the angle of the joint. For example: Tighten the muscle in your arm by making a fist, or draw in your stomach muscles. This type of exercise is most effective when the maximum contraction is maintained for five seconds and repeated ten times. This can be repeated during the day.

  These are the first type of muscle-strengthening exercises I assign after an injury. The main disadvantage with isometrics is that the muscle gains strength in one position and not through the entire range of the muscle.
- Isotonic Exercises. This is exercise that involves a change in the length of the muscle through part of its complete range against a moving or resisting force. The most common isotonic exercises are pushing weights. In this form of exercise, the load remains constant and the muscle fiber contribution varies, with the greatest gain being in the initial part of the movement of overcoming inertia.

  Isotonic exercises are especially effective when the weight or resistance is increased in following sets. This technique is known as *Progressive Resistance Exercises* (PRE), and was first described by Thomas DeLorme, M.D. Traditionally, it involves three sets of exercises with ten repetitions each. The first set is performed at resistance at one-half of one's maximum, the second set at three-quarters maximum, and the final set against maximum resistance. As you get stronger, the resistance is increased.

  PRE is relatively new. These exercises can be performed on Nautilus, Universal gym, or other similar machines to produce uniform resistance during the entire exercise.
- Isokinetic exercise is one of the new techniques in sports medicine. Isokinetic is a constant speed and constant resistance exercise. These exercises are performed on an orthotron or a Cybex machine. A Cybex is an orthotron with a computer. No matter how hard you push against the machine, the speed stays the same. The great advantage of this system is that it's difficult to exceed the capacity of your muscles and joints. It's safer, especially if you have an injured joint or muscle.

Secondly, there is a maximum involvement of the muscle fibers. I like that because it means that the entire muscle is being used throughout the exercise. I think that more of the muscle fibers that are used, the stronger the muscle becomes.

If you want to remain muscular, you must use your muscles vigorously. When they are in a cast, I have seen football players lose one inch off the circumference of their thighs in as little as ten days. Jim Rice, the heavy hitting left fielder, lost one inch off his forearm in the summer of 1980 when I applied a cast to his forearm because of a fractured wrist.

Muscle Injury and Rehabilitation

Every year millions of people who exercise or play sports sustain muscle injuries. Government specialists set the number of muscle injuries at 15 million a year. Most remain unreported. Only the severest are handled in doctors' offices or hospital emergency rooms.

I see muscle injuries everyday of my life. They are so common that they have their own jargon: bowler's thumb, halfback's hamstring, swimmer's shoulder, and tennis elbow. I can hardly recall a Boston Red Sox player who hasn't had a muscle pull, strain, or contusion during the season. Football players get muscle injuries in every game.

Muscle injuries can occur in many different parts of the muscle structure. For example:

- It may tear within its body or fleshy part. This happens in 40 percent of the cases.
- It may tear at the junction of the muscle to tendon. This happens 40 percent of the time.
- Also, the muscle may tear away from its bone origin.
- Or, the tendon which extends from the muscle may tear.
- And, lastly, the muscle may tear away from its bone insert.

I locate a muscle injury by touch. Usually, I can touch the tear even if it is deeply seated in the muscle body. I always compare the uninjured side. I make the diagnosis based on experience.

Like ligament tears, there are three grades of muscle tears or pulls—in medical parlance, these are called *strains:*

Grade I strain is merely the stretching of a few muscle fibers with minimal tearing of fibers, less than 10 percent, and without a touchable defect in the muscle.

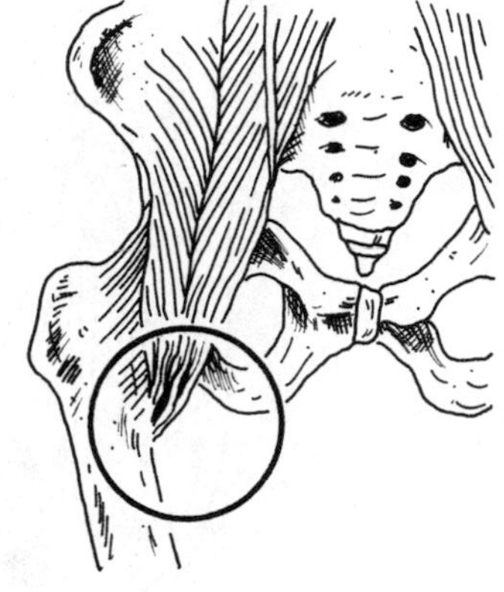

Grade I

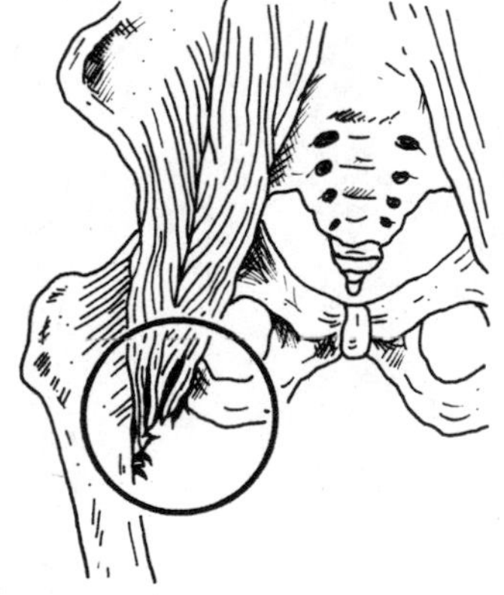

Grade II

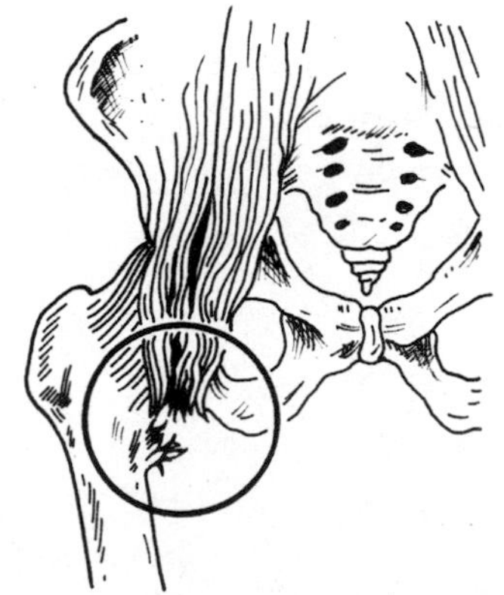

Grade III

Muscle Tears

Grade II strain is a partial tear of muscle fibers, between 10 percent to 50 percent of the fibers. I can usually touch the defect in the muscle unit.

Grade III strain is an extensive tear or complete rupture, 50 percent to 100 percent of the muscle fibers, with a large palpable depression in the muscle unit. The muscle may be torn away completely. In the grade III pull, you can't contract the muscle normally or at all.

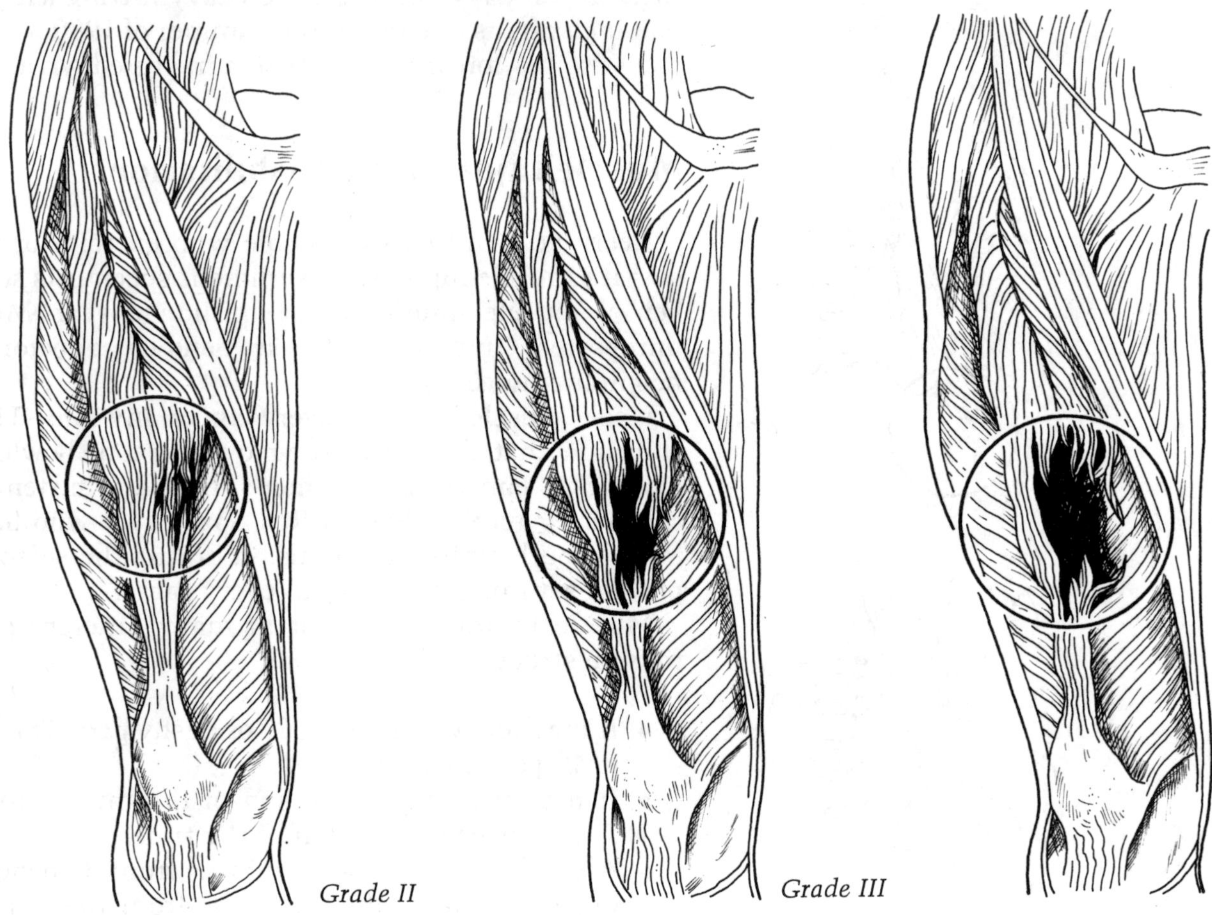

Muscle Strains

The extent of an injury is dependent on the force generated within a muscle. The strain is usually related to sudden changes of tension in the muscle. Sudden bursts of power as in sprinting in track and field, a charge to the net in tennis, a triple jump in figure skating are some examples of the kinds of athletic activity that cause strain. But I've had athletes who have had a muscle pull going up a flight of stairs. You can also strain a muscle in decelerating. The muscle is somewhat relaxed but the antagonistic muscle which is helping with the deceleration is still contracting vigorously. This is about 75 percent more common than an accelerating muscle pull.

Over the years I have uncovered certain factors that predispose you to multiple muscle strains:

- A muscle that has been previously injured and has neither properly nor completely rehabilitated. I know a doctor who runs for fun, a form of relaxation. But he has been hobbled by a hamstring strain, a pull that has been bothering him for years. He has never been to a sports therapist and he has never let it heal for more than two weeks. When he returns to running, "the hamstring goes out."
- A muscle that has been previously injured and healed and the healing occurred with contracted scar tissues that does not permit normal muscle excursion. A sudden force will reinjure the scar tissue and recreate the symptoms of a muscle strain.
- An unusual tightness within a muscle group that does not permit the normal excursion in the muscle-tendon unit.
- Improper or incomplete stretching and warming up prior to exercise or competition.
- Muscle fatigue from overexertion.
- A muscle that has been exposed to cold temperatures for an extended period of time, thereby losing its normal contracting excursion.

When I treat a tendon injury or a muscle pull, I always worry about secondary muscle atrophy and the need for rehabilitation and strengthening of the entire muscle-tendon unit. For example, in baseball players, rotator cuff tendon injuries of the shoulder are quite common. It is presumed that, in most instances, they are a result of repetitive trauma and microtears within the tendonous portion of the rotator cuff muscles. The treatment is generally directed toward quieting down the inflammation of the tendons. The muscle component of the rotator cuff injury is usually ignored while the anti-inflammatory medications are directed toward the inflamed tendons. This is one of the most common reasons for sore shoulders and decreased performance in baseball pitchers after rotator cuff probems. Pitchers usually return to pitching once the pain of the tendinitis has subsided. Because their muscles are weak, they change their pitching motion to compensate for the weakness. Their pitching style becomes obviously abnormal and their effectiveness diminishes because of a loss of speed and control. It is very important to classify the extent of the injury and to determine the exact anatomical sites of disruption. In other words, get a diagnosis. Without this information you can't start an intelligent rehabilitative program.

TREATMENT

Like all injuries, the first phase of treatment of muscle strains is RICE (See page 40).

I often provide a protective splint or support to maintain the muscle-tendon unit at its resting length. This decreases the pain and avoids either stretching or contracting of the muscle.

A stretched muscle results in a failure of the muscle to contract and therefore weakens the muscle. A shortened muscle will result in a limited excursion of the muscle and therefore an imbalance of function of the muscle and its related muscle groups. If it is an injury in the lower extremities, I recommend crutches.

As distinct from the use of surface spray local anesthetics, such as ethyl chloride, which are frequently sprayed on the skin to relieve superficial discomfort, some physicians inject muscle sprains with local anesthetics, such as Novocain or Xylocaine, and a cortisone drug. I do not use injectable drug therapy. I believe that pain provides a protective mechanism to prevent the athlete from additional injury. It also tells me which possible motions can be attempted in the early stages of post-injury.

Depending on the location and extent of the injury, RICE the muscle injury for twenty-four to seventy-two hours.

The second phase of the treatment is to assure maximum, normal lengthening and contracting (excursion) of a muscle-tendon unit. The degree of injury will determine the rate of progression of this phase. In a grade I or mild grade II muscle strain, complete excursion without pain will be possible within a few days. In a grade III strain, with or without the surgical intervention, a complete, normal excursion requires at least ten days, and sometimes many weeks.

I start phase two with limited range-of-motion exercises. I watch the response of the muscles carefully. Too much exertion can extend the injury. If it is a grade II or III sprain, I recommend that the range-of-motion exercises be started in a whirlpool or a bathtub. The warm water and exercise both increase the blood supply to the injured tissue and bring nutrients—the building blocks to heal the tissue. If possible, the earliest motion exercises should be guided by a physical therapist, trainer, or physician.

Phase three is when normal excursion returns to the muscle. At this point, progressive resistance exercises are started. I recommend light weights. Gradually increase the weights. Sometimes I prescribe the Cybex. The different techniques are explained throughout the book.

Many coaches, trainers, athletes, and doctors frequently consider the return of strength as the prime criterion for returning to competition. In my experience, this is only one factor. The key is both to regain strength and to coordinate the function of the injured muscle.

The fourth stage of rehabilitation of the integration of the injured muscle-tendon unit with the other adjacent muscles. Start with a series of very simple patterns and then progress to more complex precise patterns. For example, if a baseball player is recovering from a shoulder injury, one of the earlier stages of patterning would include complete integration of all the small muscles of the shoulder joint. This is done by throwing a baseball a short distance to a therapist or a trainer. This is a critical phase. Many individuals are unable

MUSCLE REINSTRUCTION

Mike Flanagan, one of the mainstays of the Baltimore Orioles and Cy Young award winner, sustained an isolated injury and loss of function of the infraspinatus muscle. This muscle extends from the scapula (shoulder blade) across the back of the shoulder. His performance deteriorated; his throwing pattern showed loss of velocity and loss of control. From a Cy Young award winner one year, the next he was struggling for an acceptable year. It required eight weeks to reinstruct Mike on the coordinated use of the muscles of his shoulder. Only then was he encouraged to throw a baseball some twenty feet. He was permitted to throw only for as long as his infraspinatus muscle functioned properly. Initially, this was for only seven to ten pitches. After a rest period of sixty to ninety minutes, he was able to repeat this performance. Three months of daily rehabilitation were necessary before he was able to return to throwing a baseball from a pitcher's mound to a catcher. In 1981, following an intensive rehabilitation program, he was returned to his Cy Young award condition.

to reprogram the injured muscle into an integrated pattern. Rather than spend the time to learn the reintegration, an injured athlete will develop alternative patterns and abnormal motions.

The fifth stage of rehabilitation is the preparedness for return to athletic performance. In this stage, I sometimes use a taping or bracing support. Your overall muscle agility and responsiveness should be normal. Prior to returning to competition, I review the athlete's total conditioning. Too often the focus is on the injured part and the remainder of the body is forgotten.

For college and professional athletes, I compare current performance with preinjury performance, i.e., the timing of a run, a review of a filmed performance, the throwing of a ball, the kicking of a ball, the swinging of a bat or golf club, etc. I don't believe in the adage, "playing into shape." When the athlete is deemed ready to return to competition, the return should be carefully monitored for progressive reentry. Usually it is impossible for anyone to reenter full, unrestricted competition without some progressive steps. It is at this stage that the importance of a close, consultative relationship be developed between the physician, trainer, and coach to determine the rate of reentry and the readiness of an athlete for the demands of participation and for unrestricted physical performance.

Joints and Ligaments

The body's bones form a skeletal framework that supports the soft tissue. They protect the internal organs and provide stability for the body as a whole. The skeleton supports the weight of the body and forms many levers that, when acted on by the muscles, bring about body movements. For example, your elbow joint, a body hinge, is really a junction of two bony levers, the upper arm bone and the forearm bone. Without joints, you couldn't move your hands, legs, arms, neck, spine, or mouth. My job is to restore and to improve the function of "deranged" joints.

You have three types of joints:

1. Freely movable or synovial joints. Ninety percent of the body's joints are of this category. The wrist, knee, ankle, shoulder, and hip are synovial joints.
2. Partially movable or fibrous joints. They connect bones of the leg. When you move your foot up and down, these two bones spread apart a few millimeters.
3. Immovable joints. Your body has less than ten of these joints. They are very rarely injured in sports. They hold the bones of your skull together.

This section is about synovial joints which have basic components:

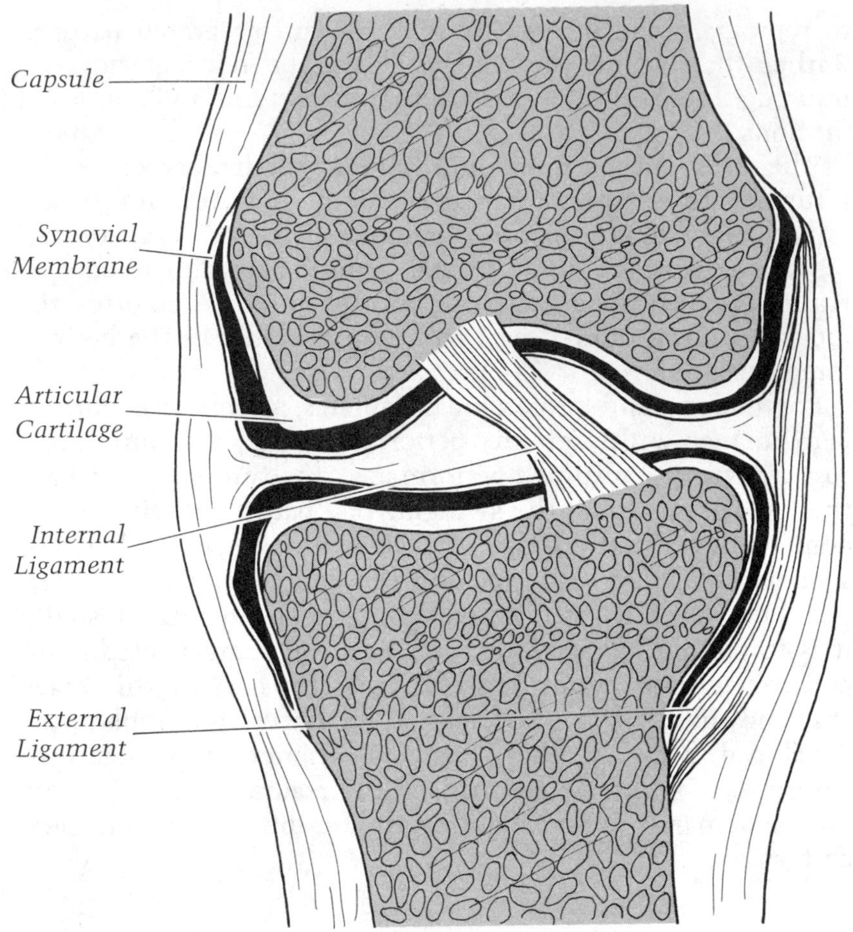

Synovial Joint

1. The end of two bones which meet to form the joint are covered with articular cartilage, a surface material much like a tread of a tire. Its strength comes from tough fibers called collagen. The joint surface cartilage is well lubricated—more slippery than well-manufactured ball bearings.

   In the mature adult, articular cartilage has no blood vessels, no nerves, and no lymphatic channels. Its living cells are nourished by joint fluid, called synovial fluid which is also extremely good lubrication. In our book, I refer to articular cartilage as joint surface cartilage, since it caps the ends of the bones. In children, this cartilage does have some blood supply and can repair itself. In adults, the cartilage repairs with scar tissue which is mechanically inferior. The tearing down of these surfaces is degenerative arthritis.
2. Every synovial joint has an outer layer which looks like a sleeve. It is composed of strong, fibrous (collagen) tissue. Ligaments are a specialized part of this sleeve and account for the primary stability of the joint. Many joints also have internal ligaments that contribute to support, such as the cruciate ligaments of the knee. The sleeve is oversized to allow for joint motion. It is nourished by blood vessels which give it the ability to repair itself after injury.

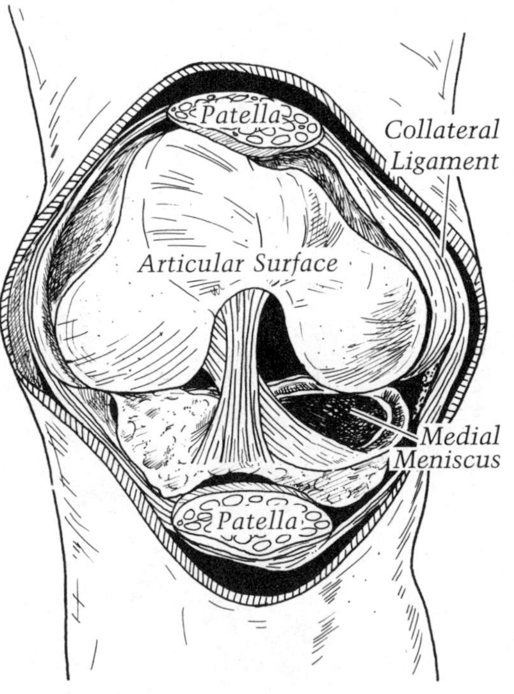

Knee Joint

Articular Cartilage of the Knee

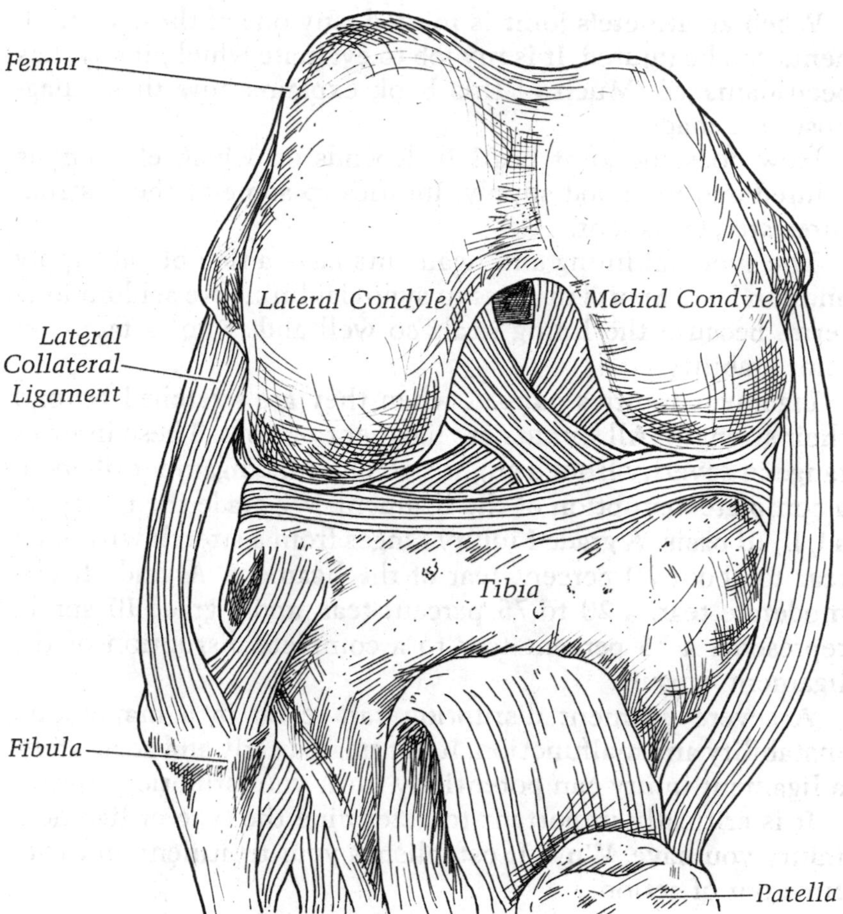

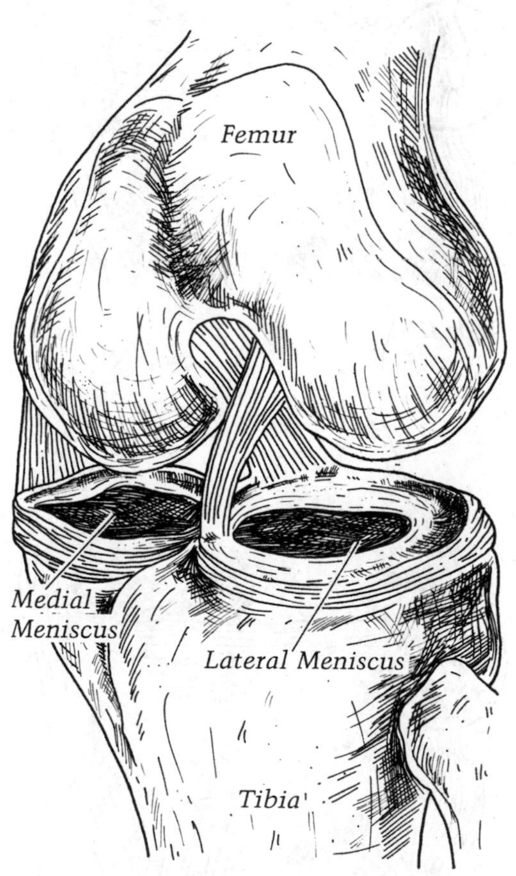

Meniscus of the Knee

3. Attached to the inner side of the sleeve is a specialized tissue lining called the synovium. This joint lining produces synovial fluid, the prime lubricant for the joint and the nutritional source for joint surface cartilage and meniscus cartilage (a washer-type structure which lies between some bone ends). Synovial fluid removes impurities from the joint. The synovial lining is an engineering marvel. It presents an enormous surface area; there are long tongues or peninsulas of synovial tissue to increase the potential for blood supply for greater nourishment, absorption, and repair. The lining acts as a two-way filter, allowing certain minerals, sugars, and proteins into the joints. Fortunately, most antibiotics pass through the synovial filter.
4. Ten percent of synovial joints have a washer-like structure between bone ends called the meniscus. Its purpose is to absorb shock, to stabilize the joint, and to spread synovial fluid. It is made out of fibro-cartilage, which is a different tissue type from joint surface cartilage, but the meniscus also has no blood supply, no nerves, and no lymphatic channels. Biologically, it can't heal itself. The knee meniscus is the most famous and most injured washer in the body.

When an athlete's joint is injured, any one of these components can be injured. It is my job to evaluate which, if any, has been damaged. Much of this book explains how these diagnoses are made.

How does the joint heal? It depends on which element is injured and its blood supply. Injuries to some of these structures are permanent.

The synovial lining and ligaments have a rich blood supply and can heal. Sports injuries to synovial lining are seldom long term, because the lining heals so well and is not a factor in joint stability.

Ligament injuries happen when they are stretched beyond their strength. All *sprains* are ligament injuries. These injuries range in severity from a slight stretch of the ligament fibers to a complete disruption of the ligament. We grade the injury on a I, II, III basis. A grade I injury ranges from a stretch without a tear to about 20 percent tear of the ligament. A grade II is a moderate tear, a 20 to 75 percent tear, and a grade III sprain represents a 75 percent tear to a complete disruption of the ligament capsule.

A severe ligament disruption can result in a major joint instability and malfunction. If not treated early and accurately, a ligament injury can potentially limit your athletic future.

It is critically important to determine the type of ligament injury you have. Why? It establishes your treatment and your time out of action.

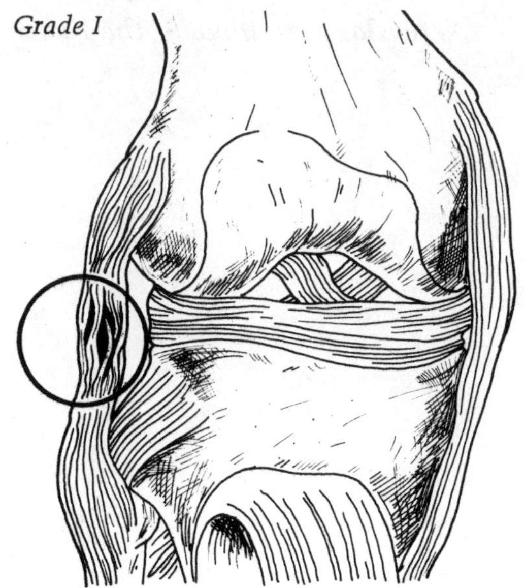

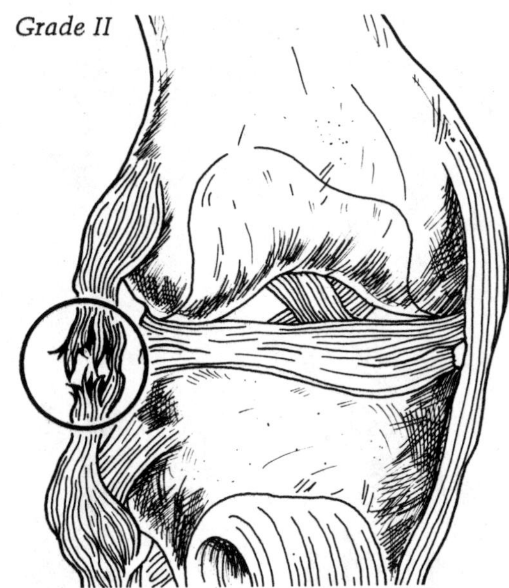

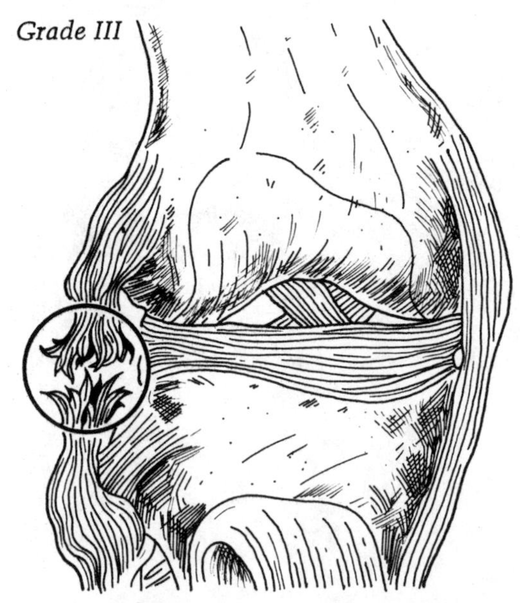

Knee Ligament Tears

TREATMENT

In general, all ligament injuries are treated with RICE. (See page 40). Grade I and most grade II injuries heal with rest, ice, and therapy. But some grade II and most grade III tears require surgery. In the lesser injuries, some fibers of the ligaments remain intact and the torn portion will heal with protection, immobilization, and rehabilitation. The completely ruptured ligament must be surgically relocated and repaired to make the joint strong again.

What is the healing time of these injuries? A grade I tear heals in five to fourteen days. A grade II tear heals in fourteen to thirty days. A grade II heals in months, often taking as long as a fracture to heal.

Ligaments heal in a four step process:

1. Blood flows into the ligament tear and the ligament itself swells up with blood and tissue fluids.
2. A blood clot forms which is mechanically weak scar tissue.
3. Gradually the scar tissue matures into strong scar tissue.
4. The damaged nerve endings of the ligaments regenerate.

The stages of treatment for a sprain must be related to the stage of injury and repair. In the initial stages of the injury, I use RICE to prevent extension of the injury, to minimize swelling, to stop internal bleeding, and to reduce pain. In addition, I immobilize the injured joint with a splint or premade immobilizer. This provides comfort and decreases the likelihood of an extension of the injury. In this phase of treatment, you should continue RICE for twenty-four to forty-eight hours.

The second phase treatment relates to the second phase of repair—the formation of a blood clot and early scar tissue. This phase of treatment starts after RICE ends. During this phase, I protect the injured part either with a compressive dressing, splints, or non-weight-bearing crutches. The swelling is reduced with the use of heat, compression, and guided motion. The guided motion stimulates the joint surfaces. I don't want the joint to stiffen and guided motion prevents healing scar tissue from encompassing other adjacent tissues, such as tendons.

The third phase following the injury, which, depending on the severity of the injury, starts in seven to twenty-one days after the injury. It is necessary to encourage the normal range of motion of the tissue into the usual length and strength of the prior-to-injury structure. It is during this phase that exercises to stimulate the return to normal muscle strength for all motions of the joint must start. At the completion of this phase, you should have a complete range of motion of the joint equal to the uninjured side. There should be no undue instability and you should have normal muscle bulk and strength of muscles controlling the joint.

The fourth phase of healing is the return of the delicate neurological control program to the joint. My objective is to reeducate carefully or to fine tune the injured joint and joint structure. The joint function must be coordinated with the other aspects of extremity activity and related to the expectations of the sport. This is the stage when a batter with a broken wrist gets his wrist speed back. In this phase, the joint has a complete range of motion with normal muscle control and the muscles are of equal or greater strength when compared with the uninjured side. The joint should not be swollen or sensitive to motion.

A trainer and a physician can adequately direct the first three phases of rehabilitation. It is during the fourth phase of rehabilitation that the injured athlete must work closely with his coach and trainer.

A final point that must be added in regard to all ligament injuries: I never inject a damaged ligament with a local anesthetic like Xylocaine or Novocain so that an athlete can return to competition with an incompletely healed ligament sprain. Because the joint can't report pain to the brain, such treatment will place the athlete at a much greater risk for a more extensive injury. However, I do, at times, use topical anesthetics to alleviate superficial discomfort.

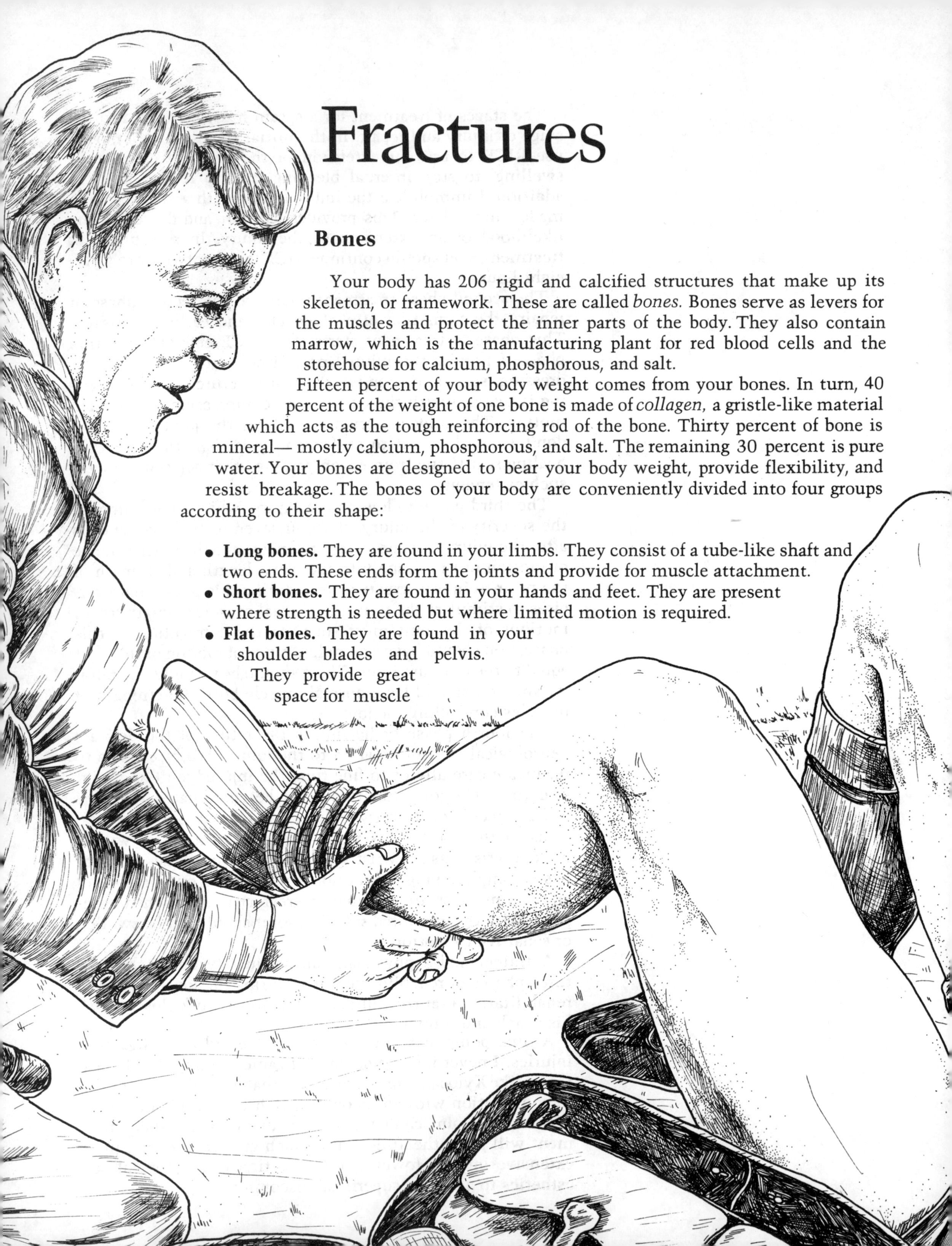

Fractures

Bones

Your body has 206 rigid and calcified structures that make up its skeleton, or framework. These are called *bones.* Bones serve as levers for the muscles and protect the inner parts of the body. They also contain marrow, which is the manufacturing plant for red blood cells and the storehouse for calcium, phosphorous, and salt.

Fifteen percent of your body weight comes from your bones. In turn, 40 percent of the weight of one bone is made of *collagen,* a gristle-like material which acts as the tough reinforcing rod of the bone. Thirty percent of bone is mineral— mostly calcium, phosphorous, and salt. The remaining 30 percent is pure water. Your bones are designed to bear your body weight, provide flexibility, and resist breakage. The bones of your body are conveniently divided into four groups according to their shape:

- **Long bones.** They are found in your limbs. They consist of a tube-like shaft and two ends. These ends form the joints and provide for muscle attachment.
- **Short bones.** They are found in your hands and feet. They are present where strength is needed but where limited motion is required.
- **Flat bones.** They are found in your shoulder blades and pelvis. They provide great space for muscle

attachment and protect the underlying organs of the abdomen.

- **Irregular bones.** They are found in the skull and in the spinal column.

The shape, length, density, and strength of your bones are determined by genes and how you use your bones and muscles. You inherit a code in your bone cells from your parents. This determines the length, width, and overall shape of your bones. If your father or mother is bowlegged, there is a good chance that you will be bowlegged. If your parents are tall, your bones are apt to be long. That is why the offspring of Scandinavians tend to be tall and those of pygmies short.

I learned about the impact of exercise or the lack of it on bone size by studying polio victims. Polio is a disease that attacks the nerve cells in the spinal cord. The result is that the nerves are unable to give the muscles of the arms and legs proper operating messages. The muscles waste away and the arm or leg becomes paralyzed. The difference is dramatic when one leg is afflicted by polio but not the other. The bones surrounded by normal muscles are strong and normal in size; the bones surrounded by paralyzed muscles are thin, brittle, and misshapen.

FRACTURES

If enough force is applied to any bone, it will break or fracture. A bone-crushing tackle, a fall on the pavement, or a repeated hard jarring of the foot can cause a fracture.

The force might be direct or indirect. Examples of *direct force* are a blow in a boxing match, a slash with a hockey stick, or a beaning with a baseball. Examples of an *indirect force* could be a fall on an outstretched hand that causes a wrist fracture or a skiing mishap in which the victim wraps a leg around a pole and breaks the lower leg bone.

There are two types of fractures:

1. A *complete fracture,* where the bone is severed.
2. A *stress fracture,* where the bone is cracked, but not separated.

Because of their variation and shape, bones have strong areas and weak areas. Fractures occur in the weak areas. For example, if a powerful force is applied to the upper arm bone (the humerus), it will break in three predictable areas: just under the ball portion, in the middle of the shaft, and just above the elbow. Other fractures are very rare. Of course, if a bone has a disease like cancer or a tumor, any part of the weakened bone can fracture or break.

Stress Fractures

CAUSES

Stress fractures are slight cracks in the bone surface. The most common sites are the bones of the feet and legs. The fractures are caused by the repetitive application of small forces, none of which alone is sufficient to break the bone.

I treated hundreds of stress fractures of the foot at Fort Dix, New Jersey, during the Vietnam War. Most of my patients were recruits under twenty-five years of age who had not been physically active in many years.

Their stress fractures were the result of strenuous army training on "lazy" bones. Within three days of arriving at Fort Dix, the recruits were forced to march twenty miles carrying heavy packs. Consider what is happening to the lower leg bone (tibia) of one of these recruits. After years of neglect, suddenly the bone is forced into activity. The marching feet pound unmercifully into the ground day after day. I call this the "battering ram effect." The surrounding muscles tug at the bone.

What happens in a case like this? The bone has no choice but to get stronger. The way the bone gets stronger is to build

STRESS FRACTURE SYNOPSIS

How Stress Fractures Develop

1. Causal event
2. Pain after exercise only and relieved by short rest
3. Pain tolerable during exercise but more marked after exertion and relieved by longer rest
4. Pain intolerable during and after exercise and partially relieved by long rest
5. Constant pain not relieved by rest

Signs of Stress Fractures

1. Point tenderness of bone (except well-shielded femur)
2. Soft-tissue swelling
3. Palpation of callus (with time)
4. Alteration of gait
5. Muscular atrophy, especially anterior tibial and gastrocnemius-soleus groups
6. Full and painless range of motion of adjacent joints
7. Painless resisted active movement of joint
8. Increased technetium 99m-diphenylmethane diisocyanate uptake indicating focal lesion
9. Hairline radiolucency, periosteal callus, or endosteal callus by x-ray; associated soft-tissue swelling

Treatment of Stress Fractures

1. Rest from running
2. Relieve symptomatic inflammation with ice and anti-inflammatory agent
3. Maintain strength (especially foot dorsum and plantar flexors)
4. Maintain cardiovascular fitness with swimming and/or biking
5. Orthotics tailored to need
6. When asymptomatic, gradually reintroduce running

SOURCE: *The Physician and Sportsmedicine.*

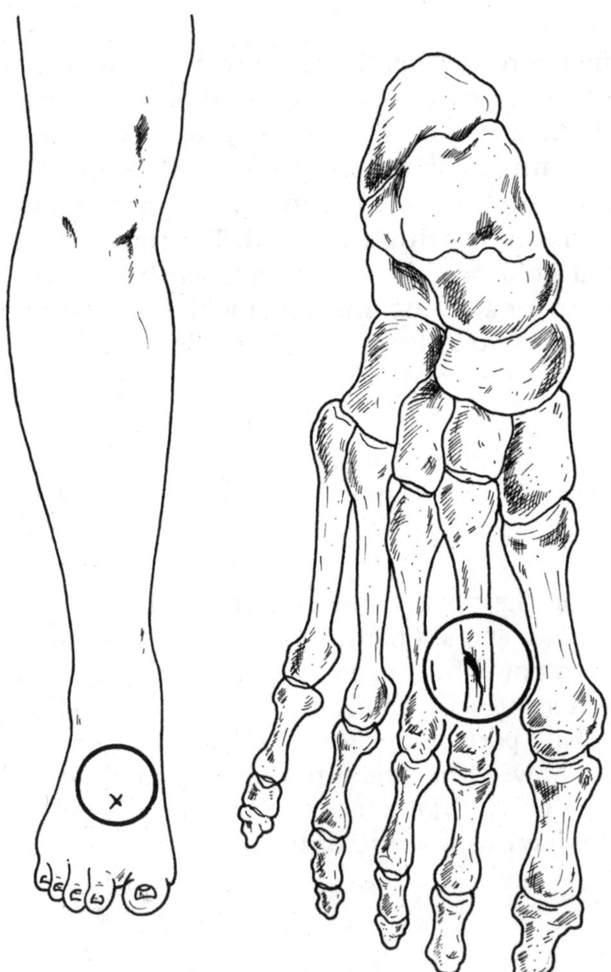

Stress Fracture of the Foot

HOW LONG YOUR STRESS FRACTURE TAKES TO HEAL	Healing time (Weeks)
Most Common Fracture Sites	
Upper arm bone, Middle Portion	6
Shoulder Blade, Coracoid Process	6
Pelvic Bones, Pubic Rami	3
Hip Bone, Central Area	12
Thighbone	
Mid Shaft	12
Lower End	6
Lower Leg Bone	
Upper End	3
Mid Shaft	6
Lower End	3
Heel Bone	
Mid Portion	6
Metatarsals	
Lower End (March Fractures)	4

new bone at the point of stress. The body calls in cells known as *osteoblasts*, bone-eating cells. They move into an area that is under great stress and start to remove old bone so that stronger new bone can fill in. However, the bone-forming cells, the *osteoblasts*, produce new bone slowly. Meanwhile, the recruit continues on the twenty-mile march every day, and finally the bone cracks. The cracks are called *stress fractures*. The osteoblast cells continue their feverish attempt to build new bone. It takes them about six weeks to heal the stress fracture.

DIAGNOSIS AND TREATMENT

How can you tell if you have a stress fracture? The pain! Nothing is quite so excruciating. The pain generally begins about two weeks after the overstressing. It takes that long for the body's biology to catch up with the event. Once the pain starts, it steadily intensifies. Sometimes you can exercise the bone, but the pain will return, usually worse than before. Many times, it keeps you awake at night.

Which bones fracture easily? Ninety-five percent of stress fractures are of the leg and foot bones. However, the upper extremities are not immune. I have treated stress fractures of the upper arm bone of champion arm wrestlers.

Besides the pain, how can you tell that you have a stress fracture? Use the finger test. A stress fracture usually hurts when you press directly on it with fingers, both from above and below. A tendon or ligament usually hurts only on pressure from one side. Also, stress fractures have sharply localized tenderness.

X-rays usually are not sensitive enough to pick up small cracks in bone. It is not until two or three weeks later—when a *callus*, a layer of new bone made from osteoblast, forms over the crack—that an x-ray diagnosis of a stress fracture can be made. By this time, if you have rested the fracture, the pain should be abated.

If you have a stress fracture, rest it and temporarily switch to another sport that will not extend the crack. For example, runners who develop stress fractures in their feet often change to riding a bicycle or swimming.

I never apply a cast to the injured area. Stress fractures heal by themselves in most cases. The immobilization caused by the cast makes the muscles smaller and weaker.

Complete Fractures

Complete fractures are usually the most painful of all athletic injuries. The jagged edges of separated bones contain a rich supply of nerves, and when they rub against each other or any tissues, they cause extreme pain. The pain and swelling can continue for weeks or even months. For that reason, I routinely administer pain medicine to patients who have complete fractures.

Complete fractures require immediate medical treatment. The sharp edges of a broken bone can cut a nerve and leave you paralyzed, can sever a blood vessel and cause you to bleed, and can cut through the skin and provide a portal of entry for germs.

While some complete fractures of the small bones of the hands and feet heal themselves, those of the large bones of the arms and legs often do not and should be checked immediately. Complete fractures of bones normally need to be aligned properly so that they can heal without complications. This often requires taping, splinting, casting, or traction. Complete breaks can take one to six months to heal, depending on the extent of the fracture, the treatment, and the absence of complications.

There are two types of complete fractures: open and closed. A *closed fracture* means that the broken bone has remained within the skin. The bone has not been exposed to the outside world. An *open fracture* means that the skin has been breached and the bone sticks out through it. Open fractures are serious business. The outside world is dirty. The bone and the wound become contaminated with the germs that live on the skin and clothes and in the atmosphere.

With an open fracture, I get my patients into a sterile operating room as soon as possible. I cleanse the contaminated bone with large amounts of salt solution. I carefully remove all dirt from the wound. Once the wound is thoroughly cleansed, I cover it with a packing gauze soaked in antibiotic material. I immobilize the broken bone with a splint or a cast. Five to

TIBIAL STRESS FRACTURE

Marty Cooksey is a national marathon champion, who came to Dr. Rob Roy McGregor, chief of podiatric medicine at Sports Medicine Resource, Inc., with a tibial stress fracture. "I have studied films of Marty's gait cycle," explains Rob Roy McGregor. "I argue that she gets tibial stress fractures because of the twisting stresses that she puts on her tibia bone. The tibia bone is designed only to take direct forces. The only way she is going to eliminate further stress fractures is to change her gait cycle slightly. I am not sure that she can do this."

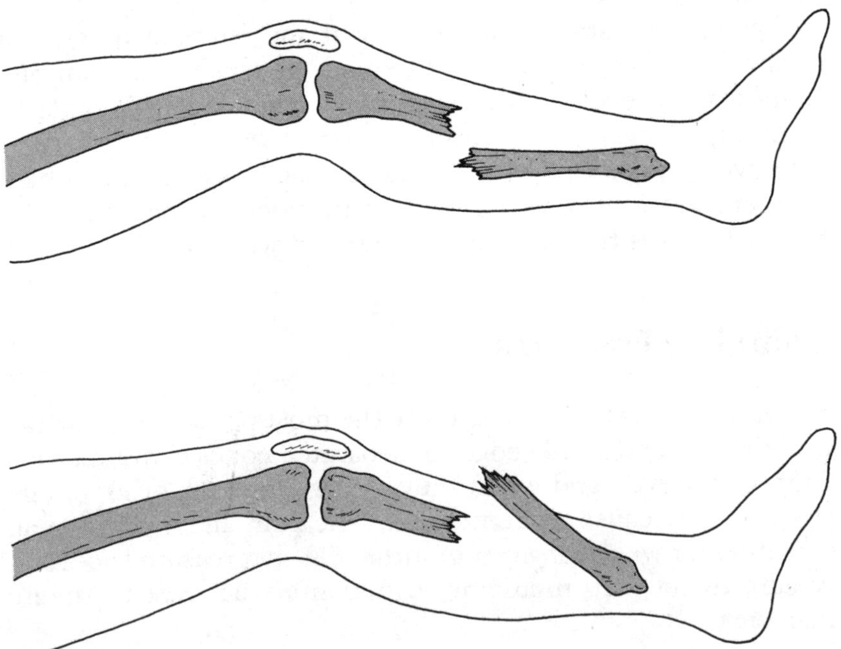

Open and Closed Fractures

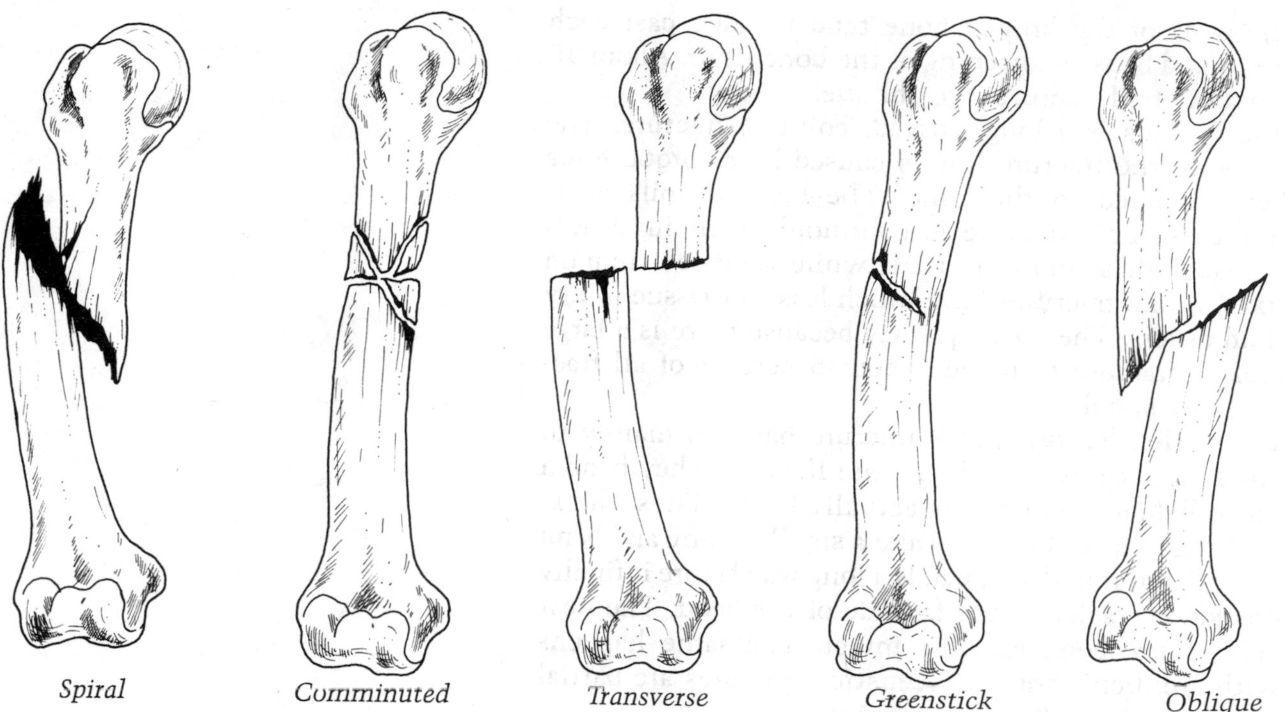

Five Types of Fractures

seven days later, I remove the antibiotic gauze in the operating room. I use a general anesthesia. Then, I close the skin with stitches. This is called a *delayed primary closure* of the skin.

When I was in the army, it was a hard and fast rule that all war wounds were treated in this fashion. No gunshot wound or flack wounds were ever stitched up in the first twenty-four hours. In general, stitching dirty wounds shut promotes the growth of bacteria that cause gas gangrene and tetanus.

Besides open and closed complete fractures, there are five ways to describe the line of a complete fracture.

- **Transverse.** The bone is broken straight across. The fracture line is at a right angle to the long axis of the bone. The edges are somewhat jagged, but the bone is broken cleanly in two. Transverse fractures are caused by a direct blow to the bone. Often, a great deal of soft-tissue injury is associated with this fracture. Transverse fractures are the "fender benders" of fractures. Sixty percent of all adult fractures are transverse.
- **Comminuted.** The fracture is not broken cleanly in two. X-rays show splinters floating free at the fracture site. If these pieces are large, they are called *butterfly fragments,* because of their shape on the x-ray. Comminuted fractures are often open fractures, with the bone sticking up through the skin. It takes tremendous force to break a bone in this fashion. About 20 percent of all adult fractures are this type.
- **Oblique.** The fracture line is on a slant. As with the transverse fracture, this, too, can be caused by direct blow, but a torque force also is involved. This is a difficult fracture for orthopedists to manage because

the ends of the broken bone tend to slide past each other and cause shortening of the bone itself. About 15 percent of all fractures are oblique.

- **Spiral.** This is a long, curved, coil-like fracture. The shape of the fracture line is caused by a torque force being applied to the bone. The bone uncoils as it fractures. This fracture is commonly seen in skiiers whose bodies rotate in a fall while their feet remain fixed. Spiral fractures have much less soft-tissue injury than others. They heal quickly because there is a large area of fracture to mend. About 5 percent of all fractures are spiral.
- **Greenstick fracture.** This fracture happens mainly to children because their bones are flexible. They bend a great distance before they actually break. Thus, the x-rays show curved bones. Take a small sapling and bend it. You will find that it bends a long way before it finally cracks. It cracks on the far side of the bend. The near side of the bend remains intact. The same happens with children's bones. Greenstick fractures are partial fractures; thus, they heal quickly.

How do fractures mend themselves? Nobody understands this process fully. It is one of the mysteries of nature. But, fracture healing has one of the highest priorities in the body. Fractures even heal in starving people.

As soon as the bone fractures, the body musters up tremendous energy and materials to heal it. All the blood vessels in and around the bone break as well. A pool of blood forms around the broken ends of the bone. In medical parlance, this phenomenon is called a *fracture hematoma (heme* means "blood;" *oma* means a "mass"). The mass of blood is a critical element in fracture healing. Within seven days, it starts to organize or becomes semi-liquid. Immature bone cells from the broken ends of the bone proliferate or reproduce in the womb-like hematoma. Within two weeks, new bone is forming from these cells. This is called *fracture callus* or immature bone. Within six weeks, the bone has regained much of its strength. The hematoma has become sticky. As the bone heals, the fracture becomes less and less painful.

Complete healing sometimes takes one year. During that period, the fracture site undergoes constant chemical and physiological change. During this time, the bone converts from immature to mature bone, then to solid bone.

As with all healing tissue in the body, fracture healing leaves a scar. The scar is visible on an x-ray. It looks like a lump of slightly irregular bone. Only if the fracture occurs in early childhood, when the bone changes tremendously in size and shape, will the scar be erased.

Bones automatically heal. The job of an orthopedist is to help them heal straight, by aligning them while the repair process goes on.

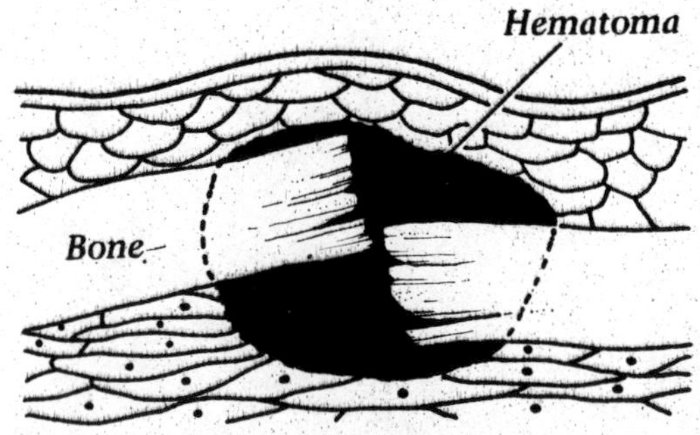

Fracture Hematoma

> **THE MOST FRACTURED ATHLETE**
>
> For ten years, I have been treating William Nayorga, a jockey who rides at Suffolk Downs. In that period of time, he has fractured his kneecap, his collarbone, his spine, his forearms, and his ankles (numerous times). Most of these injuries required surgery. Despite his multiple injuries, he is still back riding and he continues to do well in horse racing.
>
> Carter Red Rowe, M.D.
> Massachusetts General Hospital

There are four fracture mends:

- **Union.** The successful and complete healing of a bone.
- **Delayed union.** A bone which heals slower than one would ordinarily expect.
- **Nonunion.** A fracture which fails to heal.
- **Malunion.** A bone which heals crooked.

DIAGNOSIS AND TREATMENT

How do you know if you have a fractured bone? You will immediately feel like a wounded deer. It is clear that something horrible has happened. Because the fractured bone contains a rich supply of nerves and blood vessels, the injury swells and is extremely painful.

If it involves the legs, you will not be able to walk. If your

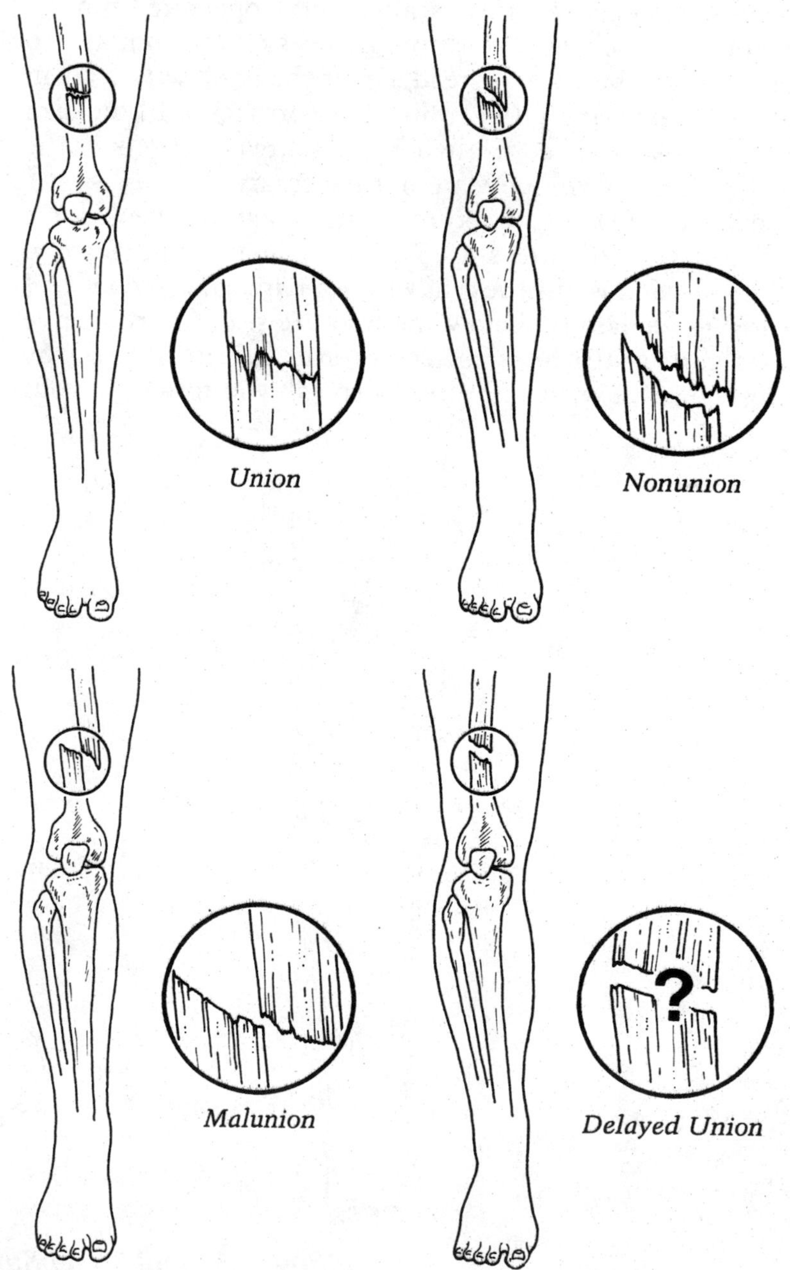

Fractures

arm is broken, you will not be able to use it. You will not want to move. If you move, the jagged edges of the separated bones will often tear tissue. Moving the fractured limb causes extreme pain.

Immobilization is important. Most emergency medical technicans know how to immobilize a fracture. In general, adhere to the old adage "Splint them as they lie." In a fractured leg, pillows can be used as splints. Wrap adhesive tape around the pillow and the fractured leg. This is called a *soft immobilization.*

Wooden splints are effective. They stop the fracture from sliding and moving. If you fracture your arm, have a sling applied to hold the arm close to your body. An Ace bandage can be used to bind the sling around your trunk. This improves the immobilization. Also, no harm can ever be done by straightening a leg or an arm parallel to the opposite limb.

If a bone is sticking out through the skin, do not try to replace it under the skin. Instead, cover the bone with a clean bandage and then apply the splint. Hemorrhaging in the arm and leg can always be controlled by pressure directly over the bleeding point. Tourniquets are not necessary.

When I examine any fracture, I immediately check for a pulse below the fracture site. This is to make sure that no major blood vessels (arteries) have been damaged. Secondly, I check for nerve damage below the fracture site. Are the nerve impulses getting to the muscles? In arm fractures, I test by pricking the palm with a needle. I also ask you to wiggle your

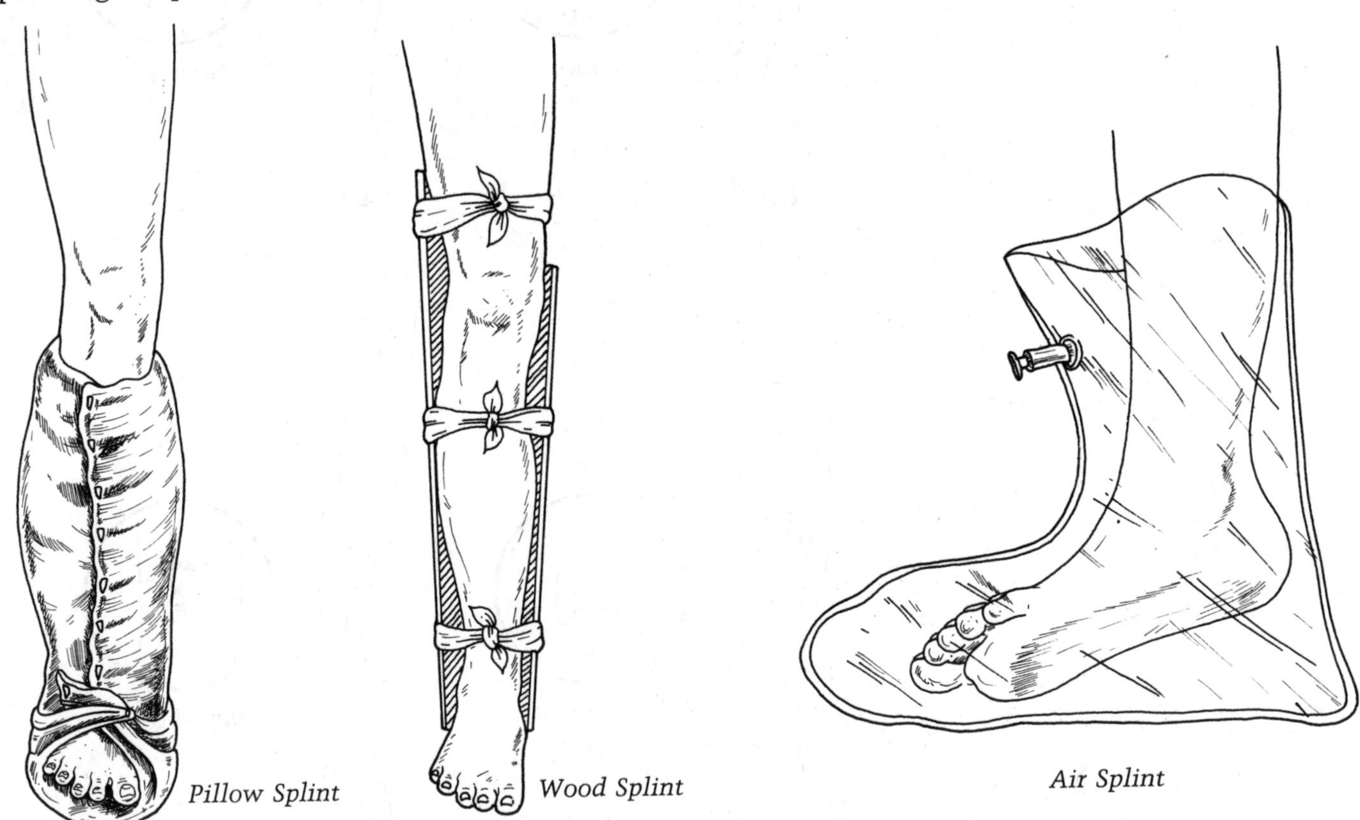

Pillow, Air, and Wood Splints

ALIGNING FRACTURES

In January 1980, I treated Steve Wiper, a fourteen-year-old hockey player who fractured his thighbone in a game. His x-rays showed a transverse fracture of the thighbone approximately four inches above the knee. The fracture was tilted at a strange angle. It was clear from the fracture and x-rays that it was impossible to reduce the fracture with my hands. The bone was too large; the muscles were in severe spasm; the angle was great.

My only choice was to take him to the operating room and give him general anesthesia to relax the muscles. I also had to put the leg in traction with twenty pounds on it. The traction reduces the muscle spasm. I do this by inserting a wire with a drill through the top of his lower leg bone. It sounds like a painful procedure, but it is not. The wire, as thick as a shoelace, is attached to an overhead pulley and the leg is drawn up with a twenty-pound weight. Only when the traction weights are less than five pounds can adhesive strips be used on the skin. If you use more than five pounds, the skin tends to peel off.

The next day, I checked Steve's x-rays. The fracture was straight. The problem now was how to hold the fracture straight. It would take at least six weeks before Steve's bone would be strong enough to allow him to put any weight on the leg without extending the fracture. The only answer was traction for six weeks.

Therefore, I kept him in bed for six weeks and then I started him on crutches. To assure complete healing of a femur fracture, which takes three months, I installed a cast to his leg for the six weeks. I placed a short-leg cast on the lower part of the leg and a plaster sleeve around the thigh. They are connected by plastic hinges about the knee so that the leg is held firmly straight, but you can walk on it. This type of cast permits the knee to move freely while the fracture heals and avoids frozen knee.

fingers. Then, I order x-rays. The patient is x-rayed in the splint or sling. The x-rays accurately reveal the scope of the fracture.

If the separated bones are sticking up through the skin, I rebandage the wound and bones with a sterile bandage and antibiotic solution. This is done in the operating room as I described earlier.

During the five days when the antibiotic packs are installed, the fracture is encased in either a half or full plaster cast to support the fractured arm or leg bone. Following skin stitching, a final cast is applied.

If the fracture is closed, I must decide whether or not the fracture needs to be straightened and how to do it. There are several methods of straightening a fracture. Sometimes, I can straighten the bone with my hands. It is a matter of touch and experience. I always check the results by x-ray. Other times, I use a traction technique whereby weights are used to pull on the fractured bone. These weights also help reduce muscle spasm around the fracture and make aligning the fracture easier. Once I have obtained a straight bone, I repeat the x-rays.

When the straightening is confirmed, I worry about maintaining the bone in proper alignment while nature heals the fracture. Some fractures require only a sling to maintain their position. Others require a half-cast or splint. Most fractures require a fully circular cast, one that goes completely around the arm or leg. The problem in maintaining the alignment stems from the reduction in swelling as the fracture heals. The cast or splint loses its supportive strength on the bone itself.

I can make casts out of plaster of paris, fiberglass, or plastic. I prefer fiberglass and plastic. They are lighter, allow better circulation of air, and are water resistant. This last quality allows the patient to shower and swim in a pool. They are easy to repair. But the one disadvantage they have is that they are tricky to apply. Often the first cast applied to a broken arm or leg must be a plaster one because it is easy to shape.

How long do you wait for a fracture to heal? It depends on the size of the broken bone, the portion of the bone that is fractured (ends or middle), and the bone's blood supply. In general, fractures heal in either three weeks, six weeks, or three months. The small bones of the fingers require only three weeks. The biggest bone in the body, the thighbone, requires three months to heal.

The middle of any bone heals slower than fractures at the end of the bone because the middle has a smaller blood supply and fewer bone-forming cells.

There are three bones that are notoriously slow healers: the carpal navicular (one of the eight wrist bones), the mid-shaft of the ulna (a forearm bone), and the mid-to-lower portion of the tibia (lower leg bone). They all have very poor blood supplies. An excellent blood supply speeds healing, bringing in necessary nutrients.

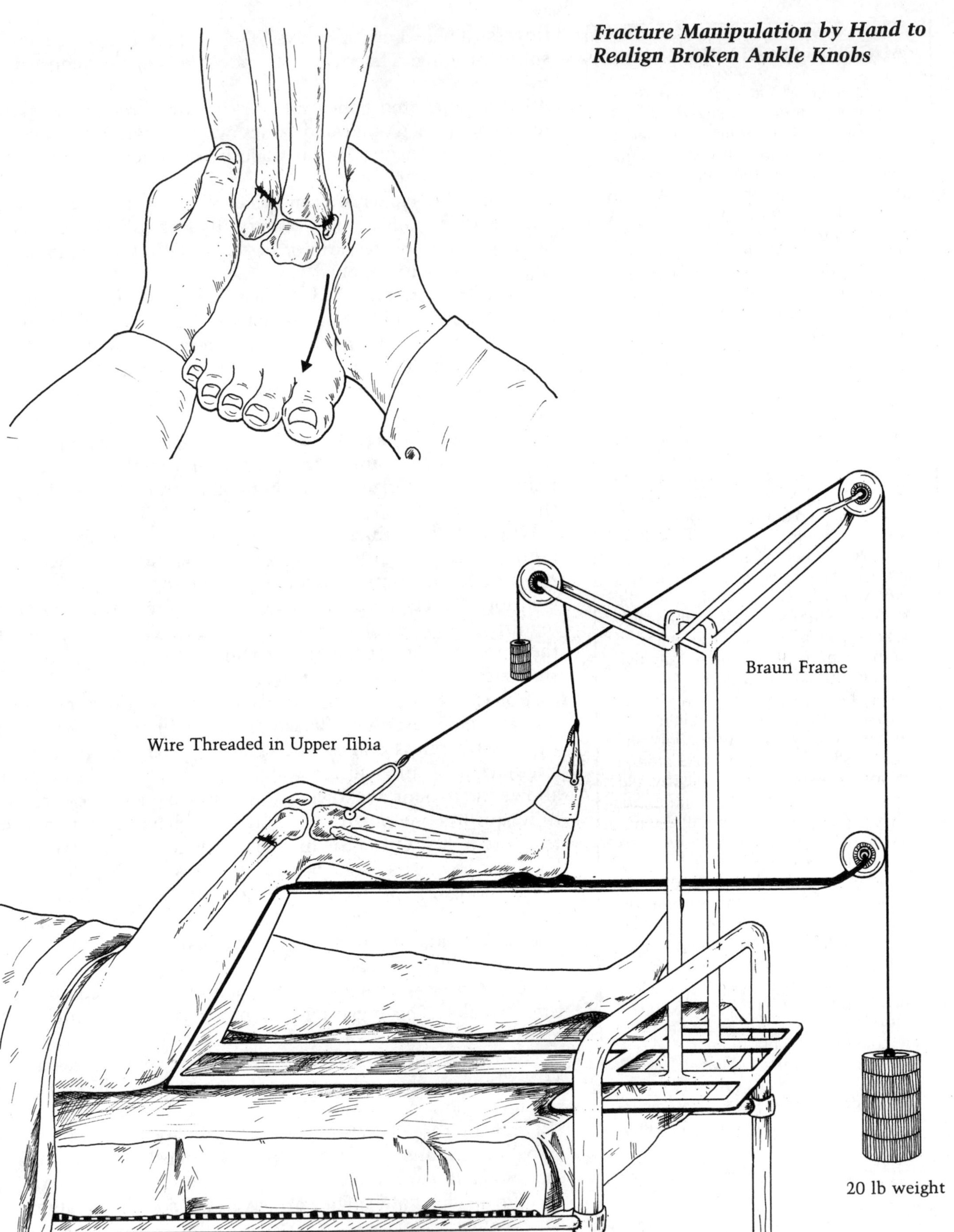

Fracture Manipulation by Hand to Realign Broken Ankle Knobs

Traction Technique to Align a Fracture

YAZ'S FRACTURED RIB

Late in the 1980 season, Carl Yastrzemski ran into the wall in right field and fractured a rib.

"Yaz had an unplaced fracture of the rib, a crack across the entire rib," says Dr. Arthur Pappas, medical director of the Boston Red Sox.

"The rib was tremendously sensitive. It even bothered him to breathe. He couldn't sleep without pain."

"After a few weeks, I was able to take batting practice," Yastrzemski says, "but I couldn't play in a game. I didn't have any fluidity. I couldn't rotate my body. Game conditions are tremendously different in any sport. My rib injury ended my season."

Fractures need to be immobilized. That is why I use a cast or an internal device such as a plate and screws to prevent the ends of the fracture from moving. The less motion, the better the chance for proper healing. Second, if the ends of the separated bones touch, the fracture heals faster. The ends do not have to be perfectly matched like pieces of a jigsaw puzzle, but at least 50 percent of the surface of the fracture ends should be touching for the quickest healing time.

When I immobilize broken bones, I make a great effort to involve as few joints as possible. In general, the joint above and the joint below a fractured bone must be immobilized for proper healing. However, there are many exceptions to this rule. For instance, the cast for a boot-top fracture of skiiers can be changed to one that allows knee motion about six weeks after the break. The initial cast follows the rule of immobilizing the knee and the ankle joint—the joint above and below the fracture. Once the fracture forms a strong callus, I can convert to a shorter cast. I do not want the knee to become stiff. The shorter cast at this point in the healing cycle does not inhibit the continued healing of the fractured bone.

Once the cast is removed, I start the patient on rehabilitation. The cast always leads to joint stiffness and wasting of the muscles. I examine many athletes who have sustained fractures and have not been properly rehabilitated. They come to my office because they cannot run at full speed or have trouble with using various joints near the fracture site. The problem is that they have not been pushed hard enough during rehabilitation. There is no medical reason to accept restricted motion of joints near a fracture. It is fair to say that it takes as long out of the cast as you were in the cast to regain full motion in your joint. The muscles around the fracture which have become smaller from disuse must be strengthened in a scientific fashion. Again, I do not settle for residual muscle weakness after fracture in athletes or fitness buffs. The muscles have only diminished in size. The muscle cells themselves have not been lost. I am confident that the entire muscle can be brought back to its normal strength. It's hard work but it's worth it.

Fracture Surgery

When do I operate on fractures? Surgery, in my opinion, is the last resort, reserved for those fractures which cannot be properly managed by nonsurgical techniques. The reason for operating is that the fractured bones will not go together well of their own accord. The mended bone should be the same length and shape as before injury. This is particularly important at the ends of bones which help to form joints. If you have a fracture which extends into one of your joints and that fracture is not perfectly restored to normal shape, the joint will deform. Its surface will become irregular and early arthritis will develop. In medical terms, this is called *traumatic arthri-*

YOUR COMPLETE FRACTURE: HOW LONG IT TAKES TO HEAL		
	Number of Weeks	
Fracture Site	**Children**	**Adults**
Finger Bones	3	3-5
Hand Bones	6	6
Wrist Bones	Rare	10, or until x-ray shows union
Forearm Bones	6-8	10-12
Upper Arm Bone		
Lower Shaft	6	8
Mid Shaft	6	8-12
Collarbone	4	6-10
Spinal Column	16	16
Pelvis	4	6
Hip Bone	6	10-12
Thighbone		
Mid Shaft	8-10	18
Lower End	6-8	12-15
Lower Leg Bone		
Upper End	6	8-10
Mid Shaft	8-10	14-20
Lower End	6	6
Heel Bone	10	12-16
Foot Bones	6	6
Toe Bones	3	3

SOURCE: *Pictorial Handbook of Fracture Treatment* by Edward Compere, M.D., Sam W. Banks, M.D., Clinton L. Compere, M.D.

tis. Any fracture that extends into a joint (an intra-articular fracture) is a strong candidate for surgery.

I also will operate on bones which would otherwise heal crooked. If a bone is not straight, it cannot function properly. For instance, the two bones in the forearm rotate around each other. If one of them is crooked, the rotation will be distorted. You will not be able to turn your palm up or turn it down correctly. Rather than accept a disability, it is best to straighten the bone surgically.

If a break of the lower leg bone shortens it to a point where you have unequal leg lengths, I would operate. Usually, the problem stems from the break not being set properly in the first two or three days. In this case, the muscles and ligaments shorten dramatically. This makes it virtually impossible—short of surgery—to get the bone to its original length.

Most fracture surgery that I perform follows this pattern. I make an incision over the bone and the bone ends are brought together. I stretch out the bones and place a plate across the fracture site. I screw the plate into the bone on either side of the fracture. This guarantees that the bone will remain straight. I remove the plate one year later.

Bone surgery always carries risks. Even though the operation takes place in a sterile operating room, the bone ends are exposed to the outside world. I always administer antibiotics to the patient during the operation and for forty-eight hours afterward. This minimizes any risk of infection. Occasionally, the metal device keeps the bone apart. This can actually retard healing. Therefore, the decision to operate should not be taken lightly. I proceed only if I feel that the advantages and gains far outweigh the risks.

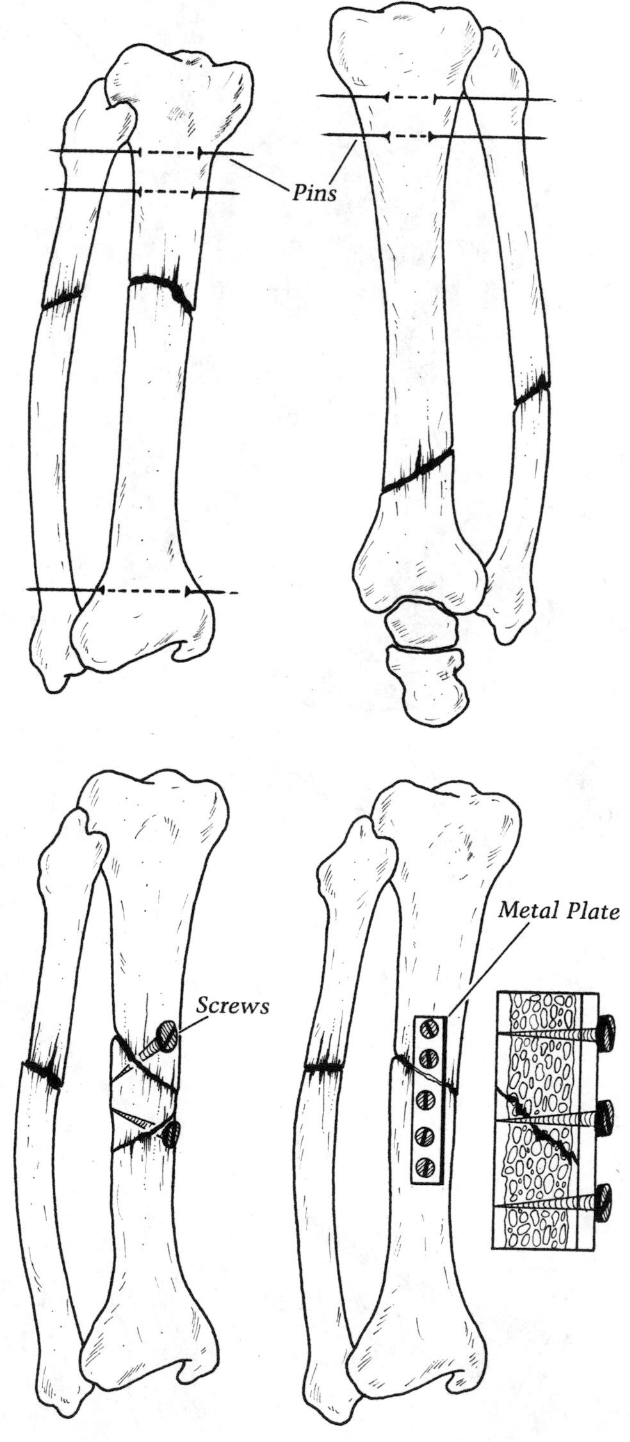

Pins, Screws and Plates for Complete Fractures

The Shoulder

The shoulder is my favorite structure. It is complex, versatile, and logical. In my opinion, it has a degree of perfection that no other anatomical body possesses. The shoulder is the only structure that can rotate 360 degrees. It is the most mobile joint in the body. It allows you to serve a tennis ball, throw a baseball, bowl, hang from a tree limb, and hug your spouse. You need a mobile shoulder to be able to use your hand effectively. Imagine having a frozen shoulder. You couldn't feed, wash, or dress yourself. Your hands would be almost useless.

Bones and Joints

The shoulder, sometimes called the shoulder girdle, is where three bones join: the collarbone (clavicle), the shoulder blade (scapula), and the upper arm bone (humerus).

The *collarbone* serves as the only bony attachment between the shoulder and the trunk itself. It attaches to the breast bone and also to the shoulder blade. The remainder of the bones of the shoulder girdle attach to the trunk only through muscles.

The collarbone is a round, dowel-like structure. When the arm is raised overhead, the collarbone turns on its long axis in swivel-like fashion. The outer end of the collarbone is attached to the shoulder blade by a small joint called the *acromioclavicular joint* (A-C joint). Because of its prominent position on the outer tip of the shoulder, it is often injured by direct blows.

The true *shoulder joint,* seated deep in the shoulder girdle, is the junction of the upper arm bone (humerus) and the shoulder blade. The upper end of the arm bone is shaped like a ball. It sits in a socket-shaped structure of the shoulder blade called the *glenoid fossa.*

All joints in the body are a trade-off between mobility and stability. The shoulder joint is an extreme example of a mobile joint with very little stability. It is the joint in the body that comes apart or dislocates most frequently. Dislocations are common in contact sports.

Finally, the shoulder blade itself is attached to the chest wall by the *scapulothoracic joint.* This is not a true joint. It has no cavity or joint lining. It is a joint only in the sense that the shoulder blade must slide over the chest wall in order for the shoulder girdle to function properly.

When you raise your arm directly overhead and the shoulder girdle is viewed from behind, the first thirty degrees of that motion is entirely in

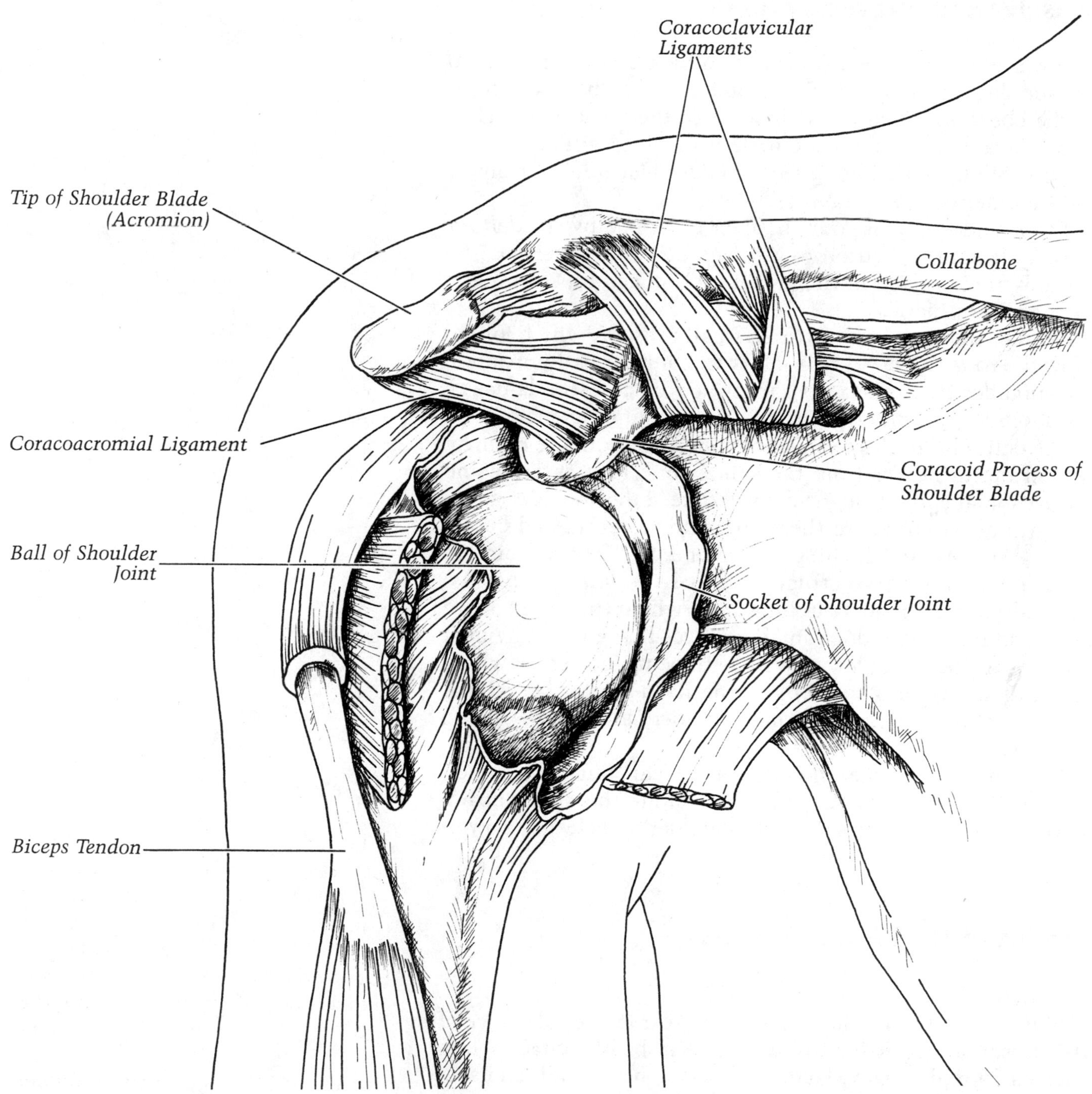

Anatomy of Shoulder

the shoulder joint. There is no motion of the shoulder blade up to that point. However, as the arm proceeds in an overhead direction, for every two degrees that the shoulder joint moves the shoulder blade rotates one degree. Finally, when the arm is directly overhead, the shoulder blade has rotated so that the socket is lying directly underneath the ball portion of the arm bone. Mechanically, this is the most stable position for the shoulder joint.

Muscles and Nerves

Powerful skeletal muscles run from the trunk of the body to the shoulder girdle itself. They come from the back and front of the chest wall and from the base of the neck itself. They form the contour and bulk of the shoulder girdle and are responsible for positioning the shoulder blade and for giving the shoulder power.

The rounded cap of your shoulder is formed by the *deltoid muscle.* It is in a position to give you powerful leverage. The deltoids are involved in the proper motion and functioning of the upper arm bone.

Deep in the shoulder are three small muscles which join to form the *rotator cuff.* Their job is to hold the ball portion of the shoulder tightly against the socket. If these are damaged, the shoulder can function at only 30 percent efficiency.

The deltoid muscles receive their message from the brain by way of motor nerves from the spinal cord. Five nerves come off the spinal cord at the neck level to feed the shoulder girdle and arm area. These are the fifth, sixth, seventh, and eighth cervical nerves and the first thoracic nerve. They combine to form a complex nerve structure called the *brachial plexus.* This is like a railroad switching yard, where the original five nerves split off to produce smaller nerves that go to individual muscles in the shoulder, as well as three major nerves which continue down into the arm to the hand area. These three major nerves—the *radial, ulna,* and *median* nerves—run to the hand.

The muscles of the shoulder girdle can fail to get their message because of injury anywhere along the switchboard. Without the nerve messages, the shoulder muscles are powerless.

Strains of the Shoulder Muscles

CAUSES

A strain of the shoulder muscle is a tear in the substance of the muscle itself. It is most common in body contact sports such as football, rugby, lacrosse, hockey, basketball, and soccer. I have also treated severe shoulder strains in weight lifters, wrestlers, and handball, squash, and baseball players.

MUSCLE STRAINS
HRINIAK'S TRIGGERS

Walter Hriniak, a coach with the Boston Red Sox and a former major league baseball player, throws four to five hundred pitches every day at batting practice. All this throwing rips and strains his shoulder muscles. The muscles seem to rip in the same spots and become heavily scarred. I call the spots "trigger points." When I put my finger directly on these trigger points, I can feel the heavily scarred tissue, which is clearly different from normal muscle tissue. They are often tender. I treat Walt with ultrasound treatments and heat packs. This seems to help him. Before pitching, he stretches and throws gently. When I press on them, Hriniak says, "That's like ringing a doorbell, Doc." After pitching, Walter whirlpools.

I have had many shoulder strains in my athletic career. This was probably because I played guard in college football, a position with a great deal of mashing of shoulders. It was not until I went to medical school that I understood why I had sustained so many strains.

First of all, I knew when my muscle strain happened. I could feel the actual ripping and tearing at the center of the muscle itself. Initially, there was very little pain. Foolishly, I ignored the injury. In the next six to twelve hours, the dull pain began. It was accompanied by muscle spasms and swelling. If I had been wiser in those days, I wouldn't have passed off the injury, which usually took at least a week to heal.

The main cause of shoulder and muscle strain is improper warm-up before competition. Warming up increases the blood supply to the muscles and raises their temperature. This, in turn, makes them more pliable and resistant to tearing. It is why you see most professional football players in pregame warm-ups hit shoulder to shoulder lightly.

The second cause of shoulder and muscle strain is that the shoulder muscle is not ready for the workload. I often see shoulder strains early in the baseball season.

Red Sox outfielder Dwight Evans was the victim of a severe shoulder strain when he tried to throw out a runner at home plate from the deepest part of center field in spring training of 1978. His arm was not ready for the strain, and the muscle tore slightly. Because of the injury he lost seven days of practice.

DIAGNOSIS AND TREATMENT

You know that you have sustained a muscle strain when the pain sets in. It is commonly accompanied by swelling, spasm, and discoloration of the skin, caused by internal bleeding. This is called "black and blue" by the layman. I can usually tell how extensive the strain on the shoulder muscle is by feeling the affected area. Small tears produce local tenderness, but if the rip is big enough I can actually touch the hole in the muscle. X-rays seldom help me.

The best treatment is rest. Your body will lay down scar tissue between the torn ends of the muscle. This process, depending on the extent of the tearing and your age, takes from two to three weeks. The decision to return to physical activity is a day-to-day event. If you return before the injury is completely healed, the chances of reinjury are greatly enhanced.

When the injury happens, apply ice immediately. This will reduce the swelling and limit the internal bleeding. Be careful not to freeze the skin. Remove the ice every twenty minutes in order to rewarm the skin. After the initial forty-eight hours, a warm bath, twice daily, or using heat packs can speed the healing process. Wet and dry heat seems equally effective. Some doctors recommend ultrasound treatment because it is a deeper penetrating form of heat.

I recommend very light exercises for strains of the shoulder muscle, called pendulum exercises. They are important in restoring mobility. Start them five days after the injury. Do them until the pain begins. Even the slightest exercise will increase the temperature of the muscle and increase the blood flow. The extra blood flow promotes healing. Remember that the older you are, the longer it will take for the injury to heal. After two weeks of healing, I recommend strength training.

PENDULUM EXERCISES

I always recommend pendulum exercises to patients with shoulder injuries. My reasoning is simple. The exercise will bring blood to the area which aids in the healing process. Secondly, because most shoulder injuries are treated with a sling, there is a tendency for the muscles to become smaller and less pliable from disuse. Lastly, if your shoulder isn't exercised daily, it has a tendency to stiffen.

Here are four pendulum exercises. Perform each of the exercises standing with your back bent forward at the trunk, using a three-pound weight. By bending, you reduce the impact of gravity. Balance your weight by placing your free hand on the arm of a chair. Do each exercise thirty times.

1. Rotate your arm in small clockwise circles. Because of stiffness, you will be able to make only small circles, perhaps twelve inches in diameter, depending on the injury.
2. Rotate your arm in counterclockwise circles. Always try to increase the size of circles.
3. Swing your arm forward and backward at a right angle to your side.
4. Swing your arm away from your body and across the front of your trunk.

As the injury heals, the mobility of your shoulder will improve.

Shoulder Separation

CAUSES

If you play collision sports you have about a 10 percent chance of getting a shoulder separation. A shoulder separation is when the collarbone separates from the shoulder blade at the acromioclavicular joint (A-C joint).

It happens when the tip of the shoulder receives a sharp blow which pushes the shoulder down and away from the collarbone. Shoulder separations are very common injuries for football players who carry the ball: ends, running backs, and quarterbacks. The runner who is protecting the ball carefully is tackled in the open field and driven to the ground on the tip of the shoulder. The separation normally occurs on the side where the ball is being carried and ironically it happens to those players who protect the ball the best.

Hockey and lacrosse players succumb to shoulder separa-

WEIGHT EXERCISES TO STRENGTHEN SHOULDER MUSCLES

Rehabilitation of injured shoulders after a period of disuse is directed at strengthening two major sets of muscles, the rotator cuff muscles and the deltoid muscle.

Here are exercises to strengthen the *rotator cuff muscles*, the small muscles form the top third of the sleeve of tissue that holds the ball-and-socket shoulder joint.

Place the elbow, bent at 90 degrees, on a table. Point your hand toward the ceiling. Using a five pound barbell, let your arm descend slowly to the table in a forward direction. It should take about three seconds. Raise the arm to the upright or ceiling position. Repeat this fifteen times. After a five-minute rest, do a second set of fifteen. After a second rest period, do a third set. As this exercise becomes easier and easier, increase the weight, but the weight should not exceed fifteen pounds.

Here are exercises to strengthen the deltoid muscle, the large muscles overlying the cap of the shoulder which give the shoulder its power. The front deltoid muscle attaches from the collarbone to the upper armbone and permits powerful arm movement in front of the body. The middle deltoid attaches from the shoulder blade to the upper arm bone, and gives the ability to move the arm away fromt he side of the body. The back deltoid muscle runs from the shoulder blade to the back of the upper armbone, and allows movement from the side of the body directly backwards. The deltoid muscle is the second strongest muscle group in the upper body. The strongest are the triceps in the back of the upper arm.

To strengthen the front deltoid muscle, hold a ten-pound barbell at your side. Lift it to shoulder level. The barbell should be right in front of you. Patients under 130 pounds should use a five-pound barbell. Hold this position for three seconds—count 100-1, 100-2, 100-3. Bring it down. Repeat 15 times. After a five-minute rest, do another set of 15. Rest five minutes. Do a third set.

To strengthen the middle deltoid muscle, the ten-pound barbell is moved directly out from the side of the body: Hold the weight at shoulder level for three seconds. Repeat fifteen times. Rest five minutes. Do two more sets with an additonal rest period.

To strengthen the back of the deltoid muscle, the barbell is moved directly backward, as if reaching for your wallet. Again, 15 repetitions are performed. After a rest, do two more sets of exercises.

For the highly competitive athlete, further gains in shoulder strength can be accomplished by the use of the Universal gym or Nautilus equipment. These systems are safer than free weights because you cannot lose control and have them fall on you; ideally, these exercixes should be supervised by a physical therapist.

In the Nautilus system, use the double chest machine, the double shoulder machine, the double pull-over machine and the military press machine. Set the weights on each machine so that your muscle fatigues on the fifteenth repetition. If you can repeat the exercise more than fifteen times, the weight is too light. If you can do only ten repetitions, the weight is too heavy. It will take some experimenting. Do only one set. One set will completely exhaust your muscles. The exercises will strengthen mainly the deltoid muscles.

In my opinion, there is a distinct important advantage for shoulder rehabilitation if there is some play or give in the weights before they are engaged. This stretches the shoulder joint before the muscle contracts. The result: a quicker return to mobility and strength.

In the Universal gym system, use the behind-the-neck pull down, the military press and the bench press. Find the weight that will fatigue the muscle at 15 repetitions. Do three sets of fifteen repetitions. Rest five minutes between sets. A Universal gym is much like free weights. It will take you three sets to completely fatigue the muscle.

tions when an athletic stick is jammed against the tip of the shoulder. The result of the separation is the same; only the manner is different.

DIAGNOSIS AND TREATMENT

A shoulder separation is painful with any motion of the A-C joint, but the pain is most severe when you lift your arm above your head.

I diagnose the injury by touch. I press firmly on the A-C joint. If there is a separation, the patient will scream bloody murder. When I have touched this injured joint, I have heard some unique locker room language. There is no other shoulder injury that will respond the same way to touch in that area.

Even with the slightest separation, the pain is persistent and dull. You will find it very difficult or impossible to find a comfortable sleeping position. That's why I always prescribe painkillers for a few days.

There are three degrees of severity of shoulder separations. The first-degree separation is a stretching of the joint. There is no physical separation of the bones and the x-ray is normal. The treatment: use of a sling for at least three days. Within a week the joint repairs itself, and if you are pain-free, you can return to your sport. The shoulder hasn't been immobilized

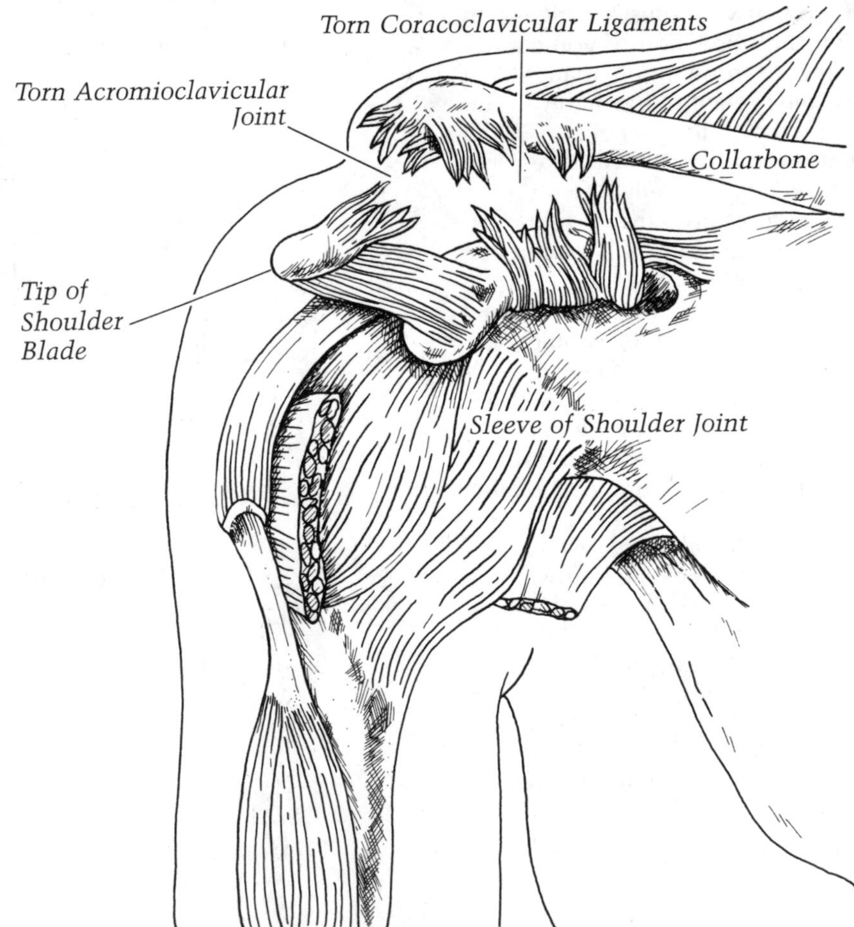

Third Degree Acromioclavicular Separation of Shoulder

long enough to have lost strength and size. A rehabilitation program is unnecessary.

A second-degree shoulder separation is a complete separation of the A-C joint, but the ligaments which attach the shoulder blade and the collarbone are still intact. However, the two bones are clearly separated. This shows up on x-rays. I usually take a second x-ray, with the patient holding a five-pound weight. This makes the separation more dramatic.

With a second-degree injury, I can feel the separation. Normally the joint is level. The end of the collarbone sticks up. The pain, swelling, and tenderness are more intense than a first-degree separation. The immediate treatment: ice, a sling to immobilize the shoulder, and painkillers. It will take two or three weeks for nature to work its wonders. During the healing period, you will be in pain. Use ice twice daily for twenty minutes. Continue for two days, then start heat treatments.

One week after the injury, I recommend pendulum exercises. This will prevent shoulder stiffness. If you don't exercise the shoulder, it could become a "frozen shoulder," a condition where the baggy tissue of the shoulder joint sleeve adheres to itself. It has the same effect as sewing your armpit shut.

Muscle-strengthening exercises should be started two weeks after the injury. When the tenderness subsides, you can return to your sport. I ask my patients to wear a rubber donut which I have designed to fit directly over the A-C joint. It is held in place with tape and skin adherent. My donuts are at least one inch thick and are available at medical supply stores.

A third-degree shoulder separation is the most severe form of the injury. The A-C joint is torn away, and the ligaments which attach the shoulder blade to the collarbone and hold the collarbone down are ripped. Almost any medical professional can recognize a third-degree separation. The skin over the collarbone is drawn tightly and discolored. When I touch it, I can feel the entire end of the collarbone. The x-ray shows the complete separation. The gap is usually one-half of an inch wide. A weighted x-ray is not necessary.

The bigger the separation, the more the swelling, tenderness, and internal bleeding. With a third-degree separation, the arm is useless. It just hangs by the side, and any movement is painful.

In my opinion, third-degree separations generally require surgical repair for two reasons:

1. The shoulder joint is so damaged that it can't repair itself. Specifically, there is no way for the bones to realign themselves. The gap is too large.
2. If you don't have surgery, there is almost 100 percent chance that your shoulder will lose some mobility. Also, the shoulder structure will be permanently weakened.

The surgical repair for a third-degree separation is an oper-

MARY'S SHOULDER

What happens if you do not get proper treatment of a third-degree shoulder separation? Take the case of Mary Kay, a twenty-three-year-old softball player. During a game in May of 1978, she slid head first into second base. She landed on the tip of her right shoulder. She experienced an immediate onset of pain. Because of the pain, she was taken to the emergency room, where x-rays showed a third-degree shoulder separation. She was given a sling and pain pills and sent home. She did not consult with me until four months after the accident. By that time, her shoulder had become deformed. Her collarbone had risen one inch above the shoulder blade. It was still tender to the touch and because of the pain, she had only 50 percent use of her shoulder. Due to disuse, many of her shoulder muscles were weak.

Because of the delay in treatment, Mary needed a complicated operation. I operated the next week. Because of the massive scarring in the area of the injury, it was impossible for me to realign the joint in the normal way. Therefore, using a small surgical saw, I removed about one inch of the outer end of the collarbone. This left a one-inch gap between the new end of the collarbone and the shoulder blade. This fills with scar tissue. The operation completely relieved her pain. There is still some disfiguration, but her shoulder works. She is even back at softball.

ation which normally takes less than an hour. There are at least 150 surgical techniques to realign and hold the collarbone to the proper position on the shoulder blade. The surgical principle is simple. I insert a pin or screw across the joint to hold the bones in proper alignment. It is the same principle as joining two pieces of wood with a screw. The pin remains implanted in the bone for six weeks. I start my patients on pendulum exercises as soon as possible—generally two weeks after surgery—to prevent shoulder joint stiffness. Gradual is the watchword.

After six weeks, I remove the pin. Now you can start your weight-lifting program. It should take you at least six more weeks to recover fully. You shouldn't return to athletic competition until the shoulder has returned to 95 percent of its strength and mobility. This can be measured precisely on a Cybex isokinetic machine.

Fractures of the Collarbone (Clavicle Fracture)

CAUSES

Fractures of the collarbone are a result of contact, usually from a fall with your arm outstretched. The force of the fall is transmitted up the arm to the collarbone, causing the fracture.

The most common site for a collarbone fracture is the

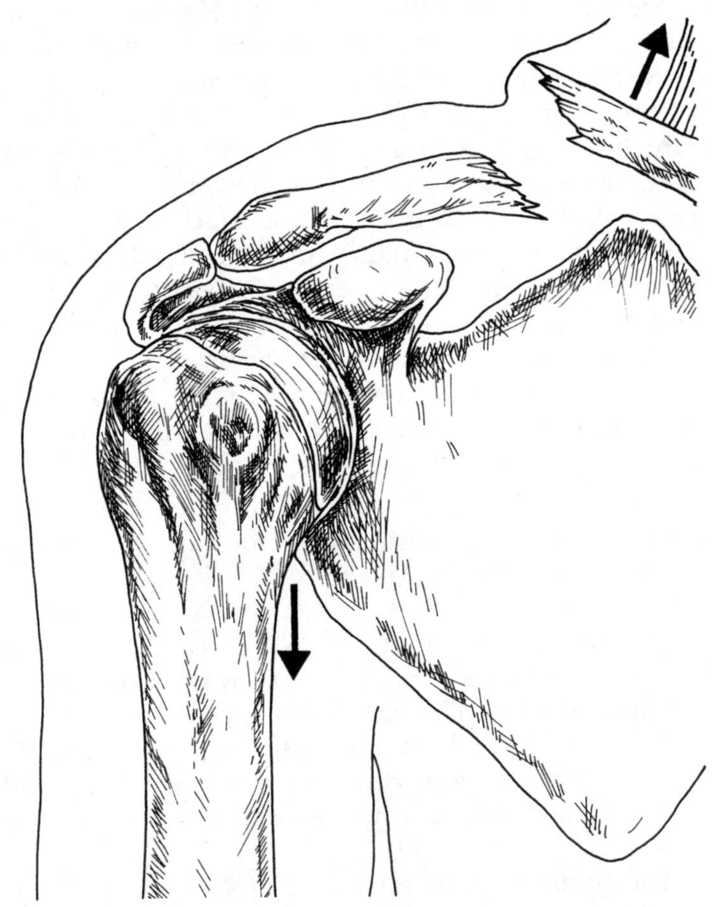

Fractured Collarbone

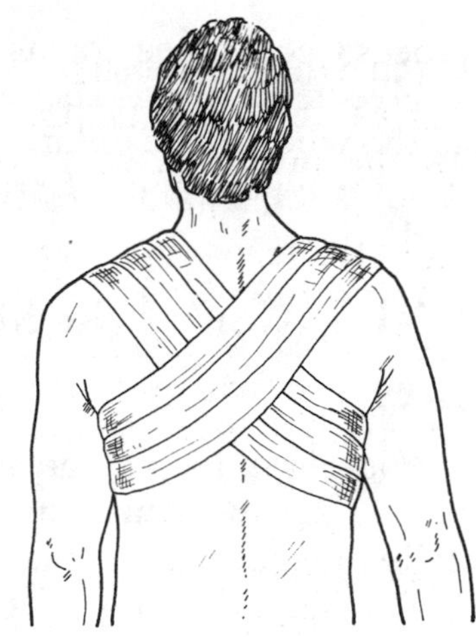

Shoulder Harness and Sling for a Fractured Clavicle

middle of the bone. Because their bones are soft, teenagers are most susceptible to fractured collarbones. Fractures on the inner and outer ends happen rarely, and I have treated such fractures only five times.

DIAGNOSIS AND TREATMENT

How do you know that you have a fractured collarbone? It is very tender and touching the point of the fracture will cause great pain. You will not be able to move your shoulder because of the pain. Within six hours, swelling starts. X-rays reveal the fracture clearly.

The treatment: Apply ice to limit swelling and internal bleeding, get rest, use painkillers and a figure-of-eight splint. This is a harness that is applied by a doctor to immobilize the shoulder. The shoulder harness is fitted about both shoulders and forces the patient to stand like a West Point cadet. It is worn all the time. In addition, I use a sling for the injured arm for the first seven to ten days. For patients under the age of twelve, the period of immobilization with the figure-of-eight splint is two to three weeks. In adults, the period is six weeks. As soon as the pain decreases, pendulum exercises should be started even though the patient is still in the splint. Weight exercises to strengthen the shoulder should not be started until there is a union of the bone, which only a doctor can determine after the splint has been removed. Once healed, the collarbone is as strong as normal and there is no susceptibility to repeated injury. Protective equipment is not needed.

Collarbone fractures almost always heal themselves. When the collarbone fractures, a pool of blood forms around the broken ends. After a few hours, the pool of blood organizes into a clot, which looks and feels like a ball of soft putty. In two weeks, this soft putty gradually stiffens as bone is produced within it. As it hardens, it becomes a bony callus, called a *fracture callus.* You can feel the lump at the site of the fracture.

While it takes only a few weeks to produce enough new bone to allow return to sports activities, the fracture healing

TED WILLIAMS' BROKEN COLLARBONE

In spring training in 1952, Ted Williams, the last batter to hit .400, tried to make a shoestring catch of a line drive.

"There is very little doubt in my mind that the collarbone was broken," says Jack Fadden, the Boston Red Sox trainer. "I convinced Ted to go back to Boston to be taken care of.

"He was operated on by a surgeon named Shortell, who pinned the collarbone together. Williams lost five weeks at the beginning of the season. But the collarbone healed perfectly."

process continues for as long as a year. In children under twelve years of age, the healing lump will gradually disappear as the whole bone enlarges with normal growth. In adults, the lump is a permanent fixture, usually visible.

Dislocation of the Shoulder Joint

CAUSES

The shoulder has great mobility because of the shoulder joint. It allows your arm to swivel, pivot, and rotate. It is this ball-and-socket joint that makes the arm so versatile.

The shoulder joint lies deeply seated in the shoulder girdle under the protective covering of the A-C joint. The joint itself has a very shallow, bony socket. The socket, however, is deepened by a gristle-like lip called a *labrum.* The lip is shaped like the Los Angeles Coliseum.

A *dislocation* of the shoulder joint happens when the ball is suddenly and forcefully moved out of the socket. Football players who arm-tackle and skiers whose bodies pass their poles are the most likely candidates for this injury. This is how I explain it to my patients: When an arm is fully extended, it becomes a lever. If the arm is hit in that extended

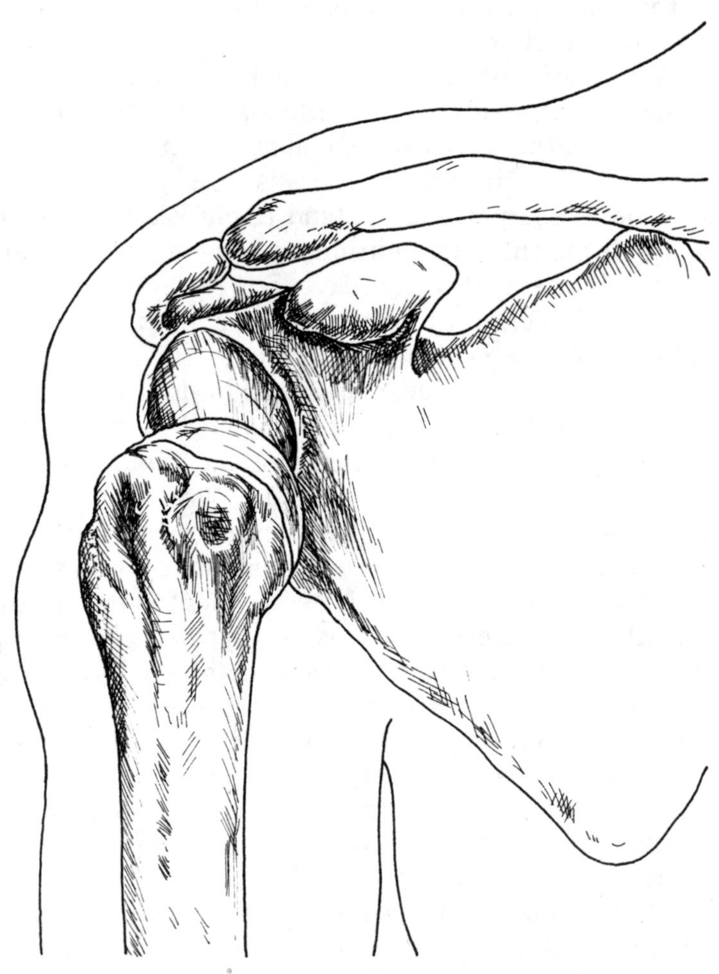

Shoulder Dislocation

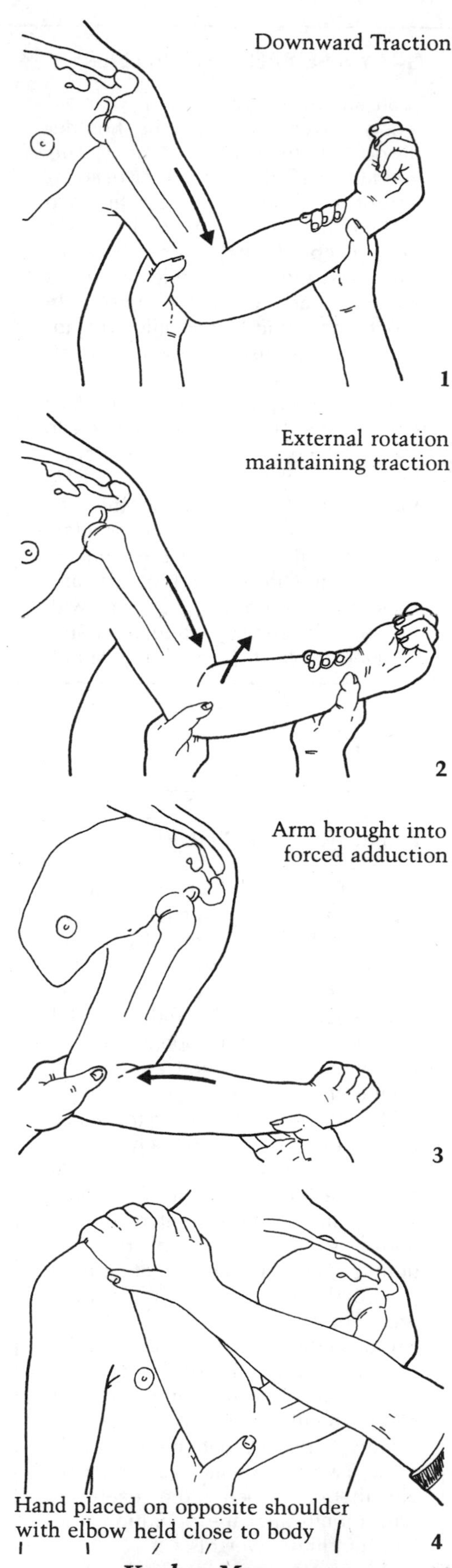

Kocher Maneuver

position, the force radiates up the arm to your shoulder joint, prying the ball out of the socket. Ninety-five percent of all dislocations are front (anterior) dislocations where the ball of the arm bone rips out through the front of the socket. It comes to rest close to the chest wall.

DIAGNOSIS AND TREATMENT

When you sustain a dislocated shoulder, you get an immediate sense that there has been a major disruption in your body. The normal contour of the shoulder is lost. The shoulder muscles go into spasm. Any movement of the arm is extremely painful. That is why you have a tendency to hold the arm close to the body. Other symptoms are tenderness and discoloration.

The x-ray of an anterior shoulder dislocation is unmistakable. The ball portion of the arm bone comes to rest directly beneath the lip of the shoulder blade. The empty socket is clearly visible. Occasionally, a fracture of the head of the arm bone will accompany the dislocation. It is important for your doctor to look for this fracture. A fractured ball might prevent relocation.

See a physician immediately. He will put the ball back into the socket. The sooner you get to him, the less the shoulder muscle will spasm. It is the spasm of the shoulder muscles which makes it difficult to relocate the shoulder. Meanwhile the only thing you can do for yourself is to ice the injured part.

There are several methods of relocating a dislocation. The oldest method is one developed by Hippocrates, the Greek physician and author of the Hippocratic Oath. The ancient physician gently placed his foot in the armpit, then grasped the arm and pulled it directly downward. After two or three minutes of steady traction, the muscles in spasm around the shoulder joint relaxed. Gently, he turned the ball of the joint back into its socket.

The second method, my favorite, is called the *Kocher maneuver.* I leave the elbow at the side of the body and rotate the hand away from the body. This stretches the muscle in spasm. I gently raise the upper arm when it is relaxed, by moving the elbow away from the body. I can feel the ball return to the socket. The patient is immediately relieved.

The third method is a variation of the first two. I place the patient on the examining table face down. I let the patient's arm drape over the side of the table. I then lash a ten-pound weight onto the arm. This has the same traction effect as the first two methods. After two or three minutes, the shoulder muscles in spasm relax. Then I can guide the ball back into the socket.

Most dislocated shoulders get reset in hospital emergency rooms. In 1970, I got lots of experience with dislocations at Fort Dix, New Jersey. That was the year I was drafted, and I had just finished my internship at the Massachusetts General Hospital. During my first week at Fort Dix, we had a rash of dislocations, mainly in recruits from New York City and Philadelphia.

I used the Kocher maneuver to reset most of the dislocations. I liked this method because it took some finesse. I also enjoyed doing the Hippocratic method, but it seemed to me less than professional to take off my military boot and stick my foot into a soldier's armpit.

Once I had to turn to the ancient Greek physician's method. A recruit who was a big strapping former college football player was brought in with a shoulder dislocation. I tried the Kocher maneuver, but his huge muscles would not relax. I considered a general anesthetic, but the patient had just eaten. I was running out of options, so I laid him on a stretcher, placed my foot into his armpit, and started to pull. I am strong, but after five minutes of steady pulling, I was not making any progress. I called for another doctor. We both started pulling on this poor soul's arm. The other doctor stood directly behind me and a nurse tied us together with bed sheets. After a few minutes, the muscles relaxed and I was able to relocate the shoulder. One doctor jested: "The new Laurel and Hardy." The patient said: "Doc, I thought you were going to pull my arm off."

In my first three weeks at Fort Dix, I relocated about twenty shoulder joints. The other doctors, many of whom had witnessed my "Laurel and Hardy" antics, sent me all their dislocation cases. It was my first experience in orthopedics.

I wondered why so many recruits were dislocating their shoulders, so I asked a few discreet questions. Apparently, one battalion commander was overzealous with his charges. He had his troops cross elevated ladders six times a day. The ladders, called "monkey bars" by the soldiers, are about eight feet off the ground. The soldiers, supported only by their arms, swing from bar to bar. The ladders are about thirty feet long. The dislocations would happen when a soldier lost control—a hand would slip reaching for the next bar—and all the force would be put on one shoulder. Most of the recruits who suffered dislocations were big or overweight. Thin soldiers, even those with skinny arms, would swing across the ladders without difficulty. Though their arms were small, they did not have much weight to support.

When I was a resident in orthopedics at the Massachusetts General Hospital, I witnessed the ultimate method of relocating a shoulder. A college football player was brought in with a dislocation. I prepared myself for my "Laurel and Hardy" act, but the doctor in charge of the emergency room took him to the operating room to administer general anesthesia. The player's entire body went limp in a few minutes, including the shoulder in spasm. The doctor relocated the shoulder with hardly any effort.

"Why did you give him the anesthesia?" I asked.

"Did you see the shoulder muscles on that guy?" the doctor asked. "It would have taken three of us to relax his muscle."

GRAY'S SLIPPERY SHOULDER

Susan Gray, a twenty-four-year-old tennis player, had her right shoulder dislocated twenty-five times. This condition is called recurrent dislocation of the shoulder. For her first two or three dislocations, she had a doctor put the shoulder back in place. However, now the joint is so loose, she is an expert at replacing it herself. In 1979, she came to my office for advice and to discuss possible treatment.

The maneuver that I like best to determine if a shoulder dislocates is to stand behind the patient and support the arm in a throwing position. I hold the weight of the arm so that the muscles of the shoulder can relax. With the other hand, I push on the back of the upper part of the arm bone, the ball portion. I did this with her shoulder and she let out a scream.

"Doc, you dislocated my shoul-

BANKART PROCEDURE

I prefer the Bankart procedure to repair recurrent dislocation of the shoulder. In fact, I wrote a paper on this particular procedure with Doctors Carter Rowe and Dinesh Patel, both of Massachusetts General Hospital. The purpose of this operation is to reinforce the weakened area on the front of the shoulder joint. I do this by drilling three small holes in the socket bone. I use the holes to anchor stitches which I attach to the sleeve-like capsule against the front of the bony socket. In time, the joint regains its normal strength. The Bankart procedure tends to give the patient far more shoulder mobility than most other procedures.

Our study found that four out of five patients regain full mobility after surgery. One key to regaining motion is early rehabilitation. Three days after surgery, I get my patients out of the sling. Two weeks after surgery, I begin them on pendulum exercises. One month after the operation, they can start lifting weights.

der."

Sheepishly, I agreed. Quickly, I put her down on the examining table and used the Kocher maneuver to relocate her shoulder.

"I have to do something about my shoulder," she said. "Last summer, during a swim, I dislocated it about 200 yards from shore; suddenly, I thought I was going to drown. Fortunately, I was swimming with my boyfriend, and he was able to support me until I could work the shoulder back in place."

"How did you manage that in the water?" I asked with surprise.

"If your shoulder has been out as many times as mine, you get it back in under any circumstances."

Two weeks later, I operated on her. She has never dislocated the shoulder again.

Once relocated, the shoulder should be immobilized in a sling. The sling should be rigged so that the arm cannot be moved away from the front of the body. The purpose of this is to prevent outward rotations of the arm, which could cause another dislocation or slow the healing process. The sling should be worn for three weeks. Use ice on the shoulder twice a day for twenty minutes during the first two days. Then use heat packs. After the three-week healing period, start the pendulum exercises. Then, as soon as the pendulum exercises can be done without pain, start the weight exercises.

Once you dislocate your shoulder, you have an almost 95 percent chance of dislocating it again. That's what the medical literature shows. The reason for this high repeat rate lies in the nature of the injury. The damaged lip of the shoulder joint is much like knee cartilage in that it has no blood supply; without a blood supply, the lip has no way to heal itself. Thus a dislocation is a permanent injury. I explain it to my patients this way: "When the shoulder dislocates, the ball portion damages the front part of the joint lip. It is very much like a wrecking ball blasting out one side of the Coliseum wall."

If you don't want to be a candidate for a second, third, or fourth dislocation, there is a surgical alternative. This involves a choice of any of five separate operations—the Bankart Procedure, the Puttyplat Procedure, the Bristow Procedure, the Magnusen-Stack Procedure, and the Dutoy Staple Procedure—to stabilize the shoulder and return full motion to it. All these operations are designed to hold the ball in the socket, and they are 97 percent successful.

After the operation, about one out of four patients has some restriction of external rotation of the shoulder. That means that they can't cock the shoulder properly up to the throwing position. For quarterbacks and pitchers, it could mean a loss in throwing velocity and accuracy.

Partial Shoulder Dislocation (Shoulder Subluxation)

CAUSES

Sometimes the ball of the upper arm bone is leveraged out of the socket for an instant, but the muscles in the front of the shoulder are strong enough to keep the ball from coming out completely. This partial dislocation is called *shoulder subluxation.* Subluxations are caused by the same type of accidents that create total dislocations.

DIAGNOSIS AND TREATMENT

Normally, a victim of a partial shoulder dislocation will have a sense of the shoulder being separated and then coming back together again. The subluxation will be extremely painful for thirty seconds, then the pain will rapidly diminish. Within four to six hours, pain and stiffness will return and the shoulder will lose mobility. But within five to seven days the pain often disappears.

X-rays do not help diagnose the problem. They show that the bones are aligned properly. Also, there is no great swelling or tenderness. Since a partial dislocation does almost as much damage as a full dislocation—the lip of the socket has been permanently damaged—I am anxious to diagnose it.

Carter Rowe, the eminent Boston shoulder surgeon, taught me an excellent method of diagnosing the problem. I stand behind the patient and grasp the wrist of the affected arm, then I lift it into the throwing position and support the arm so that the shoulder muscles relax. With my other hand, I push on the back of the ball of the upper arm bone. If the front of the shoulder joint has been damaged, the patient will scream. The apprehension is tremendous. They feel that the ball is going to jump out of its socket again.

The treatment for acute subluxation: ice for twenty minutes daily during the first two days, then heat treatments after that. I usually put the arm in a sling for three days. This ensures rest. As the pain decreases, start pendulum exercises. Two weeks later, start weight training.

If permanent damage has been done to the lip in the front of the joint, the shoulder will continue to slide in and out. You feel this shifting as you try to throw or use your arm in a throwing position. Since the damage is permanent, surgery is the only answer.

Rarely is damage done to the back of the lip. I've only seen it occur three times, most recently in Red Sox outfielder Tom Poquette.

As in recurrent shoulder dislocation, I personally favor the Bankart Repair. In my hands it gives the best chance for full return of shoulder motion after surgery. Other orthopedic surgeons are equally satisfied with the other surgical procedures.

Rotator Cuff Tendinitis (Shoulder Impingement Syndrome)

CAUSES

Three small muscles and their tendons—the supraspinatus, the infraspinatus, and the teres minor—tightly hold the ball and socket of the shoulder joint together. They are called the *rotator cuff muscles*, and along with their tendons run from the top of the shoulder blade to the top of the arm bone.

Players of racquet sports and baseball pitchers—those who raise their arms over their heads—often irritate these muscles and tendons. When the arm is in an upright position, the rotator cuff tendons tend to rub against the bony undersurface of the shoulder blade, causing a slight tearing and inflammation. We call this *impingement*. The problem, in nine out of

TOM POQUETTE'S SHOULDER

In mid-season 1979, the Boston Red Sox traded for outfielder Tom Poquette. Poquette played very effectively until he was injured in September. He batted .331 and led the club with a .345 average on the road.

The injury to Tom's shoulder happened on an attempted pickoff play at second base. Tom lunged back to the bag to avoid the tag. His outstretched elbow jammed into the bag, driving the upper arm back into the shoulder. He felt immediate pain, but he continued to play. Poquette, a quiet man, didn't complain about the pain. As the season ended he returned home and throughout the winter did not comment on any unusual discomfort. But, in the 1980 spring training his pain started to return.

"When Dr. Southmayd and I examined Poquette in spring training, it was evident that he had major spasms in the posterior aspect of the shoulder," recalls Dr. Arthur Pappas, medical director of the Boston Red Sox. "We ordered a battery of tests, x-rays, and a shoulder arthrogram. We did the tests between March and May. We couldn't identify the exact problem."

"There's something that catches; there's something that gives," Poquette complained.

"We did an arthrotomogram (a test which helps visualize the shoulder) on the back area of his shoulder. This is a usual test in an unusual body area. Ninety percent of shoulder injuries are in the front part of the shoulder—the anterior part. His symptoms indicated a problem in the back of the shoulder—the posterior—and this was confirmed by the arthrotomogram."

Tom underwent an arthroscopic examination of his shoulder, using a small telescope-type instrument to look into the shoulder joint. The findings confirmed the arthrotomogram and showed a T-shaped tear in the posterior glenoid labrum which looked somewhat like a torn cartilage in the knee. This injury had never been reported in a baseball player before. The shoulder was surgically approached from the back, and the torn labrum removed. Tom worked on a "rehab" program and is fully prepared for his return to professional baseball.

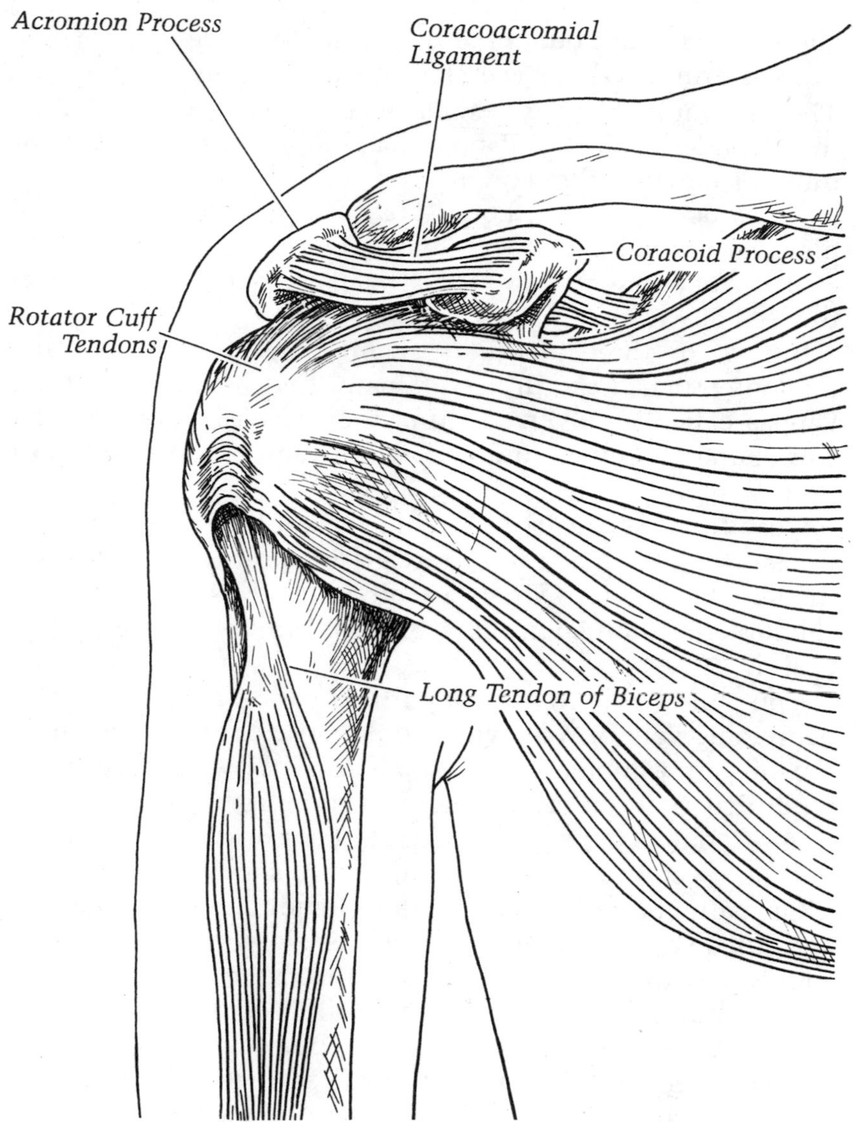

Rotator Cuff Muscles and Tendons

ten patients, is due to shoulder overuse. A single blow brings on the problem in the tenth patient.

Once tendinitis starts in the rotator cuff tendons, the tendons swell, leaving even less room between the upper surface of the tendons and the overhanging shoulder blade. The irritation creates more irritation. Picture a rope being drawn across a craggy rock again and again; it will fray. That is what is happening in your shoulder.

DIAGNOSIS AND TREATMENT

Like most shoulder injuries, rotator cuff tendinitis can be diagnosed by touch. By rotating the arm backward and away from the body, the rotator cuff tendons become exposed from under the shoulder blade. Touching them usually will make a patient cry out.

The second method of diagnosing this problem was taught to me by Dr. Arthur Pappas, the medical director of the Boston Red Sox. In 1978, I watched him examine pitcher Luis Tiant's shoulder. He put Louie on a rubdown table face down. He let Louie's arm dangle at the side of the table. In this position, Tiant's rotator cuff muscles relaxed and became exposed under the shoulder blade. Dr. Pappas' hand followed the rotator cuff muscles down to the tendons.

"How does that feel, Louie?" Dr. Pappas asked. Tiant took his cigar out of his mouth: "Nasty, Doc, nasty."

"How about swinging your arm in a circle?" Pappas asked the Red Sox hurler.

In two or three positions, Tiant winced as his swollen tendons touched the undersurface of his shoulder blade. Dr. Pappas called me over. "I think that he has tendinitis. If it were bursitis, Louie would be in constant pain in every position."

Despite the severe irritation, it is very difficult for me to feel the swelling because the irritated tendons are so deeply set in the shoulder cavity. X-rays are often of little use. Only when calcium has formed from a previous injury do they prove useful.

You will be able to use your arm at your side, but will have pain using your arm above shoulder level. The pain is on the top of your shoulder. Almost all patients with rotator cuff tendinitis have difficulty finding a comfortable sleeping position. If you deeply sever or tear the rotator cuff tendons, you will not even be able to lift your arm to shoulder height.

On patients with severe or longstanding pain, I normally perform a shoulder *arthrogram,* a procedure where I inject the shoulder with dye under local anesthesia. I do this with an x-ray machine called a *fluoroscope.* A fluoroscope is something like the x-ray machines used in the shoe store years ago. You put your feet into the machine to see if the new shoes fit properly. A fluoroscope operates the same way. Basically, it is a continuous movie of the joint. Using the fluoroscope, I inject the shoulder joint with 15 milliliters of renographin, an iodine dye. I watch the joint sleeve fill with the dye. I move the shoulder up and down ten times so that the dye can reach every nook and cranny of the shoulder joint. Doctors call this "milking." If there is a complete tear, the dye will squirt into the overlying bursa sac. The bigger the tear, the faster the dye escapes into the sac.

In cases where there is not a complete tear, but only roughening of the tendon, the dye outlines that roughening.

When the shoulder first becomes painful, ice it for twenty minutes after use. This will relieve some irritation. Rewarm the skin. Ice it down at least two times more. Ice twice a day for two more days. Then use heat treatments after that.

The best treatment is rest. Inflammation will usually heal itself. Allow the shoulder a week off. After the rest period, I recommend four exercises to strengthen the rotator cuff muscles and tendons. They help about 95 percent of people with small tears. A Cybex machine is also effective in strengthening these muscles.

BILL LEE'S ROTATOR CUFF TENDINITIS

In 1978, Bill Lee was the only starting left-handed pitcher with the Boston Red Sox. In June, he developed pain on the top of his left shoulder. With the tips of his finger, he would touch the top outer edge of his shoulder.

"That's exactly the spot, Doc."

Bill, a knowledgeable athlete, understood the problem. Dr. Pappas and I showed him an anatomic chart. We explained that he was experiencing pain on the supraspinatus tendon, one of the rotator cuff tendons—a common ailment for pitchers.

Despite ice treatments and anti-inflammatory medication, Bill's pain persisted.

"Why is it taking so long to heal?" he asked.

"It's an anatomy problem," I explained. "The small blood vessels that nourish your tendon start out in the muscle over the shoulder blade and then must traverse along the tendon.

HISLE'S ROTATOR CUFF

Larry Hisle, the all-star outfielder of the Milwaukee Brewers, has struggled with rotator cuff injury to the shoulder of his throwing arm for one and a half years. He first injured his shoulder in April of 1979 when he made a hard throw to the infield in a game against the Baltimore Orioles. A diagnosis of rotator cuff injury was made by a number of physicians. He was started on rest, weight training, and other exercises. When the shoulder didn't respond to that therapy, surgery was recommended. Hisle was reluctant to undergo the surgery. One factor in his decision was the fate of Hisle's long-time friend, Don Gullett, a New York Yankee pitcher who underwent surgery for rotator cuff repair in 1978 and has never recovered from the operation. He has been on the injured reserve list since the operation.

Hisle started the 1980 season feel-

That's a long way to go. Whenever we have this situation in the body, it takes a long time for injuries to heal. The greater the blood supply, the faster the body heals."

Bill, known for his feisty nature, accepted my explanation. In three weeks, he started to pitch again, but he pitched with pain. He refused cortisone injections.

His pain persisted at a low-grade level throughout the season. As soon as the season was over, I tested his pitching arm on the Cybex isokinetic machine. I found he had experienced a 20 percent loss of strength in the shoulder muscles. His rotator cuff muscles were especially weak.

Lee worked all winter long on the Cybex machine to regain full strength in the rotator cuff muscles. The following year, he was traded to the Montreal Expos and had an outstanding season. He told us later that his rotator cuff tendons were pain-free all season.

ing much improved. However, in a game in May, while sliding into second base, he braced himself with his right arm. He felt his shoulder tear. Dr. Frank Jobe of Englewood, California, and Paul Jacobs, the Brewer's team physician, performed an arthroscopy of Hisle's shoulder and it showed a complete rotator cuff tear. His tendon tear was repaired with open surgery of the shoulder. At the same time, Dr. Jobe removed the undersurface of the acromion and the coracoacromial ligament. The rotator cuff was found to be damaged in its substance and had not pulled away from the bone.

During the first month of rehabilitation, Hisle's therapy was limited to a small range of motion exercises and gentle physical therapy. He then started lifting weights and swinging a bat and is now able to throw the ball again with force.

For patients with nagging and persistent pain, I prescribe oral anti-inflammatory medicine for a few weeks. In 80 percent of the cases, it soothes the pain. If the pain continues after fourteen days, I inject a cortisone-like drug into the area around the tendon. The drug I prefer is called Depomedrol. It is similar to a timed-release capsule, and like other cortisones of its type takes forty-eight hours to start working. The full power of the injection does not reach maximum potency until it has been in the body for five days.

I carefully warn patients of this potential hazard. "Do not go back to sports because your shoulder feels better," I admonish. "Depomedrol is like Sneaky Pete. It makes you feel good, then it drops you off the edge of a cliff."

Three weeks after the injection, the tendinitis should have disappeared. Occasionally, the Depomedrol therapy has to be readministered, but I never give more than three injections.

In those patients whose tendon is badly roughened or completely torn, the severe pain will return again and again. When all conservative measures fail, you should consider an operation. It will relieve the pain and increase shoulder mobility. It is not a complicated procedure, and the stay in the hospital is only two days. In 95 percent of the patients, the operation gives complete and permanent relief from pain.

Surgery eliminates the mechanical problem in the shoulder. If there is any tendon tearing, I stitch them. I remove a slice of bone from the undersurface of the shoulder to give the tendon more room to operate. Always, I remove a tight ligament called the coracoacromial ligament, which pushes down on the tendon. In three months, you can be back at your sport, but only after you have experienced a return of 95 percent of normal strength and mobility to the shoulder.

"Swimmer's shoulder" is a special type of rotator cuff tendinitis. It is due to the mechanical impingement of the rotator cuff tendons against the coracoacromial arch of the shoulder blade. It is common in swimmers who reach out with their arms: free stylers, back strokers, and butterfly competitors. These three strokes require similar use of the shoulder.

The more you swim, the more likely you are to get swimmer's shoulder. A study by the National Athletic Health Institute evaluated 137 of the country's best swimmers. Forty-three percent had experienced shoulder pain due to swimmer's shoulder. Eighty-one percent of the swimmer's shoulder sufferers reported that hand paddling made their shoulder pain worse. The purpose of using hand paddles is to increase hand resistance through the water, increasing the load on muscles of the upper arms and shoulders. In theory, this exercise should strengthen the swimmer's arms. However, for the injured swimmers, it only makes the impingement worse. Treatment includes stretching exercises, rest, ice therapy after workouts, oral anti-inflammatory agents, and the judicious use of cortisone injections. Surgery is rare and should be used only as a last resort.

EXERCISES TO STRENGTHEN THE ROTATOR CUFF MUSCLES AND TENDONS

Here are four exercises to strengthen the rotator cuff muscles and tendons. Do them twice a day to rehabilitate and prevent rotator cuff tendinitis. They take a half-hour to perform with your back and heels touching the wall. This will keep your body straight and isolate the arm during the following exercises.

1. Let your hand hang by your side. Your palm should be facing inward. Bring your arm forward slowly from your side as if you were goose-stepping. But, continue the motion until your arm is over your head. Keep your elbow straight. Throughout the motion, the thumb should be pointing to the ceiling. Do not rotate your wrist. At the end, your thumb should be touching the wall. Your arm should be next to your ear. During the exercise, count 100-1, 100-2, 100-3. Hold this position for one second. Repeat ten times. Rest three minutes. Repeat ten times. Rest. Do a third set. If you experience pain, stop.

Do the same exercise with weights. Start with two pounds. If you experience pain, stop. Return to the exercise without the weight.

Increase the weights two pounds at a time. If the pain returns, stop. Reduce the weights.

The goal is to find the exact weight which you can lift only ten times without pain. Ideally, that means as hard as you try, you cannot make the eleventh repetition. In weight-lifting parlance, this is called "lifting to failure."

2. Stand sideways. Lean against the wall. Your shoulder and the side of your shoe should touch the wall. This is only for stability. You are going to exercise the opposite shoulder. Your palms should be facing inward. Raise your arm sideways and away from the body. Keep your elbow straight. Do not raise your arm above your shoulder. The movement should take two seconds or less. Count 100-1, 100-2. Hold for one second. Repeat ten times. Rest three minutes. Repeat ten times. Rest. Do a third set.

Add the weights gradually. The goal is to find the weight that you can lift ten times to failure without pain.

3. Lie on a table on your side. With your elbow bent ninety degrees, place the free arm across your chest. Keep your elbow firmly against your rib cage. With your palm facing inward, raise the forearm and hand off the chest and stomach until the forearm is parallel to the table top. This movement should take less than two seconds. Count 100-1, 100-2. Hold for one second. Let your forearm fall. Repeat ten times. Rest. Do two more sets with a rest.

Add weights gradually. The goal is to find the exact weight that you can lift ten times to failure without pain.

4. Lie on a table on your side. This time, the arm to be exercised is the one that is underneath you. Your elbow should be at right angles. The back of your arm should be flat on the table and your palm should be facing the ceiling. Move your forearm and hand across your chest. At the end of the exercises, your fingers should be pointing directly to the ceiling. The movement should take less than two seconds. Hold for one second. Let your forearm fall. Repeat ten times. Rest three minutes. Repeat the exercise. Rest. Do ten more repetitions.

Peter Stone and Donald Labourr, Sports Therapists, Sports Conditioning Center, Chestnut Hill, Massachusetts.

WAYNE GARLAND'S ROTATOR CUFF

Wayne Garland, who signed a $2-million, ten-year contract with the Cleveland Indians in 1976, is another pitcher who has had a tear of the rotator cuff tendons. Frank Jobe and Robert Kerlan, sports medicine physicians from Los Angeles, diagnosed Wayne's problem in May 1978. He was operated on the same month. It took Garland almost a year to return to pitching.

"Wayne is a very dedicated, hard worker," says Jim Warfield, the trainer for the Cleveland Indians. "I wish people knew how hard Wayne worked to make his comeback. Wayne and I almost were living together during his rehabilitation.

"We did a lot of range-of-motion and stretching exercises. Initially, I just wanted to get his arm moving. We did small smalls—moving the arms in small circles. Very gradually, we started to use some weights. Dr. Jobe was in constant contact by phone, and supervised the program."

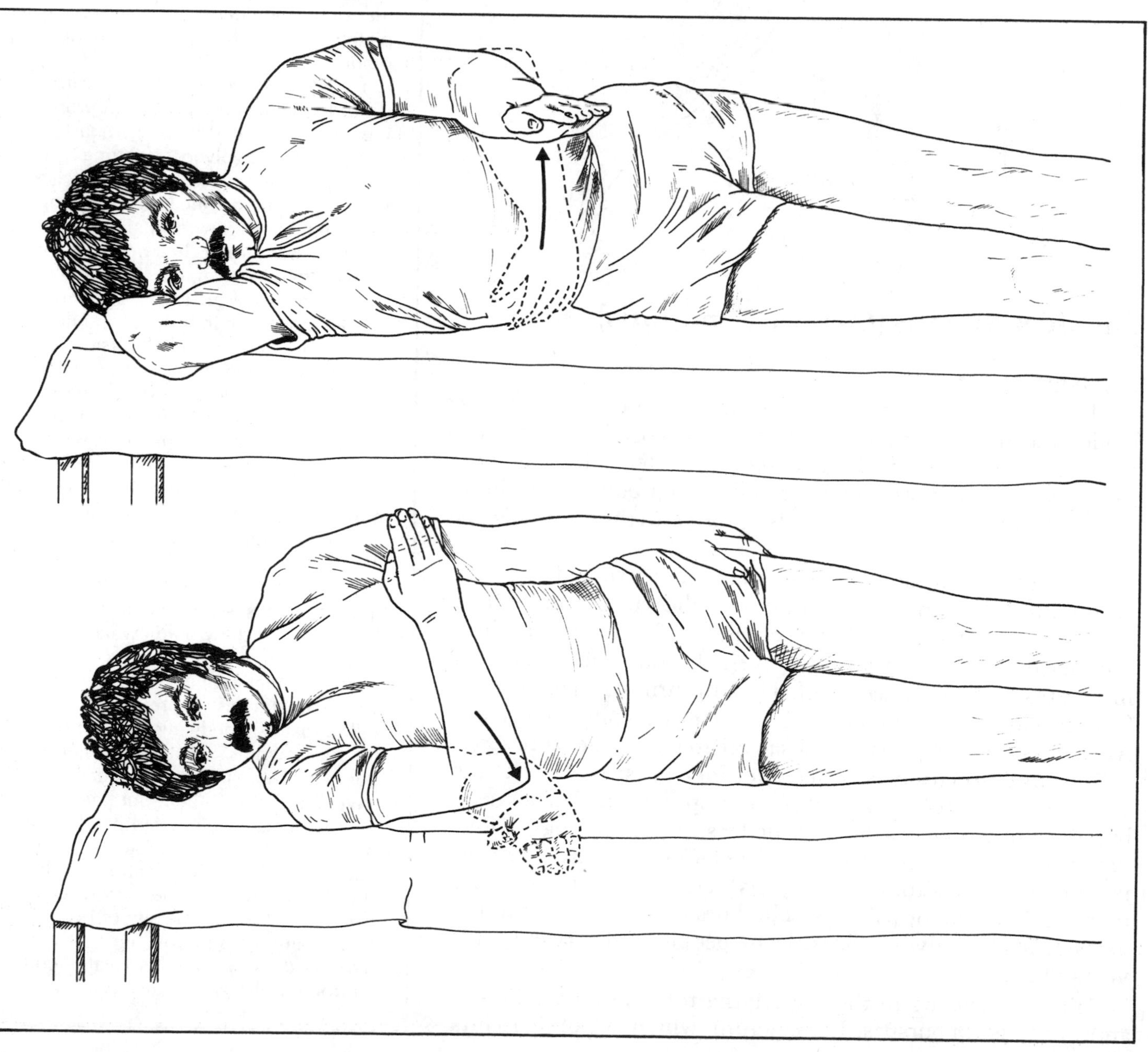

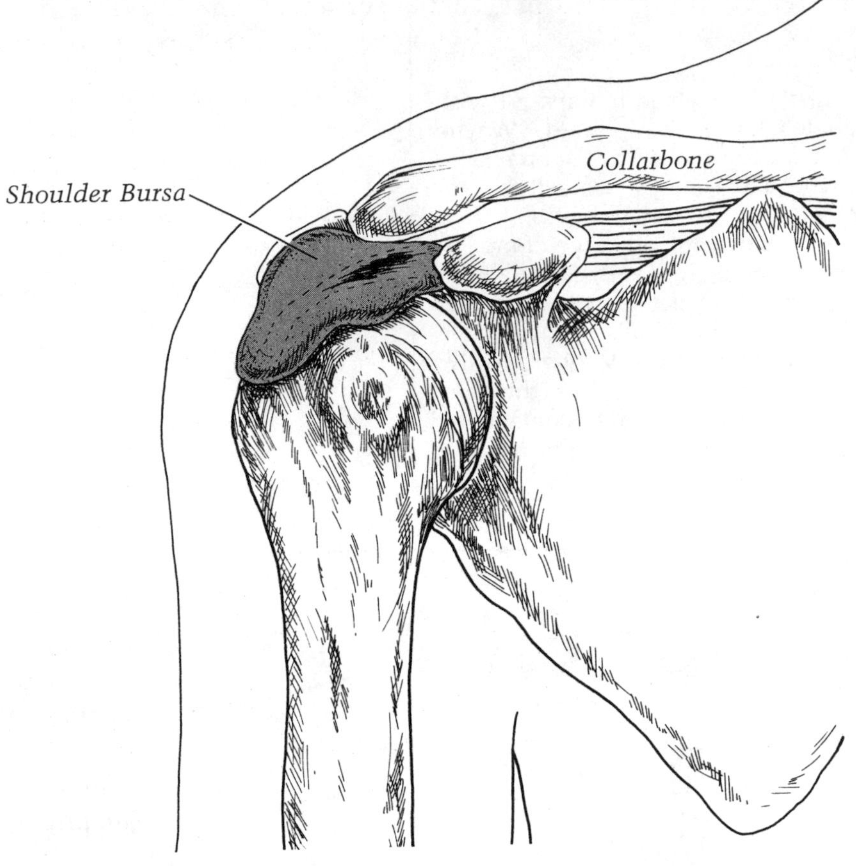

Shoulder Bursa Sac

Shoulder Bursitis (Calcific Tendinitis)

CAUSES

The *bursa sac* lies between the underside of the shoulder blade and the rotator cuff tendons. Its purpose is to lubricate the rotator cuff tendons. It looks like a deflated balloon and its lining is composed of specialized cells that produce joint fluid, called *synovial fluid.* The fluid also lubricates the sac as well as the tendons.

When the bursa sac becomes inflamed or irritated, it starts producing extra amounts of fluid, and the sac expands. Once this process starts, it is self-perpetuating. The increase in fluid production leads to more tension in the sac, which leads to more pressure on the sac itself. This, in turn, stimulates more fluid production. The end result is a badly swollen bursa sac. Any pressure on the expanded sac creates extreme pain.

This inflammation is called *shoulder bursitis*. Its victims are players of racquet and throwing sports, but I have seen bursitis in patients who cut hedges, paint ceilings, move furniture, and shovel snow. In 95 percent of the cases, bursitis is an overuse syndrome. It happens mainly to people who are part-time players or painters. The bursa sacs cannot take the extra load. I rarely see bursitis in people under twenty-five years of age.

Calcium deposits in the underlying rotator cuff tendons is another cause of bursitis. The calcium, which is like sharp bits

SHOULDER IMPINGEMENT SYNDROME

George Jessup, a forty-six-year-old educator, was an outstanding athlete at Springfield College. He had been both a football player and a gymnast. Throughout his collegiate career, he had sustained several shoulder injuries, including an A-C separation of the right shoulder, which he completely recovered from within three months.

Twenty-five years later, Jessup's shoulder pain returned. He was working as a high school gymnastics coach. This involved "spotting" of the gymnasts during their exercise routines. Spotting means supporting the gymnasts in mid-air. Thus, George was using his arms at or above shoulder level in a very forceful and strenuous way. In 1975, he noticed a small pain in his right shoulder which became steadily worse. The pain was located over the top of the shoulder. It would intensify when he used his arm above shoulder level. The pain was devastating in his work, but it also interfered with activities that he enjoyed doing, like swimming, chopping wood, and painting around his house. I judged that he lost 40 percent of his shoulder mobility. Lastly, George was unable to get a decent night's sleep. Every time he lifted his arm, he would awaken with pain.

For eighteen months, I tried conservative treatments. I prescribed anti-inflammatory medication by mouth for four months. No help. I injected cortisone twice into his shoulder. This gave temporary relief, but the pain stormed back. He needed painkilling pills to control his symptoms. Time and again, my examination of his shoulder showed tenderness directly over his A-C joint, the site of his college injury. In addition, he had tenderness directly over his rotator cuff tendons. I discovered this by moving

of sea coral, forms in the rotator cuff tendon area and pushes into the bursa sac. The bursa sac reacts violently to the irritation and starts producing fluid.

DIAGNOSIS AND TREATMENT

Shoulder bursitis starts slowly. First you sense an uneasy feeling about your shoulder. Over the next six- to twelve-hour period, the pain intensifies. I have examined many bursitis patients in the middle of the night in desperate pain. The pain is on the top of your shoulder.

If you get bursitis, you will be most comfortable by holding your arm directly at the side of the body with your forearm across the stomach. This is the same position that people with shoulder dislocation favor. However, in bursitis, the contour of the shoulder remains normal. The other difference is that the arm can be rotated without much discomfort. However, if it is moved away from the side of the body, the patient screams in pain. This movement causes the top of the arm bone to push against the expanded bursa sac.

In an acute bursitis patient, I can feel the enlarged bursa sac. I run my fingers along the top of the shoulder blade and then I touch out, just off the shoulder blade. There is a distinct swelling in this area. The swelling is not present on the opposite shoulder.

Apply ice immediately. It is extremely helpful in bursitis, because it shrinks the bursa sac. Continue for twelve hours. Ice for thirty minutes; rewarm for fifteen minutes. Be careful not to freeze the skin. Do not use heat; it tends to make the sac more tense and increase the pain.

The fastest way to relieve acute bursitis is to inject the bursa sac with cortisone. By placing the cortisone directly in the sac, the inflammation is quickly controlled and fluid production stops. Within twelve hours, the painful swelling disappears.

I inject the cortisone from the back of the shoulder area under the corner of the shoulder blade. I numb the skin with Novocain. I then place a one and a quarter-inch needle directly into the bursa sac, which runs parallel to the undersurface of the shoulder blade. I avoid placing the needle directly into the rotator cuff tendon below the bursa sac. Injecting cortisone into tendons weakens them and can cause damage.

After the injection, I always give anti-inflammatory medication orally. It quiets the irritation and swelling over time. If patients feel squeamish about taking a cortisone shot, they can get relief with these pills, but it takes at least two days for them to work. I place the patient in a sling to rest the arm while the pills are acting.

If the cortisone shot is not administered early, strong analgesic or painkilling pills must be taken. As soon as the pain and tenderness start to subside, begin the pendulum exercises. Do them twice daily, thirty repetitions for each exercise. Continue them for one week. This is to prevent the onset of stiffness in the shoulder joint and a frozen shoulder.

his arm in full circles. In certain positions, the tendons were rubbing under the shoulder bone. The x-rays showed a narrowing of the A-C joint with a few bony spurs present—a possible sign of early traumatic arthritis of the joint. The x-rays did not show any calcium deposit, so I ruled out bursitis.

George and I talked about surgery. I explained that surgery could help him, but I could not guarantee how much it would help. I knew that if he continued to bite the bullet and bear the pain the rotator cuff tendons would gradually fray more and more. Eventually, they would tear completely, the result being a permanently disabled shoulder. George opted for the surgery.

When I opened his shoulder, I found severe arthritis in the A-C joint. To cure this, I removed an inch of the collarbone and reshaped its outer edge. This is called a *resectional arthroplasty*. Secondly, the rotator cuff tendons were being restricted by the coracoacromial ligament. Therefore, I removed this ligament which does not attach to a muscle or another ligament. Finally, I removed a lifesaver thickness piece of bone from the undersurface of the shoulder blade, the acromion process. This gives the rotator cuff tendons more room to move.

George's recovery was speedy. I started him on pendulum exercises two weeks after surgery. A month after surgery he began weight-lifting exercises, and within two months his shoulder was back to normal strength. At that point, he resumed his coaching duties.

George occasionally has stiffness in the shoulder, especially if he chops wood. He will occasionally have night pain. However, his shoulder has improved from 60 percent to 95 percent of its full functioning capacity.

Cortisone is effective on calcium deposits, too. It shrinks them. Calcium deposits are rarely removed surgically. I have only done the operation three times in patients with recurring bursitis. The calcium deposits ranged in size from as small as a dime to as large as a fifty-cent piece.

The surgery is straightforward. An incision is made over the top of the shoulder. The deltoid muscle is split apart about one inch. This allows access to the bursa sac and the calcium deposit. The entire bursa sac is removed. A new one grows back in six to eight weeks. The calcium deposit is scraped off the rotator cuff tendons. The deposit is a mixture of firm little granules and toothpaste-like material, which sits on the tendon. I dig it out of the tendon and sew the tendon up. If there are any rough edges on the undersurface of the shoulder blade, a small slice of the bone is removed. The ligament which pushes down on the rotator cuff tendon is also removed at the same time. Surgery really has two purposes here, to remove the calcium deposit and to relieve any pressure on the tendons, because it is the pressure that causes the calcium deposits.

After the surgery, the arm is kept in a sling for approximately ten days and pendulum exercises are started one week after surgery. Shoulder-strengthening exercises are begun two weeks after the operation. In six to eight weeks, the patient can return to full activity, including sports. I have never seen a calcium deposit return after surgery.

Surgery on calcium deposits was performed frequently prior to the discovery of cortisone. There was no choice then but to operate. Ernest Emery Codman, in his book *The Shoulder,* published in 1932, was the first to describe the nature of the calcium deposit and the surgery which often relieves the pain. This book is not highly technical and makes excellent reading for those who wish to understand shoulder function in more detail.

CODMAN'S CALCIUM

Ernest Emery Codman, an eccentric surgeon who practiced in Boston in the 1920s and 1930s, was the first surgeon to insist that patients be interviewed in a systematic way after their operations. Because of his insistence on follow-up studies and his eccentricity, Codman was highly criticized by his colleagues at the Massachusetts General Hospital, and at one point in his career he was dismissed from the staff of that hospital.

Codman's book, *The Shoulder,* is one of the most enjoyable medical books I have ever read. The first fifty pages are a social critique of Boston medicine at that time. Codman even drew his own cynical cartoons. The one I like best depicts golden eggs

Biceps Tendinitis

CAUSES

The *biceps* muscle—the large, strong muscle in the front of the upper arm—divides into two major sections. The larger section of the bicep, called the *long head* of the muscle, is connected to the shoulder by a long tendon. This tendon passes through a groove in the upper end of the arm bone, called the *bicepital groove.* That is the point where tendinitis strikes. Beyond the groove, the tendon enters the shoulder joint and attaches to the top of the shoulder socket.

The biceps tendon is held in the groove of the upper arm bone by the *intertubercular ligament.* If this ligament is injured or stretched, the tendon slides in the groove. The sliding irritates the tendon, and the result is *biceps tendinitis.*

A second cause of biceps tendinitis may be a deformity in

BICEPS TENDINITIS

In the fall of 1978, Suzanne Laurie, a seventeen-year-old gymnast, completed a somersault exercise. Her arms were fully extended in the crucifixion position. As she started to rise, another gymnast ran into her arm. The force jerked her arm up and back. Suzanne heard and felt something snap. After the accident, her shoulder ached all the time, but the pain intensified when she used it. Handstands and overhead exercises on the uneven bars were impossible for her. She told me that the pain was coming from the front of the joint and described a snapping sensation in her shoulder.

Suzanne Laurie had a classic case of biceps tendinitis. She had tender-

being rolled down Beacon Hill toward the Massachusetts General Hospital. There surgeons scoop up the eggs and carry them away. Codman's view was that only Boston's affluent were being well cared for, while the poor were being ignored.

It was this same type of critical thinking which made him a first-class medical pioneer. Codman was the first to recognize that calcium deposits often irritated the bursa and caused shoulder bursitis. Initially, he removed the calcium surgically. Later, he used x-ray treatments to shrink the deposits. That method was also successful. Much later, the large doses of the radiation from the x-ray were proven to produce cancer.

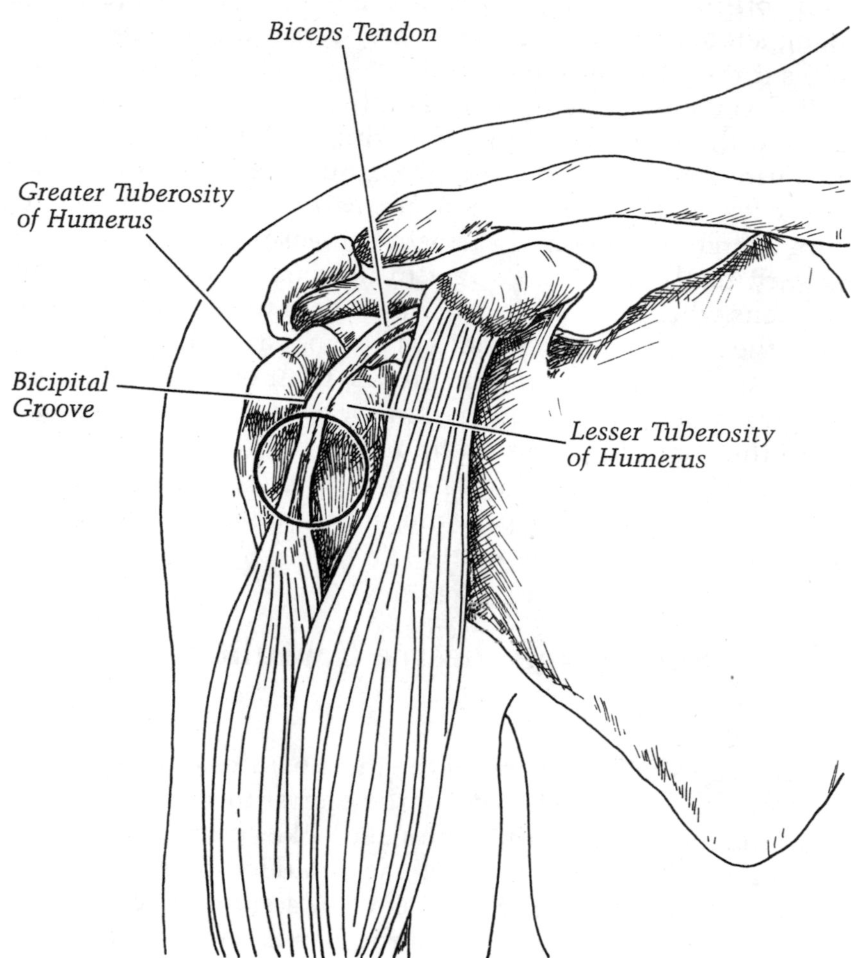

The Bicipital Groove

ness directly over the tendon, but had a full range of motion of the shoulder joint. When I put her arm in an overhead position, I could feel a shifting sensation in the area of the biceps tendon. She had no tenderness in any other part of her shoulder.

I told Suzanne and her parents that the tendon had to be rested for nine months. The alternative was an operation, which they chose because the girl was at the height of her athletic career. She underwent surgery in February 1979. I moved her tendon next door to the coracoid process. She underwent a full rehabilitation program and by the summer of 1979 was able to compete with the U.S. National gymnastics team in Europe.

the groove itself. The groove can be too shallow or have rough edges. In any overhead position, the tendon is forced to curve, but the groove is fixed. If the tendon is not firmly held down, it will slide out.

Biceps tendinitis is a thrower's ailment. I find it in players of racquet sports and baseball, and in gymnasts.

DIAGNOSIS AND TREATMENT

Like most shoulder ailments, biceps tendinitis is painful, especially in the throwing or serving position. I can touch the irritated groove and compare it with the groove on the other arm. I can feel the swelling. The pain is on the front of your shoulder.

For simple cases of biceps tendinitis, I recommend ice and rest. Ice the affected area for twenty minutes twice a day for three days. Start the pendulum exercises as soon as the pain decreases. Usually, the irritation disappears in a week. If you are a tennis player, gradually work your arm back into shape by playing "long ball" or volleyball, but don't exert yourself. If you put too much pressure on the tendon too soon, the tendinitis will flare up again.

For the more severe cases of biceps tendinitis, I prescribe

oral anti-inflammatory medication and two weeks of rest. The icing should continue twice a day. Perform pendulum exercises if they are not too painful.

If this course does not alleviate the pain, I inject cortisone around the tendon. Normally, within twenty-four hours, the irritation will start to abate. Continue the ice treatment for two days, and follow it with heat treatments. Start the pendulum exercises as soon as the pain decreases.

Very rarely does biceps tendinitis require surgery. It only happens when the tendon will not stay in the bicipital groove, and the irritation is chronic. The operation is simple and interesting. I make an incision in the front of the shoulder and surgically detach the tendon from the shoulder joint. I reattach it to the coracoid process, a lip in the front of the shoulder. I line it up with the short head of the biceps muscle. The operation alleviates the pain and, almost unbelievably, there is no loss of strength or mobility.

Frozen Shoulder (Adhesive Capsulitis)

CAUSES

A *frozen shoulder* is one of the most painful and disabling of conditions, yet it is also one of the easiest to avoid. What is it that freezes? The culprit is the sleeve-like structure that holds the ball-and-socket portion of the shoulder joint together. It operates very much like the sleeve of a shirt. When the arm lifts, the sleeve tightens. When the arm is down at the side of the body, the sleeve is floppy. The sleeve of the shoulder joint operates in exactly the same way.

When the sleeve is not stretched out several times each day, it has a tendency to adhere to itself with tiny adhesions. These look like small strings of scar tissue. Normally, the sleeve gets its exercise. Everyone raises his or her arm above the shoulder level each day. You do this in sleep, you do it getting plates down from a high shelf, and you do it getting dressed in the morning. This allows the sleeve to stay pliable and healthy.

The problem arises when you have a painful shoulder condition, such as tendinitis or bursitis, and you do not move your shoulder for a week or more. Some adhesions form in the sleeve. If the shoulder is not moved for two to three weeks, then these adhesions will become very dense and strong. They will shrink the size of the joint sleeve, with the result that the shoulder cannot move freely and a frozen shoulder develops. It has the same effect as sewing up the armpit. Frozen shoulder is one of the few ailments that results from a lack of use, rather than overuse. That is why I feel so strongly about shoulder exercise being initiated early.

DIAGNOSIS AND TREATMENT

How do I tell if a patient has a frozen shoulder? I do this in two ways. First, the patient will bitterly complain of shoulder

WALL WALKERS

This exercise stretches the shoulder sleeve to its maximum:

Line up a chair against the wall. The side of the chair should be at right angles to the wall. The arm with the frozen shoulder should be closest to the wall.

With that arm, reach up and touch the wall. At the maximum height, your arm should be rubbing against your ear.

Now, with your fingers, walk up the wall as far as your shoulder will let you.

Mark the point with a fingerprint or a pencil mark. Wall-walk fifteen more times. Wall-walk twice daily.

Each day, attempt to walk one inch higher. Continue for three weeks.

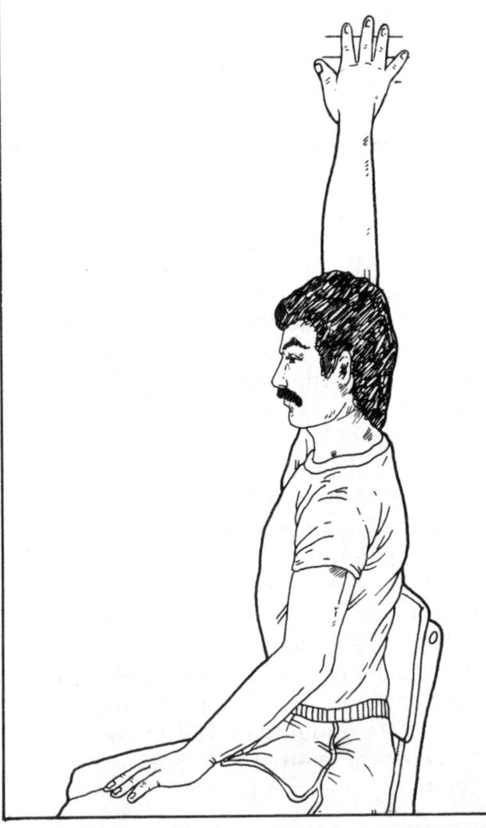

PULLEY EXERCISES

These simple arm-lifting exercises are designed to inhibit the reformation of the small adhesions in your shoulder.

You can build a simple pulley exerciser. The materials you need: a small pulley, some clothesline, and two four-inch dowels for handles. Attach the pulley to a ceiling or, if the weather permits, to an overhanging tree branch. Thread the pulley with the clothesline and attach the dowels to each end of the rope.

1. Sit directly under your homemade pulley and grasp each handle with your hands. Lift your arm with the frozen shoulder as high as it will go. Repeat thirty times.
2. Place your chair a foot behind the pulley. With the good arm, lift the arm with the frozen shoulder thirty times. This gives the shoulder some forward movement.
3. Put your chair a foot in front of the pulley. Lift the frozen shoulder thirty times. This gives the shoulder some backward movement.

Do the three pulley exercises twice a day for three to four weeks.

pain. It hurts to move the shoulder in any direction. You cannot put your arm into a coat; you cannot snap your bra; you cannot comb your hair. If you attempt to move beyond the area of restricted motion, you produce severe, jabbing pain. One patient described it as "a sword going into the shoulder." In addition to the pain, I notice a symmetrical decrease in shoulder motion. The arm cannot be extended to its fullest in any direction. No other condition produces the same degree of limitation.

What can you do to prevent frozen shoulder? Exercise the shoulder, even though it's painful. I recommend pendulum exercises for all shoulder injuries, no matter what their origin. If possible, these exercises should be done twice a day.

Prompt treatment is needed for frozen shoulder. I start with ice, aspirin, and exercise therapy. If that does not work, I add oral anti-inflammatory medicine. A cortisone injection should be considered. The key to treatment is to keep the shoulder moving. Even though you experience pain, continue to use it for daily activities. Fight the pain; it will not harm you.

If you have not used your shoulder (sleeve) for a few months, therapy will not free the adhesions, which have become very strong and stiff. I favor breaking the adhesions in the shoulder under general anesthesia. That means you must be hospitalized. While you are asleep, I inject the shoulder joint with 10 milliliters of salt water and 80 milligrams of Depomedrol, a cortisone-like drug. The joint accepts the salt water, which is used because sterile water would upset the mineral balance of joints. The joint sleeve expands and stretches with the solution. The muscles entirely relax, and the joint can be moved manually. I firmly grasp the upper end of the upper arm bone about two inches below the tip of the shoulder. I then push on the tip of the shoulder blade. This stretches the joint, and the adhesions give way. The whole procedure takes about forty-five seconds. There is an immediate and dramatic increase in the motion of the shoulder. To reduce swelling, the shoulder is iced.

Physical therapy starts the next day with pendulum exercises. In addition, there are two other exercises that I like for rehabilitation: wall climbing and pulley exercises.

Does frozen shoulder ever require surgery? In my opinion no. I have never operated for frozen shoulder. I firmly believe that shoulder manipulation under anesthesia and very careful physical therapy following the manipulation will allow the shoulder to become supple. Breaking of the adhesions surgically would itself tend to cause scarring and would be self-defeating. I have never had to remanipulate a frozen shoulder.

Diabetic patients seem to develop frozen shoulder more quickly than others. If you have diabetes and develop shoulder pain, get a diagnosis quickly and start treatment immediately. Women are more likely to get frozen shoulder, too. Science cannot explain why these groups are more susceptible to this ailment than others.

THERE ARE ALMOST 180,000 SHOULDER INJURIES IN THE UNITED STATES PER YEAR **1979**

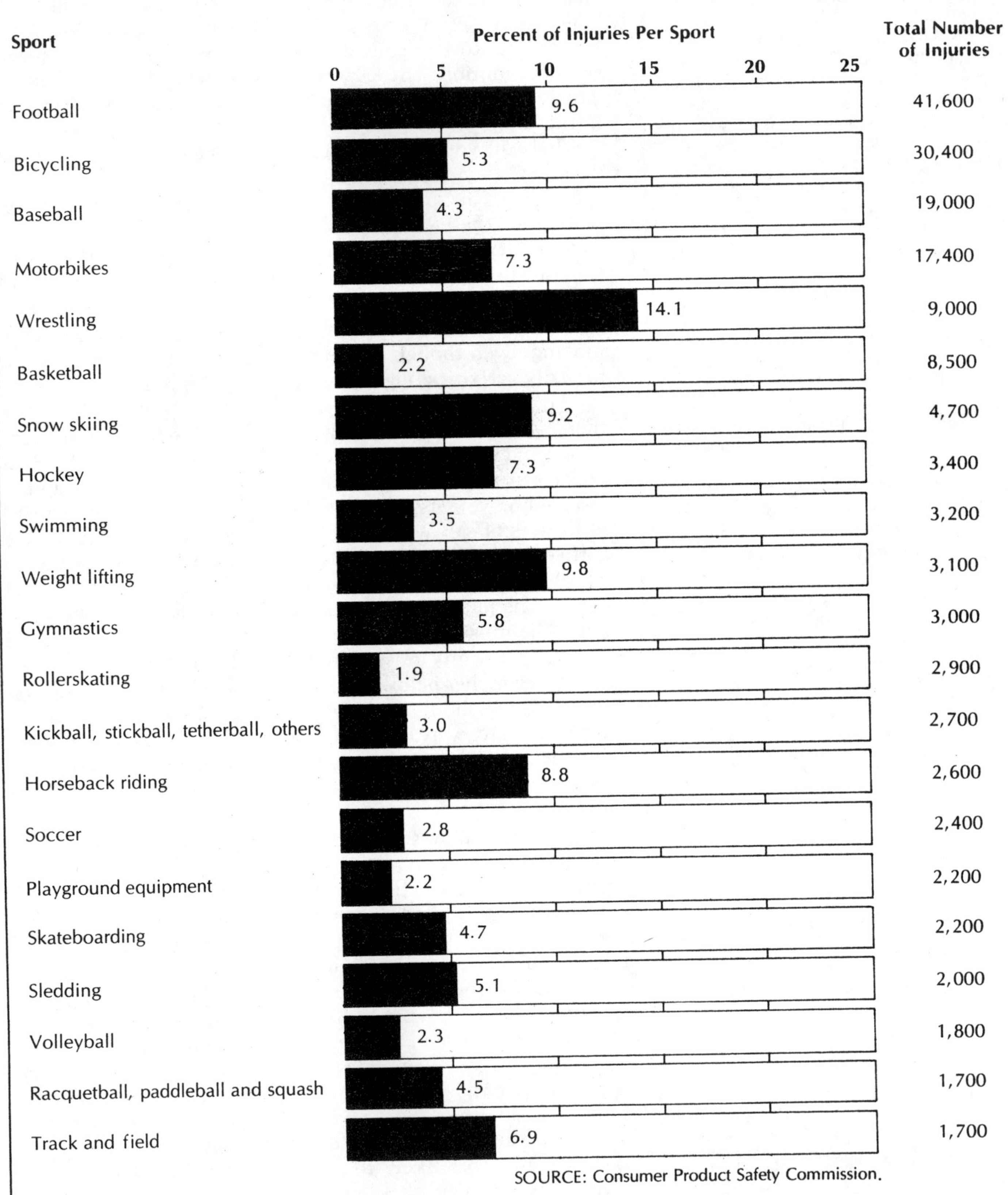

SOURCE: Consumer Product Safety Commission.

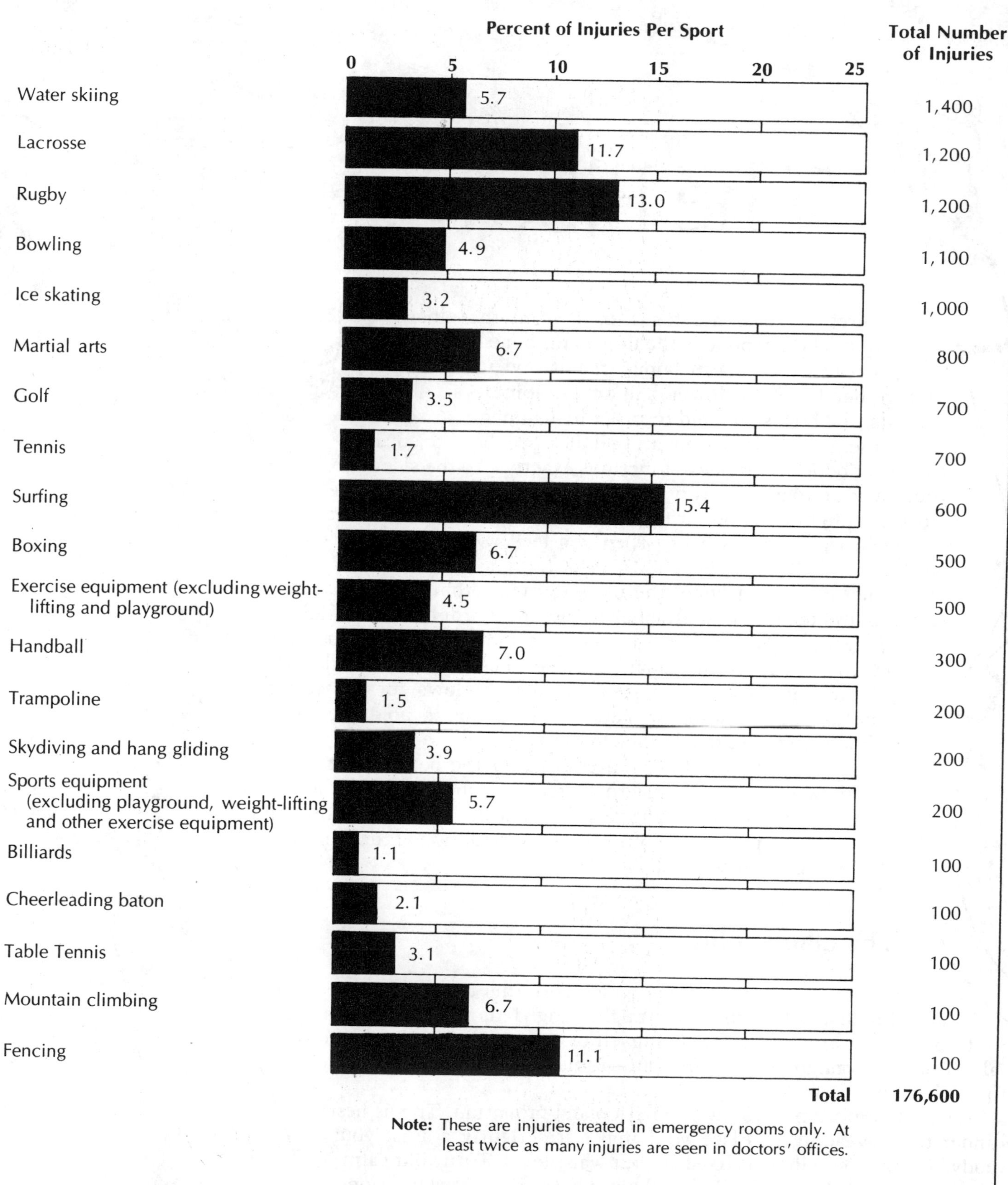

Note: These are injuries treated in emergency rooms only. At least twice as many injuries are seen in doctors' offices.

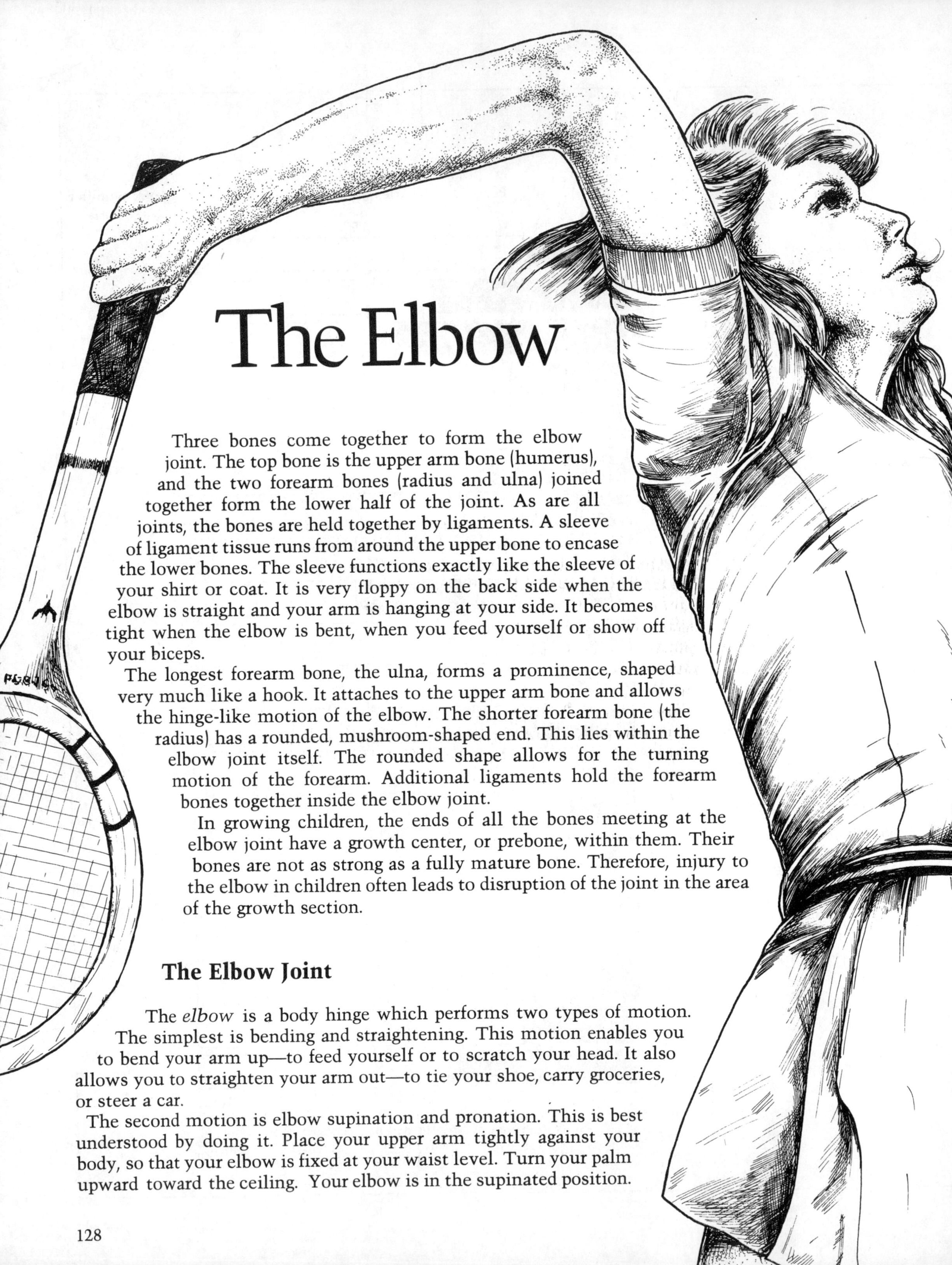

The Elbow

Three bones come together to form the elbow joint. The top bone is the upper arm bone (humerus), and the two forearm bones (radius and ulna) joined together form the lower half of the joint. As are all joints, the bones are held together by ligaments. A sleeve of ligament tissue runs from around the upper bone to encase the lower bones. The sleeve functions exactly like the sleeve of your shirt or coat. It is very floppy on the back side when the elbow is straight and your arm is hanging at your side. It becomes tight when the elbow is bent, when you feed yourself or show off your biceps.

The longest forearm bone, the ulna, forms a prominence, shaped very much like a hook. It attaches to the upper arm bone and allows the hinge-like motion of the elbow. The shorter forearm bone (the radius) has a rounded, mushroom-shaped end. This lies within the elbow joint itself. The rounded shape allows for the turning motion of the forearm. Additional ligaments hold the forearm bones together inside the elbow joint.

In growing children, the ends of all the bones meeting at the elbow joint have a growth center, or prebone, within them. Their bones are not as strong as a fully mature bone. Therefore, injury to the elbow in children often leads to disruption of the joint in the area of the growth section.

The Elbow Joint

The *elbow* is a body hinge which performs two types of motion. The simplest is bending and straightening. This motion enables you to bend your arm up—to feed yourself or to scratch your head. It also allows you to straighten your arm out—to tie your shoe, carry groceries, or steer a car.

The second motion is elbow supination and pronation. This is best understood by doing it. Place your upper arm tightly against your body, so that your elbow is fixed at your waist level. Turn your palm upward toward the ceiling. Your elbow is in the supinated position.

Front View

Ulna
Humerus
Radius

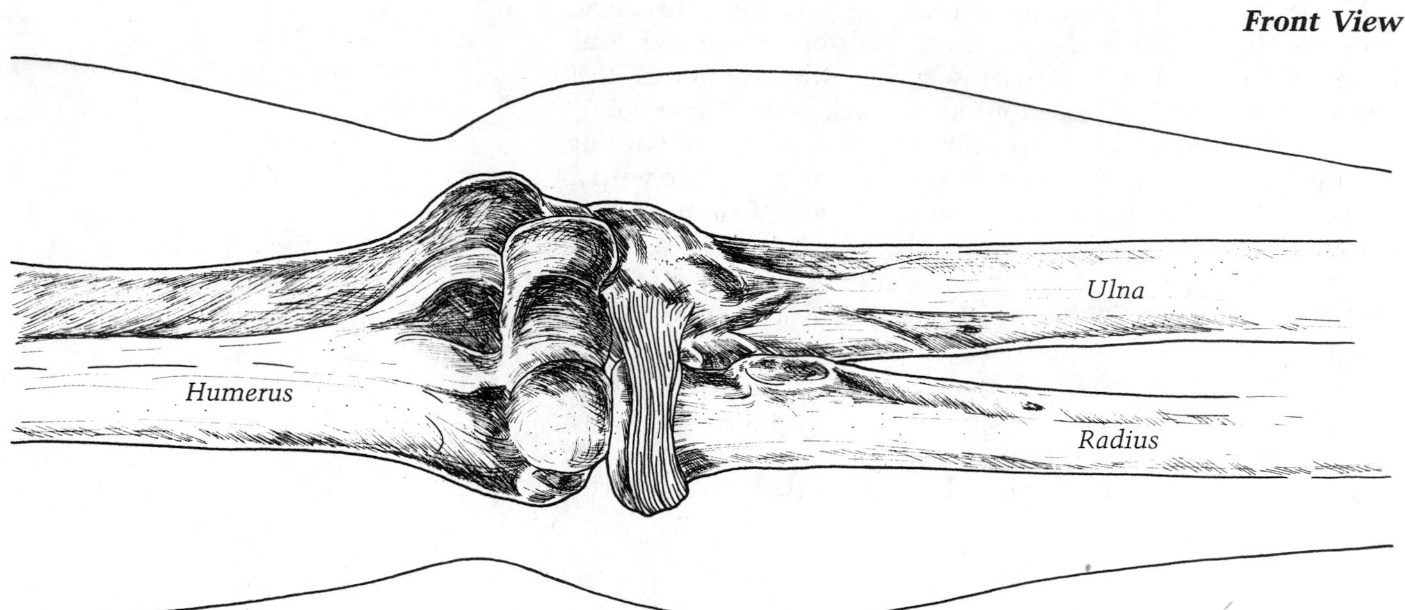

Side View

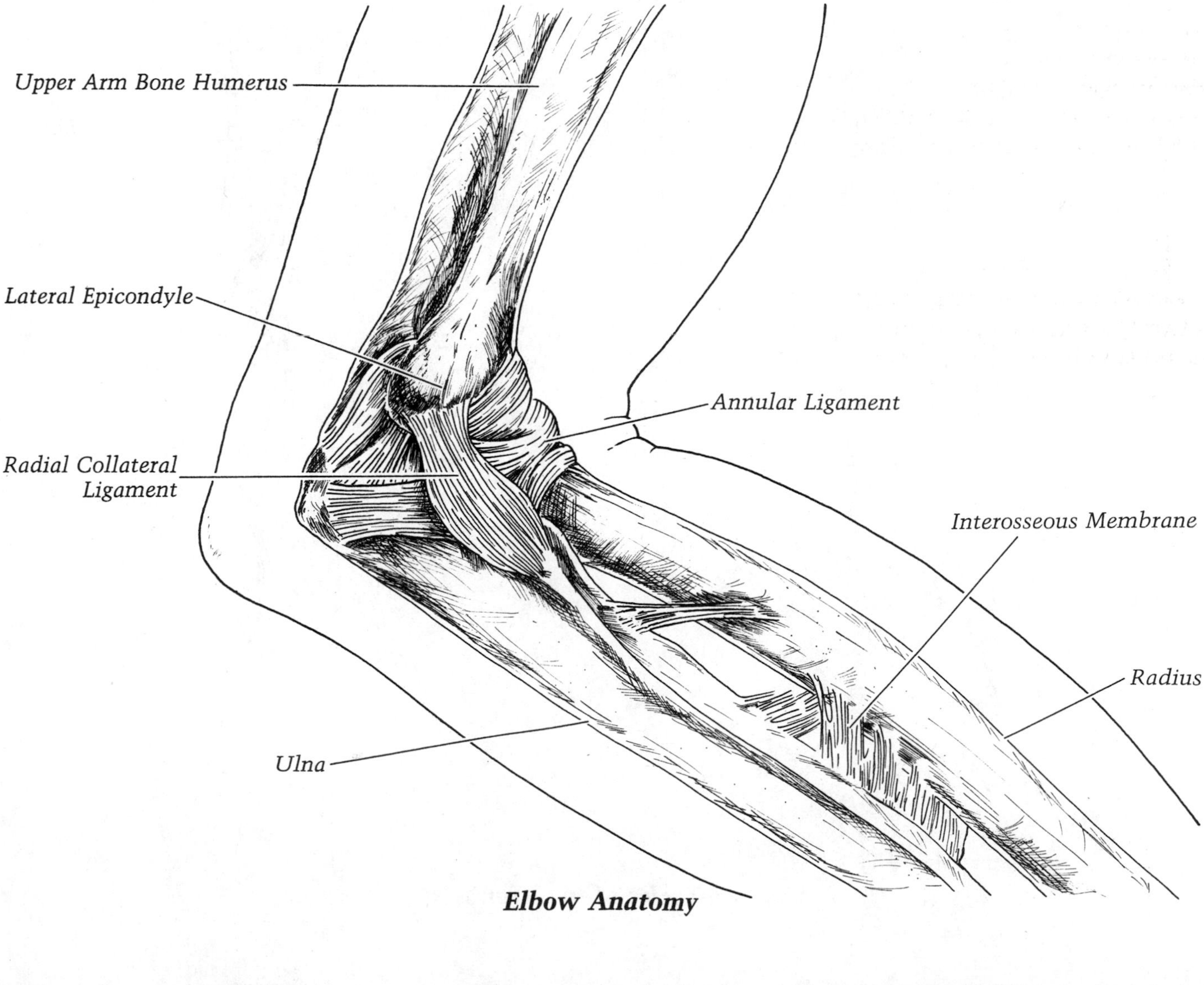

Upper Arm Bone Humerus
Lateral Epicondyle
Annular Ligament
Radial Collateral Ligament
Interosseous Membrane
Radius
Ulna

Elbow Anatomy

Now rotate your forearm so that the palm of your hand faces the floor. This is the pronated position. If you use your thumb as a pointer, you will see the forearm rotates 180 degrees. The rotational motion takes place at the elbow joint, and it is this movement that allows you to collect change at the supermarket with the palm of your hand up or write with a pen with the palm of your hand facing down. Without a good elbow (and shoulder joint), your hand would be almost useless.

THE MUSCLES AND NERVES OF THE ARM

The arm is one of the great body parts. It is versatile, powerful, and mobile. It allows baseball players to strike home runs, violinists to play symphonies, and eye surgeons to remove cataracts. This, and more, is all possible because of the

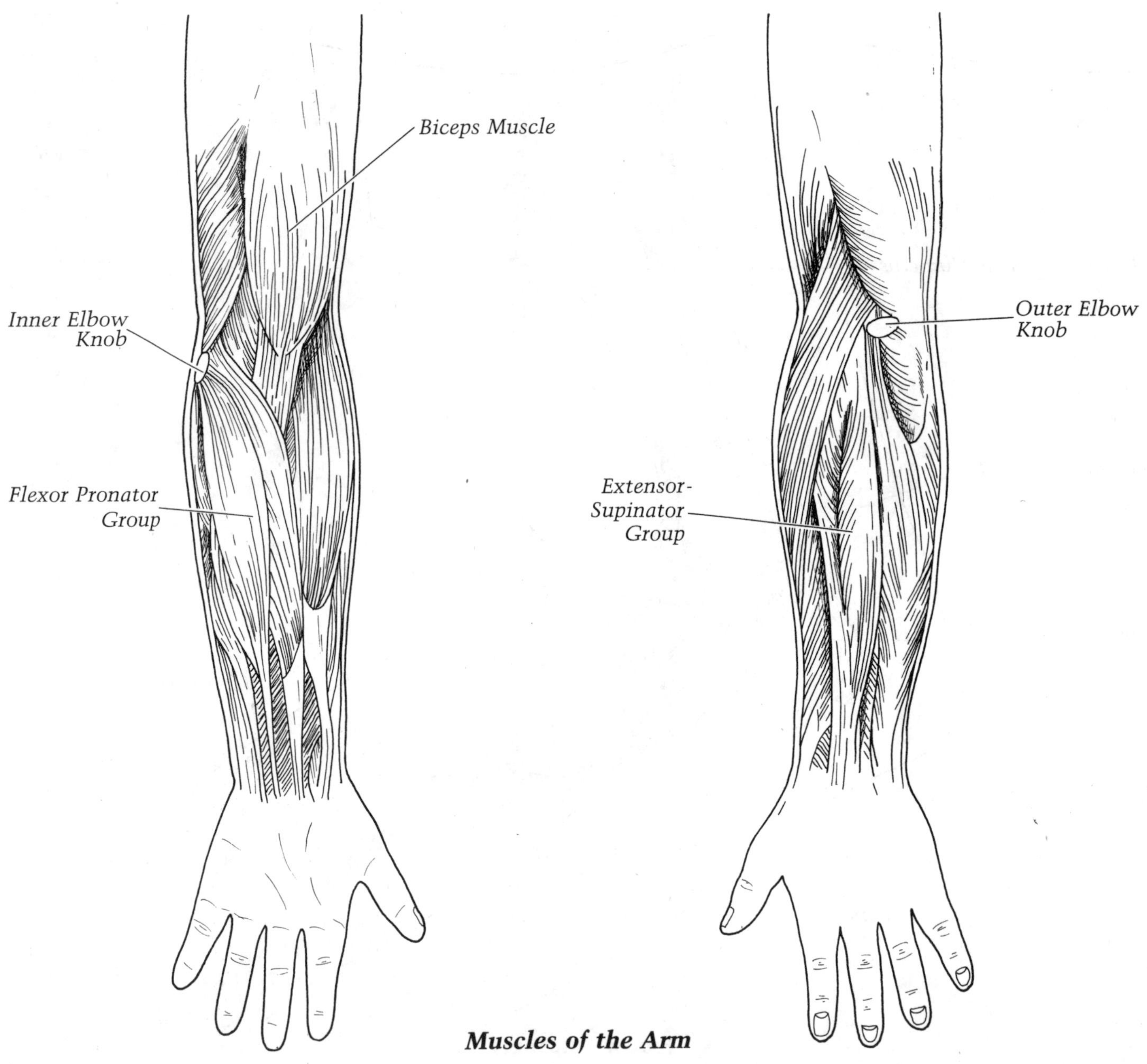

Muscles of the Arm

marvelous way the arm is constructed.

The upper arm muscles, the biceps and triceps, which join at the elbow, are so arranged to give the forearm tremendous power to thrust and bend. The *biceps muscle* attaches to the radius bone of the forearm at the elbow joint. It gives the elbow the ability to bend the arm with power and the wrist to rotate in a counterclockwise direction. The *triceps muscle* in the back part of the upper arm attaches to the ulna bone of the forearm. It allows the elbow to straighten with power.

You have two sets of muscles in the forearm. The forearm muscles on the back portion of the arm (if your palms face the floor, they would be on the top of the forearm) straighten the wrist. They attach on a bony knob on the outside edge of the elbow. This knob is called the *lateral humeral epicondyle.* This is the point where most tennis players get tennis elbow, a stress syndrome of these forearm muscles.

The front forearm muscles give the wrist the power to snap inwardly. This muscle is attached to a bony knob on the inner elbow. This is called the *medial humeral epicondyle.* It's where most pitchers and throwers get pitcher's elbow, a common overuse syndrome of these muscles.

The nerves that direct these muscles are branches of the brachial plexus. There are three major nerves in the arm at the elbow level: the *radial nerve,* which lies on the back and outer aspect of the elbow; the *median nerve,* which lies on the front of the elbow; and the *ulnar nerve,* which lies on the inner side of the elbow. The ulnar nerve is what most people call the "crazy bone." It is very prominent and is covered only by skin.

Each nerve continues on to the hand, where it picks up sensation and relays it to the brain. Therefore, numbness in a specific portion of the hand allows your doctor to diagnose difficulty with one of the major nerves of the arm. Tingling in the hand is also a sign of nerve trouble.

Elbow Sprains

CAUSES

An *elbow sprain* is a partial tearing or stretching of the ligament or sleeve of tissue (capsule) that holds the elbow together. These injuries are common, especially in football, hockey, and gymnastics. Arm tackling in football is the number one elbow sprainer.

It is far more common to sprain your elbow by having it quickly forced straight out; bending the elbow quickly toward the body doesn't seem to sprain it. When force is put on the elbow, fortunately, it does not often spring apart, but rather the ligaments and the capsule along the front of the elbow joint tear. The stretching and ripping cause tearing of the blood vessels which nourish the elbow ligaments and lead to hemorrhage in the elbow joint.

Elbow sprains run the gamut from mild sprains to complete

dislocation—when the elbow does, in fact, come apart. The degree of the sprain depends on the force applied to the elbow and its duration.

DIAGNOSIS AND TREATMENT

Thirty minutes after the injury, the elbow starts to swell. Because blood accumulates from the torn blood vessels, the elbow starts to feel tight and painful. If you don't apply ice early, the elbow continues to swell and gets stiffer and stiffer. You will find it very painful to try to straighten the elbow.

What should you do if you sprain your elbow? Apply ice for thirty minutes and allow the skin to rewarm for fifteen minutes. Continue for four hours. The elbow should be reevaluated the morning following injury. This is the best time to diagnose how severe the sprain is. By the amount of swelling, tenderness, and muscle spasm it's clear whether you have a small, moderate, or a large tear in the ligaments and joint capsule. For mild and moderate sprains, two to three days of warm baths or whirlpools is sufficient to help heal the ligament. When 95 percent motion returns to the elbow, you can return to athletic competition.

For insurance, I like to place you in extension-stop tape apparatus. I shave portions of skin above and below the elbow. Next, I apply several wraps of tape on the upper arm and several wraps of tape on the mid-forearm. This is the foundation. With the elbow bent just ten or fifteen degrees, I lay down strips of tape between the two foundations. I anchor these pieces of tape with more tape wrapped around the foundations. Since the tape acts as a check rein, you can't fully straighten your arm. This prevents repeat stress on the injured area. You should wear the tape restraint for two weeks.

For severe sprains, ten days to two weeks is usually necessary for the elbow to quiet down and regain 95 percent of motion. Warm baths or whirlpools on a daily basis should be taken. After the warm baths, reinstall the check-rein tape apparatus. It requires a full six weeks from the time of injury for this type of ligament capsule injury to repair itself. In the more severe injuries, it is very important to do strengthening exercises to allow full return of muscle power to the arm. I recommend elbow crankers.

Elbow Hyperextension Strapping

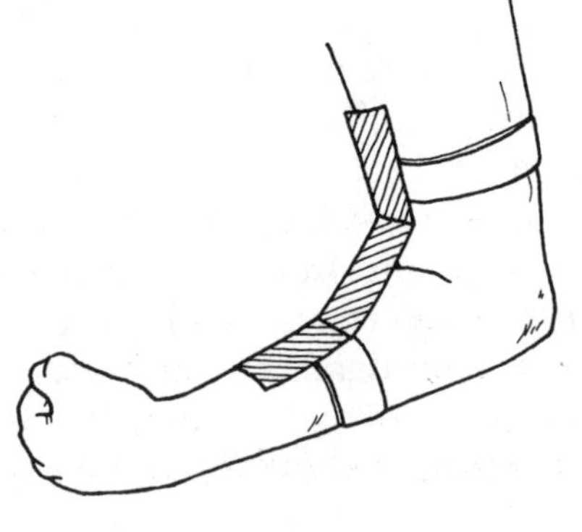

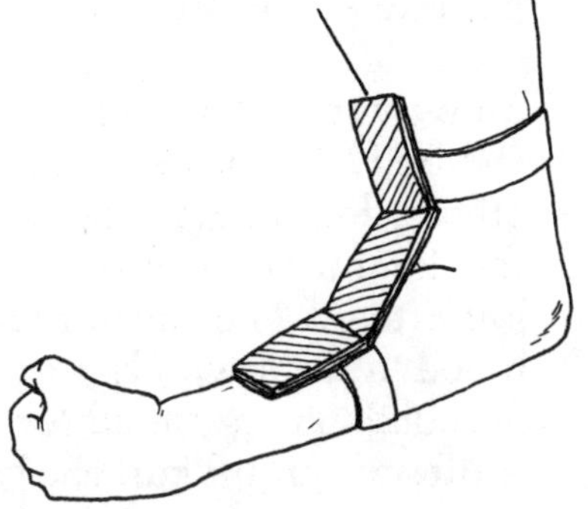

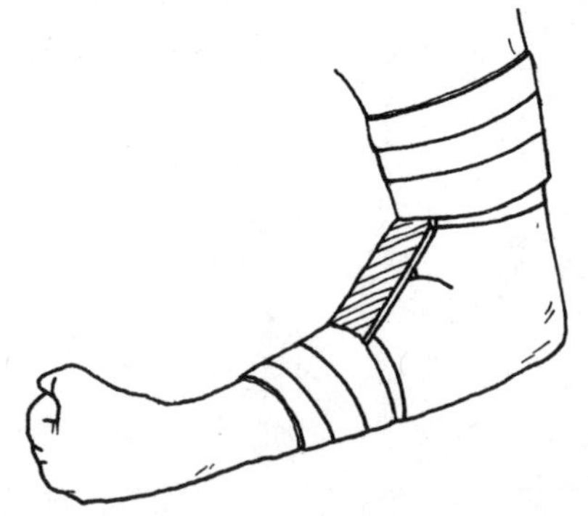

Elbow Bursitis (Dart Thrower's Elbow)

CAUSES

Why, when you fall on your elbow, does the skin scrape instead of break open? Feel your elbow. The skin is mobile. Nature has provided a *bursa sac* between the skin and the bone to make the skin slide easily. As in the shoulder, this sac looks very much like a deflated balloon. The lining of the sac produces fluid which lubricates the walls of the structure and provides a low friction barrier between the skin and bone.

If you fall directly on the tip of the elbow, you can injure the bursa sac. A blow or repeated falls irritate the walls of the sac. The irritation, called *elbow bursitis,* leads to excessive production of bursa sac fluid. This is the same synovial fluid that is found within joints and around sliding tendons. The extra fluid distends the sac and it looks very much like a small egg, hanging off the tip of the elbow bone. It is unmistakable. The bloating produces discomfort. The awkward lump can disturb your arm coordination.

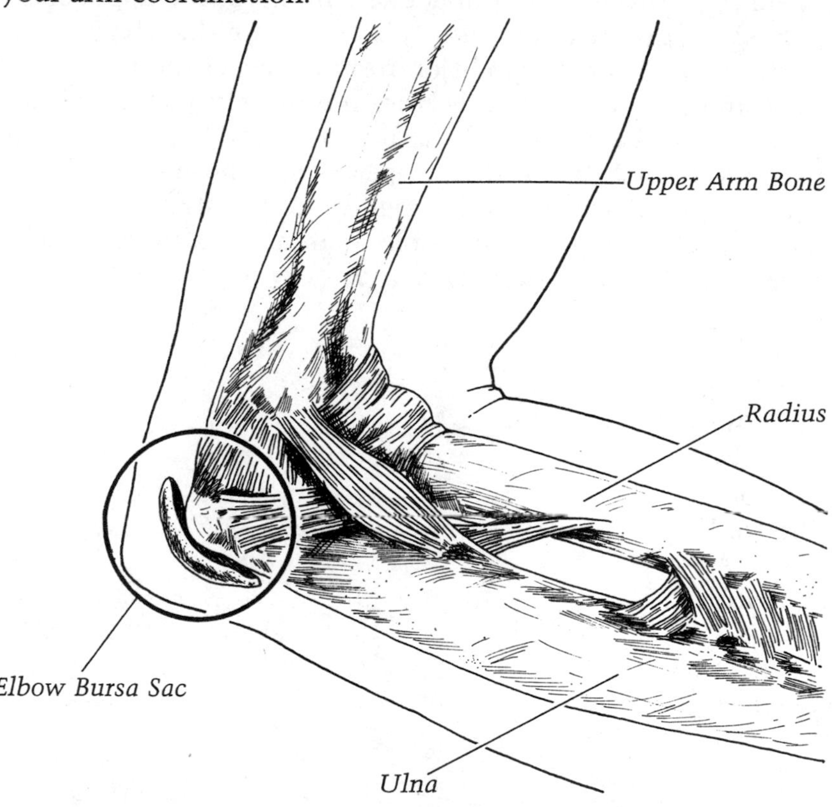

Elbow Bursitis

DIAGNOSIS AND TREATMENT

Initial treatment for elbow bursitis is to apply ice twice daily for twenty minutes. Heat tends to make the bursa sac swell more. Two aspirins at meals relieves discomfort and decreases the irritation. Continue the ice and aspirin for one week. If this does not bring about shrinking, I prescribe stronger anti-inflammatory medicine by mouth. Relief should come on the

ELBOW CRANKERS

Support the elbow on the edge of table.

Straighten the elbow slowly until it stops.

Hold position three seconds.

Bring your forearm toward the shoulder. It is the same position as "making a muscle."

Hold this position for three seconds.

Repeat 30 times.

Do these exercises three times a day for two weeks.

After this two-week period (one month after surgery), using the same exercises, add a two-pound weight.

You can use an ankle weight instead of a barbell.

fifth day. The medication should be taken for two weeks.

If medication alone does not bring relief, then cortisone injection is indicated. I inject local anesthesia into the skin directly over the tense bursa sac. This numbs the area. An empty syringe is placed into the sac and the fluid is drained away. The fluid is yellow, the color of straw, and is extremely slippery. In fact, it is better than any artificial lubricant on the market. Once this fluid is drained out, I inject the cortisone into the bursa sac. The cortisone quickly lessens the inflammation. Rarely, do I repeat the cortisone injection. If the bursa sac remains constantly inflamed and tender, I surgically remove it. This can be done under local anesthesia and involves a simple operation. The body will actually grow a new uninflamed bursa sac after this removal.

DART THROWER'S ELBOW

Dr. Robert E. Leach, noted sports medicine orthopedic surgeon in Boston, described this entity in 1979. He related a case involving a twenty-year-old college student who had thrown darts for three hours one night. The next morning, he woke up with his arm stiff and painful at the elbow area. He was seen at the Boston University orthopedic clinic and was found to have a swollen, hot olecranon bursitis of the dart-throwing elbow. The irritation even extended into his forearm. Fluid was removed from his bursa sac with a needle and only anti-inflammatory medication was used. Two weeks after treatment, his symptoms had disappeared. With this in mind, the term "dart thrower's elbow" was coined.

Tennis Elbow (Lateral Humeral Epicondylitis)

Tennis elbow is the number one ailment of the upper body. With 32 million regular tennis players, more than 10 million are afflicted with this condition from time to time.

The more you play tennis, the more likely you are to get tennis elbow. A new study by James D. Priest, Vic Braden, and Susan Gerberlich found that 45 percent of those players who played daily suffered from the condition. Pain developed in 33 percent of those who played three to four times a week and in 25 percent of those who played once or twice a week.

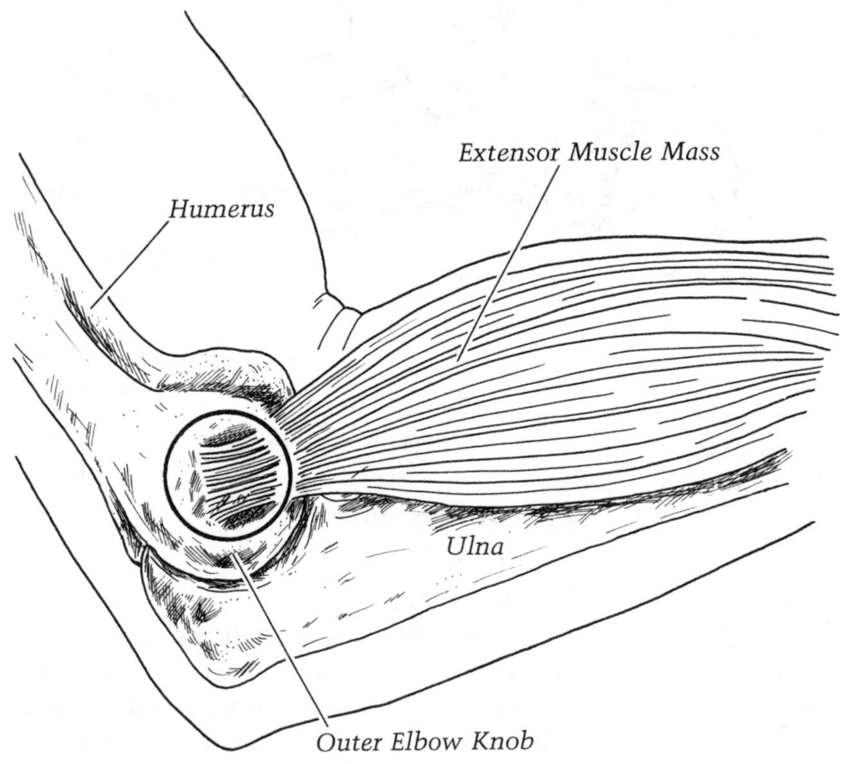

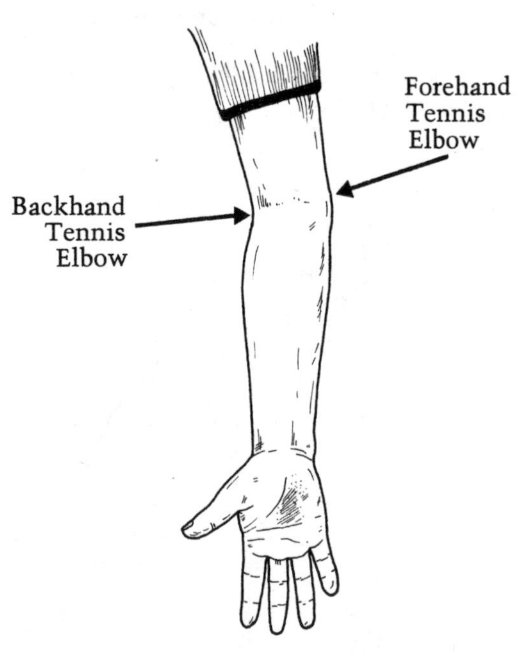

Tennis Elbow

PETROCELLI'S TENNIS ELBOW

Rico Petrocelli, the veteran shortstop for the Boston Red Sox, was a victim of tennis elbow. In 1969, a year after setting an American League record for home runs by a shortstop, he reported pain on the outer knob of his elbow. It sounded like tennis elbow, but I had never heard of a baseball player having this particular ailment. I had a long chat with Rico. He revealed that in the off season, he was playing squash four times a week. The puzzle was clear. The tennis elbow came from squash and not baseball.

Petrocelli's tennis elbow threatened his career. The pain was so severe he was unable to throw effectively; it did not bother his batting stroke.

His symptoms were classic. He was unable to lift objects—even the weight of a water glass—without severe pain. His elbow became stiff in the morning after use, and he had pain when he tried to straighten it.

In desperation, Rico asked for surgery. Dr. Thomas Quigley, then the orthopedic consultant for the Boston Red Sox, performed the surgery on Rico's throwing arm. It was done in the off season and he had plenty of time to recover. He began range-of-motion and progressed to strength exercises. At spring training, he worked his way back into shape and played the following season without pain.

CAUSES

Tennis elbow is caused by excessive strain on the muscles of the forearm which attach at the elbow. These muscles produce forward and backward movement at the wrist. The pain is usually slightly below the elbow attachments. But the pain can radiate into the forearm, too.

Tennis players suffer from two types of tennis elbow:

- Backhand tennis elbow, common in novice or weekend players, comes from hitting backhand strokes incorrectly. A backhand stresses the muscles that straighten the wrist. These muscles, called the extensor supinators, also allow you to extend the fingers and rotate the palm up. They attach on the outer side of the elbow. *Ninety percent of tennis elbow sufferers get this type of injury.*
- Forehand tennis elbow, common in professional players, stems from the wrist snap in booming serves. Serving strains the muscles that bend the wrist down and allow you to make a fist. These muscles, called the flexor pronators, attach on the inner side of the elbow; consequently, the pain occurs at this site. *Ten percent of players get this type of elbow pain.*

Tennis elbow is not limited to tennis players. I have treated plumbers, mechanics, surgeons, bowlers, pitchers, and factory workers with tennis elbow. It can happen to anyone who fires his wrist in a powerful way.

People who try to hit the ball with a wrist movement, rather than using their entire shoulder and arm, put tremendous force on the muscles of the forearm. Hitting a tennis ball traveling at thirty miles an hour is equal to lifting a fifty-pound weight. Thus, the average match for a player with good skills is the equivalent of doing fifty curls with a fifty-pound weight.

Where does this force go? The impact radiates from the racquet to the forearm muscles and ends up at the elbow tendon, especally if you hit the ball only with the forearm. The easiest way to distribute the impact of the stroke is to use a backhand that employs the entire upper arm. The shoulder muscles are stronger than the forearm muscles and will help carry the load.

Here are some additional factors that put added stress on muscles, tendons, and joints and may result in tennis elbow:

Using too heavy a racquet. The heavier the racquet, the more stress on the arm.

Playing on grass or cement. The ball bounces off these surfaces with a greater velocity, hitting the racquet with a greater force, which is transmitted to the elbow.

Using heavy balls. The heavier the ball, the greater the force against the racquet.

Having too much tension on the strings. If the strings are too tight, they do not give sufficiently when the ball hits and a greater force is transmitted to the elbow.

Using an oversized grip. If the racquet is held insecurely, the ball will be hit with a wobbly stroke and the elbow absorbs the shock.

Almost Half of Tennis Buffs Who Play Daily Get Tennis Elbow

Frequency of Play	Percent with Elbow Pain
Daily	45
3-4 times per week	33
1-2 times per week	26
2-3 times per month	7
Once per month or less	9

BEST TREATMENT FOR TENNIS ELBOW

Treatment	Percent Who Said That Treatment Helped
Changing tennis stroke	100
Stretching and strengthening exercises	89
Wearing elbow brace or support	84
Using aspirin	84
Rest	83
Cortisone or steroid injection	75
Heat treatment	74
Butazolidin injection (horse pill)	71
Cold treatments	62
Ultrasound treatments	52

Most Painful Strokes (For Tennis Elbow Sufferers)	Percent*
Backhand	38
Serve	25
Forehand	24
Backhand volley	7
Overhead smash	4
Forehand volley	3

*Because of rounding, components do not add up to 100 percent.

LOCATION OF ELBOW PAIN

		Percent**
Lateral epicondyle	(Attachment of the forearm muscle and tendon in outer arm about one inch below the elbow)	75
Lateral muscle mass	(Below outer side of elbow)	17
Medial epicondyle	(Attachment of the forearm muscle and tendon on arm)	10
Other		8

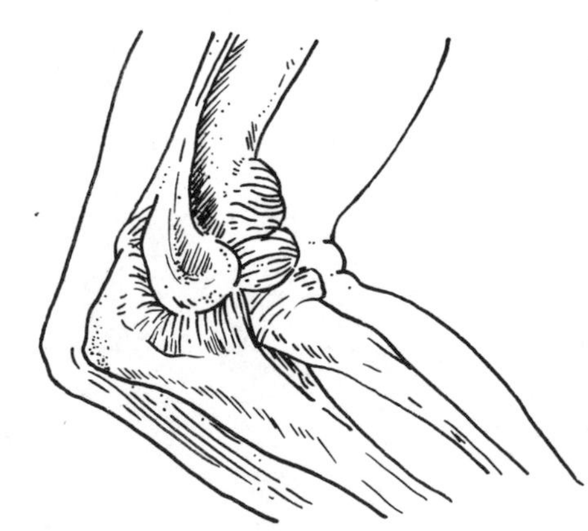

**Components add to more than 100 percent, because participants in the survey experience pain more than one place.

SOURCE: James D. Priest, Vic Braden, and Susan Goodwin Gerberlich. *The Physician and Sportsmedicine*, April and May, 1980.

The more you play, the greater the chance of elbow pain and injury. That is a conclusion of the Priest, Braden, and Gerberlich study. It is common sense, too, but for some reason tennis players do not view their muscles in the same light as other athletes.

For example, every time you play a vigorous game of tennis, your forearm muscles and tendons tear slightly and rip. The healing process causes the muscles to shorten. That is why your arm feels stiff in the morning. It takes twenty-four to forty-eight hours for the muscles to heal. If you play everyday, the muscles and tendons do not have time to repair. Even if you play four times a week, the muscles do not recover. If the muscles are under stress all the time, they will break down.

Most top runners work out hard twice a week. Mark Cameron, America's best Olympic-style weight lifter, lifts heavy weights twice a week. But the average tennis buff plays hard four times a week—full out. For the tennis enthusiast, there is no such thing as an easy game, easy serve, easy shot. Tennis is very competitive.

For my patients, I compare the tear in the tendon, tennis elbow, to a small rip in your skin. If it is allowed two to three weeks to heal, it heals in a solid fashion, no problems or pain. If the cut is ripped open every two to three days—it becomes inflamed, turns red, and weeps. This same interruption of normal healing happens in your elbow every time you play tennis. Even daily activities—opening your car door—interrupt normal healing.

One reason that tennis elbow is slow to heal has to do with your anatomy. The forearm muscles are attached to what most physicians call a tendon. But in reality, they are attached to a muscle origin. A true tendon slides. It has a sheath and is bathed in synovial fluid, which both nourishes and lubricates. The muscle origin of the forearm muscles has none of these.

In a few instances, I have operated on the muscle origin. It is like cutting into gritty tissue. It has the worst blood supply of any structure in the body. The incision gives forth only a drop or two of blood. It is this infinitesimal blood supply which slows the healing process when the muscle origin is ripped or torn.

The older you are, the slower the healing process. That is why the Priest, Braden, and Gerberlich study found that players more than forty years old had a significantly higher incidence of elbow pain than younger players. For example, 41 percent of men aged forty-one to forty-five reported elbow pain, compared to 27 percent of men aged thirty-six to forty.

Besides age, the number of years that you have played is another factor. I have seen this. Operating on the muscle origin, I have found scar tissue heaped on scar tissue. Thus, the damage to the muscle origin is cumulative. The Priest, Braden, and Gerberlich study found that male players with a history of elbow pain had played for slightly less than ten years. Those without pain had played only for 6.4 years.

DIAGNOSIS AND TREATMENT

In 90 percent of tennis elbow sufferers, the pain is at the point slightly below where the muscle origin attaches at the outer knob.

I can feel the tenderness. Touching the inflamed spot brings on pain. Next, I ask you to raise and bend your arm so that the hand touches the shoulder. It is the position you use should you flex your biceps. Straightening out your elbow brings on the pain at the elbow. With your elbow straight, I ask you to cock your wrist up. This is especially painful. Sometimes, I push gently against your hand as you attempt to cock up your wrist. This activates the muscles on the back of the forearm and pulls on the injured tendon. In my opinion, the pain produced with the elbow straight out and the wrist cocked up is the most reliable sign of tennis elbow. This duplicates the position your wrist is in as you make a backhand shot or try to pick up a heavy object.

I usually confirm the diagnosis with the "chair test." I ask you to pick up a light chair with the ailing arm. Most cannot. The pain is too great.

Finally, I take x-rays of the elbow. I am looking for bone chips or a bone tumor in the elbow, which sometimes causes pain only on the outer side of the elbow. While these cases are rare, they should be ruled out with x-ray. Most of the time, the x-rays are negative. Occasionally you will see a small calcium deposit approximately one-half an inch from the outer elbow bone. This is very much like the calcium deposit in the pitcher's elbow, but less frequent.

The best treatment for tennis elbow is rest. I do not have anything in my medical bag that can beat it. If you continue to play tennis, you will extend your injury.

Be careful not to use the affected arm. Avoid opening car doors, carrying a briefcase, or lifting milk cartons. Use the other hand. Daily use can interfere with healing. Ice the elbow twice daily for twenty minutes for two to three days. Then, start heat treatments twice daily for twenty minutes. I recommend two to three weeks of rest. The longer you wait, the better the chances that you will heal completely. It is only guess work. As far as I know, there is no scientific study on the healing process of this muscle origin. Take two aspirin with each meal. The aspirin reduces the inflammation and the pain.

Tied with "rest" for the number one treatment in my book, is "go see a tennis coach." By learning to hit the ball with your whole arm instead of your elbow, you reduce the strain on the muscle origin. The Priest, Braden, and Gerberlich study found that 100 percent of those surveyed said that changing the tennis stroke was helpful treatment.

I have never had tennis elbow. But, I understand the problem. I am a squash player. When I have chased the ball into a corner and hit it only with the forearm muscles, I have felt the stress on the outer elbow area. I call it "coming up short." It

EXERCISES FOR TENNIS ELBOW

When you first develop pain in your elbow, stop playing tennis. After waiting two to seven days for the pain to disappear, start the following exercises. Perform each twice a day.

Lay your arm flat on a table, letting your hand extend over the edge, your palm facing up. With a five-pound weight, flex your wrist ten times. Do two more sets.

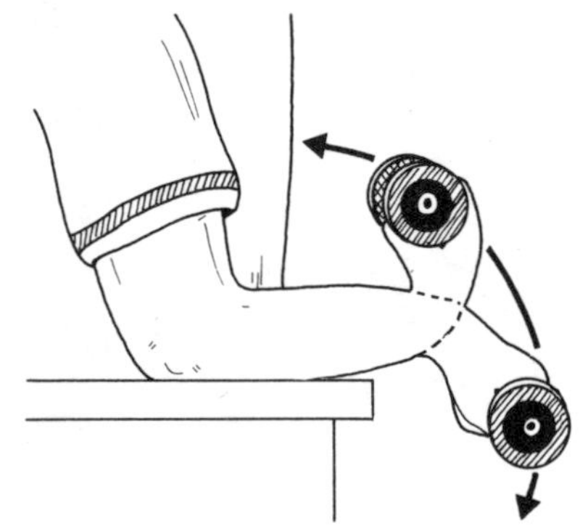

Lay your arm flat on a table, letting your hand extend over the edge, your palm facing down. With a five-pound weight, flex your wrist ten times. Do two more sets.

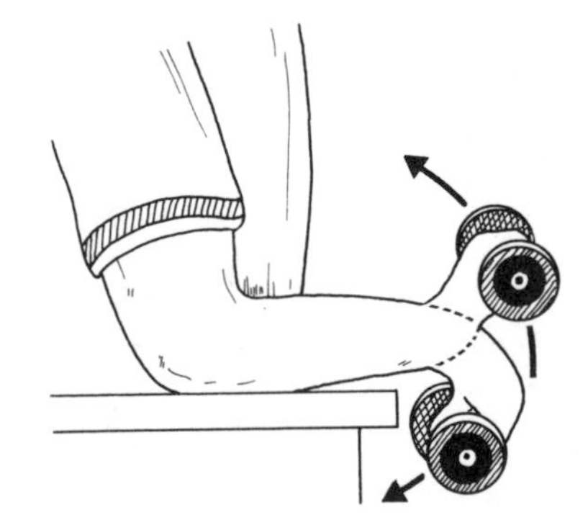

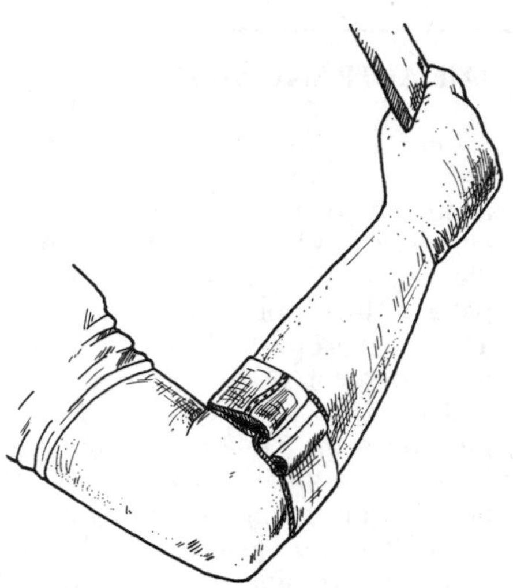

Tennis Elbow Strap

happens in tennis players who get caught at the net. The ball is smashed directly at their bodies; they cannot set themselves. They end up taking a half-swing. It is more instinct than anything else.

I have never treated a single tennis player who regularly stretches and strengthens the forearm muscles and tendons. Tennis players have not gotten the message yet—that the muscles damage slightly when they are stressed, and when they heal they are shorter. It is the same principle as a skin tear. The healing process draws the skin together. The drawing tightens the skin.

Most other athletes have the message. I have never been in the Boston Red Sox's clubhouse before a game when players were not stretching their muscles. Before marathons, I see prerace stretching.

For stretching and strengthening, I recommend wrist curls with a five-pound weight. Do not start them until the pain disappears. Perform them twice a day. It will take you ten minutes a day.

Personally, I do not stretch with weights just before playing squash or golf for a physiological reason. The weighted exercises make my muscles swell. I feel pumped up and I lose my touch. I recommend doing the curls at least two hours before playing. Ideally, it is best to do them first thing in the morning.

Always volley before you play. This allows the muscles to fill up with blood and become more pliable. The more pliable they are, the less likely they are to tear.

I always recommend the first three treatments: rest, tennis lessons, and stretching and strengthening. They are simple, painless, and they work.

Another such remedy is the tennis elbow strap, fitted with Velcro hooks. The hooks make it adjustable. It should be worn about one and a half inches below the outer knob on the forearm. It should be at least one and a half inches wide. The strap acts like a shock absorber. In the army, I learned that a tennis elbow victim could pick up a chair if he wore the strap. Without it, the pain was too great. After two weeks of wrist curls, you can return to tennis—with your tennis elbow strap.

The strap should not be made out of elastic material, because then it acts like a tourniquet, damming up the blood in the hand. As you play, loosen the strap. The exercised muscle swells with blood. The swelling varies from individual to individual. I have had very good results with the tennis elbow strap.

If all these treatments fail, I prescribe an oral anti-inflammatory medicine. The pills are taken three times a day with meals. It takes five days to reduce the pain. If the medication is going to be effective, it will happen within a two-week period.

As a last resort, I inject cortisone into the muscle origin. This is the only tendon-like structure that I inject directly. I do it on the outer side near the tendon and bone. My reason: The tendon does not slide. Thus, the chances of rupturing are slim when compared with the Achilles tendon. Secondly, the inflammation is so acute that it cannot heal itself. With the cortisone, I am trying to break the inflammation cycle. Lastly, the cortisone might weaken the muscle origin slightly, which may allow it to pull away gently from the bony anchorage. This seems to reduce the tension and, thus, relieve the pain.

Because there is not much room to maneuver, injecting cortisone into this muscle origin is more complex than injecting other areas. I use Novocain to numb the area. Additionally, I mix Novocain with the cortisone in the same syringe. I make a single entry through the skin, but I reposition the needle four or five times to spread it around. I want to be sure to get down to one level. Rubbing the area with your thumb helps spread it out. This can be a painful injection because there is very little room for the cortisone and Novocain to spread out. Thus, the fluid creates pressure in an area that is already inflamed. This tension causes pain.

When the Novocain wears off, I prescribe painkillers. Icing helps reduce swelling from the shot. I recommend three hours of elbow icing. The skin should be rewarmed for five to seven minutes every half-hour. The following day, ice it for twenty minutes, twice daily.

The cortisone takes forty-eight hours to start working. In five days, you should be feeling close to normal. In two weeks, the pain should be gone. At this point, I start tennis elbow patients on wrist curls, wearing the tennis elbow strap. If the pain continues, I give a second shot a month after the first.

Rarely is a third shot necessary. Three shots are my limit. One out of fifty tennis elbow sufferers whom I examine need an operation. Frankly, most of my patients consult me as a last resort. They have been through other therapy with other physicians. But, now their pain is chronic and constant.

The operation that I prefer is called the Bosworth Procedure, invented by David Bosworth, an orthopedic surgeon at St. Luke's Hospital in New York City. The purpose of the operation is to relieve the tautness of the muscle origin. I cut the muscle origin away from its bony anchorage. When released, it literally jumps away from the bone. It is then stitched securely to its new position, a bit lower down on the forearm. The forearm is not weakened. I send a tendon sample to the pathology laboratory. The pathologist usually finds fighter cells, a reflection of long-standing irritation.

After surgery, the patient wears an immobilization splint for two weeks. This supports the elbow at a ninety-degree bend. When the splint is removed, range-of-motion exercises called elbow crankers are started immediately. After exercising, the splint is replaced. The elbow, more than any other joint, loves to get stiff. Nobody knows why.

One month after surgery, the splint is discarded and weight exercises are started. I recommend curls with a two-pound weight. I let my patients return to full sports activities when

THE AUFRANC METHOD

Otto E. Aufranc, the dean of Boston orthopedic surgeons, had an unusual approach to the treatment of tennis elbow. It was his theory that the muscle origin had to be weakened for the patient to improve. In order to achieve this, he took his patients to the basement of his office building and let them hang on a cold water pipe from the ceiling until the pain became too great. It was his contention that this would allow slight tearing of the muscle origin and, thus, relieve the tension. He swore by this treatment. His approach parallels the cortisone treatment and the surgical release of the muscle origin as an effective cure of the problem. Aufranc's method was a primitive, but effective cure.

> **SEVEN CHIPS**
>
> "I have seven chips in my elbow," says Rudy May, of the New York Yankees. "The chips are on the outside of the elbow. My doctors diagnosed the chips as an epicondylitis condition. A strained epicondylar tendon.
>
> "The condition causes me great pain, especialy when I throw a curve ball, but I still go on pitching. Why? Because I am going good now, and the team has a chance to win the pennant.
>
> "After I pitch, Barry Weinberg, one of the great trainers for this team, ices down my elbow and he also puts some Kapsolin on it. This helps relieve some of the swelling."

the muscle strength reaches 95 percent of the opposite arm. It takes about three months, and I insist that they not play any sports until then. I have never had to reoperate on anyone for this problem.

Loose Bodies of the Elbow (Joint Mouse)

CAUSES

The throwing motion places tremendous stress on the elbow. As you move forward from the cocking phase of throwing into the acceleration phase, the stress focuses on the elbow joint. The inner part of the elbow wants to fly apart. On the other hand, the bones that meet at the outer part of the elbow smash together. This mashing of the outer elbow is called *valgus stress* (outward stress). The round radial head of the forearm acts as a battering ram against the outer part of the upper arm bone.

For reasons that are not clear, some young throwing athletes drive these two bones together with such force that they damage the surface of the bones. This damage looks very much like a divot taken from the fairway with a golf club. A piece of the joint surface (articular cartilage) along with a small sliver of underlying bone loosens from the surface of the joint itself. The loose piece is called *osteochondritis dissecans* or as we call it, Panner's disease. Dissecans can be painful.

Over time, the piece can break free of the joint surface and become loose in the joint. We call this a "joint mouse" because it is small, white, and can run freely around the joint. A joint mouse acts like a marble in a gearwork. When it falls

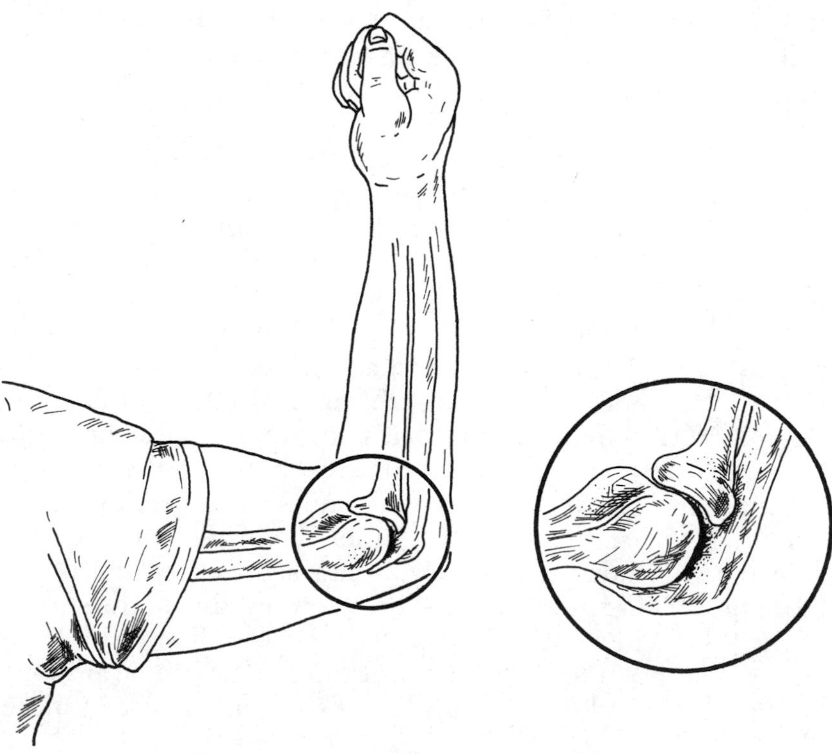

Osteochondritis Dissecans

into the teeth of the gear, the joint locks in one position. Until the mouse moves, the elbow joint is frozen.

Osteochondritis dissecans starts with pain over the outer aspect of the elbow. The pain occurs only when throwing. The pain does not radiate up or down the arm.

The rough surface or the loose pieces irritate the joints. To combat the inflammation, the joint produces extra fluid, which distends the joint sleeve. The joint becomes hard to bend. It is very much like trying to bend a sausage.

Once the loose piece breaks free and starts to move in the joint, it is a game of Russian roulette. The joint can suddenly freeze in the middle of a throw. This is excruciatingly painful. The muscles around the joint go into spasm and the joint swells quickly. It usually takes several minutes for the joint to unlock.

In both the dissecans problem and the joint mice, the patient loses the ability to straighten the elbow. Why this happens is unclear. I always compare arms. The ability to bend the arm up and the rotational motion of the forearm are normal.

When examining the joint, I feel for tenderness on the outside edge of the arm bone. This is not over the knob where tennis elbow hurts, but rather further forward toward the joint itself.

I always take x-rays of both elbows. The comparison view of the normal elbow is especially important if the patient is not fully grown, because all growth centers or prebones are still present and make the x-ray hard to interpret.

The x-ray shows a crater in the upper arm bone. Sitting within this crater is a small fleck of bone which is attached to the piece that has broken free. The joint surface itself—bone cartilage—is not visible on plain x-rays. The dissecans, or the joint mouse, which is both bone and cartilage, is four to five times larger than the bone crater. If the loose piece has fallen out of its crater, it wanders in the joint as a loose body.

There are two techniques for examining the surface of the joint. One way is to inject dye into the joint. The dye will coat the surface of the joint. We call this an *arthrogram. Arthro* means "joint" in Greek. Infrequently, the dye runs down along the edges of the loose piece and outlines its contour.

A better technique is to use an arthroscope to make the diagnosis. An arthroscope is a small telescope. I can place this in the joint under regional anesthesia. Only the arm is put to sleep. I fill the elbow joint with a saline fluid. The arthroscope is introduced into the joint through a tiny skin incision. Looking around the joint is very much like peering into a fishbowl. The saline water magnifies the inside of the joint.

In osteochondritis dissecans, I can see the outline of the loose piece lying in its crater. There is wrinkling of the surface of the joint which ordinarily is smooth as a billiard ball. If the crater is empty, I look for the joint mice. This can be arduous, because I do not know if I am looking for one mouse or more.

HOBSON'S CHIPS

Butch Hobson, a major league third baseman and former University of Alabama quarterback, had chronic elbow pain all his athletic career. It started when he was seventeen. In 1977, his second year with the Red Sox, his elbow swelled after throwing hard.

In the spring training camp of 1978, detailed x-rays showed three loose pieces in the elbow joint. Butch could now feel the pieces. Occasionally, when throwing, his elbow locked in odd positions, sometimes in the overhead throwing position, sometimes at three-quarter arm. It was like dropping a marble into a free-wheeling winch. As the season wore on, Butch's elbow deteriorated. He started making more and more throwing errors. Often, the ball would end up in the first-base dugout.

Early in the season, Dr. Arthur Pappas, the medical director of the Red Sox, and I discussed the possibility of surgery with Hobson. However, in May of that year, Butch was leading the American League in home runs and RBIs and decided that he would rather wait on the surgery until after the season.

As soon as the season ended, Hobson underwent surgery. Dr. Pappas and I removed the three marble-sized pieces from his elbow. We also trimmed away some small sharp bone spurs in the joint and flushed out the joint with saline fluid. His recovery was smooth. Three months after the operation, he began lifting weights. In the first three months of 1979, he gradually worked his arm back into throwing shape. He returned to the regular lineup in May and had an outstanding season.

In Dr. Pappas' opinion, the loose pieces in Butch's elbow were formed early in his athletic career. Repeated injury to his elbow during football, especially falling on artificial turf, and the stress of throwing compounded his difficulty.

While Hobson's elbow will never be normal anatomically, it is functioning normally. It is one example of how professional athletes can function at a high level of play without normal joints.

Hobson is an incredibly determined and dedicated athlete.

"I have never once heard him use the word 'pain,' " says Dr. Pappas. "Perhaps the best words to describe the way Butch plays are 'with courage.' "

If the fragment is lying securely in its crater and is not likely to be falling free, rest is the best treatment. Do not throw for at least three months. The fragment might reattach itself to its base. I repeat x-rays in three months. If pain is still present and the x-rays are unchanged, I recommend an additional three months of rest.

If time is the answer, the dissecans should heal in six months. If after six months, the athlete is not able to throw competitively, or for fun, surgery is the only answer. I surgically remove the loose piece from the crater. I drill the crater itself with a small bone drill. A blood clot forms in the crater and eventually converts to scar tissue—tough and pain-free. The surgery restores normal elbow function.

After surgery, I immobilize the elbow for two weeks until the skin heals. Then, I start the patient on elbow crankers. Crankers with weights start one month after surgery. Three months after surgery, you can return to sports.

If the patient comes into the office with joint mice already in the elbow, surgery is likely. Leaving a loose body in the elbow will gradually grind up the surface of the joint. This eventually leads to arthritis. Removal of a loose piece from the elbow is a straightforward operation. I make an incision over the outer aspect of the elbow joint. I find and remove the loose body. I wash out the joint with lots of salt solution to make certain no other loose pieces remain. In this case, x-rays are taken in the operating room. This makes it easier to find and remove them. The same post-operative course is followed as in dissecans.

Radial Head Fracture

CAUSES

When you lose your balance in sports, it's natural to stick out your arm to break your fall. This is how most fractures of the radial bone happen. When you land on your outstretched arm, the force moves up the radius bone in the forearm. This bone is very large and strong in the wrist area, but becomes small in the elbow. In this end of the bone it looks like a mushroom or rounded disc. That's why it is called the "radial head."

The *radial head* lies just beneath the skin approximately one inch below the outer knob of the elbow itself. Feel it on your own elbow. Start by touching the outer elbow knob. Move about one inch down your arm and move your forearm back and forth. You will feel a smooth round structure slipping beneath your finger. That's your radial head. It occupies an area inside the elbow joint. When it breaks, the pieces often fly apart; it's like a dish smashing on a cement floor. When the pieces heal, the shape of this structure is permanently deformed. A radial head fracture is a serious and painful problem.

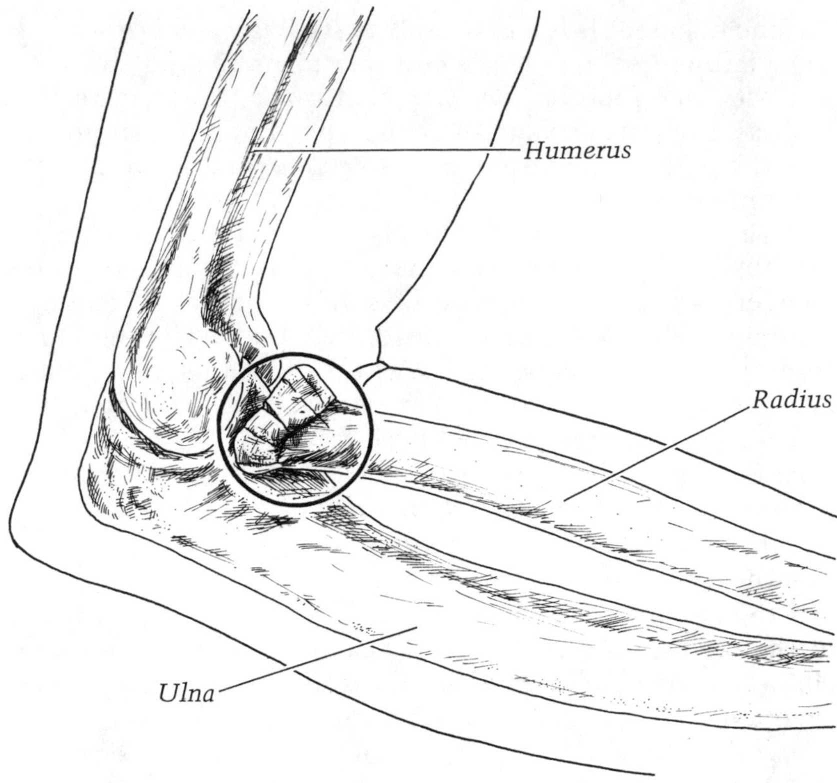

Radial Head Fracture

DIAGNOSIS AND TREATMENT

How do you know if you have broken your radial head? You feel pain immediately on the outer side of your elbow. In the first thirty minutes after injury, the pain decreases. You begin to feel a lot better; you might even feel normal. However, in the next two hours the elbow becomes steadily more painful and stiff. When the radial head fractures or breaks, blood pours out from the fracture into the elbow joint. The more blood in the elbow joint, the greater your discomfort. You get a pumped-up feeling, and the elbow becomes very stiff. Bending your elbow is awkward. The pain caused by swelling in the joint is intense, worse than the pain of the fracture. If left untreated, within twenty-four hours, you will not be able to use your elbow at all.

All elbow joints with a fractured radial head swell with blood. They all look the same. They swell on the outer aspect of the joint. I always compare the normal joint with the injured one.

If you have a fracture, you will hold the arm at about a ninety-degree bend. You will be reluctant to bend it up and down. If I touch the outside face of the joint even lightly, you will scream. If I try to turn the forearm palm up or palm down, you will resist. That's because I am turning the broken bone in the joint. X-rays show if the bone is only cracked or shattered.

Fractures that do not break apart, and therefore maintain their normal shape, are called *undisplaced fractures.* In the radial head, it is like dropping a plate on the floor and having it only crack. A cracked bone heals. A shattered one can't. The

"I BROKE MY ELBOW IN THE DARK"

Don't ever walk in the dark. I did and I sustained the worst injury of my career.

I walked into the kitchen to return a couple of bowls to the sink. Because my hands were filled, I couldn't turn on the light. I stumbled and fell over my dog. I went down like a lead balloon, landing on my right hand and my left elbow.

My elbow was sore from the fall, but I wasn't concerned. I went to bed. About 5 A.M., I woke up with a throbbing, painful arm. My wife rushed me to the hospital where my elbow and arm were x-rayed. I had fractured my elbow and a forearm bone. I left the hospital with two casts on. I was supposed to be pitching that night in Oakland. What a surprise for the manager!

Pitcher Rudy May,
New York Yankees

main problem with an undisplaced fracture is bleeding into the joint. The pain can be easily relieved by aspirating the joint. This means removing fluid from a joint with a needle. I numb the skin over the outer aspect of the elbow with Novocain. After meticulously cleaning the skin, I introduce a needle into the elbow joint. Using the syringe on the needle, I draw out the blood in the joint.

This relieves the tense feeling in the elbow and eliminates 80 percent of the pain. Next, I place the elbow in a posterior plaster splint. This is a half-cast which extends down the back of the arm to the wrist. The splint places the elbow at a ninety-degree angle. You rest in the splint for only one week.

Because the elbow tends to get stiff, I start you on elbow crankers within five days. An undisplaced fracture takes six weeks to heal solidly. By then, you should be back to full motion in the joint and regain full power in the muscles about the elbow. Because the radial bone is not blown apart, the chance of arthritis is minimal. The elbow should heal as "good as new."

Fractures where bones shatter are called displaced fractures. Think of it as a plate shattering into large pieces which scatter. A displaced fracture is a serious problem. If this fracture is allowed to knit together in a deformed position, you will have a deformed elbow joint and will lose some of your mobility. The fact that the radial head has been blasted apart is unmistakable on the x-ray. In this instance, I favor surgery. The surgery involves removal of both the radial head and it's broken pieces. I do this through an incision over the outer aspect of the joint. The radial head is easily accessible. After I remove the radial head, I wash the joint carefully so that no loose fragments are left behind. These pieces could become joint mice.

Don't you miss your radial head? Yes, without a doubt. It would be much better if you had a normal radial head and had never sustained a fracture. But the bone has been shattered. It must be removed in order to get maximum return to normal function from this elbow. Surgery offers you the best chance. It allows you to play tennis, bowl or ski, but you cannot expect to play major league baseball.

What takes the place of a radial head? Some scar tissue fills in the gap between the remaining bone and the upper arm bone (humerus). This provides a good buffer between the bones and allows excellent functional return.

After the operation, I put the athlete in an immobilization support for one week. This can be either a plaster half-cast down the back of the arm to the wrist or a prefabricated elbow splint. The advantage of the elbow splint is that it is lined with soft material and is a bit kinder to the skin. It also has Velcro fasteners across the front, which makes it easily removable and replaceable.

One week after surgery, I start you on elbow crankers to regain motion in the elbow. If I don't, I am afraid that the

TED WILLIAMS' FRACTURE

In the 1956 All-Star game, Ted Williams, one of the all-time best hitters in baseball, crashed heavily into the outfield wall. His right elbow smashed into the fence. Immediately, Williams's fell to the ground writhing in pain. He was holding his throwing arm.

X-rays revealed a fracture of the radial head, but only part of the head was displaced. If the fracture had been allowed to heal untreated, Williams' career would have ended in 1956.

Surgery was performed to remove only the broken portion of the radial head and leave the rest. It was felt this procedure would give Ted the best chance of returning to major league play.

After the operation, Ted's recovery was smooth. He spent the winter in Florida with Jack Fadden, the Red Sox trainer, working daily on elbow exercises. They worked especially hard on rotational exercises of the forearm to improve the pivoting of the arm as well as muscle-strengthening exercises for the upper arm and the forearm and wrist. The effort covered six months.

It worked. At that time, it was one of the great medical comeback stories in baseball, and the next season Williams won the American League batting crown.

elbow will stiffen. Three weeks after the motion exercises are started or one month after surgery, you can start weight-lifting exercises—elbow crankers with weights. I like to wait at least two months from the time of surgery before I okay you for athletic activity. Three months should be allowed between surgery and the return to contact sports.

Ulnar Neuritis

CAUSES

Have you ever hit the inner part of your elbow and felt an uncomfortable electric-type sensation? Some people call this hitting the "funny bone." You have not hit a bone at all. You have hit the *ulnar nerve,* one of the three major nerves in the arm. It is located on the inner knob of your elbow just behind the point where the inner forearm muscles attach. The nerve itself is larger than a telephone cord. It functions like a two-way wire. It brings messages from the brain to half the muscles in the forearm. It also returns sensation messages from the palm of your hand, the little finger, and half of the ring finger (next to the little finger) to the brain. If this part of your hand touches a hot stove, the message is sent to the brain by way of the ulnar nerve.

The problem arises when the nerve stretches or becomes

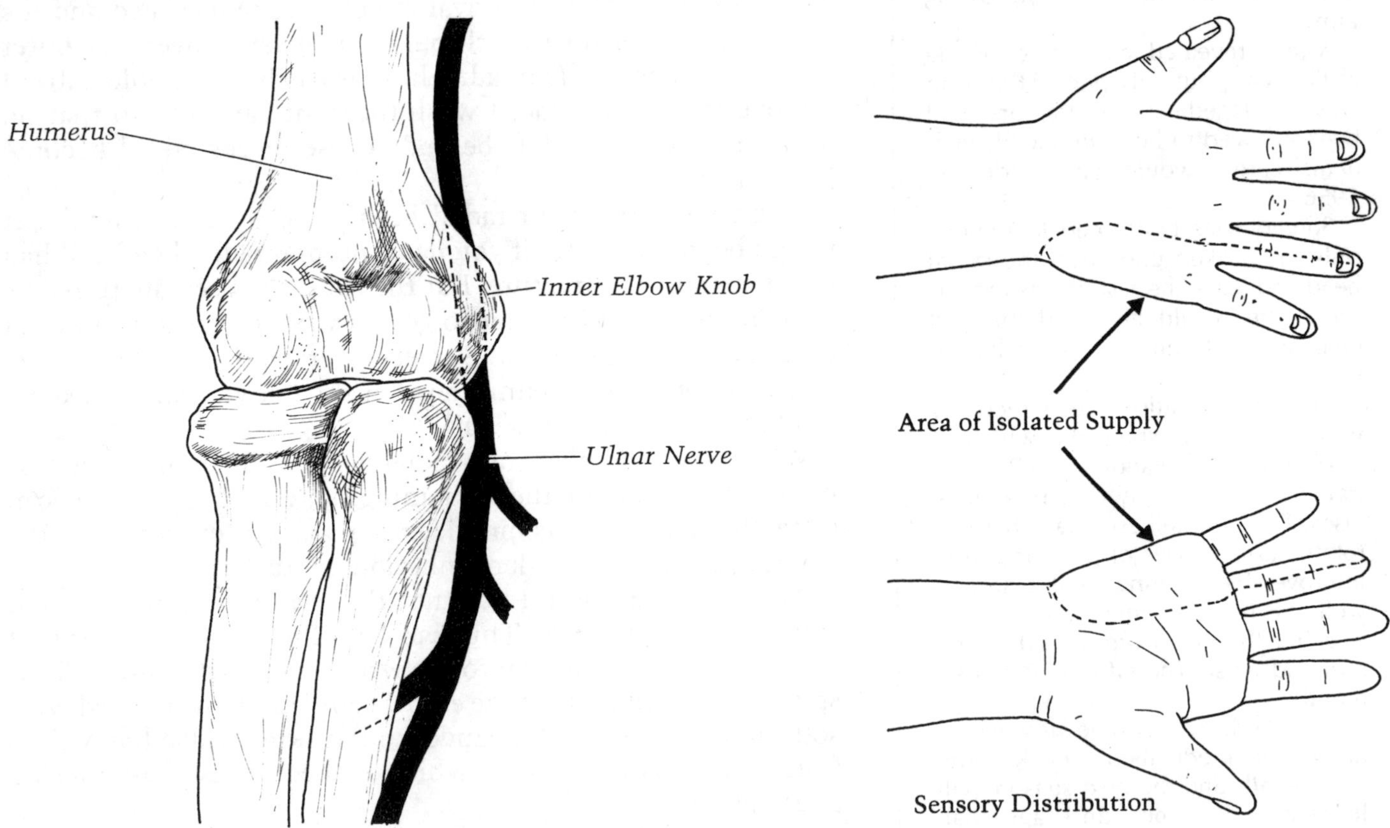

Ulnar Nerve at the Elbow

loose. You can be born with a loose nerve or you can develop it from throwing or from racquet sports. Every time you forcefully throw a ball or hit with a racquet, you twang the nerve and irritate it. The irritation called *neuritis,* causes pain at the inner elbow.

DIAGNOSIS AND TREATMENT

My patients with ulnar neuritis describe the onset as "sneaky." They will feel discomfort on the inner side of the elbow after a long tennis or golf match. The discomfort goes away in a few days only to return a week or two later. This is stage I.

The pain radiates down the arm, from the elbow. I call this stage II.

Stage III occurs when the irritation interrupts normal nerve function. This means that you feel pins and needles or even complete numbness in the little finger and half the ring finger. It feels as if your arm has fallen asleep.

At this point, you will not be able to play sports effectively. Stage III scares people and they seek medical advice. Getting to stage III usually takes two to three months.

Stage IV is permanent numbness. The nerve stops functioning. However, I have never seen stage IV.

I diagnose ulnar neuritis by pinpointing the discomfort. Taking the elbow in my hand, I touch the outer and the inner elbow knobs. The bones and the tendons are not sensitive. By using my index finger, I can reach just behind the inner elbow knob and find the cord-like nerve. This is approximately the diameter of a pencil. In ulnar neuritis, it is very mobile. I can actually twang it. This causes the patient to protest strongly. Even touching the area causes pain. Sometimes the pain can radiate down the forearm. In severe cases, light tapping on the nerve produces pins and needles in the fourth and fifth fingers. We call this a *positive Tinel's sign.* This simply means that the neuritis is acute, and the nerve is very sensitive. I always compare the normal nerve on the other elbow. Tapping on this nerve shouldn't produce the same discomfort.

Some patients use an elastic bandage or elbow sleeve to try and reduce the pain. However, it does not help. The elastic is not strong enough, nor can it put enough pressure on the nerve to hold it in place.

I always refer the patient to a neurologist. Neurologists specialize in problems of the brain and nervous system. I ask them for a nerve conduction study to determine the degree of damage to the nerve. A nerve conduction study is an electrical test, which is very much like testing the flow of electricity in a wire. A small electrode needle is inserted in the nerve above the elbow. A second electrode needle is placed in the nerve below the elbow. The upper needle is charged and sends shocks down the nerve. The stimulus is collected by the second needle. By timing the charge between the two electrodes, the neurologist learns how long it took for the signal to get from one point to the other. A normal nerve conducts a message at a rate of fifty-five meters per second. This means that if you have a nerve 180 feet long, the message would normally be transmitted in one second. An irritated nerve slows the signal. Readings below forty meters per second indicate that the nerve is so inflamed it is not able to conduct messages at a normal speed.

Why do I order a nerve conduction study? Sometimes a pinched nerve in the neck produces numbness in the fourth and fifth fingers. I want to be sure.

Once the diagnosis is made, treatment begins. Rest is the number one treatment. Refrain from heavy use of the arm for four to six weeks. Sometimes, I prescribe anti-inflammatory medicines by mouth. This reduces the inflammation. In 50 percent of the cases, this treatment provides complete relief. The patient can return to activity gradually over the next month.

In the other 50 percent, the neuritis is chronic. With rest, the inflammation will disappear; with new activity, it returns. The nerve is just too mobile. For these people, I recommend an operation or a year of rest. The purpose of the operation is to move the nerve from the side of the front of the arm. The new position relaxes the nerve. I suture the nerve to the new position below the biceps muscle. The body accepts the new positioning without side effects.

I keep patients in an elbow splint which holds the elbow at a ninety-degree angle for two weeks. Then I start the patient on elbow crankers. Four weeks after surgery, the patient adds a two-pound weight to the elbow crankers. Heavier weights are used as tolerated.

I have enjoyed a 100 percent success rate with this operation in terms of relieving nerve pain. The nerve is positioned on the front side of the elbow and does not slip back into its original position. My patients go back to full sports activity.

Others have had the same success rate. A recent paper by Wilson Del Pizzo, Frank Jobe, and Lyle Norwood—physicians from the National Athletic Health Institute—found that the three major league baseball players who were operated on for ulnar neuritis have returned to their preoperative level of play.

Little League/Pitcher's Elbow

CAUSES

Almost every Red Sox player, at one time or another, has had pain on the inner knob of the elbow of his throwing arm.

When x-rayed, their elbows all show evidence of wear and tear. This ailment is called *Little League elbow* in growing children and *pitcher's elbow* in adults. Think of this ailment as an inside tennis elbow.

The pain is caused by the throwing motion. When you

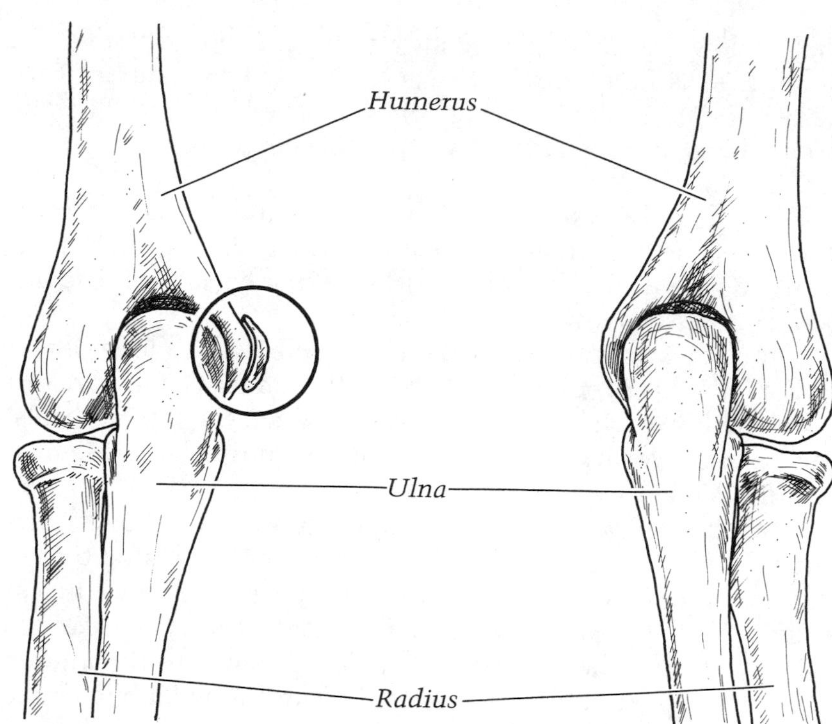

Little League / Pitcher's Elbow

throw a baseball, the final release of the ball involves a powerful downward motion of the wrist and fingers. The muscles—flexor pronator group—which power the motion, all lie on the front part of the forearm. The force of the throw radiates up the arm to the weakest portion of that muscle group, the attachment at the inner knob of the elbow, called the *medial epicondyle.*

With repetitive use of the throwing muscles, tremendous stress eventually produces a small rip or tear in the muscle origin close to the attachment to the bone.

In growing children, the muscle origin is attached to a growth center, or a prebone, which is not as strong as bone itself. Thus, Little League elbow is stress to the growth center tending to pull it away from the main trunk of the bone. In adults, the ailment is right at the muscle attachment and the bone.

PITCHER'S ELBOW

In 1979, Carlton Fisk, the Boston Red Sox catcher, played only one-third of the season because of pain on the inner aspect of the elbow of his throwing arm. He had a classic case of pitcher's elbow.

Initially, the pain started when he threw out a runner trying to steal second base. Immediately, he felt acute pain, which persisted.

X-rays showed an old calcium deposit just below the bony prominence of the inner elbow. The calcium was a sure sign that the muscle origin had been injured previously.

Fisk had one of the most stubborn cases of pitcher's elbow I have ever treated. Anti-inflammatory medicine and cortisone injection were not effective. He tried strengthening exercises within the limits of his discomfort.

But, the pain persisted. He was unable to pick up a bat without pain. He sat out the last part of the season. His pitcher's elbow required an additional six months of healing time before he could effectively throw again at the start of the 1980 season.

DIAGNOSIS AND TREATMENT

Little League or pitcher's elbow usually starts slowly. The pain gradually builds over a two- or three-day period. Rarely does it start abruptly after a specific hard pitch or throw.

Biologically, the pain originates from the tear in the muscle origin. The larger the tear—the greater the discomfort. As the pain sets in, you will not be able to throw hard or accurately. When you wake up, you will notice a stiffness and pain. The pain often radiates down the forearm two or three inches. Some are unable to straighten the elbow fully.

I diagnose the problem by touch. I can put my fingers directly on the injury and determine the nature of the problem. In adults, the tenderness is found approximately one-half inch below the medial epicondyle. In growing athletes, the tenderness is found directly over the medial epicondyle. Tenderness is not found anywhere else in the elbow area. There is no evidence of any difficulty with the nerve function in the forearm or hand.

X-rays are always taken of the elbow to make this type of diagnosis. They allow comparison with the normal, non-throwing elbow. In the adult, with the first episode of discomfort, the x-rays show no abnormality.

Sometimes, if you have had pitcher's elbow for a long time, I find a calcium deposit, lying one-half inch down the forearm from the bone. It shows up on x-rays. When the tendon heals, the calcium forms in some people and not others. Medical scientists do not know why.

Pitcher's elbow and Little League elbow are treated differently. For pitcher's elbow, the best treatment is to stop throwing for at least ten days. As the tear starts to heal with scar tissue, start "long toss"—throwing the ball gently with a high arc. Long toss is helpful because it brings blood to the injured area without extending the injury. Do not throw hard for three weeks. Like tennis elbow, pitcher's elbow is a slow healer. The longer you hold off from pitching or throwing, the better are your chances of full recovery.

If the pain continues, I prescribe anti-inflammatory medication by mouth. If that does not help, I inject cortisone. However, the cortisone does not speed the healing.

Because you cannot throw hard for two to three weeks, your arm must be restrengthened with weight-lifting activities. In that period, you have some muscle atrophy. Without the restrengthening, your arm will be vulnerable to reinjury.

Charlie Moss, Red Sox trainer, likes to use paste-like linament rubbed into the elbow approximately thirty minutes before a player takes the field for throwing activities. He recommends a long sleeve jersey during the healing time. This keeps the elbow warm. Ice is applied to the inner aspect of the elbow for twenty minutes at the end of the game or practice.

Little League elbow is managed differently. Until your elbow loses its tenderness, all throwing is stopped. The tenderness often takes six to nine weeks to disappear. If you throw before the elbow is fully healed, the pain immediately returns. Anti-inflammatory medication does not work well in growing athletes. They don't respond well to it. I never inject cortisone into children because it has been shown to interrupt growth in the injected areas.

In children, the x-rays of Little League elbow show a gap between the growth center and the main trunk of the bone. If the separation is greater than one-quarter of an inch, there is a serious displacement of the growth center. Always compare the gap to that in the other elbow.

Six weeks after the onset of the ailment, I x-ray the patient again. If the x-rays show only a small gap between the growth center and bone, and the inflammation has abated, then throwing may be resumed. Again, "long toss" and a strengthening program should be started prior to return to full velocity throwing.

If the second x-ray still shows a gap of more than one-quarter of an inch displacement, then surgery should be considered. Displacement of this magnitude will lead to inability of the bone to mature. Permanent weakness results. Abnormal anatomy such as this leads to abnormal function, and you will never participate in throwing activities at a competitive level.

The purpose of the surgery is to restore the growth center to its normal position. I do this with two threaded wires. These wires are left intact for six weeks, and then removed under local anesthesia. A complete physical therapy program follows to regain strength and motion in the elbow. Often, throwing is not resumed for six months.

CHUCK RAINEY'S MUSCLE TEAR

Some injuries just happen without explanation. Take the case of Chuck Rainey, a starting pitcher for the Boston Red Sox in 1980. Chuck reported to spring training in excellent shape, according to his coaches and the team physicians.

In 1980, he started off the year strongly, compiling an 8-3 record. In early July, while pitching against the Baltimore Orioles, Chuck threw a curve ball and felt something pop on the inner side of the elbow.

He was taken out of the game. Charlie Moss, the Boston Red Sox trainer, iced the elbow. The next day Chuck still had pain on the inside of the elbow.

"Chuck had experienced a partial tear of the flexor muscles," explains Dr. Arthur Pappas, the medical director of the Boston Red Sox. "It was a tear in the muscle itself near the inner elbow knob.

"Unfortunately it was in one of the muscles that count for a pitcher. It is amazing how something as apparently simple as a partially torn muscle can be so damaging. The tear ended Rainey's season. There's *nothing* in a medical way to hasten the healing. Nature has its own timetable.

"It was a very frustrating season for Rainey and the Red Sox medical staff."

MORE THAN 134,000 PEOPLE SUSTAIN ELBOW INJURIES EVERY YEAR **1979**

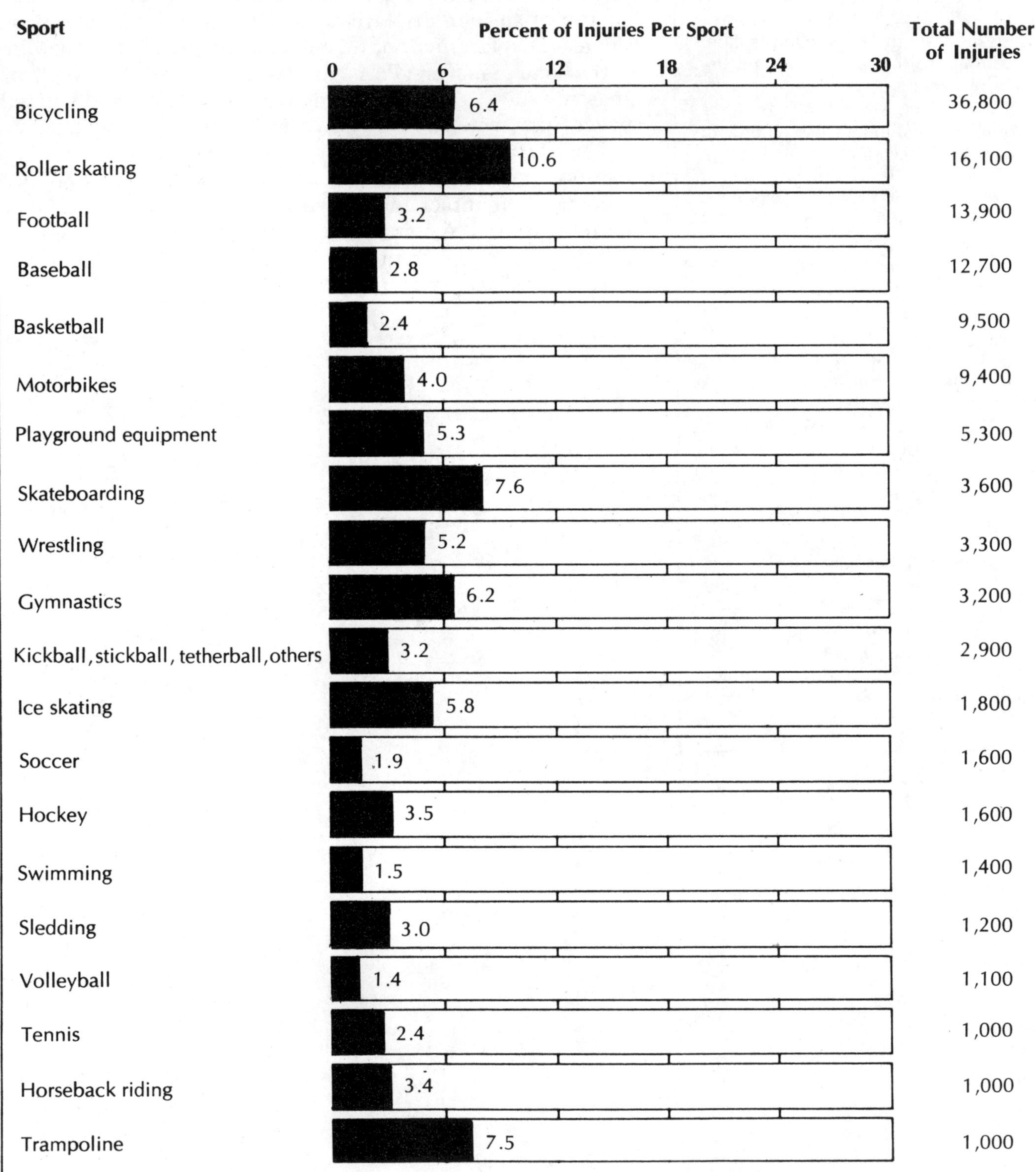

SOURCE: Consumer Product Safety Commission.

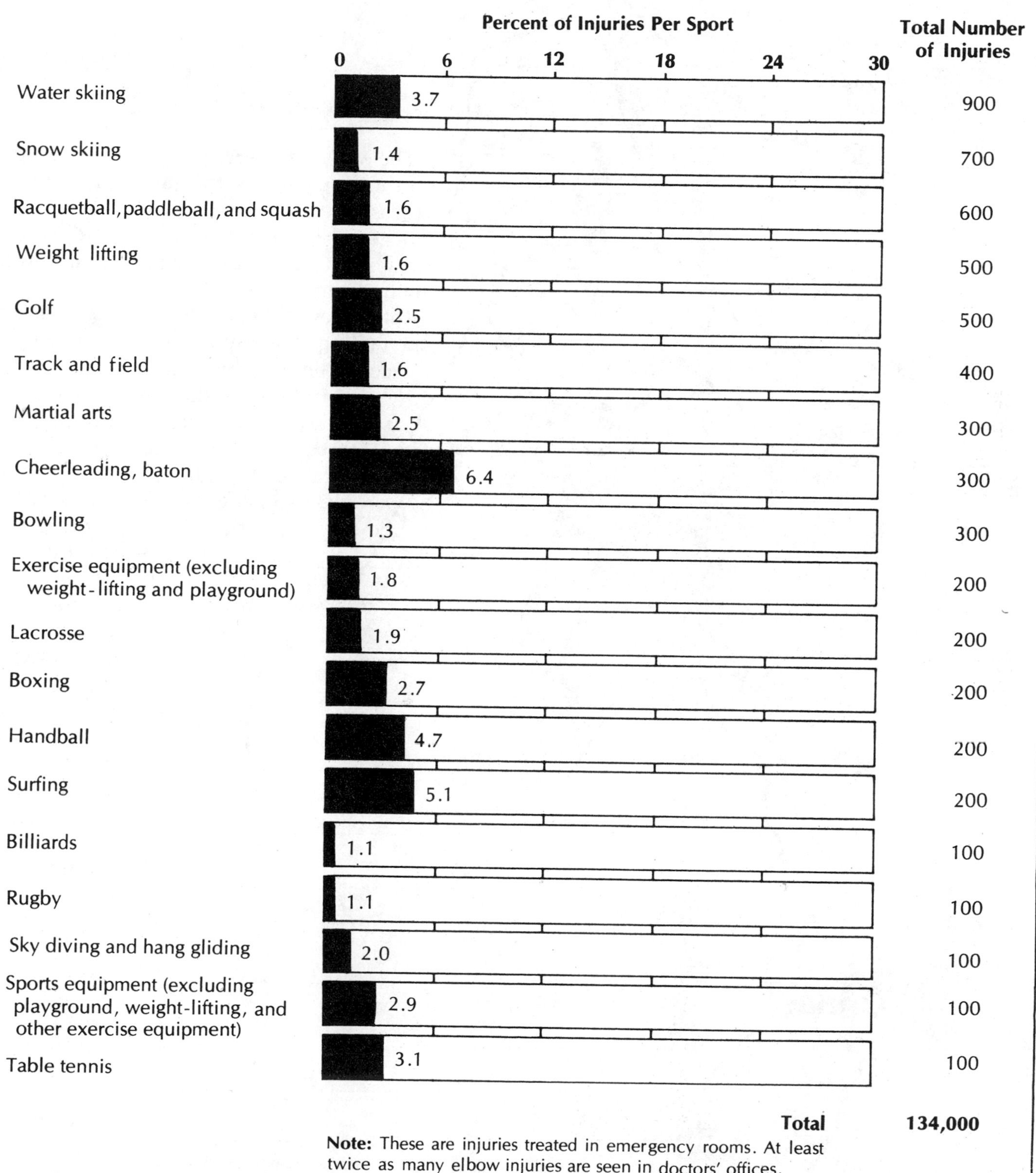
Percent of Injuries Per Sport
Total Number of Injuries
0
6
12
18
24
30
Water skiing 3.7 900
Snow skiing 1.4 700
Racquetball, paddleball, and squash 1.6 600
Weight lifting 1.6 500
Golf 2.5 500
Track and field 1.6 400
Martial arts 2.5 300
Cheerleading, baton 6.4 300
Bowling 1.3 300
Exercise equipment (excluding weight-lifting and playground) 1.8 200
Lacrosse 1.9 200
Boxing 2.7 200
Handball 4.7 200
Surfing 5.1 200
Billiards 1.1 100
Rugby 1.1 100
Sky diving and hang gliding 2.0 100
Sports equipment (excluding playground, weight-lifting, and other exercise equipment) 2.9 100
Table tennis 3.1 100
Total 134,000
Note: These are injuries treated in emergency rooms. At least twice as many elbow injuries are seen in doctors' offices.

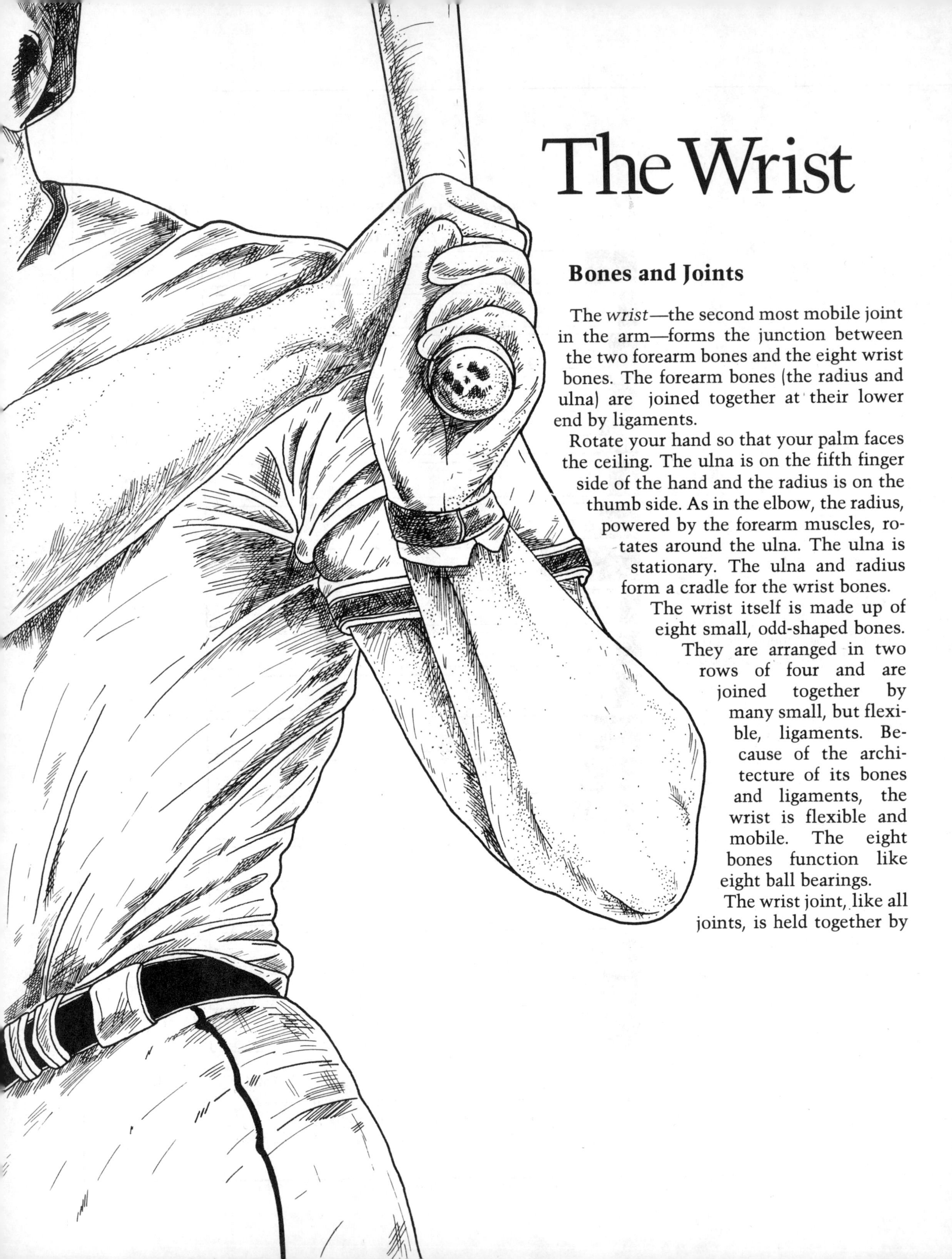

The Wrist

Bones and Joints

The *wrist*—the second most mobile joint in the arm—forms the junction between the two forearm bones and the eight wrist bones. The forearm bones (the radius and ulna) are joined together at their lower end by ligaments.

Rotate your hand so that your palm faces the ceiling. The ulna is on the fifth finger side of the hand and the radius is on the thumb side. As in the elbow, the radius, powered by the forearm muscles, rotates around the ulna. The ulna is stationary. The ulna and radius form a cradle for the wrist bones.

The wrist itself is made up of eight small, odd-shaped bones. They are arranged in two rows of four and are joined together by many small, but flexible, ligaments. Because of the architecture of its bones and ligaments, the wrist is flexible and mobile. The eight bones function like eight ball bearings.

The wrist joint, like all joints, is held together by

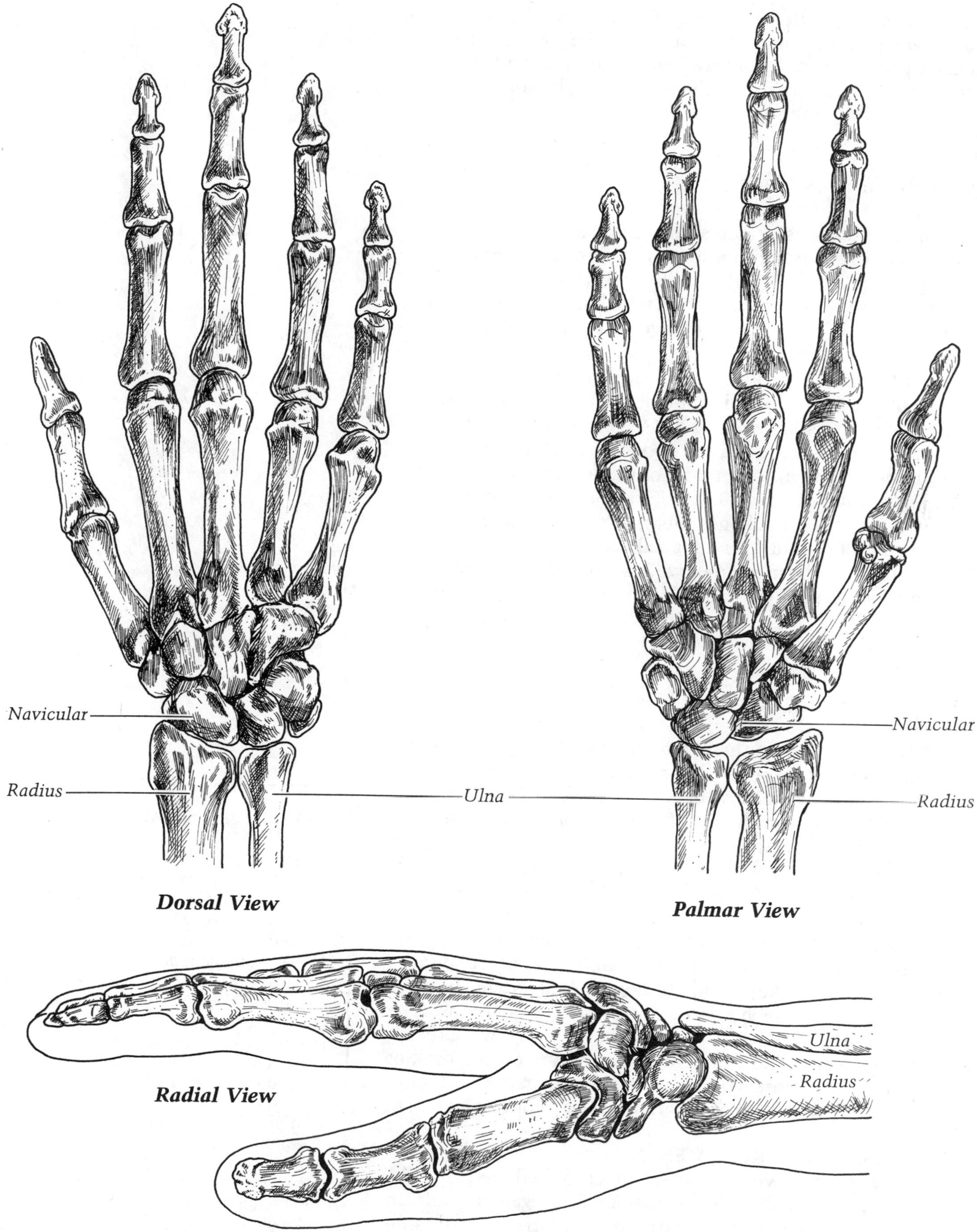

Navicular

Radius

Ulna

Dorsal View

Navicular

Radius

Palmar View

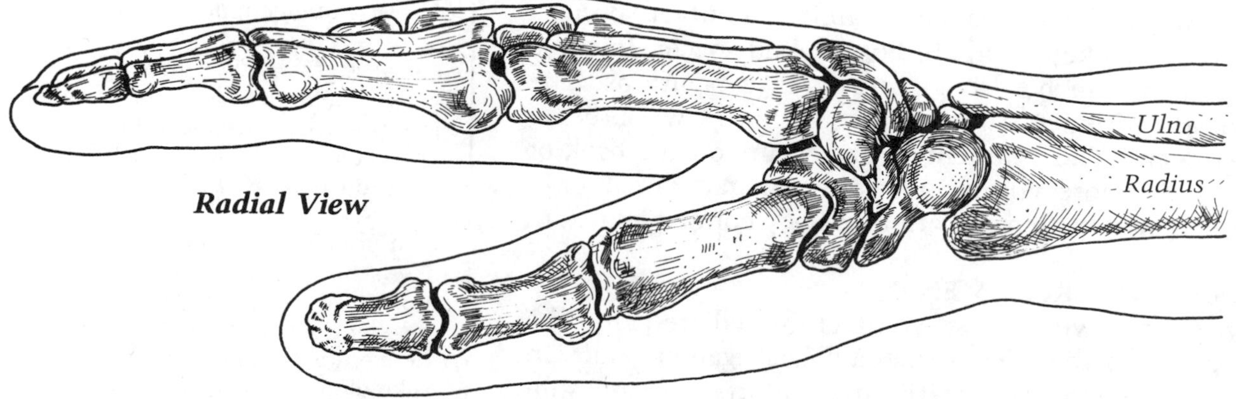

Radial View

Anatomy of the Wrist

a flexible sleeve called a *capsule.* The capsule is loose enough to allow equal movement back and forth and side to side. The wrist itself can rotate slightly, due to the shape of the eight wrist bones. In my opinion, the wrist is a felicitous combination of mobility and stability.

Muscles and Nerves

Four large tendons, attached to the forearm muscles, powerfully move the wrist in four directions: up, down, and side to side. They stabilize the wrist for heavy lifting. If these tendons are injured, you cannot play sports which use a club, bat, stick, or racquet. A second group of five forearm muscle tendons controls the fingers.

All wrist and hand movement is directed by the three major nerves of the arm: the radial, ulnar, and median nerves. The *radial nerve* runs along the radius bone in the forearm, the *ulnar nerve* runs along the ulna bone in the forearm, and the *median nerve* lies in the middle of the forearm and runs into the hand through a canal. The floor and two sides of the canal are formed by the wrist bones themselves. The top of the canal is formed by a stout ligament called the *transverse carpal ligament.* Thus, the median nerve is surrounded on three sides by bone and on the fourth side by an unyielding ligament.

Wrist Sprain

CAUSES

All the structures of the wrist—the two forearm bones and the eight wrist bones—are connected by ligaments. In addition, the joint capsule is made up of ligament-like material.

A *sprain* is a tearing of a ligament, or joint capsule. It happens when the wrist is forced into a position which exceeds its normal range of motion. This extends the ligaments beyond their fixed length and they rip. The tear generally occurs in the middle of the ligament.

Ligament injuries (sprains) are far more common than fractures in the wrist. Sprains happen mainly in contact sports—rugby, lacrosse, hockey, football, boxing, and soccer. In these sports, the wrist can be driven into strange positions. The most common wrist sprains happen with the wrist being forced downward. The ligament injury occurs on the back of the wrist. It is more common in men than women only because men participate more often in body contact sports.

DIAGNOSIS AND TREATMENT

When you sprain your wrist, you feel immediate pain. The pain is centered in the area of the damaged ligament. Fifteen to thirty minutes after this initial discomfort, the pain subsides, but then it gradually intensifies over the next three

HARGROVE'S SPRAINED WRIST

Early in the 1980 season, Mike Hargrove, the Cleveland Indian first baseman, suffered from a sprained wrist which occurred when he was trying to check his swing.

"After the team doctor, Earl Brightman, examined Mike's wrist, I iced the wrist for twenty minutes," says Jim Warfield, the Cleveland trainer. "I iced the wrist twice the following day. Three days after the injury I started Mike on contrast baths. That means we used alternating hot and cold whirlpool baths. I taped his wrist with a wrist and hand-type strapping. It is a football type of strap which gave him excellent wrist support. Because of the ice, contrast bath, and the taping, Mike didn't miss any action."

FRONT WRIST CURLS

Lay your arm flat on a table, letting your hand extend over the edge, your palm facing up. With a five-pound weight, flex your wrist ten times. Repeat three times.

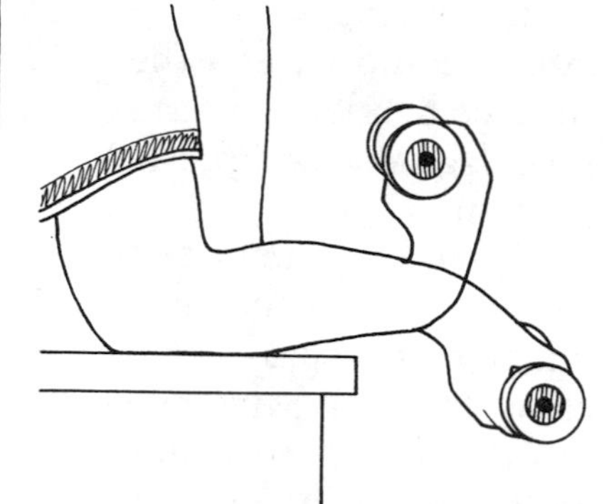

BACK WRIST CURLS

Lay your arm flat on a table, letting your hand extend over the edge, your palm facing down. With a five-pound weight, flex your wrist ten times. Repeat three times.

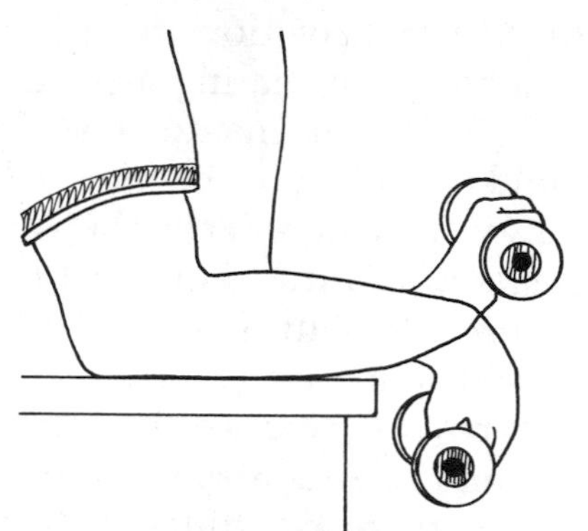

YAZ'S WRIST SPRAINS

Carl Yastrzemski has one of the hardest swings in baseball. He is the only American Leaguer ever to combine 3000 base hits and 400 home runs. In August of 1978, Yaz swung at a fastball which suddenly changed directions. He tried to check his swing, but his bat was already moving at full velocity. All the muscles in his forearm were desperately trying to hold the swing back. His wrist, the fulcrum, was caught in the middle of the two forces.

Something had to give. It was Yaz's left wrist. He felt a tearing sensation on the thumb side of the wrist. After a few innings, Yaz experienced severe pain in the wrist, and was removed from the game.

I took x-rays, but there was no evidence of bone damage. I had Yaz ice the wrist for twenty-four hours. That night, I examined him again and found very little tenderness, but he complained of pain when he moved his wrist to the side. I made him a half-splint out of plaster of paris. He continued the icing and elevated the wrist on a pillow in bed.

Yaz explained to me that he had had wrist problems six or seven times during his baseball career. "I guess I swing too hard," he said sadly.

Despite the icing, Yaz's wrist swelled the next day. This was a sign that many blood vessels were broken and that the blood was oozing into the wound. It was tender to touch. He was unable to move the wrist without pain. I explained that no matter how tough he was mentally the pain would be so great that he would have to sit out for at least one week. He was unhappy.

A seven-day layoff for a severely sprained wrist is minimal. With tape support of the wrist, I was hoping he could get back quickly.

I made a splint to rest the wrist. He continued icing for another forty-eight hours and then I started him on warm whirlpool baths. When he had his wrist and arm completely in the whirlpool, I instructed him to flex his wrist up and down and from side to side. I did not want his wrist to stiffen. After the whirlpool treatments, he returned to a wrist splint. In a week, the swelling and tenderness had diminished greatly.

Charlie Moss, the Red Sox trainer, devised a special wrist support that wrapped very tightly around Yaz's thumb and his wrist. The purpose of the support was to hold the wrist steady and protect the injured ligament as he swung his bat. In fact, at one point, the tape was so tight that it pressed on the thumb and made it numb.

This "little sprain" caused Yaz to lose ten days during a race for the pennant. In four weeks, the ligament injury healed completely. He was then able to discard the tape support and was able to use his regular wrist wrap for protection.

Yaz's wrist was an example of how an athlete can return to the field of play before an injury is completely healed and pain-free.

Conventional orthopedic treatment of this kind of sprained wrist would have been to use the plaster splint for three to four weeks and then to have started physical therapy. Our treatment was possible only because we could reexamine Yaz's wrist every day for damage.

Our "unconventional" treatment program got Yaz three more weeks of baseball in 1978. His wrist healed completely over the winter and he has had no difficulty with it since.

hours. Swelling starts one hour after injury. As a general rule, the more the swelling, the worse the sprain. If the injury is severe, it turns black and blue in six to twelve hours—a sign of internal bleeding. The discoloration is confined to the injured area.

The best treatment for a sprain is rest. That is why even for a simple wrist sprain I immobilize the wrist with a half-cast or splint. Almost any movement of the wrist—writing, thrashing in your sleep, or dialing a phone number—could extend the injury. The splint ensures rest. I use an Ace bandage to bind and hold the splint to the forearm. The bandage also compresses the wrist, which keeps the swelling down.

I make the splint out of *casting tape,* which looks and feels like gauze tape, but is covered with particles of plaster of paris. It is pliable and takes twelve layers to make a firm splint. I cut and shape the tape so that it fits across the full palm and thumb and runs across the inner part of the wrist, stretching approximately two-thirds of the way up the forearm. The top of the hand and upper forearm are not covered. The next step is to immerse the tape in water, which activates the plaster of paris. The splint follows the contour of the hand and forearm perfectly. I do not use prefabricated splints; they are uncomfortable and do not fit as well. It is like the difference between the custom-made and an off-the-rack suit.

As the splint takes shape, I remove it and line it with a soft, felt-like material called webril to protect the skin from chafing.

While the plaster of paris is hardening, it undergoes a chemical reaction and gives off heat. I cock the wrist up about ten degrees as the plaster splint hardens. This will put the wrist in a position of rest. It also allows use of the fingers. A good trainer can make a splint.

I usually reexamine the wrist the next day to check the swelling. The more severe the pain, swelling, and tenderness, the more severe the sprain. A slight sprain will heal itself in one week. A severe sprain requires three to six weeks of rest. The length of time for healing depends upon how much scar tissue must form between the ends of the ripped ligaments. A partially torn ligament will heal much more quickly than a completely torn ligament. Given enough time, they will all heal.

Slight sprains should be immobilized in a plaster splint for

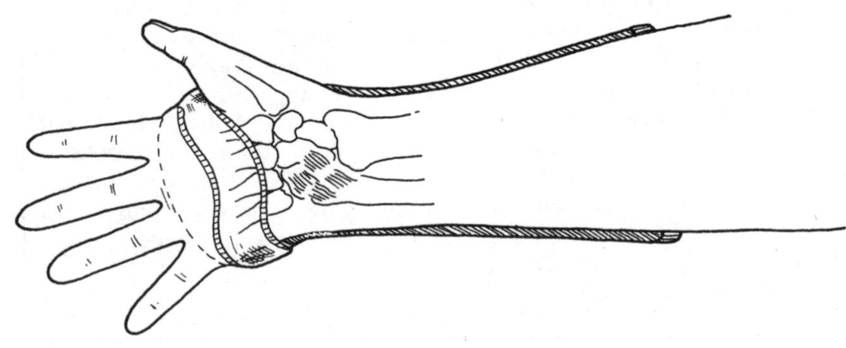

Half-Cast For a Sprained Wrist

three days. On the fourth day, start doing wrist curls. Do not use weights. Do a set of thirty, two times a day. In one week, you can return to sports.

To protect your wrist from reinjury, tape it with adhesive. The tape is like a soft splint; it prevents the wrist from being jammed forward or backward easily. Wrap the wrist with a prewrap to protect the skin. Then apply the adhesive tape. It should be at least one and a half inches wide. Extend your fingers fully. This prevents you from wrapping the adhesive too tightly around the tendons.

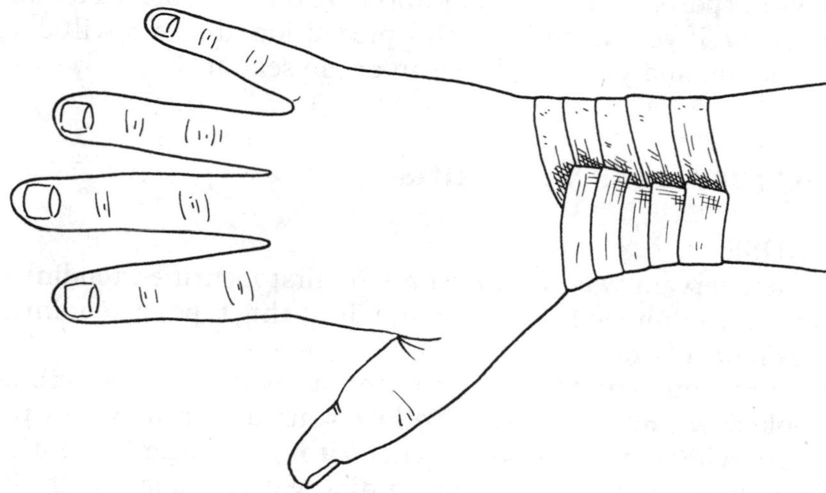

Wrist Taping

If you are like me, you will need some help in wrapping the adhesive. When I played football at Harvard in the early sixties, I always taped my wrists. It was the only way to prevent constant sprains. Taping wrists (and ankles) on college and professional teams in many sports—soccer, football, baseball, and hockey—is commonplace.

More severe sprains require one to two weeks of splint support. However, the splint should still be removed at the end of three days, and gentle motion exercises—wrist curls—should be initiated immediately.

The wrist, like the elbow, freezes up quickly.

Do the wrist curls in water. A washbasin is fine, and the wrist can be immersed. A buoyant wrist requires much less muscular effort.

Only when the swelling is down and the area is much less tender can wrist curls with weights be started. Use a three-pound barbell to start. You can return to athletic competition when your wrist has regained 95 percent of its normal motion and strength. I estimate this by asking you to squeeze a spring-type hand exerciser. I compare performance of the injured wrist to the uninjured one.

When you return to sports, tape your wrist for at least one month. If you return too early to activity, you interrupt normal healing. This can eventually produce an inflamed ligament, and the wrist pain becomes chronic. I reset the plaster splint to reduce the inflammation and allow proper healing for the ligament.

Secondly, I recommend taking two aspirin with each meal for ten days. If the aspirin does not relieve the pain, I prescribe stronger oral anti-inflammatory medication. Twelve times in my career I have injected cortisone into the inflamed area. I injected it into the most tender and painful areas. After the injection, I used a wrist splint for three days. Then I started the wrist curls.

If wrist sprains are not treated properly, they can be nagging, disabling problems. You are far better off losing seven to ten days of sports competition by allowing the wrist sprain to heal properly. If you do not take this precaution, the pain will drag on and on and you could lose an entire season.

DeQuervain's Tendinitis

CAUSES

DeQuervain was the surgeon who first identified tendinitis on the thumb side of the wrist. Thus, this type of tendinitis was named after him.

Turn your wrist so that the thumb points to the ceiling. Look down at your wrist. Follow your thumb down to the point where it joins the wrist. With your thumb pointing directly up, note that it forms a ninety-degree angle with the wrist. There are two tendons which must turn this ninety-degree angle to get to your thumb. They are called the *abductor pollicis longus*—which attaches at the base of the thumb bone—and the *extensor pollicis brevis* tendon—which attaches to the knuckle of the thumb. Without these tendon attachments, you could not thumb your nose!

Like all sliding tendons, these are enclosed in a tunnel, or sheath. The sheaths are attached to the forearm bone and keep the tendons from bowstringing out away from the bone when you point your thumb up or down. They also stabilize the tendon against the bone (like other sheaths of the wrist) and function like a pulley.

With heavy use of the thumb in throwing and racquet sports, these tendons can become irritated. The irritation happens at the right-angle turn. When the tendons get irritated, the sheath produces extra synovial fluid, the liquid which lubricates and feeds the tendon. The sheath cannot expand to accept this extra fluid, with the result that the fluid pushes hard on the tendon and it becomes inflamed. Over time, this extra fluid presses hard against the sheath and causes it to thicken. Then, there is even less room for the tendon, and the tendinitis becomes chronic. The inflammation creates more inflammation.

Sometimes, patients with DeQuervain's tendinitis have extra tendons in the sheath. It is a fluke of nature. The irritation is caused by the extra tendons.

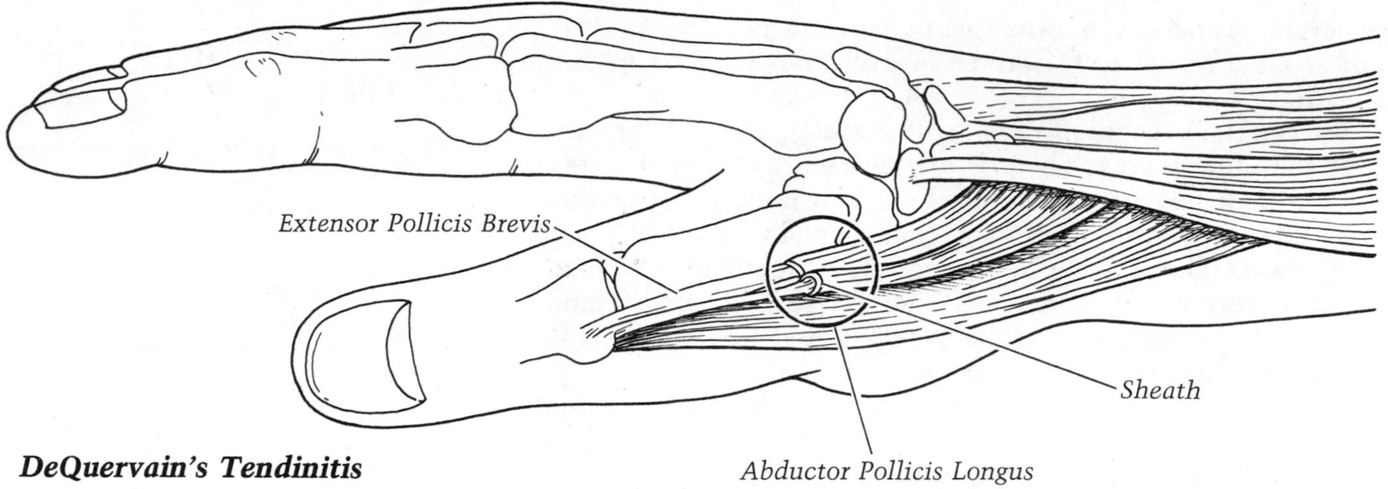

DeQuervain's Tendinitis

DIAGNOSIS AND TREATMENT

DeQuervain's tendinitis develops gradually. The pain is localized on the thumb side of the wrist and is dull and aching in nature. It radiates two to three inches up the forearm and down into the thumb itself, but there will be no numbness or tingling. You will not be able to thumb your nose without discomfort. If you try to put your thumb and fifth finger together, the pain will be almost unbearable.

I listen carefully to the patient's complaint. I then feel along the thumb tendons. I start at the tip of the thumb and work toward the base. When I get to the forearm bone at the base of the thumb, I usually notice some thickening and enlargement. This is the swollen sheath. Pushing on this inflamed area causes pain. I ask the patient to cock the thumb up tightly at a ninety-degree angle. This hurts. Moving the thumb over the fifth finger is very painful. If I push on the thumb in a bent position, the patient screams. We call this a positive Finkelstein's test, and it is a sure sign of DeQuervain's tendinitis.

Are x-rays helpful? Not really. In fact, unless there has been a specific injury which preceded the onset of tendinitis, there is no need to x-ray at all.

The initial treatment: Ice the thumb twice daily for twenty minutes. Take two aspirin with each meal. I make up a thumb splint—out of orthoplast, a synthetic plastic which softens with hot water and hardens as it cools—to rest the thumb. Without the splint, daily activities keep the tendinitis acute. The ice, aspirin, and the splint are continued for one week.

If the thumb is still tender after a week, I prescribe stronger anti-inflammatory medicine orally. I leave the splint on for another week. If the pain does not disappear, I inject cortisone into the tendon sheath, *not* the tendon. The technique I use is to mix the cortisone with Novocain in one syringe. The splint is then continued for five more days, and by the end of this time the patient is usually symptom-free.

Occasionally, the tendinitis recurs. I usually try another cortisone injection. But if the tendinitis persists despite these injections, I suspect that the sheath has multiple tendons. It is not at all unusual to find three or even four tendons in one sheath.

The surgical technique I favor is a zigzag incision directly over the tendon area. There is a large skin nerve in the area which must be carefully avoided prior to finding the tendon sheath. The sheath is then found and opened with a small pair of scissors. This gives the tendons more room in which to work. It relieves the pressure, and recovery is rapid. Hand function is encouraged within five days, and complete recovery can be expected in four weeks.

Carpal Navicular Fracture

CAUSES

The word *navicular* in Greek means "canoe-shaped." One of the eight wrist bones is, in fact, shaped like a small canoe. It lies in the thumb side of the wrist. It is in the first row of the two rows of eight wrist bones, and it lies up against the end of the forearm bone, the radius.

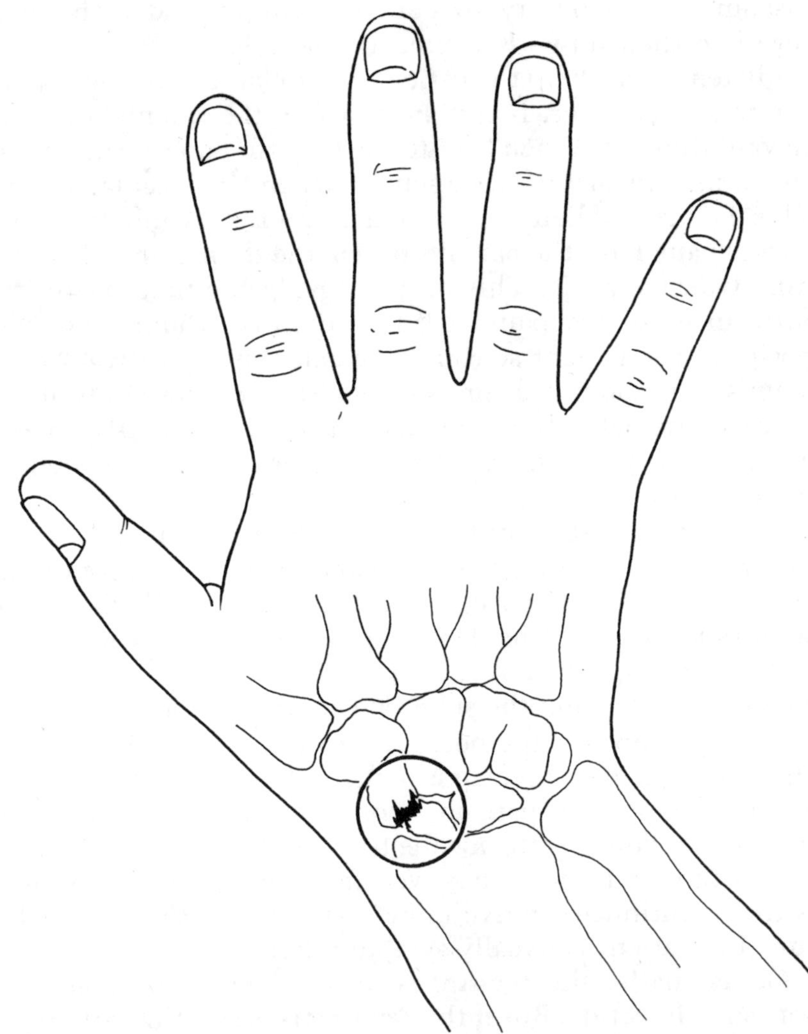

Carpal Navicular Fracture

When you fall directly on your outstretched hand, the force comes through the palm area in the wrist. This drives the navicular heavily against the radius. An athlete who loses balance and topples to the ground instinctively puts out a hand to break the fall. When this happens, there is a good chance of breaking a wrist bone. The force produced when one hits the ground is like hitting the wrist with a hammer. The bone usually breaks in the middle, the part we call the waist.

This can be a very sneaky injury. When you first fall, the thumb side of your wrist will become painful immediately. In the next four to six hours the pain lessens, only to return with more intensity later. You assume you have sprained your wrist, but you can't be sure. After several days when it does not improve, you fear a break and must seek out professional help.

ELMORE'S WRIST

In 1973, Elmore Smith of the old Buffalo Braves basketball team fractured his carpal navicular bone. Like many wrist injuries, it required an extensive series of x-rays by Dr. Steven Joyce to prove the finding.

"We treated him with ice, rest and compression," declares Ray Melchiorre, the former trainer. "When a diagnosis was finally made, the doctor built a splint for Smith. Believe it or not, Elmore, a defensive player and rebounder, was able to play effectively wearing the splint. It took eight weeks before the navicular healed completely and the splint was removed."

DIAGNOSIS AND TREATMENT

The diagnosis of a carpal navicular fracture involves a high index of suspicion. I must always consider a fracture possible when a player complains of a sore wrist after falling on it. The location of the pain is often significant. I ask where the wrist hurts. The patient points to the base of the thumb, and says the wrist has been stiff and mobility limited. The patient will also say the pain has not abated despite the passage of several days.

Examination of the wrist shows that cocking up (dorsiflexion) and bending down (palmar flexion) are both limited. This is because the broken bone releases blood into the wrist joint. The wrist is pumped up with blood, and therefore cannot move freely. Side-to-side motion will also be limited.

Occasionally, there will be a black-and-blue area directly over the thumb side of the wrist. There will always be some mild swelling in the area of the broken bone. The bone itself can be felt if you position the hand correctly. I shake hands with the patient in such a manner that the thumb is moving away from the wrist. I can now easily feel the end of the forearm bone, the radius. This is the most prominent bone on that side of the wrist.

You move your finger just over the end of the radius and the next bone in line is the navicular. Pushing on other bony prominences in the wrist area does not cause pain but pushing this area makes the patient very uneasy. This part of the wrist is called the "anatomic snuff box," because early users of snuff placed the powder here. By tensing up your thumb, you can make a small cradle in this area, bounded by the thumb tendons. It is a large enough space to hold a small amount of snuff. Tenderness in the anatomic snuff box is the key to this diagnosis.

X-rays must also be taken: front-to-back, side-to-side, and angled views. These three positions are necessary to visualize properly the wrist bones. The sneaky part of this diagnosis comes with x-ray interpretation. Often, the initial x-rays are not positive. A crack cannot be seen because the bone is so dense. You must wait for at least two weeks to determine whether or not the bone has been fractured. In two weeks time, bone disappears around the fracture site and the crack can be seen. Therefore, if the initial x-ray is negative, you still must protect the wrist and keep the player from using it until a later diagnosis is made.

I prefer a splint immobilization for protecting the wrist. My preference is to use the synthetic plastic-like material called orthoplast to fabricate this splint. The material is cut with a large scissors and then heated with a hair dryer or warm water. It is then molded around the thumb and carried across the wrist and halfway up the forearm. The splint can be held in place with Velcro straps or an Ace bandage. This immobilizes the thumb and wrist area.

Repeat x-rays are taken in two weeks. A navicular series is performed on the x-rays. This involves six to ten different views of the carpal navicular. The patient rotates his wrist on the x-ray film so that many different angles are obtained with which to view the bone. This gives better diagnostic accuracy than the original three x-rays. If no fracture line is seen with the navicular series, then the patient may return to full activity and no rehabilitation should be necessary.

If a fracture is seen, continued casting is imperative. The carpal navicular is one of the three bones in the body that is slowest to heal after fracture. This is because it has a blood supply which enters only one end of the bone. Therefore, the half of the bone away from the blood supply is cut off from its food, making the healing process very slow. Three months is the usual time that a cast has to be used to allow proper immobilization for healing of a fracture of the carpal navicular.

A cast called a *thumb spica* is utilized. This is a short-arm cast which stops below the elbow but extends down over the thumb. The tip of the thumb may be left free so that it can wiggle and continue to be useful in daily activities, such as writing. The cast may be made out of plaster, plastic, or fiberglass. All three give predictable and reliable immobilization.

I remove the patient from the cast after three months. A new navicular series is taken. In 80 percent of the cases, the bone will be healed. I can tell this on the x-ray because little lines of new bone cross the crack in the bone. This is new healing bone which has united or healed the fracture. If these

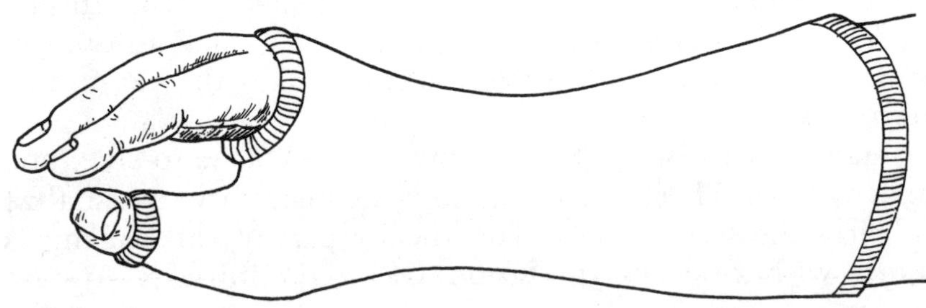

Thumb Spica Cast

bands are not seen crossing the fracture line, the patient is replaced in a cast for an additional three months. In these patients, the cast is removed once again and new x-rays are taken. In 50 percent of these cases, the fracture is still not healed.

In 10 percent of the initially injured patients, the bone just does not heal despite proper immobilization in a cast. These patients must be operated on to achieve healing of the fracture. The operation involves taking a small bone graft from the nearby forearm bone. This introduces new live bone cells across the fracture line by hollowing out a trough, the center of which is the fracture itself. Again, three months in a cast is necessary after surgery.

What happens if these fractures are ignored? Arthritis of the wrist sets in. When I was in the service at Fort Dix, New Jersey, I saw many young men from remote areas where medical care was not the best. These recruits would describe having fallen on their wrists three or four years prior to their induction in the army. They would have seen no medical personnel whatsoever, assuming they had a sprained wrist. In fact, they had a fractured carpal navicular. In two to three intervening years, they had used their wrists, despite pain. By the time I x-rayed, we could detect arthritis in the wrist joint. This led me to conclude that these fractures must be aggressively dealt with and healing of the fracture must be achieved. Otherwise, the patient is condemned to an uncertain and painful future.

JIM RICE'S FRACTURED WRIST

Jim Rice was batting against the California Angels in late June of 1980 when an inside pitch struck him on his left wrist.

"I felt his wrist at the end of the inning," relates Dr. Arthur Pappas, medical director of the Boston Red Sox. "I knew it was sensitive, but Jim wanted to continue playing."

His next time up, which was about thirty minutes after his injury, he hit a ball that just missed being a home run by about two feet in left centerfield.

After the missed homer, Dr. Pappas asked Rice if he wanted x-rays after the game. "I'm feeling fine, just a little sensitive where I was hit, Doc," Rice said.

"Later in the evening his sensitivity increased," says Dr. Pappas. "I felt it was important for him to get x-rays, which we did. That was about three hours after the injury. He was hurting more by that time, and this was not unusual.

"The x-rays revealed a fracture of the ulnar styloid, which is the tip of the ulna on the outside of the wrist. It was not a severe fracture. For a football or hockey player this would be a nothing injury. They would be playing within days, but for a power hitter like Jim Rice the fracture was in a critical spot. As he swings, he's closing down on his left wrist and rotating his left wrist over. If he's going to pull the ball, which is most of the power for right-handed hitters, he really has to be able to roll that wrist over.

"I placed his wrist in a splint initially for two days so I could observe the swelling. Also, I wanted him to ice the wrist. Once I felt that the swelling was well controlled, I then applied a short-arm cast, holding his wrist and fingers in the position that would be most beneficial for him.

"Jim was ready to return after five weeks of immobilization in a cast and use of gradual physical therapy, but I found that he still had difficulty for another six weeks.

"For a man who has really powerful forearms, it was obvious his left arm was mighty thin. I kept him on ball-squeezing exercises and exercises to maintain mobility and strength, but it still required considerably more rehabilitation."

Those eleven weeks accounted for the majority of the 1980 season. It wasn't until early September that Jim started feeling more comfortable at bat, and his record shows it. In the month of September he was named Outstanding Player of the American League. He finished the season with fury, batting over .290 and hitting twenty-five homers.

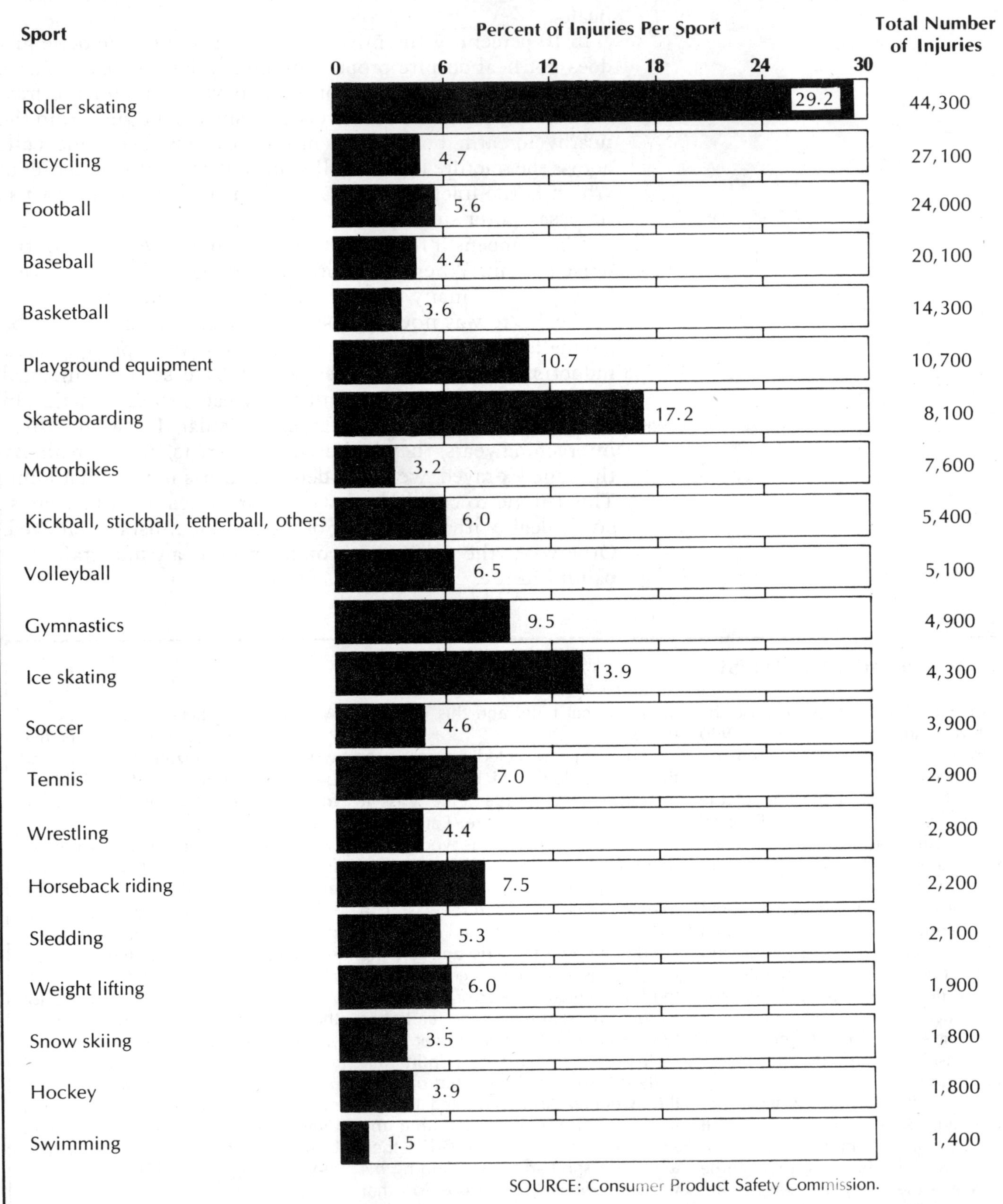

THERE ARE ALMOST 205,000 WRIST INJURIES IN THE UNITED STATES PER YEAR
1979
Sport
Percent of Injuries Per Sport
Total Number of Injuries
0
6
12
18
24
30
Roller skating
29.2
44,300
Bicycling
4.7
27,100
Football
5.6
24,000
Baseball
4.4
20,100
Basketball
3.6
14,300
Playground equipment
10.7
10,700
Skateboarding
17.2
8,100
Motorbikes
3.2
7,600
Kickball, stickball, tetherball, others
6.0
5,400
Volleyball
6.5
5,100
Gymnastics
9.5
4,900
Ice skating
13.9
4,300
Soccer
4.6
3,900
Tennis
7.0
2,900
Wrestling
4.4
2,800
Horseback riding
7.5
2,200
Sledding
5.3
2,100
Weight lifting
6.0
1,900
Snow skiing
3.5
1,800
Hockey
3.9
1,800
Swimming
1.5
1,400
SOURCE: Consumer Product Safety Commission.

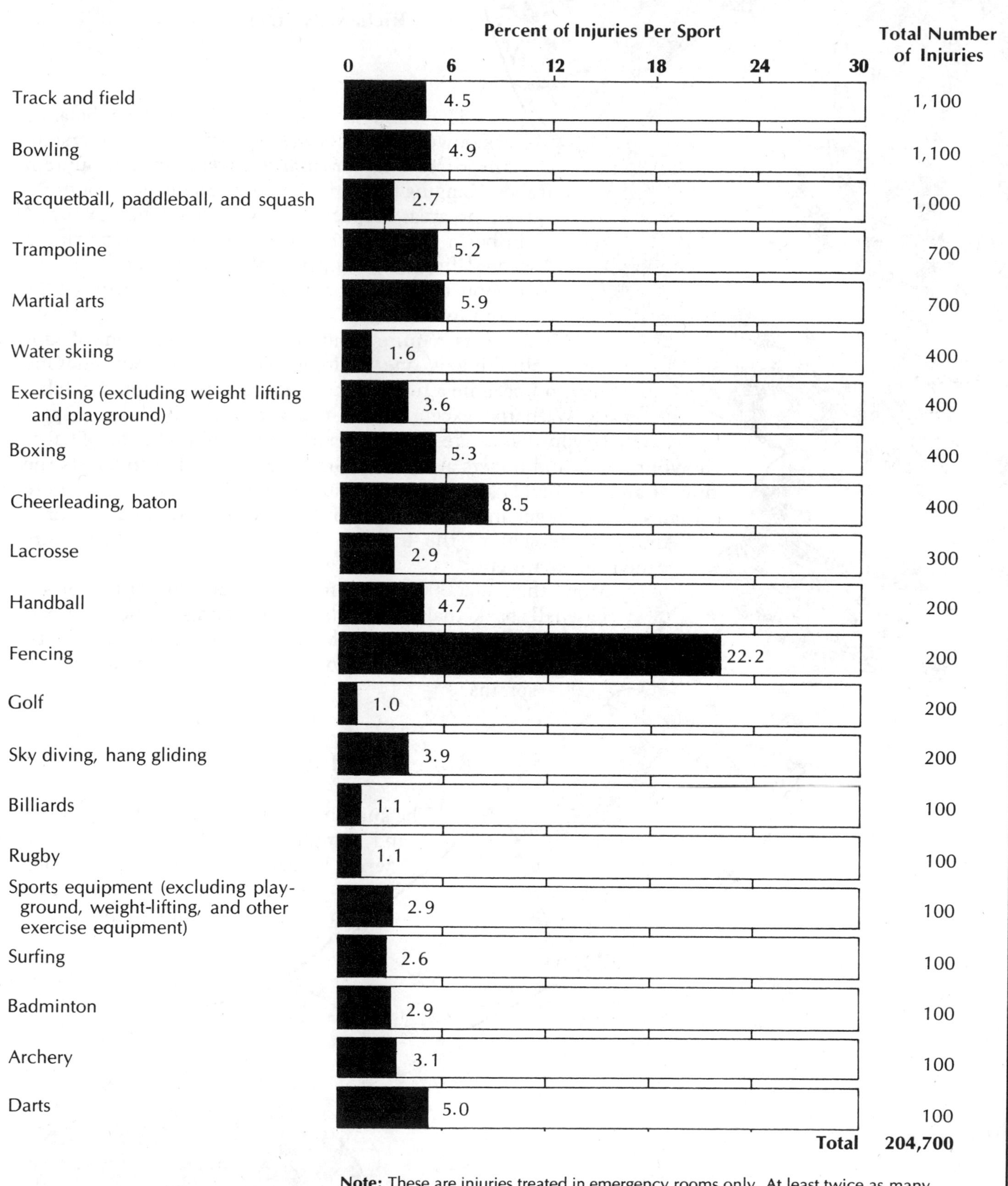

Percent of Injuries Per Sport
0
6
12
18
24
30
Total Number of Injuries
Track and field
4.5
1,100
Bowling
4.9
1,100
Racquetball, paddleball, and squash
2.7
1,000
Trampoline
5.2
700
Martial arts
5.9
700
Water skiing
1.6
400
Exercising (excluding weight lifting and playground)
3.6
400
Boxing
5.3
400
Cheerleading, baton
8.5
400
Lacrosse
2.9
300
Handball
4.7
200
Fencing
22.2
200
Golf
1.0
200
Sky diving, hang gliding
3.9
200
Billiards
1.1
100
Rugby
1.1
100
Sports equipment (excluding playground, weight-lifting, and other exercise equipment)
2.9
100
Surfing
2.6
100
Badminton
2.9
100
Archery
3.1
100
Darts
5.0
100
Total 204,700
Note: These are injuries treated in emergency rooms only. At least twice as many injuries are seen in doctors' offices.

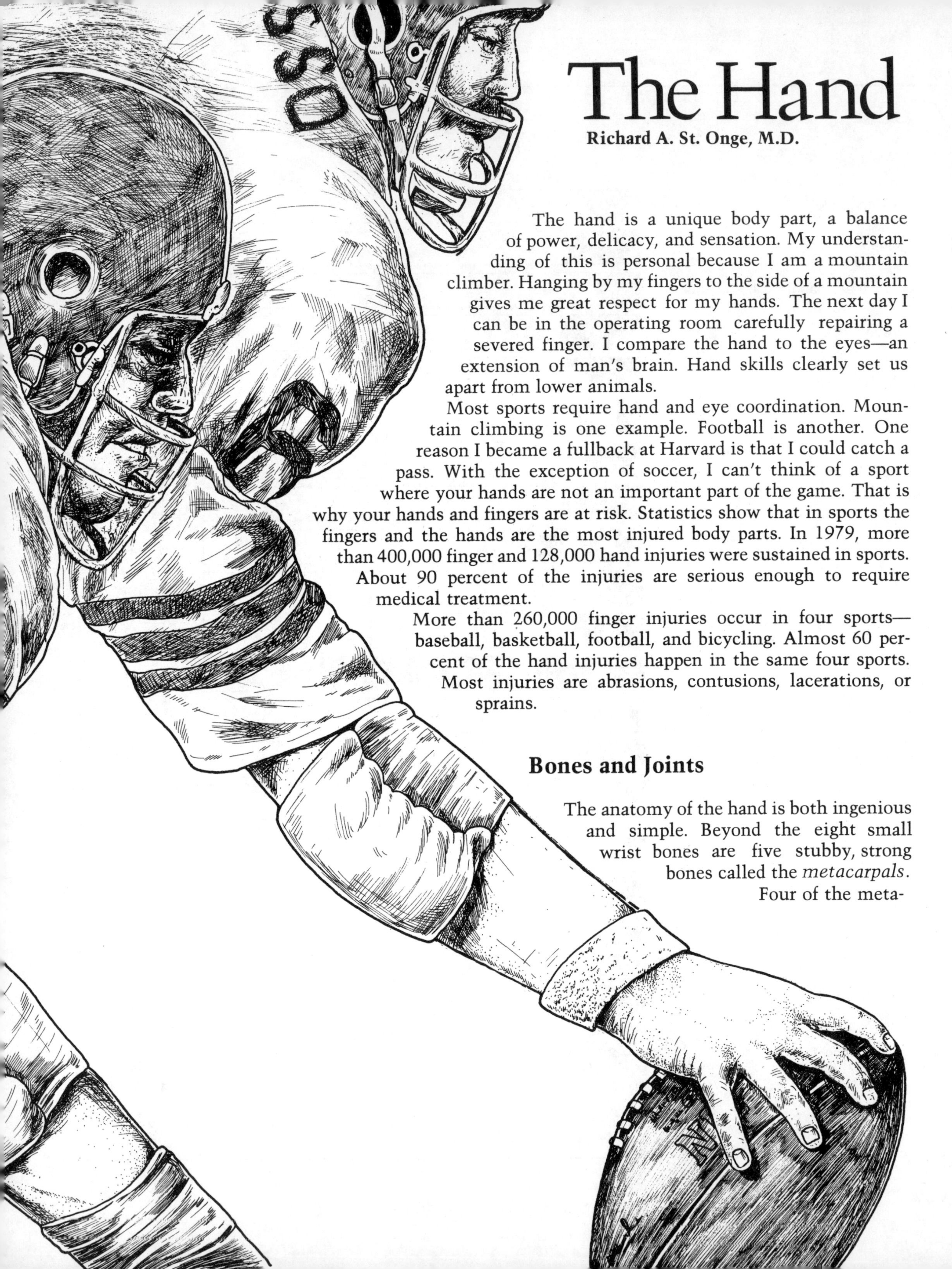

The Hand

Richard A. St. Onge, M.D.

The hand is a unique body part, a balance of power, delicacy, and sensation. My understanding of this is personal because I am a mountain climber. Hanging by my fingers to the side of a mountain gives me great respect for my hands. The next day I can be in the operating room carefully repairing a severed finger. I compare the hand to the eyes—an extension of man's brain. Hand skills clearly set us apart from lower animals.

Most sports require hand and eye coordination. Mountain climbing is one example. Football is another. One reason I became a fullback at Harvard is that I could catch a pass. With the exception of soccer, I can't think of a sport where your hands are not an important part of the game. That is why your hands and fingers are at risk. Statistics show that in sports the fingers and the hands are the most injured body parts. In 1979, more than 400,000 finger and 128,000 hand injuries were sustained in sports. About 90 percent of the injuries are serious enough to require medical treatment.

More than 260,000 finger injuries occur in four sports—baseball, basketball, football, and bicycling. Almost 60 percent of the hand injuries happen in the same four sports. Most injuries are abrasions, contusions, lacerations, or sprains.

Bones and Joints

The anatomy of the hand is both ingenious and simple. Beyond the eight small wrist bones are five stubby, strong bones called the *metacarpals*. Four of the meta-

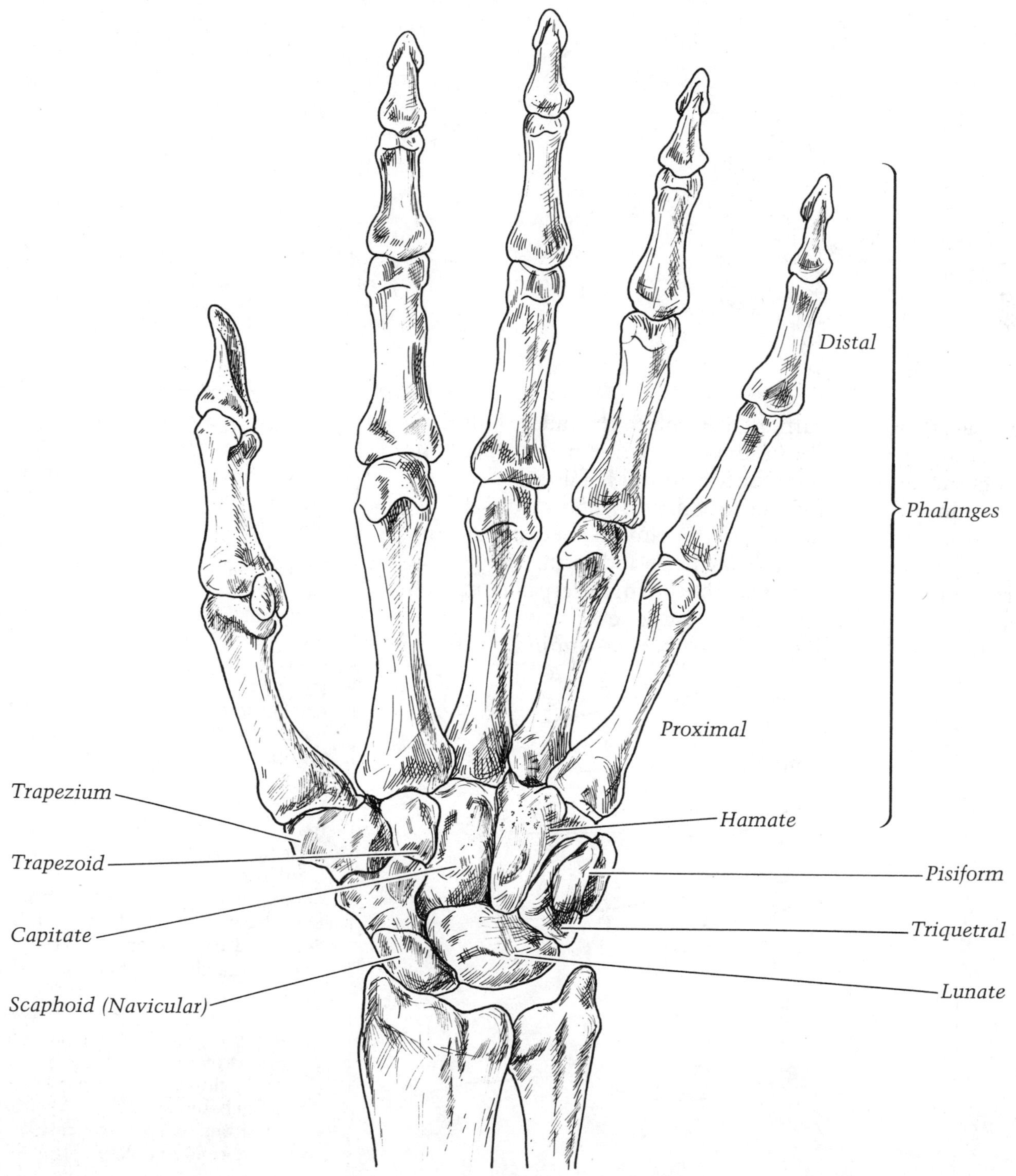

Palmar View of Left Hand

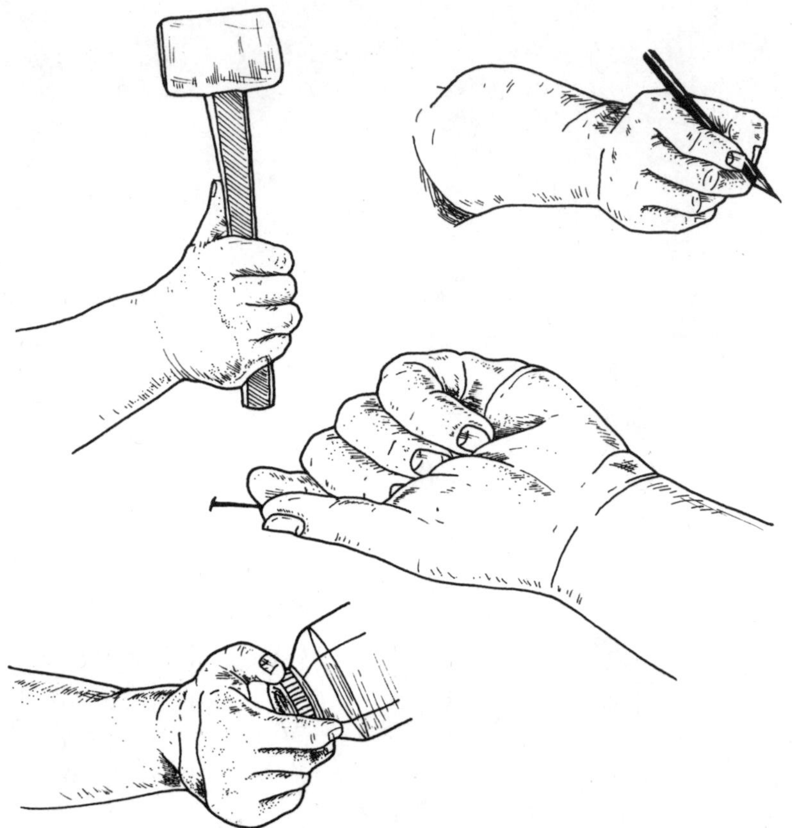

Hand Functions

carpal bones form the palm and attach to the index, long, ring, and fifth fingers. The thumb has a separate metacarpal bone, which is the shortest and stoutest in the hand.

The thumb has incredible mobility, which results from a unique joint at the bottom of the thumb metacarpal bone. This saddle-shaped joint allows back-and-forth, side-to-side, and rotational motion of the thumb metacarpal bone. The saddle itself is one of the eight wrist bones.

The thumb is composed of two bones—similar to the big

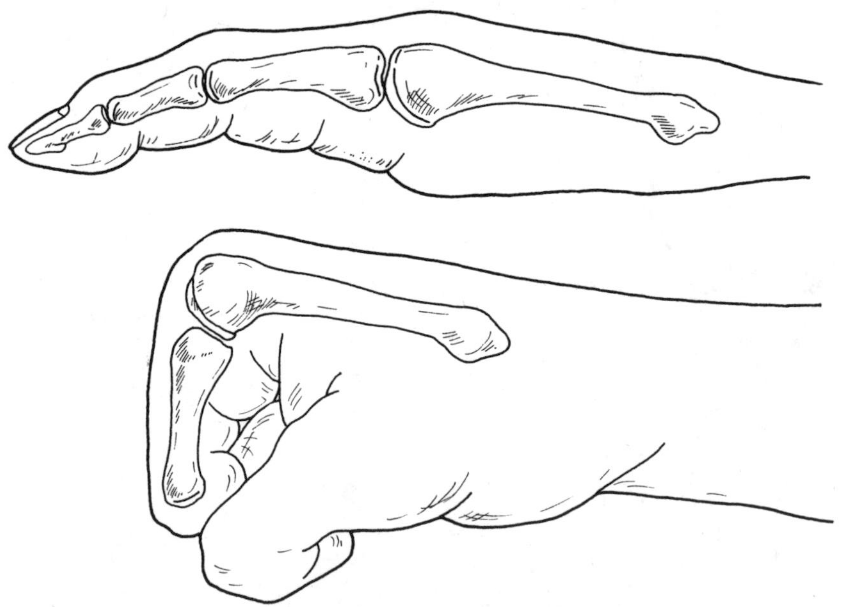

Side View of Index Finger

A CATCHER'S HAND

No athlete's hand takes more beatings than a major league catcher's. Big league pitchers can throw a baseball between 90 and 100 miles an hour. What do catchers wear in their gloves?

"I usually wear a golf glove inside my mitt, with a quarter-inch piece of foam rubber between the glove and the palm," said Carlton Fisk, all-star catcher formerly with the Boston Red Sox. "Usually after a game, my catching hand is red and swollen, but by the next day it heals.

"To minimize the injury to my hand, I try to catch the balls in the upper part of the glove. I am a one-handed catcher, and I catch the balls behind the plate, like a first baseman would catch the balls at first base. Second, I try to cradle the ball, like trying to catch an egg without breaking it.

"I have very little doubt that some of the blood vessels in my hand have been destroyed. I find that on cold days, my fingers get extremely cold when I'm behind the plate. My fingers are blue and white. The rest of my hand is normal in color."

STAPLETON'S THUMB

The first time I met Dave Stapleton was in a hospital. I asked him what was wrong with his thumb.

"Every time I hit the ball," he said, "I get a shock in my thumb."

"How is it now?" I asked.

"It feels just fine!" he answered.

Well, I told him that he shouldn't expect me to operate on his thumb if it was feeling fine and suggested that he go home. I didn't think we should risk ruining his career.

"What do you mean, go home? I just flew up here from Alabama today . . . My career is no good with this thumb," he complained.

I said that he had nothing to lose by going home and getting someone to throw baseballs at him, and that if hitting bothered him he could call me.

In three weeks, he called back.

"I need the operation," he said. "I can't play baseball."

When I opened his thumb, I found a large, thickened nerve. It was inflamed and swollen. There are a number of things I could have done. One was to trim off the scar tissue. Another was to remove the nerve completely. But Dave would lose sensation in the tip of his thumb, and that might end his baseball career. My third alternative was to move the nerve to a new position. The principle is to cover the nerve with a body padding—muscle or fat. In Dave's case, I covered the nerve with a small muscle in the hand. This operation is like moving the ulnar nerve.

Dave Stapleton's recovery is now history. In his first year in the majors, he was one of the top batters in the American League.

Lewis Millender,
Boston Red Sox consultant
and Boston-based hand surgeon

toe. The four other fingers have three bones. All the finger and thumb bones, called *phalanges*, are held together by joints. These joints, smaller than many other body joints, have a synovial lining and are enclosed by a gristle-like sleeve of tissue—the *joint capsule*. Ligaments are specialized strands of tissue in the sleeve. The ends of the bones are covered with joint surface cartilage.

Examine your own index finger. As you bend your finger down, you will see that each of the finger joints can bend almost ninety degrees. As you straighten your finger, the joints are brought back to a fully straight position. The ligaments of your knuckle joints are the strongest in the hand.

Muscles, Tendons, and Nerves

The hand is powered by muscles in the forearm and in the hand itself. The forearm muscles are connected to the hand by tendons that run across the wrist and into the hand. They are called the *extrinsic* hand muscles. The numerous small muscles that originate in the hand are called the *intrinsic* hand muscles. The two muscle groups work together to produce both the power and the delicacy of hand function. Acting together, they allow you to straighten your fingers or make a fist. The tendons that go to the fingers live inside tubes that provide lubrication and protection. The tendons on the back of the hand do not have a true sheath, but rather slide just beneath the skin. The thumb has eight separate muscles and

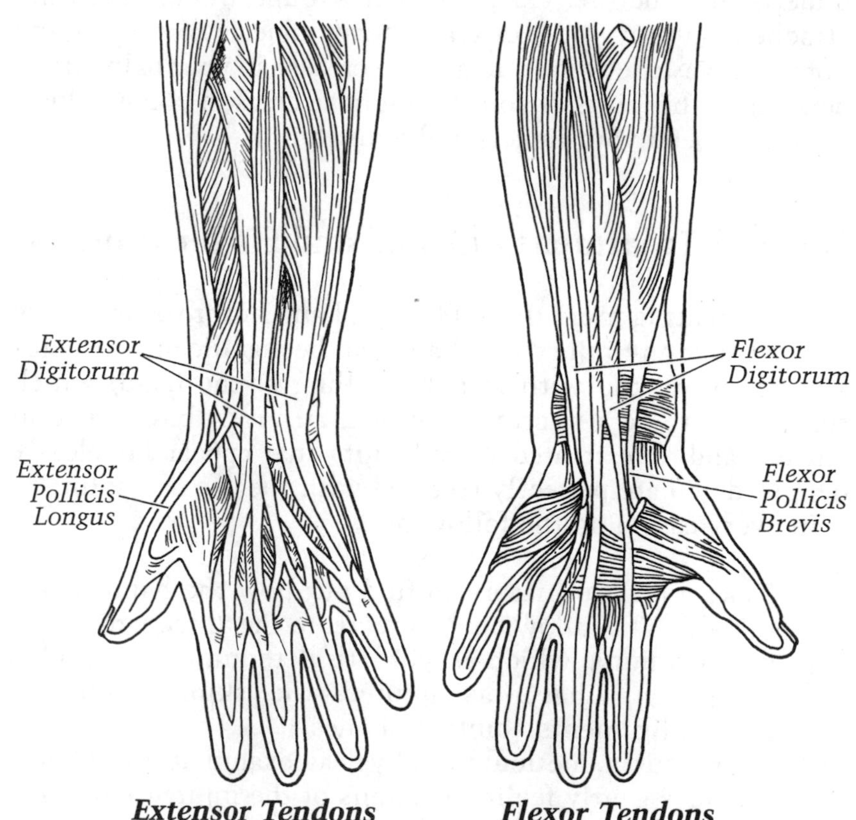

Extensor Tendons ***Flexor Tendons***

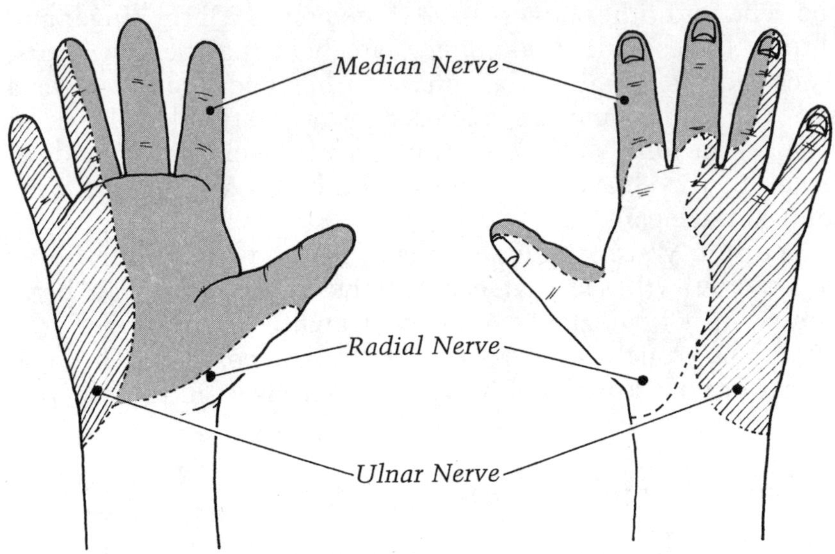

Nerve Distribution of the Hand

tendons, which make it extremely versatile.

All the hand muscles are controlled by nerves that bring messages from the brain. The same nerves carry messages to the brain from the skin. This highly specialized system is basically provided by the three major nerves of the arm—the *radial*, the *median*, and the *ulnar*. The nerves of the hand lie close to the skin and therefore are vulnerable to injury.

The skin that covers the hand is highly specialized. The skin on the back of the hand is very loose and moves in all directions without difficulty. It is not tightly attached to the underlying structure. The palm skin is tough, thick, and firmly attached, so that you can grasp heavy objects forcefully and not tear away the skin. The muscle creases in your palm allow bending of the joints. If you did not have creases, you would be unable to bend your fingers and your thumb.

General Treatment of Hand and Finger Injuries

From treating more than 3000 hand and finger patients, I've learned that any injury to a hand can be significant. Athletes have come to me with minor swelling of the hand, which turned out to be a fracture. I've seen an insect bite infect an entire hand. I've treated finger joints that appear hopelessly deranged and are perfectly restored with the use of a splint.

In every case, I do the following:

1. Take a careful history to find out how the hand was injured. This allows me to recreate the "mechanism" of the injury, which is vital to determine the hand component that was injured—the skin, tendons, nerves, ligaments, joints, or bones.
2. Carry out a meticulous physical examination. This includes gently feeling portions of the injured hand. I

A SPIDER BITE

In 1979, I was in Nepal to climb Mt. Himal Chuli. I was called by the American ambassador and asked to see an American trekker, who was in the hospital. "What's the matter with him?" I asked.

"He is scheduled to have his hand amputated," said the American ambassador. "I understand you are a hand surgeon."

"That's right. I'll be happy to take a look at him."

At the hospital I found a young Californian, who had apparently been living in the back mountains of Nepal. He told me that he had been bitten by a spider on the end of his finger. The finger had become infected and then secondarily infected. The disease was so rapid and the sanitary conditions so minimal that his whole hand had become massively infected.

The infection spread deep into the tendon spaces and the finger fascia. The Nepalese doctors had opened up all the hand areas to let the infection out. They had made longitudinal incisions. Unfortunately, this kind of incision doesn't work very well on the hand. It causes a lot of disability and scarring.

His hand was a total disaster, with draining pus and exposed flexor tendons and nerve vessels. I spent almost a whole day operating on the hand, providing skin coverage with grafts. This experience impressed on me the severity of a relatively minor injury in a primitive setting.

test sensation in the hand and fingers with a pin and a feather.

3. Ask the patient to move the injured part. If it won't move, a tendon may have been cut.
4. X-ray the hand and fingers. Hand and finger fractures are common.

The immediate treatment for all hand and finger injuries is RICE (rest, ice, compression, and elevation).

I always build a supportive splint for hand and finger injuries—even for simple strains. My reason: It is almost impossible to rest the hand and fingers. Daily use—picking up a carton of milk or carrying a briefcase—can extend the injury. The splint helps reduce the swelling.

I make the splint from gauze impregnated with plaster of paris. It is the same kind of half-cast used for wrist sprains. The plaster of paris is covered with a soft cotton padding dressing called webril. The plaster of paris is dipped into water, and the splint is applied to the palm of the hand. I can design the splint to extend beyond the fingertips or well up on the forearm. The splint is held in place with a three-inch Ace bandage. Before I apply the splint, I cover any cuts with a gauze pad and antiseptic ointment. As the splint hardens, I bend the wrist back approximately fifteen degrees. This "rest" position is the most comfortable.

If the splint is to be worn for more than two days, it is important to have the fingers bent slightly at all of their joints with the splint in place. Never splint the fingers out straight for more than a few hours, because they quickly stiffen in that position.

Finger Sprains and Dislocations

CAUSES

The bones of the finger joints are held together by a gristle-like sleeve called the joint capsule. Specialized strips within the capsule are ligaments. Injuries to these ligaments are similar to ligament injuries in other joints.

When the finger ligament is stretched beyond its strength, it tears, or sprains. In sports circles, finger sprains are called jammed fingers. They are common in football, baseball, mountain climbing, and skiing. I sprained my hand ligaments in football by catching hard passes.

Ligament sprains may be classified as follows: a grade I sprain is a 25 percent tearing of the ligament; a grade II sprain is a 25 to 75 percent tearing of the ligament; and a grade III sprain is a complete tearing of the ligament.

Sometimes a finger ligament is so strong and tough that the force of an injury does not rip the ligament. Instead, a small piece of bone is pulled off the finger bone at the end of the ligament. Called an *avulsion fracture*, it is important because it represents a complete ligamentous disruption, or a third-degree sprain.

DIAGNOSIS AND TREATMENT

All sprains cause swelling of the joint, restricted motion of the joint, and black-and-blue discoloration. The worse the injury, the faster the symptoms develop. As the swelling increases, stiffness sets in.

If you think you have sprained your finger, start RICE (rest, ice, compression, elevation) immediately. Do not wait for the swelling to start.

I always like to feel the sprained joint. I first ask you to move the joint to find out how much motion has been lost. Next, I take the two bones that form the joint in my hands. I gently try to pry the joint apart. If the two bones move easily away from each other, I know the ligaments have been completely torn (a third-degree sprain). The ultimate third-degree sprain is a complete dislocation of the joint, or a complete separation of the bones. A dislocation is the most severe form of third-degree sprain. You have to tear the ligaments completely to dislocate the joint. Finger injuries should always be x-rayed. X-rays tell me if you have a fracture or an avulsion in addition to the sprain. At least three x-rays of the finger should be taken for an accurate diagnosis.

I place first- and second-degree sprains in aluminum splints. They should be elevated for the first twenty-four hours of injury. Ice the sprain for twenty minutes, three times a day, to help minimize swelling. After the first forty-eight hours, start heat treatments. Soak your finger in hot water. Within five days, start gentle motion exercises of the injured joint and continue for five days.

You can return to sports that do not require precision use of the hand within a week. Tape the injured finger to the normal finger next to it. The tape should be applied to the flat portion of the finger bone and not over the joints themselves. I call it the "human splint." The splint should be worn for two to three weeks.

Swelling from first- and second-degree sprains can last for many weeks. I sprained the ring finger on my left hand in my senior year in college football. The joint was so swelled that I had to buy an oversized wedding ring when I was married the following June. It took two years for the joint to slowly return to normal size. I had to get the ring refitted. My point is that it takes a long time for finger-joint swelling to recede.

A third-degree sprain means that the ligament has been torn severely. As in other joints in the body, it is important that finger ligaments return to their normal length so that you can use your fingers properly. This will often require surgical intervention.

A common third-degree sprain is called *gamekeeper's thumb*. It is a ligament tear at the base of the thumb, where the thumb meets the metacarpal bone. Gamekeeper's thumb,

FINGER DISLOCATION

One easy way to dislocate your finger is to catch a football thrown by a strong quarterback. In 1980, Randy Burke dislocated his finger trying to catch a bullet pass from quarterback Bert Jones of the Baltimore Colts.

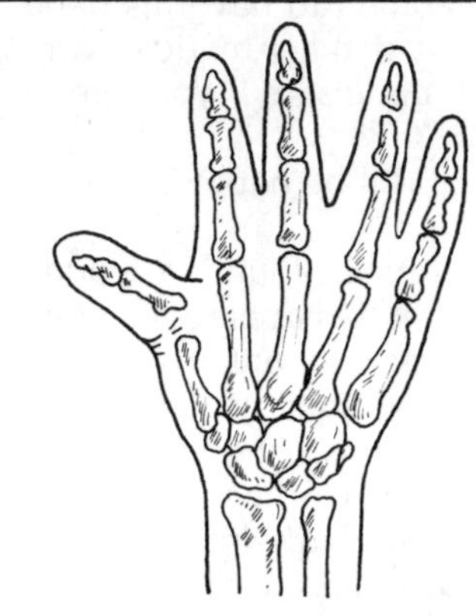

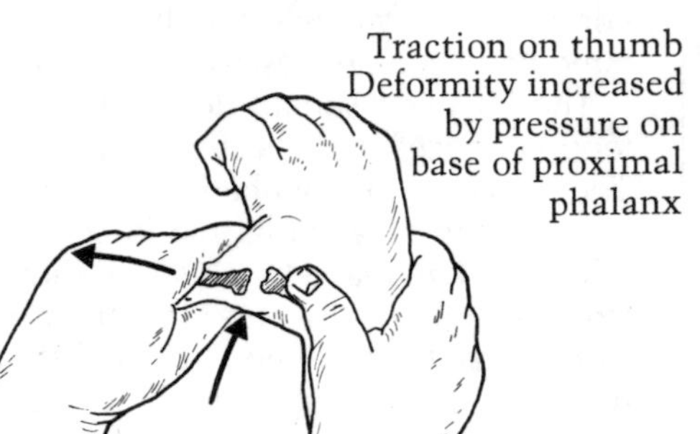

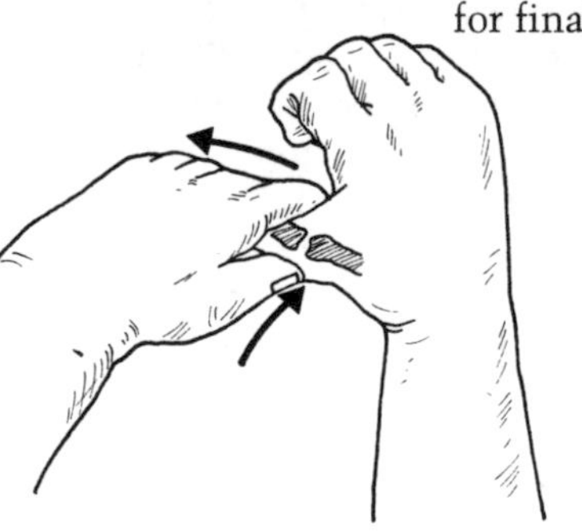

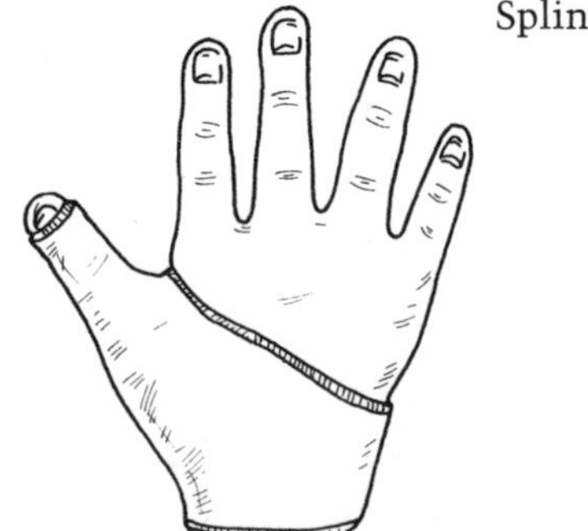

Treatment for Thumb Dislocation

TIANT'S FINGER

"In 1978, I dislocated my finger in spring training against the Detroit Tigers," said Luis Tiant, who pitched for the New York Yankees in 1979 and 1980. "I pitched and . . . boom, the ball came back like a cannon. To protect myself, I put up my bare hand. What a mistake! The ball caught the index finger of my pitching hand. One look at the finger and I thought that my career was in big trouble."

Dr. Arthur Pappas, the medical director of the Red Sox, described Tiant's injury medically. "The index finger of the pitching hand was obviously deformed. From the first joint, the proximal interphalangeal joint, the finger was going off at a forty-five-degree angle. If you looked at it from the side, it was z-shaped.

"Luis, usually a very jolly fellow, had a long face. I could see from his eyes that he was concerned. The index finger is one of the main fingers controlling the ball. Without a good index finger, his career was in jeopardy.

"I drove him to Winter Haven Hospital for x-rays, which revealed a finger dislocation, not a fracture. The ligaments holding the first finger joint together had been torn apart. To relocate the finger, I used a local anesthetic. I said to Luis, 'I am going to inject you with a very small needle. It will feel like a mosquito bite.' When I injected the anesthetic, he told me that it stung more like a bumblebee than a mosquito. When the finger began to numb, I was able to set it straight. I put the finger in a flexed position with a splint.

"Luis was lucky. Often, a joint with that amount of disruption can't be straightened without surgery. The finger was gradually restored to normal strength, motion, and stability with a series of exercises with rubber bands and balls to squeeze. Also, I recommended warm, moist heat for the finger.

"It took the finger about four to six weeks to heal. Luis tells me that he still has some sensitivity in the finger joint, but, fortunately, he has been able to pitch without major discomfort."

a common ski injury, happens when you fall with your ski pole handle between your thumb and index finger. The fall forces the thumb away from the index finger and rips the ligament that is designed to restrict this motion. If it is a third-degree sprain, the thumb moves freely away from the thumb metacarpal without resistance. The joint is completely disrupted. Surgery is necessary to repair the damage.

When I was studying hand surgery in Scotland, I learned how this sprain got its name. In the nineteenth century in Scotland, gamekeepers on large game reserves caught rabbits in their snares. To put the "small wee beasties" out of their misery, they snapped their necks in guillotine fashion between the thumb and the index finger. This caused outward stress on the thumb and ruptured the thumb ligament.

The object of the operation on gamekeeper's thumb is to reattach the ligament to the thumb bone. Following surgery, the finger or thumb is immobilized in a cast for six weeks.

If the third-degree finger injury is not treated initially, the finger and thumb become chronically unstable. At a later date a reconstructive surgical procedure is required, in which a new ligament is created from a forearm tendon. This, however, is a much more difficult operation than early ligament repair, and the results are less perfect.

Therapy after finger sprains should include warm water soaks three times daily. When the hand is in the water, the fingers should be moved in unison in an attempt to regain full motion in the injured joint. Spring-type hand squeezers should also be used to regain strength in the hand. A ball of Silly Putty can be substituted for the hand squeezer. The motion and strength exercises should be continued until 95 percent of motion and 95 percent of strength have returned to the hand. Even after the operation, it is not necessary to sit out from sports activities until the 95 percent mark is achieved. It depends on the sport and your position. A quarterback will have to sit out longer with a finger sprain on his throwing hand than would a defensive linebacker. Soccer players can return to duty sooner than baseball players. Finger problems are individualized problems.

What about finger dislocations? They are extreme examples of complete ligament tears. Because the finger joint is crooked, they are easy to diagnose. I always take x-rays of dislocations to rule out an accompanying fracture. As soon as possible, I freeze the finger by injecting Novocain into the base with a very small needle. Once the finger is numb, I can gently pop the joint back into its proper position. In medical terms, this is called reducing the dislocation.

Once the joint is straight, I always x-ray the finger again to be sure it is lined up properly. Sometimes the finger bones do not align correctly. One cause of misalignment is a piece of torn ligament stuck between the joint. This condition often requires surgery.

Ice the dislocation for twenty minutes, three times a day,

for two days. The splint compresses the dislocation. The ice and the splint minimize the swelling. The splint is left on for three weeks. Then start warm finger baths and finger exercises. Three to six months may be required for full motion to return to a finger joint that has been dislocated.

Hand Contusions

CAUSES

Even though the hand is a fine, delicate instrument of precision movement, it is often used in sports as a battering ram or hammer. I used it that way in football. It is an instrument of violence in boxing, karate, and other contact sports. The result is that the hand commonly receives severe bruises and contusions.

The contusions occur when your hand strikes a blunt object; usually the back of the hand suffers. The force of the blow breaks blood vessels, and blood pours forth beneath the skin, forming a large black-and-blue (ecchymotic) area.

DIAGNOSIS AND TREATMENT

If you hit your hand hard, the chances are that you will develop a hand contusion. Start RICE to reduce the swelling and internal bleeding. Sometimes a plaster splint helps to immobilize the hand. Compression is achieved with an Ace bandage.

The ice, elevation, and splint should be continued for the first forty-eight hours. Then start range-of-motion exercises. I tell my patients to move their fingers in lukewarm water for fifteen minutes, three times daily. This encourages circulation, which helps clear away the old blood and brings nutrients to the contusion.

Because the hand joints have a tendency to swell and become stiff, the simple range-of-motion exercises are important.

Finger Fractures

CAUSES

Most finger fractures happen when fingers are hit hard—with a bat, ball, baseball, or hockey stick.

If you fracture your finger, you will know it immediately. The onset of pain is quick and sudden. It will be no secret.

DIAGNOSIS AND TREATMENT

Swelling starts within two hours. A black-and-blue hue, from internal bleeding—often appears within three hours of injury. You will be unable to use the hand effectively because of pain and swelling. I always x-ray to confirm the diagnosis. Two views of the finger should be taken—ninety degrees apart. This ensures that no fractures are missed. If there is any

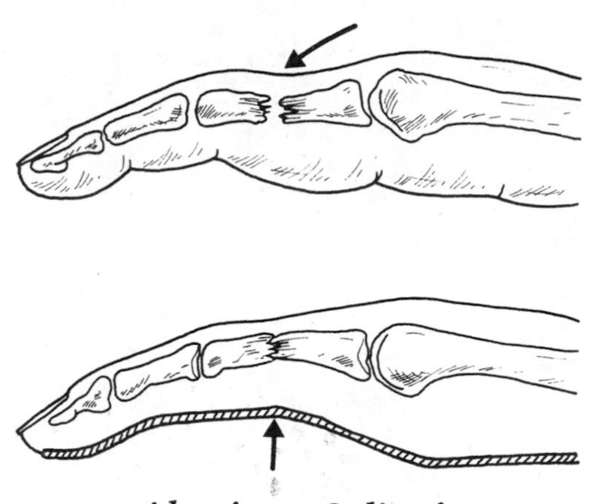
Aluminum Splint for Fractured Finger

doubt about the diagnosis, I order additional x-rays.

A finger fracture usually can easily be managed with an aluminum finger splint. These splints come in long strips, with a soft, foam rubber backing, and can be cut to the appropriate length. The foam rubber goes against the skin of the finger, usually on the palm. I bend the splint so that the fingers are immobilized, with each finger joint bent approximately thirty degrees toward the palm. This is a relaxed, comfortable position for a finger. Three weeks are required for healing finger fractures.

If the fracture cannot be reduced or straightened, that usually means surgery. I straighten the finger with a small metal wire. Healing time is usually three weeks; the wire is then removed.

As with all joint fractures, it is important to restore the normal architecture of the joint. Surgery is often necessary. An irregular joint produces arthritis at an early age and limits the function of the hand.

Boxer's Fractures

CAUSES

The most common fracture of the hand occurs at the base of the little finger—at the far end of the fifth metacarpal bone. This fracture is known as a *boxer's fracture* because it is caused by striking the fifth knuckle against a "hard" head or other unyielding object. Ironically, I have treated very few boxers with this fracture because they wear padded gloves. Boxer's fractures are more often seen in hockey players who drop their gloves to fight.

DIAGNOSIS AND TREATMENT

Because the bone usually fractures toward its far end—near the knuckle—that is where I look for this injury. The fracture often causes the knuckle to misalign with the others. It slides down toward the palm. The fracture often causes a great deal of swelling and pain, and sometimes the area becomes black and blue. You will not be able to make a fist because of the pain. You will be unable to move the fifth finger effectively. The sensation of the finger should not be disturbed; x-rays clearly demonstrate the fracture.

I straighten the bone. In medical terms, this is called a closed reduction of the fracture. Once the bone is straight, I design a finger splint, incorporated into a short-arm cast. It is important to maintain the fractured bone as straight as possible. If the knuckle heals tipped down into the palm, it can create residual pain. The bone sticks into the palm itself and acts like an irritating thorn.

The bone does not have to heal perfectly straight to give an excellent functional result, because the fourth and fifth metacarpal bones are flexible. Examine your hands and wiggle the

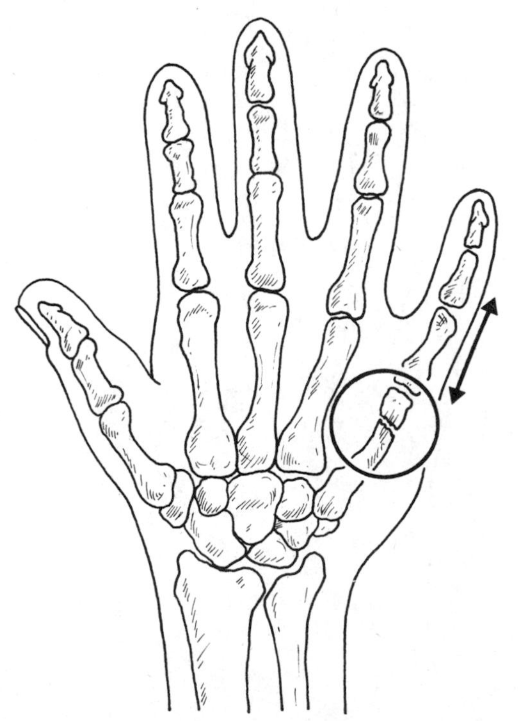
Boxer's Fracture

metacarpal bones and knuckles of the fourth and fifth fingers. See how much more mobile they are than your index finger. Therefore, if there is a slight deformity of the fifth metacarpal bone, its extreme mobility more than compensates for the deformity, thus providing a margin for error.

You must wear the splint for six weeks—the normal healing time. After I remove the splint, you start finger exercises. If the finger is especially stiff, start in water. The exercises are simple. Rotate the finger in circles for a few minutes in each direction. Bend the finger back and forth thirty times. Rest—then do two more sets.

Bennett's Fractures

CAUSES

A fracture at the base of the thumb metacarpal is called a *Bennett's fracture*. It is common in basketball and skiing. Named after E. H. Bennett, the Irish surgeon who first described it, the fracture occurs when the small, powerful muscles of the thumb tear a triangular fragment off the thumb bone or when the thumb is forced back powerfully.

DIAGNOSIS AND TREATMENT

When a patient tells me, "I hit my thumb against a blunt object," I suspect Bennett's fracture. Usually any movement of the thumb produces excruciating pain.

With this fracture there is always swelling and discoloration at the very bottom of the thumb metacarpal, about one inch below the wrist joint. It is where your thumb bends at the bottom. X-rays reveal that the thumb metacarpal bone is broken at its bottom end and that a triangular-shaped piece has split away from the main body of the bone.

Because of the strength of the thumb muscles, it is very hard to reduce, or straighten, this fracture. When I can reduce the fracture, I always build a thumb spica, a cast especially made to support the thumb.

Often I have to straighten the thumb surgically and reattach the broken piece of bone. I make the incision at the fracture site. The main bone is realigned with the triangular fracture fragment. A small steel wire is then placed across the two bone pieces to hold them together. The thumb is immobilized in a thumb spica cast, which is left on for six weeks. At the end of that time the metal wire is removed from the fracture area, and physical therapy is started. Within three months of the fracture, full use of the hand is restored.

Therapy after this fracture has two goals—to return motion to the stiffened joints and to strengthen the thumb muscles. Soak your hand in warm water for fifteen minutes three times a day. Move your thumb in all directions in the water. Try to touch your thumb tip to the tip of your little finger. Do this for seven days.

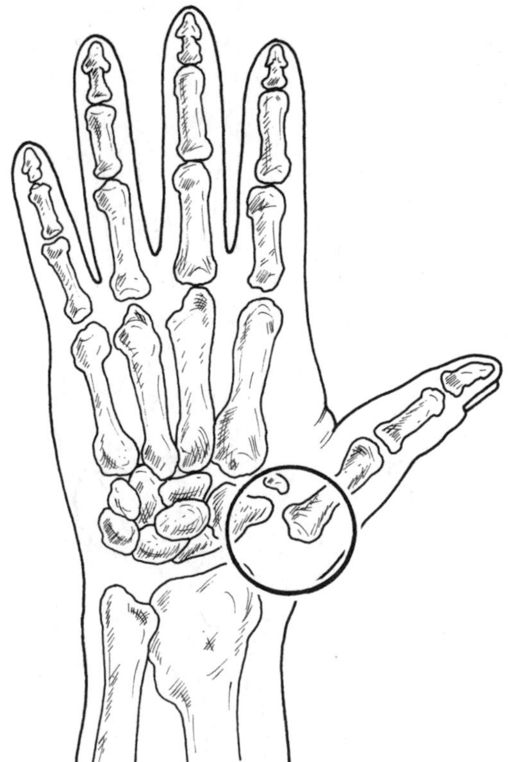

Bennett's Fracture

Get a spring-type hand squeezer, which all sporting goods stores have. Squeeze it as many times as you can, three times a day, to build hand strength. A ball of Silly Putty is also a good hand strengthener; squeeze it until fatigue sets in.

Tendons of the Hand

The hand tendons are extensions of the muscles in the forearm and the hand. More than fifteen tendons extend from the forearm muscles to the hand. One set of tendons bends the fingers; another straightens them.

The tendon arrangement in the hand is complex. On the palm side, each finger has two sets of tendons. One set allows you to bend the fingertip joint. The other set allows you to bend the middle joint—the joint between the tip and the knuckle. The two tendon systems on the palm give both superb power (a strong grip) and fine control. The palm tendons are enclosed in sheaths, or tubes, which are lined with

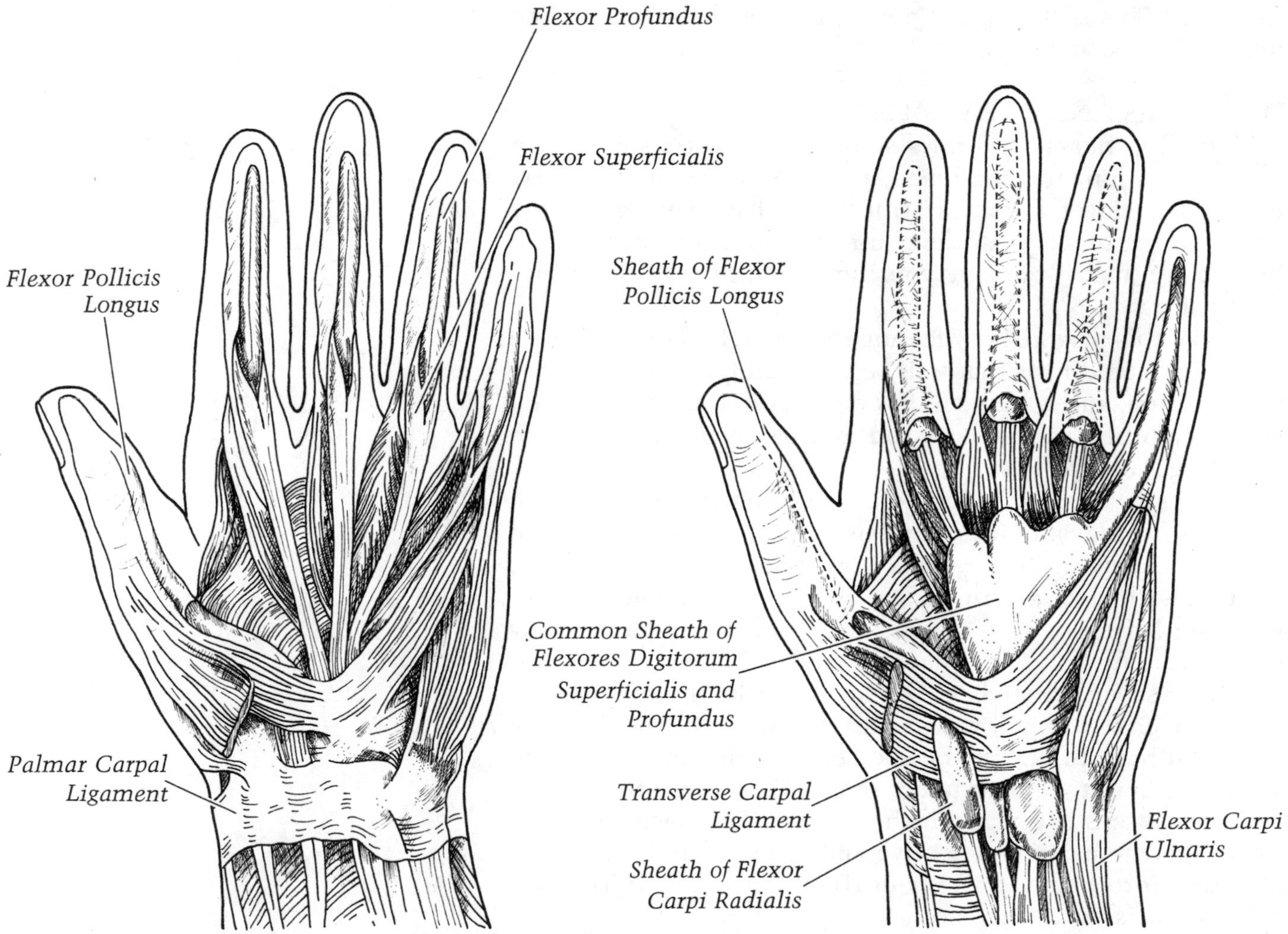

Flexor Tendons in the Palm and Fingers

Synovial Sheaths of Flexor Tendons

synovial tissue. The tissue produces fluid to lubricate and feed the tendon. The palm tendons are called the *hand flexor tendons*.

The tendons on the back of your hand, which straighten the fingers, are called the *hand extensor tendons*. They are far less numerous and are weaker because their function does not require strength. These tendons do not have sheaths; they slide in the fatty tissue under the skin.

Flexor Tendinitis of the Hand

CAUSES

Because the hand flexor tendons are used constantly during sports activities, they absorb a great deal of wear and tear. It is not uncommon for these tendons, which direct the index and long fingers, to become irritated. The irritation is called *flexor tendinitis*. When the irritation sets in, the lining of the tendon tube produces extra lubricating fluid, which causes swelling of the tube and a full feeling in the tendon area. The athlete will notice this as stiffness and difficulty in bending the finger into the palm. This is a common problem for pitchers, hockey and tennis players, and golfers.

DIAGNOSIS AND TREATMENT

Normally, flexor tendinitis starts with a little soreness in the index or long finger after a game. Over a five- to seven-day period, the finger becomes stiffer. You have pain along the course of the tendon—in the palm and into your finger. Finally, you are unable to bring your finger all the way down into your palm.

I make this diagnosis by feeling the hand. I ask you to make a fist. With tendinitis that will be very painful. I feel along the tendon itself, and it feels full. If I push hard, you will feel pain. If the tendinitis is very bad, you will be unable to straighten your finger completely.

The best treatment is rest; an aluminum splint is used for three to five days. During that time the extra synovial fluid usually disappears. I prescribe anti-inflammatory medicine, starting with eight aspirins a day for five days. If your tendinitis is not better, I switch to a stronger oral anti-inflammatory medicine.

If, after seven to ten days, the tendinitis has not completely healed, I give you a cortisone injection. This can be tricky because I have to put the needle between the tendon and the sheath; I do not want to inject the cortisone into the tendon itself because it can weaken the tendon. The cortisone takes forty-eight hours to be effective and generally five days for a full cure. Fortunately, only about 10 percent of tendinitis cases require cortisone injections.

Tendon Laceration

A tendon laceration is a severe injury that occurs commonly in industrial accidents. Most often a sharp object, such as a knife, cuts the tendon. Fortunately, sharp objects are seldom used in sporting activities.

What makes these injuries difficult to treat is the complexity of the tendon arrangement in the palm. If a tendon is severed and has to be stitched together, scar tissue inevitably results. Because the tendon lives in a sheath, it will not slide properly. Second, it can shrink in healing. I have operated on only a few athletes with this problem.

Tendon Avulsion Injury

CAUSES

Once in a while the finger tendon rips away from the finger bone, taking a piece of bone. This is called a *tendon avulsion injury*. In baseball the most common tendon avulsion injury is *mallet finger*. When you are hit on the end of the finger bone with a hard ball, the finger is driven toward the palm. The tendon that attaches near the fingernail completely tears away. The problem with a mallet finger and other finger avulsions is that you lose one of the guidewires in a two-guidewire system. One side of the joint has a strong tendon pulling, while the opposing partner has been completely stripped away. The result is that you may be able to bend the finger but not straighten it, or vice versa.

A tendon avulsion injury can occur on the bottom of the last finger joint. This occurs mainly in football players and is called a *jersey finger*. When defensive players try to tackle ball carriers by grabbing their jerseys, there is a sudden yank on the end of the finger that pulls the flexor tendon away from the bone. In this case it is impossible for the athlete to bend the tip of the finger down because the tendon that ordinarily supplies this power has been pulled away. The player can straighten the joint, however, without difficulty.

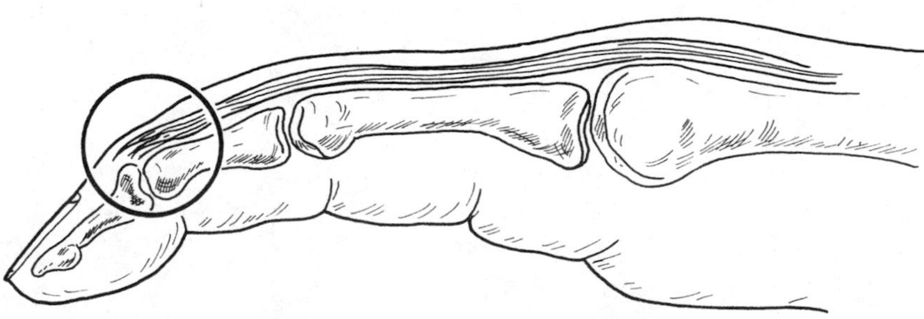

Baseball Finger

DIAGNOSIS AND TREATMENT

"Please bend and straighten your finger," I say. Normally, if you have finger tendon avulsion injury, you can bend or straighten the finger, but you cannot do both.

I always take x-rays to determine if, in fact, a piece of bone has pulled away or if just a tendon has ripped.

The treatment for mallet finger is to apply a short aluminum splint to support and straighten the joint. The splint should not extend to the next finger joint when you are treating finger injuries. I do not immobilize normal joints. The splint should be worn for two to three months. In 95 percent of the cases, the splint does the job.

If, after three months, the finger still droops, surgery is indicated to restitch the tendon to the bone. After the operation you wear a splint for three weeks. Then, start finger exercises in a bowl of warm water. Flex the finger thirty times and rest. Do two more sets daily.

The treatment for jersey finger almost always requires surgery. The reason is that the tendon on the palm side of the finger is more powerful and tends to recoil into the finger. Therefore, simply applying a splint will not bring the torn end of the tendon into proximity with the bone surface, where it belongs. The surgery opens the palm side of the finger so that the tendon can be located. The tendon is then brought back down into the end finger bone and attached with a special wire stitch. The finger is encased in a splint for at least four weeks. The wire stitch is removed under local anesthesia. The tendon is allowed two more weeks to heal before physical therapy is started.

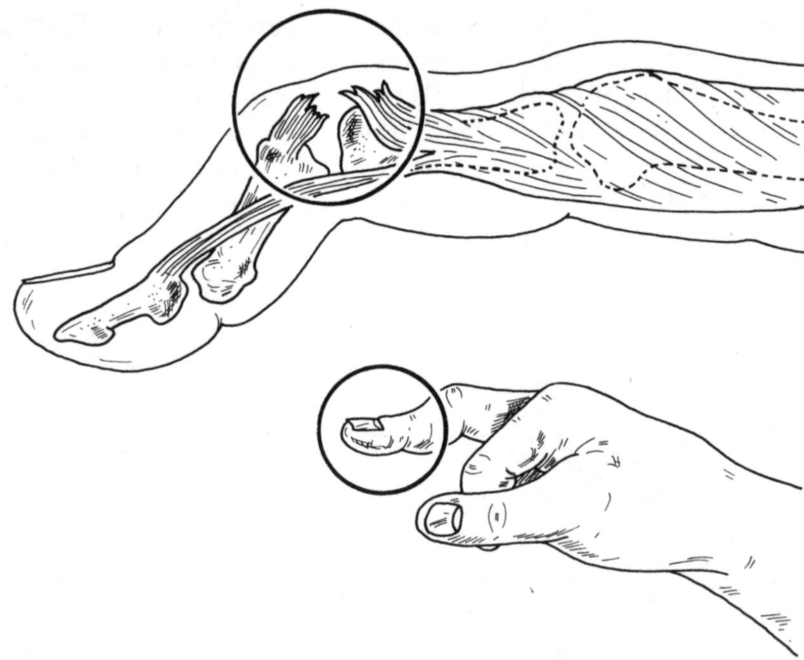

Boutonniére Deformity

Another tendon avulsion injury common in sports is called a *boutonniére deformity*. It occurs at the proximal interphalangeal joint—the next joint out from the knuckle. This, too, involves a direct blow to the top surface of the finger, which suddenly drives the joint down into a bent position. The tendon on the top surface of the joint is ripped away from the middle finger bone. You will not be able to straighten this joint normally; it remains in a permanently bent position.

The diagnosis is made by the position of the finger. You can see that the first finger joint is cocked up, and it is impossible for you to straighten your joint fully. You can, however, bend the finger into the palm.

A finger splint that holds the middle joint straight doesn't often work because the tendon will not heal directly back to the bone. Instead, a long strip of scar tissue results, which leads to permanent weakening of the joint.

For this reason surgery is often indicated to stitch the tendon directly back to the bone where it was originally inserted. The operated finger is supported in a splint for four weeks, and then physical therapy can begin. Use warm water exercises and a hand squeezer to regain motion and strength in the finger.

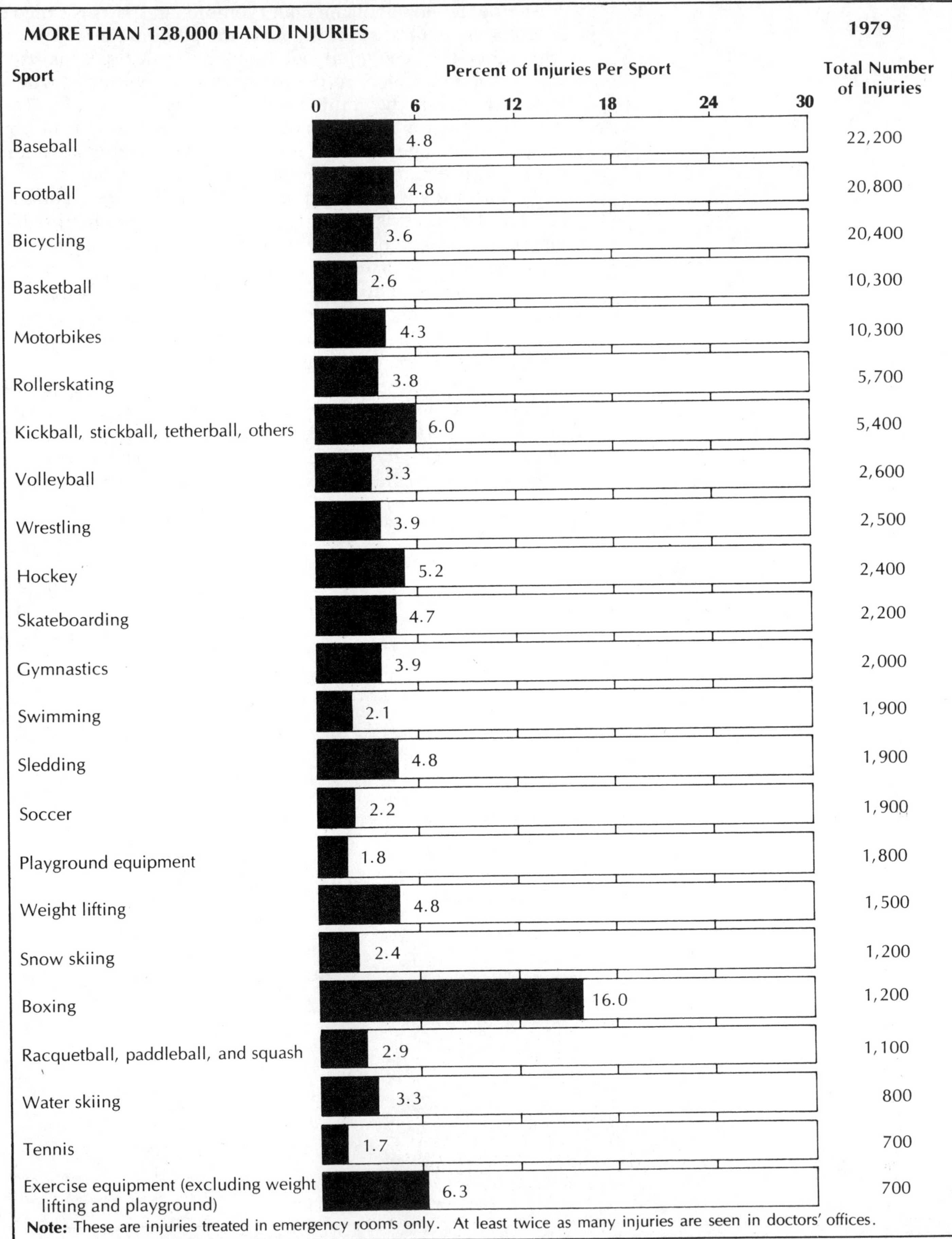

Note: These are injuries treated in emergency rooms only. At least twice as many injuries are seen in doctors' offices.

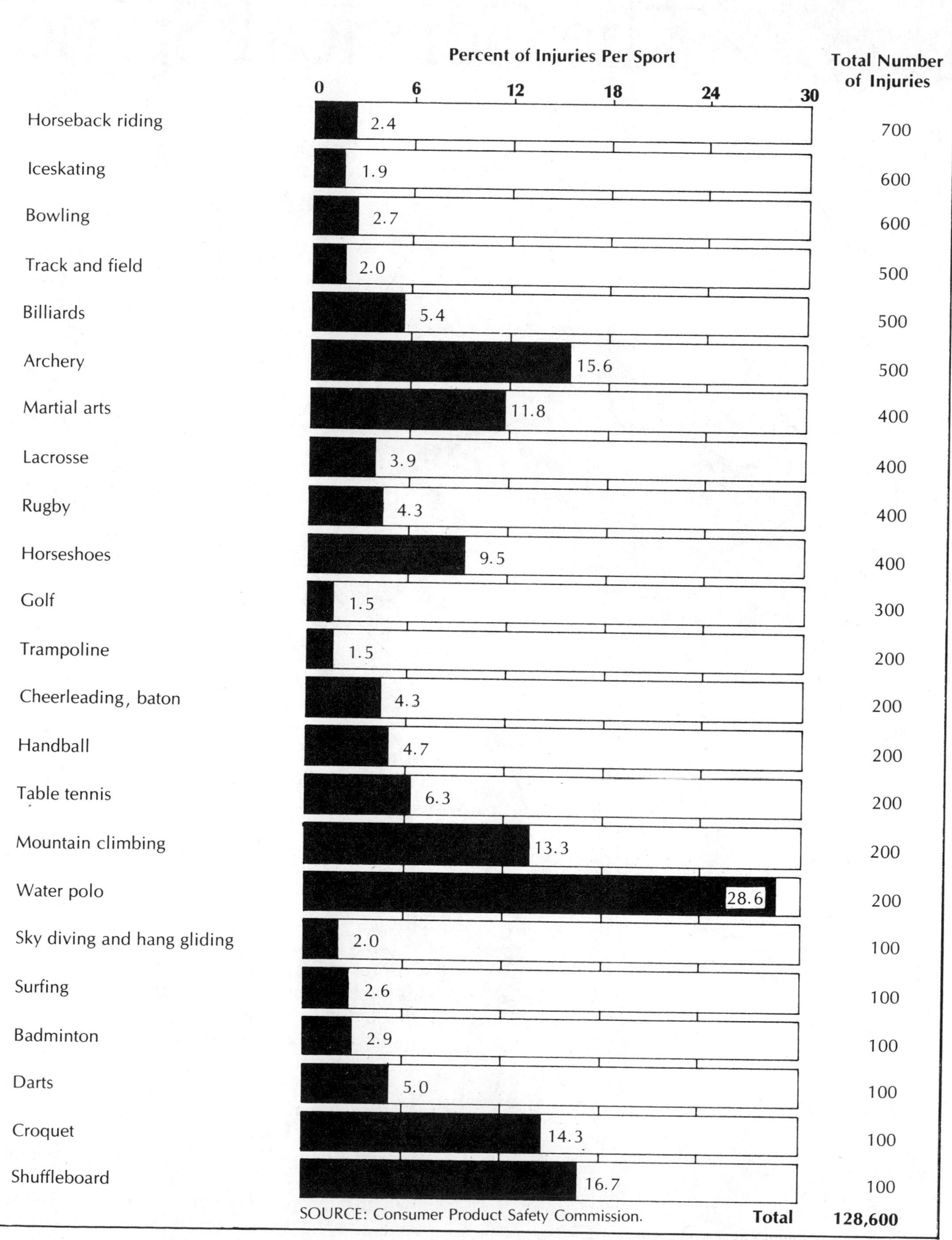

SOURCE: Consumer Product Safety Commission.

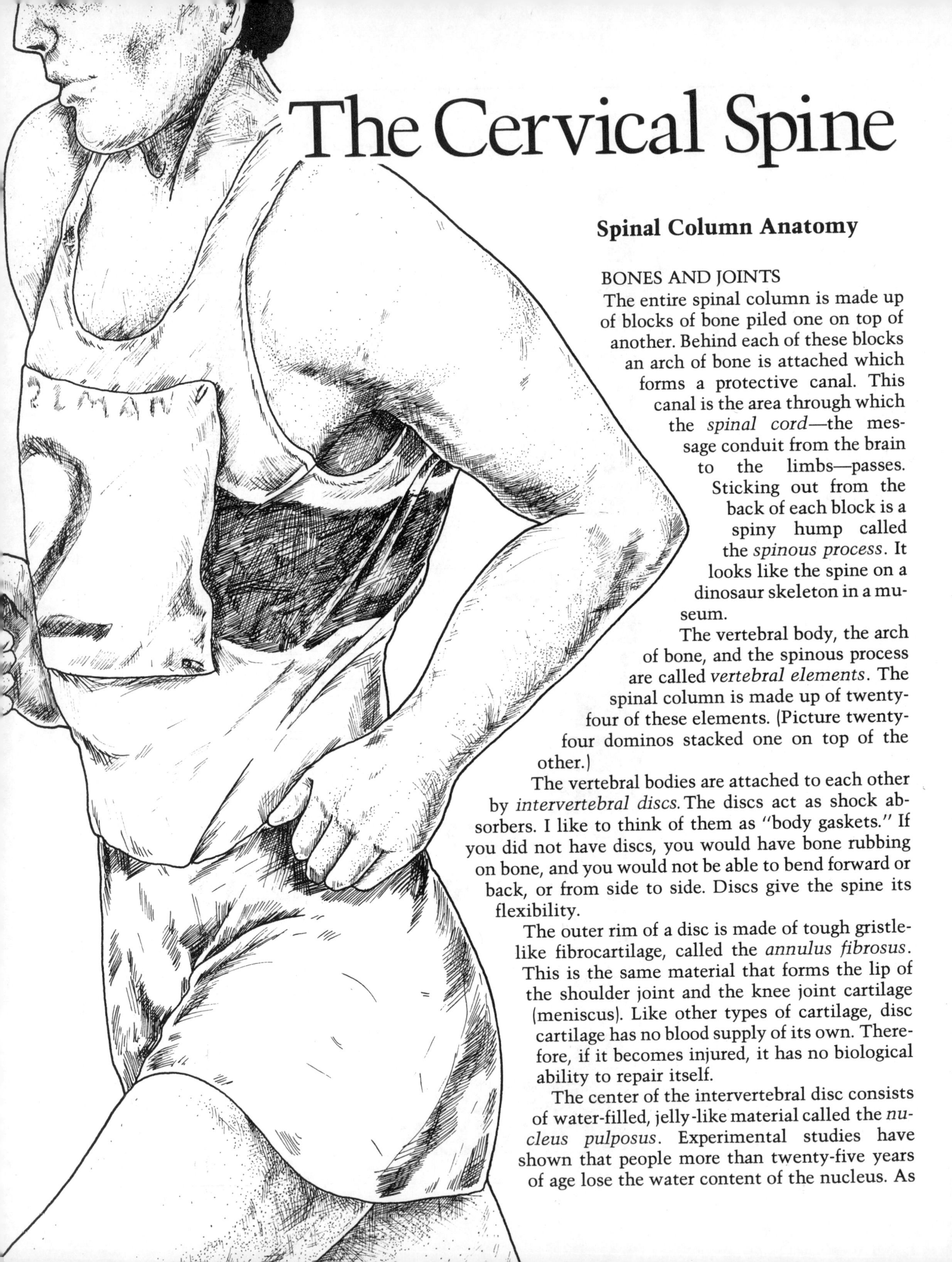

The Cervical Spine

Spinal Column Anatomy

BONES AND JOINTS

The entire spinal column is made up of blocks of bone piled one on top of another. Behind each of these blocks an arch of bone is attached which forms a protective canal. This canal is the area through which the *spinal cord*—the message conduit from the brain to the limbs—passes. Sticking out from the back of each block is a spiny hump called the *spinous process*. It looks like the spine on a dinosaur skeleton in a museum.

The vertebral body, the arch of bone, and the spinous process are called *vertebral elements*. The spinal column is made up of twenty-four of these elements. (Picture twenty-four dominos stacked one on top of the other.)

The vertebral bodies are attached to each other by *intervertebral discs.* The discs act as shock absorbers. I like to think of them as "body gaskets." If you did not have discs, you would have bone rubbing on bone, and you would not be able to bend forward or back, or from side to side. Discs give the spine its flexibility.

The outer rim of a disc is made of tough gristle-like fibrocartilage, called the *annulus fibrosus*. This is the same material that forms the lip of the shoulder joint and the knee joint cartilage (meniscus). Like other types of cartilage, disc cartilage has no blood supply of its own. Therefore, if it becomes injured, it has no biological ability to repair itself.

The center of the intervertebral disc consists of water-filled, jelly-like material called the *nucleus pulposus*. Experimental studies have shown that people more than twenty-five years of age lose the water content of the nucleus. As

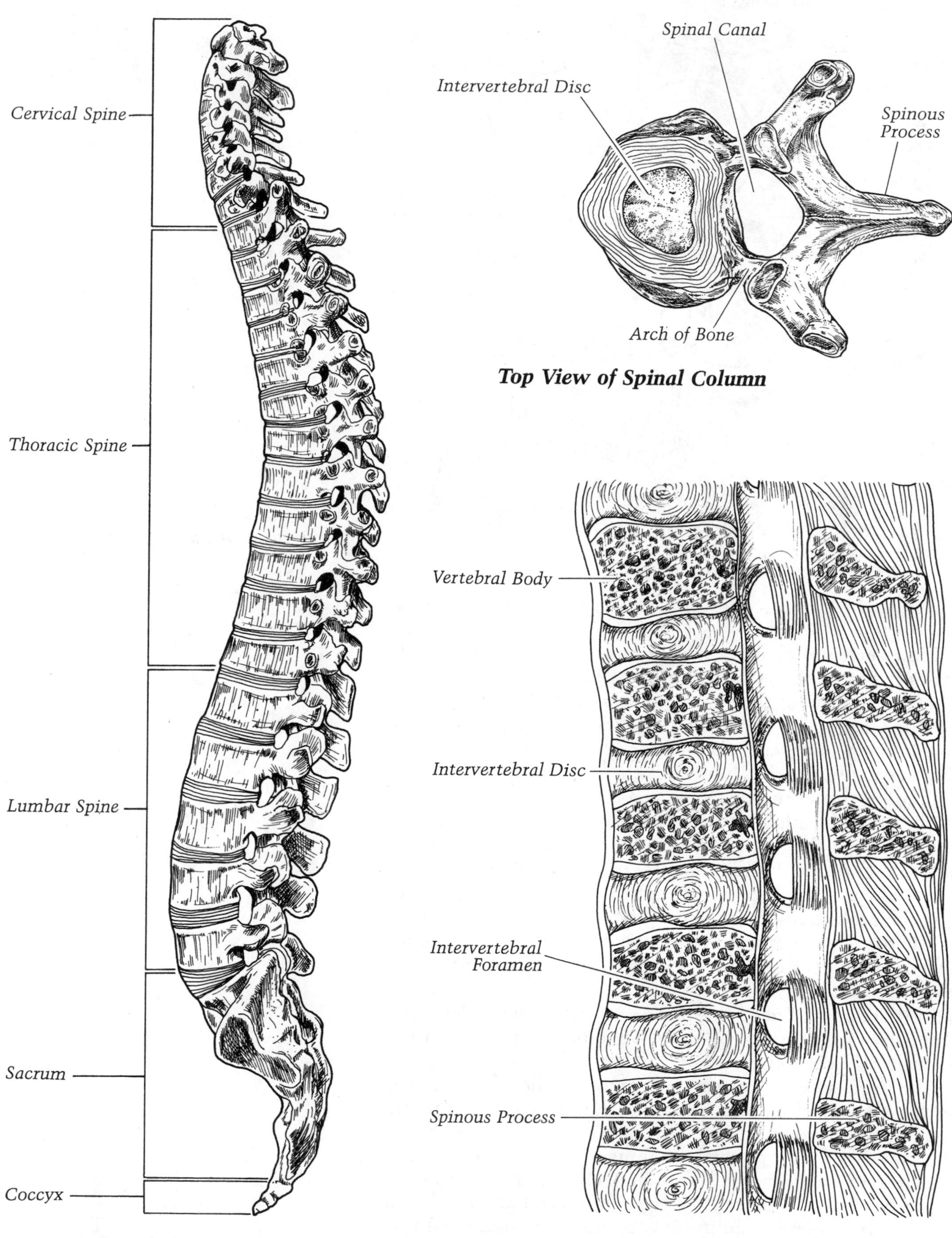

Top View of Spinal Column

Side View of Spinal Column

Sectional Side View of Spinal Column

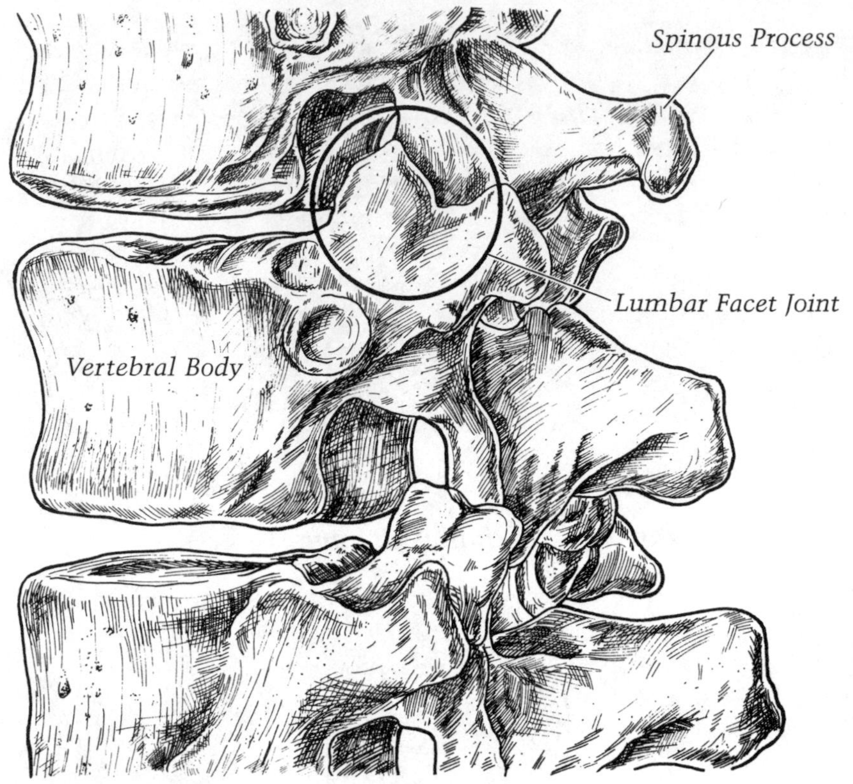

Facet Joint of Lumbar Spine

a result, the disc becomes dry and brittle, and damage can occur.

Each vertebral element is also attached to its neighbor by way of two joints, which lie just in back of the bony arch. These joints, about the size of your knuckles, are called *facet joints*. They also add to the mobility and flexibility of the spine. Facet joints are true joints. They have a joint sleeve, they have a lining that lubricates them, and they can become stiff and arthritic through wear and tear.

The stout ligaments that attach the arches to one another help stabilize the spine. They are yellow and are called *ligamentum flavum*. In addition, there are ligaments that run between the spinous processes called *intraspinous ligaments*. It is their job to resist excessive forward bending.

The top seven vertebral elements, called the *cervical spine*, support your neck. They are shaped to give you mobility. Test your neck mobility. Put your chin on your chest. Look at the ceiling. Turn your head side to side. Notice how much forward and backward motion you have. This motion is all occurring in your cervical spine.

The next twelve vertebral elements, called the *thoracic spine*, all have ribs attached. The thoracic spine forms the back anchor for your chest. This part of the spine along with the ribs helps to protect your lungs, because the rib strut attaches to each of these vertebral elements. The thoracic spine has only 25 percent of the mobility of the neck area. Because it has less mobility, it gets much less wear and tear. Thus, it seldom causes trouble.

The bottom five vertebral elements, known as the *lumbar spine*, are the strongest and largest vertebral bodies. Their structure corresponds to the tip of a crowbar. When you lift, the lumbar area absorbs most of the punishment. The mobility of the lumbar portion of the spine is midway between the neck and the thoracic spine. Wear and tear on the facet joints in the lumbar area is a frequent source of pain in the low back area. Secondly, because this portion of the spine is the last formed in embryologic development, congenital anomalies of the lumbar spine are much more common than those in other parts of the spinal column. For instance, it is not rare for people to have only four mobile lumbar elements instead of five, making their spine less mobile and flexible.

MUSCLES AND NERVES

The spinal cord, a massive trunk of nerves, runs down the spinal column from the brain to the very bottom of the tailbone. The spinal cord is the body's main telephone center to the limbs. It is protected by the bone of the spinal canal. At every vertebral element, a pair of nerves branches off from the spinal cord. They come out between the vertebral elements from holes called *foramina*. These holes lie just in front of the facet joints and just in back of the vertebral body.

The nerves run to the arms, trunk, and legs. They carry messages from the brain to the muscles and bring back sensa-

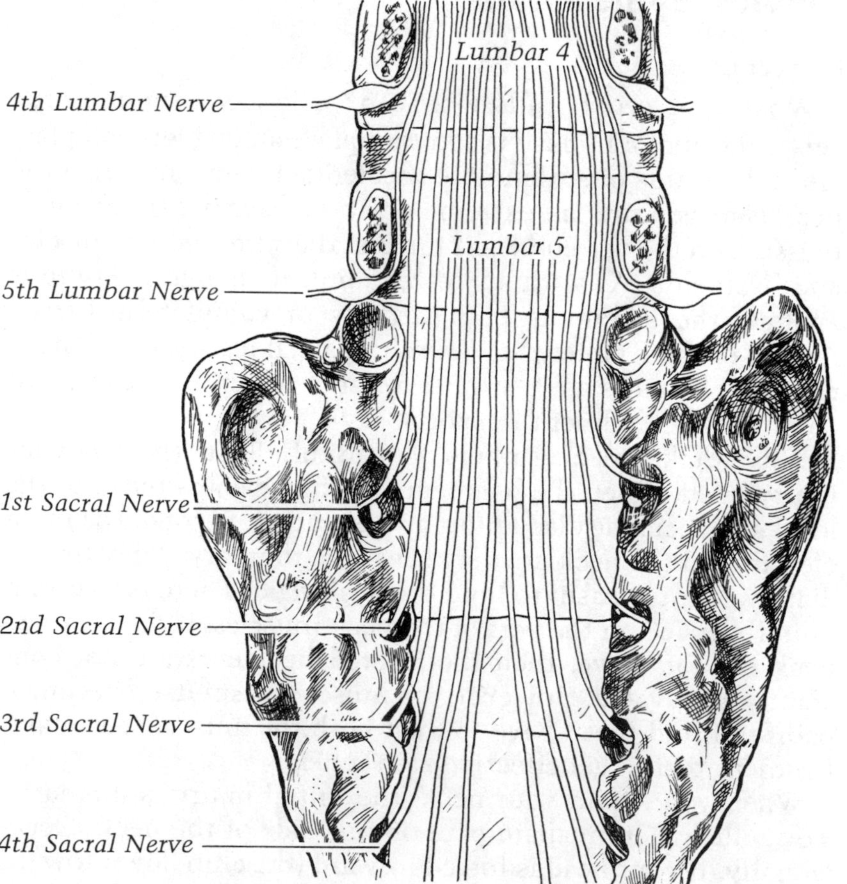

Nerves of the Spine

tion messages from the limbs. The nerves are vulnerable to pressure as they exit from the spine. Through the bony holes (foramina), a slipped intervertebral disc can push on them and produce pain and pressure. Spurs or arthritis can change the shape of the facet joints. This can build up and apply pressure to the nerves.

Think of the vertebral column as the mast of a ship. Four sets of muscles keep the mast up straight and allow it to move properly. The front guidewires are the abdominal muscles. While these do not attach directly to the spine, they are important in keeping it strong from the front side.

As you lift heavy weights, notice that you tighten your abdominal muscles. You are also pushing your abdominal organs back against the spine. This resistive pressure helps stabilize the spine.

The paraspinal muscles which run up and down either side of the spine are the main guidewires. They are tremendously strong. You can see these bulging on either side of the mid-portion of the spine itself. Additional muscles, called the *quadratus lumborum*, lie between the ribs and the pelvis and support the lower back area.

Specialized muscles called the *paracervical muscles* support the neck, but also allow it to turn and bend in its many different positions.

Cervical Spine Sprain

CAUSES

What most people call "whiplash" is in fact a sprain of the cervical spine. Whiplash is commonplace among football players and victims of automobile accidents. It happens when the neck is driven into an extreme position, exerting tremendous pressure on the vertebral column and the surrounding muscles and ligaments. As with any other joint, if enough pressure is applied, the ligaments tear partially or completely. Partial tearing of the ligaments is a *sprain*. Partial tearing of the neck muscle elements is a *strain*. Sprains are by far the most common neck ailment in sports.

What makes neck sprains so painful? First, the ligament injury itself. When the ligaments rip, blood pours out into the local area, causing discomfort and irritation. Second, the muscles around the neck are very sensitive to injury. They immediately go into spasm to hold the neck rigid. There is a certain "intelligence" on the part of the neck muscles. If the sprained neck cannot move, then the injured ligaments are not constantly aggravated. However, the muscle spasm itself becomes painful. I find that I have to treat both the muscle spasm and ligament tear to get effective pain relief.

When you sprain your neck, the initial injury is dramatic. You will feel immediate pain on one side of the neck. Occasionally, if your head is forced forward (the chin down toward the chest) both sides of the neck will sprain. I have never treated a sprain to the front of the neck.

After thirty minutes, the initial pain subsides and your neck feels good. After two to three hours, the blood from the tear leaks into the injured area and the discomfort begins. First you feel a dull aching at the injury point. Eventually, the dull ache turns into a very sharp pain. At the same time, muscles in the entire neck go into spasm. The spasms limit the amount of turning in your neck. The morning after the injury, you are in agony. Often, you have to cock your head off to one side to find a comfortable position.

DIAGNOSIS AND TREATMENT

When I examine you, I find you cannot move your neck very much. I ask you to bend your head from side to side and try to touch your ear to your shoulder. Next, I ask you to rotate your head and look over your shoulder. Finally, I ask you to put your chin on your chest and then to look up at the ceiling. The type of motion restriction tells me where the injury lies. I never push on the head to test neck motion.

Next, I feel along the muscles of the back of the neck. I search for tender areas that throb. That is the location of the sprain. I always check the reflexes of the tendons in the arms. If the reflexes are normal, the injury does not involve a nerve. I take x-rays of the neck. The x-rays usually show only that the neck is pulled off to one side, confirming the ligament injury and the muscle spasm.

Initial treatment includes painkiller pills (analgesics), muscle-relaxer pills, and a neck collar. The medication should be continued for at least five days. A collar must be worn until the pain abates. The neck collar acts as a splint so that you cannot reinjure the ligament with any sharp involuntary or unexpected motion of the neck or head. I recommend only daytime wearing of the collar because I find that patients cannot comfortably wear the collar in bed.

Forty-eight hours after the sprain, I start you on heat application. An electric heating pad, hydrocollator pack, or hot towels all do the job. Apply the heat for thirty minutes at least three times daily. Continue until the pain and muscle spasm abate and full motion returns to your neck.

Simple sprains heal in three to five days. In moderate sprains, the healing process is seven to ten days. Severe sprains require three weeks to heal.

It is not dangerous to return to sports activities before you have the full neck motion, but you will not be as effective because you cannot turn your head fully. To protect the neck, football players who have had neck sprains should use a horse collar between the head and the shoulder pads throughout the entire season.

Muscle strengthening exercises soon after the injury are not helpful. In fact, they may cause reinjury. However, in the off season the muscles about the neck should be strengthened. A

SOCCER NECK

Can you sprain your neck without actually being hit? It happened to Beth Reilly, a twelve-year-old soccer player. She came into my office holding her head tilted off to one side. I examined her and took x-rays of her neck. There was no doubt that she had a neck sprain. I asked Beth's mother what had happened. She said that she really wasn't sure, but that the neck pain and headaches had developed after a recent soccer match. We puzzled over the problem together.

I started Beth on heat treatments and gave her a neck collar for support. The next day, her mother called back elated.

"I've discovered the answer to Beth's problem. Because she can throw a soccer ball farther than any player on the team, the coach has her throw in all the out-of-bound plays. To get more distance on the ball, she brings her hands way behind her neck and then whips the ball onto the field. At the same time, she whiplashes her neck."

Beth's syndrome isn't described anywhere in the medical literature. I call it "soccer neck." After her sprain healed, she strengthened her neck with weight exercises. Beth's mother says that her daughter's neck has not caused her any further problems.

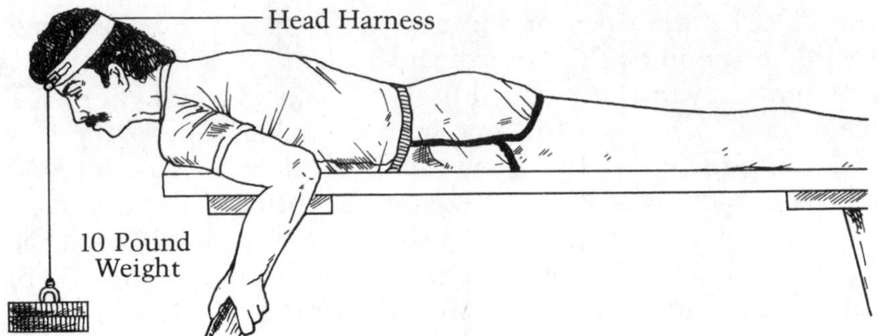

Neck Strengthening Harness

head-type harness with weight plates can be used or a Nautilus neck machine. In my opinion, neck bridges with bench-pressing-type weight lifting puts too much stress on the neck and should be avoided.

Cervical Burners (Cervical Nerve Stretch)

CAUSES

When tackling, some football and rugby players get an electric burning sensation down their arms. As they tackle, the head is driven in one direction and the shoulder in another. The result is that the cervical nerves which come off the spinal cord and out through the neck are stretched. Because of the burning sensation, stretching of the cervical nerve is called a "burner" in football circles. In my college football days, I had burners. They are common in people with my body type: short-necked and heavily muscled.

DIAGNOSIS AND TREATMENT

Should you worry if you have experienced a burner or if it has happened two or three times? An occasional occurrence is not significant, and you don't have to be concerned. However, if the stretching happens time after time, some bleeding develops around the nerve at the point where it exits from the neck. This bleeding can turn into scar tissue, and then the nerve might become imbedded in the scar. This makes it less flexible and therefore even more prone to burners. It can be a vicious circle.

The easiest and best treatment is to give the player a horse collar or foam rubber ring to wear between the shoulder pads and his neck. This essentially makes it impossible for the head to be driven far away from the shoulder. Therefore, the nerve cannot be dramatically stretched. In 90 percent of cases, the collar is totally effective.

If the collar does not work, a cantilevered extension can be placed on the shoulder pad. The purpose of the shoulder pad is to direct the force of tackling from the tip of the shoulder pad to the center of the trunk where the body is stronger. A shoulder pad that fits well does not actually touch the shoul-

PREVENTION OF CERVICAL SPRAINS IN FOOTBALL

In 1972, football trainer Red Romo of the United States Naval Academy and Dr. William Clancy devised a method to prevent neck sprains in football. They connected a semi-elastic strap one and a half inches wide and nine inches long from the back of the helmet to the shoulder pads. With this strap in place, they theorized, it would be impossible to drive the head forward—the movement that creates neck sprains.

Detailed records were kept on 135 Navy football players. Ninety-three players used the strap and forty-two did not. They found that neck sprains were twice as common in those players who did not use the strap. Perhaps even more significant was the fact that 70 percent of players who did not wear the neck strap and who had had previous neck difficulty were reinjured, whereas only 26 percent of those using the strap sustained an additional injury.

MY CERVICAL BURNERS

In my sophomore year in college, I started experiencing burning pains down my right arm. I had them in high school, too, but never with such frequency. In a practice for the third game of the season, I made a routine tackle. Suddenly, my whole right arm was on fire. My right hand felt like it was being stuck with pins and needles.

Dr. Quigley, the Harvard team

physician, checked me. "This is a cervical nerve stretch injury," he said to the medical students who accompanied him. "We see these in football, especially in linemen. When this young man tackled, his neck was driven to the left and his right shoulder was depressed downward. This put great stretch on the top nerve in the brachial plexus—the fifth cervical nerve." Dr. Quigley looked to me for confirmation. I nodded.

He took out a small pin and started to prick the skin on the outer aspect of my shoulder. I was amazed that I had no feeling whatsoever. Next, he asked me to hold my arm over my head. I could not do it. He told me and the students that my deltoid muscle had been weakened as a result of the nerve stretch. "The fifth cervical nerve gives the deltoid muscle its messages," he explained.

Next morning at breakfast, I could not lift a spoon to my mouth. It gave me an unforgettable anatomy lesson. If you cannot elevate your shoulder with your deltoid muscle, you cannot get a spoon into your mouth.

I returned to the fieldhouse for a repeat examination the next day. Once again, Dr. Quigley confirmed the findings of a stretch injury on the fifth cervical nerve. "If I send you to a neurosurgeon or a neurologist, he probably would not understand how frequently this happens in football players, and how quickly you can improve. They will put you out for the season. I will rest you three to four days. You will probably improve on your own."

Sure enough, the nerve returned to full function in four to five days. The sensation in my arm and hand returned first, and then the power in the muscle. I began lifting weights as soon as the muscle would tolerate it, and within two weeks I had regained my strength. I have always been grateful to Dr. Quigley for assuming the responsibility for the treatment of this particular problem. If he had sent me to a nerve specialist, I would have missed the whole season. Since that time, I have had no difficulty with this muscle or nerve. This is a good example of how treating sports injuries differs from treating injuries to nonathletes.

der, but rather sits above it. As you tackle, the force never hits the shoulder, it comes instead into the center of the shoulder pad in the trunk area. By placing an extra bar on top of the shoulder pad, still more of this force can be directed toward the middle of the body and away from the shoulder. If neither of these techniques brings about cessation of the burners, you should seriously consider giving up football. If the nerve becomes heavily scarred, it loses its pliability and this could diminish your ability to play tennis and golf or other recreational sports later in life. Personally, I consider that too high a price to pay.

Acute Cervical Disc Disease

CAUSES

It is called a "slipped disc" in everyday parlance, yet the word "slipped" conjures up the wrong mental picture. It is not as if your disc slips out between the vertebral bodies as one unit. What in fact happens is that the disc breaks into distinct pieces.

Your disc fragments for two reasons. First, as you get older, it naturally loses water and dries up. A brittle disc is much more likely to divide. Second, you have normal wear and tear.

Your disc may be undergoing fragmentation long before it produces clinical pain. At some point, during the course of this biologic event, symptoms will be produced. It is for this reason that the acute cervical disc can come on with very little warning and without significant injury. You might just be turning your head to look at your tennis serve or quickly snapping your neck as you are bowling. Even sneezing or coughing is sometimes enough to dislodge a disc fragment.

The problem is brought on when one of the fragments changes its position very quickly and goes out beyond the normal confines of the disc itself into the area occupied by the spinal nerves. In this way, it applies direct pressure to the spinal nerve, in a manner similar to pliers squeezing a telephone cord. The more the pressure, the more trouble the nerve has conducting its messages. In addition to direct pressure, there is always swelling and inflammation around the nerve after the disc fragment makes its assault. This intensifies pressure on the nerve.

The disc fragment almost always protrudes on one side, touching the nerve that runs into the arm. Thus, people with cervical disc disease always have neck pain and arm pain, but only in one arm. Pain in both arms is very rare.

Pressure on nerves leads to distorted nerve function. Arm reflexes slow down. Sensation signals coming back into the brain via that nerve are disturbed as well. Mild pressure causes dysesthesia, an uncomfortable feeling. Moderate pressure brings about a pins-and-needles sensation, and severe pressure produces numbness.

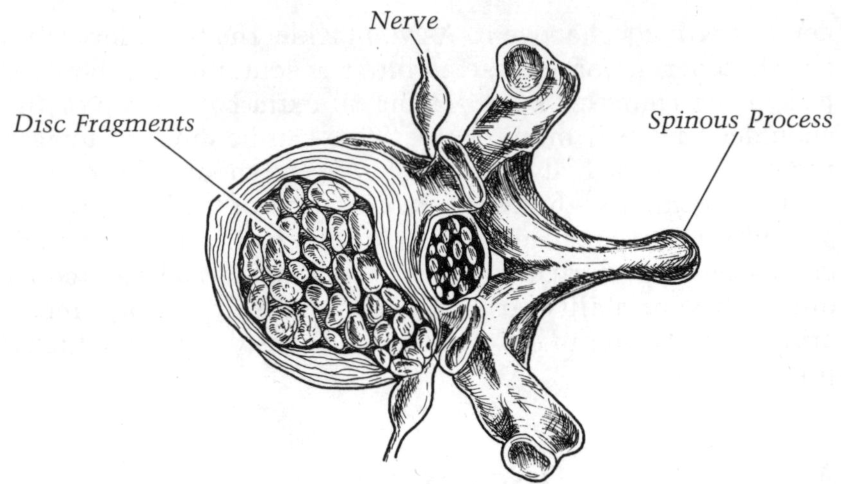

Fragmentation of Disc

DIAGNOSIS AND TREATMENT

The onset of pain from a slipped disc is sudden and excruciating. You have a difficult time finding a comfortable position. When it happens, the neck muscles immediately go into spasm. Within a few hours, a dull aching pain extends down the arm. Often, you will notice a weak feeling in the arm as well as some loss of sensation.

I can tell which disc has slipped by identifying the nerve under pressure. I do this by testing all the reflexes in both arms, as well as muscle strength and skin sensation. I test the reflexes of the biceps, the triceps, and the brachio-radialis muscle with a reflex hammer. I test muscle strength by resisting the patient as he or she contracts specific muscles in the arm. I test for sensation in the arm and hand by pricking the skin with a needle. Often, you have a tenderness to touch on the affected side of the neck.

The most common neck discs to cause nerve trouble are those that lie between the fifth and sixth and the sixth and seventh cervical vertebral bodies. These are the lowest two in the neck.

Because a disc problem has often been going on for a while prior to the acute episode, there is usually some narrowing between the vertebral bodies, corresponding to the fragmented disc. As the disc fragments, it no longer can hold the blocks of bone apart in a normal way. As a result, the blocks move closer together.

The narrowing can be clearly seen on an x-ray, especially in the side view. The involved disc will be narrower than the disc space above it or below it. X-rays confirm what I have found in the physical examination. Regular x-rays also rule out the possibility of a rare neck problem such as an infection or tumor.

How do I treat a fragmented disc? I immediately prescribe a cervical collar which is tall enough to splint the neck. Then I start you on analgesic medication, usually a strong narcotic pill and muscle relaxant such as Valium. Lastly, I recommend the application of an electric heating pad for thirty minutes, three times a day. The hydrocollator pack is also effective.

After two days, I reexamine you. Then I start you on cervical traction. You have a choice of going to a physical therapy unit or obtaining a home cervical traction unit which hangs over your door.

Your head is placed in a halter traction device. The halter is attached to a rope which runs directly up above your head to a pulley and then down to a weight. I like to start with eight pounds of weight. The traction pulls the vertebral bodies apart and opens up the disc space. This aids the process by which the slipped disc fragment can recede to its original position.

Many patients with this disorder look forward very much to their time in traction. I recommend traction for twenty minutes twice a day. Usually, it requires two to three weeks of daily treatment to heal. If you stop treatment too quickly, you run the risk of reintroducing the condition. Unfortunately, cervical disc problems tend to recur. But two or three years may elapse between attacks. It is unusual to have just one episode. However, once you have been through it, you know the early symptoms. You can avoid a full-blown attack by starting therapy quickly.

Occasionally, surgery is necessary. If my patient has been under active treatment for a cervical disc for six weeks and is no better, then I consider surgery.

Before any surgery, I have my patient examined by a neurologist, and often an *electromyogram* is performed. This is an electrical test which tells whether the nerve that runs to a certain muscle is under pressure or not. If it is, the electrical signal which travels down to that muscle, via a small needle

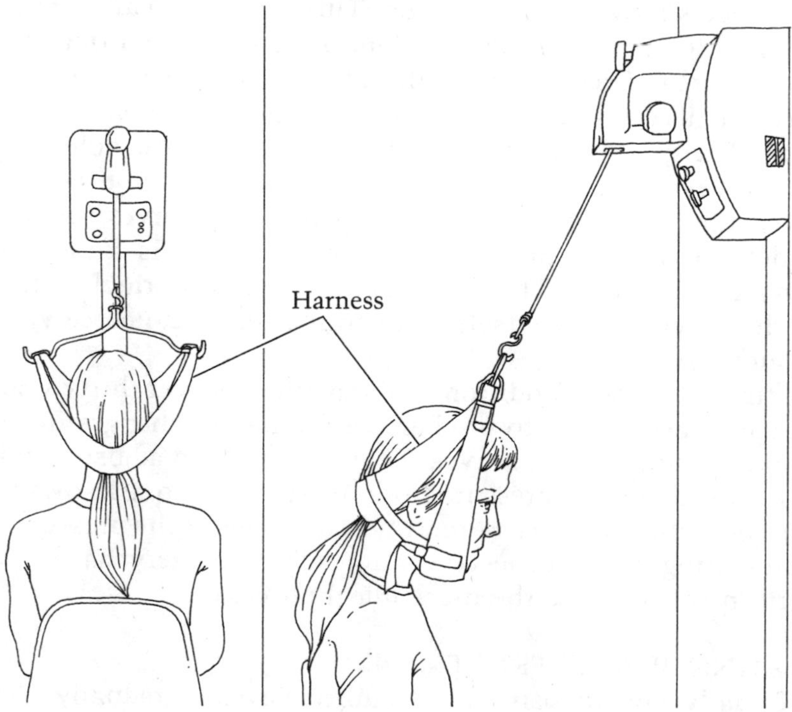

Cervical Traction

placed in the muscle, will register abnormally. Since certain cervical nerves go to specific muscles, an abnormality in one particular muscle will tell which nerve is affected, and therefore which disc is pushing on it.

In addition, I order a second test, called a *myelogram*, before surgery. This is a dye test that outlines the fluid-filled sac in which the spinal cord is encased. The sac looks very much like a long party balloon, and it cannot be seen on regular x-rays. The dye is introduced with a small needle. If a slipped disc fragment is present and pushing on one of the spinal nerves, it can be seen on a myelogram as a dent in the fluid column. The sac looks as if it has been punched. This gives me the exact size of the slipped disc fragment and the disc from which it originates.

If the dent is very large and the patient has not improved in six weeks, then it is logical to remove the disc fragment surgically. This is like taking a stone out of your shoe. The surgery requires a hospitalization of only two to three days. A cervical or neck collar is worn for three months following surgery. Fortunately, only a few people need this surgery, and it is very effective.

Cervical Spondylosis—Cervical Radiculitis

CAUSES

If your neck gets hit hard repeatedly in contact sports, eventually you will have to pay a price. The repeated trauma to the neck produces wear-and-tear changes in the cervical spine. It starts as small injuries to the discs in the lower part of the neck. As the discs lose their height and plumpness, the vertebral bodies move closer together. This process is called *cervical spondylosis*, and leads to the formation of spurs around the injured area. These spurs build up and start pushing on the cervical nerves that run out through the bony holes in the neck. The spurs decrease the circumference of these holes, and the nerves are caught in a pinch. More importantly, this leads to an uncomfortable radiation of pain along nerves that are under pressure. A pinched nerve in the neck is known as *cervical radiculitis*. The spurs cut down on the normal motion of the cervical spine itself. Spur formation is called *cervical spondylosis*.

I find these two conditions only in athletes over thirty years of age. It takes time to produce wear-and-tear changes in the neck. Amazingly, the nerves seem to be able to adjust to this steady onslaught of pressure from the spurs. No one knows for sure how this happens. Sometimes, however, if the pressure is unrelenting and the nerve cannot adjust, surgery has to be performed to reduce the mechanical pressure.

DIAGNOSIS AND TREATMENT

Usually, spondylosis and radiculitis develop gradually. You

will start to feel a deep aching sensation in your neck, which becomes worse over a period of two to three weeks. As the pain intensifies, it spreads along the shoulder to the top of the shoulder and the shoulder blade. Sometimes, it can come into the chest area and mimic a heart attack. The pain is constant and aching in nature; it is not a burning sensation. Your neck will be stiff and it will be difficult to find a comfortable position to sleep. Because you cannot use your neck fully and your arm is in pain, it will also be difficult to play sports. Using your arm tends to pull on the nerves and twang them like guitar strings.

Most of the time you do not lose sensation in the neck and shoulder. Because there is usually no great pressure on the nerve, muscle weakness is also not a problem. But often, when you grasp the muscle that is supplied by the pressured nerve, it will be tender to touch. Medical scientists cannot explain this phenomenon.

Once in a great while, a patient experiences tremendous pressure on the nerve. In this case muscle weakness, slow reflexes, and abnormality of sensation will occur. This makes the diagnosis easier—but the treatment harder. Properly managed, it takes ten to fourteen days for these attacks to abate. If your neck is painful, stop exercising. I immediately start you on anti-inflammatory medication. I find that something more than aspirin must be used. This should diminish the inflammation around the nerve root where the spurs are pressing. I prescribe a narcotic painkiller only 40 percent of the time. Often a milder analgesic will do.

The second most effective treatment is cervical traction for twenty minutes twice daily. Because you usually wake up in the morning with a stiff neck, I recommend a cervical pillow. This is a very low, flat pillow which is shaped like a large kidney bean. The indentation fits into the nape of the neck. This seems to reduce morning stiffness.

If you use traction first thing in the morning, it tends to stretch the neck out and make it more comfortable during the day. By the end of the afternoon, your neck often feels quite stiff, sore, and painful, and that will be an ideal time to use the traction again.

If you have great problems moving your neck, I prescribe a cervical collar. But in half the cases, it is not necessary.

Sometimes, the radiating pain localizes in a very small area, mainly on the top corner of the shoulder blade. In medical terms, these are called *trigger points*. If I touch a trigger point, the patient usually jumps off the examining table. A single cortisone injection into one of these points can completely eliminate discomfort, and for some reason it reduces the neck pain. I almost always inject cortisone into trigger points.

Radiculitis seems to come in successive bouts, often months apart. Patients become alert to early symptoms. By getting a head start on medication, the collar, and traction, you can avoid a full-blown episode. After three to four years of off-and-on pain, the condition seems to disappear.

Occasionally, the pressure on the nerve builds to such a point that surgical relief is necessary. If surgery is performed, I feel that a fusion of the vertebral bodies—an operation which makes a permanent bony connection between them—should be carried out as well as removal of the disc and the offending spurs. This makes the operation a bit more difficult and time-consuming. It also requires that you wear a cervical collar for three months after surgery. However, the fusion brings about a stable and pain-free vertebral element.

A Cervical Fracture or Dislocation

CAUSES

A cervical fracture or dislocation is known as a "broken neck." It is a devastating injury because it can cause paralysis of both the arms and legs (called *quadriplegia*).

When I was nine, I saw my first broken neck. It is an injury I will never forget. My brother was playing football for Hebron Academy in Maine. Toward the end of the game, one of Hebron's linebackers made a hard tackle at the line of scrimmage. There was a big pileup. Everyone got up from the pile except the linebacker. He remained flat on his back in the middle of the field.

I still can picture the trainers and doctor evaluating this injured player. He moved neither his arms nor his legs. The medical team did not remove his helmet, but they did remove his shoes and socks. The doctor started to stick pins into the soles of his feet. Why did the player not withdraw his foot? Didn't he feel the pain? He was removed from the field on a large board and carried into the nearby fieldhouse.

After the game, I visited the fieldhouse to see my brother. The locker room was in a deathly silence. The injured linebacker was lying on one of the rubdown tables. His eyes were open, but he was not moving.

The doctor phoned his parents to tell them about the accident. Next, the doctor explained to the coach that his spinal cord was probably severed and that was why he had lost all movement in his arms and legs. As I contemplated this type of injury for the first time in my life, I felt a sense of shock and horror. It had never occurred to me that a person might have to live his remaining years without ever moving his arms and legs again.

A broken neck is rare in sports. It is far more common in motorcycle, trampoline, and diving accidents. Children diving into a shallow pool may suffer broken necks. In organized sports, they occur almost exclusively in football.

In many cases cervical bones will break or ligaments tear when a severe force is applied to the top of the head, driving the neck forward and down. We call this a *flexion injury*. In football, it happens when a tackler tackles with his head down and his neck bent forward. The injury occurs when the blow

OLD TIMER'S FOOTBALL NECK

Recently, a college teammate, now a famous politician, came to my office complaining of neck and shoulder pain. He said it had become stiff and sore in the last three weeks. "The pain is in the lower part of my neck and radiates into my right shoulder, all the way down into the middle finger," he said. "My finger feels like pins and needles. I can't turn my head enough to back the car out of the driveway. I haven't had a decent night's sleep in weeks."

I asked him if he remembered any special tackle or injury in football that might have produced damage in the neck area. He said that he couldn't recall a single particular incident. However, as we both knew, he had taken many shots to the head and neck during eight years of football in both high school and college.

I examined his neck, and he did have restricted motion and spasm in the neck muscles on the right side. His reflexes showed pressure on the seventh cervical nerve as it exited from the right side of his neck. This explained the tickling sensation in the long finger of his right hand. X-rays of his neck showed that he had had an old football injury to the disc between the sixth and seventh cervical vertebral bodies.

I explained to him that many ex-football players have neck x-rays which suggest they are much older than their true age.

"Well, that's great. You're really very encouraging," he said. "Are you telling me that I have a seventy-year-old neck? What are you going to do to make me better?"

"Well, don't be too concerned," I replied. "Fortunately, 95 percent of people with your kind of neck have only occasional pain. Nature seems to be good to us on this one and it eventually settles down and causes no trouble whatsoever. It's just the price that you are paying for an athletic youth. Remember that the body never forgets the beatings that it took in the past."

He was relieved. I think it made sense to him that this was the case. I treated him with cervical traction and anti-inflammatory medication. The painful episode quieted down in three weeks.

CERVICAL DISC SURGERY

I was a surgical intern at Massachusetts General Hospital in 1968 when I saw my first neck disc operation. I had evaluated the patient before surgery. He was a college football player who had hurt his neck in a preseason scrimmage two years before this hospitalization. He explained that he had made a tackle and had had the onset of severe pain down his right arm all the way into his right thumb. During the physical examination, I discovered that the reflex in his right biceps muscle was diminished, and he had numbness over the thumb and index finger of his right hand. I examined his x-rays and they showed that he had narrowing of the disc between the fifth and sixth vertebral bodies. I concluded that he had a slipped disc and there was a piece of disc pressing on the sixth nerve out of his neck and pain ran into his right arm. His myelogram confirmed this impression. He was all set for the operating room the next morning.

As a surgical intern, you are placed in a situation that is totally new to you. I had read about this operation and had reviewed the anatomy. I knew that I would just be holding a retractor, but I was anxious to see exactly how this operation was done technically. The patient was placed on the operating table facing up. The back was carefully scrubbed and painted with an antiseptic iodine fluid. Robert Boyd, the surgeon, explained that the incision had to be placed three fingerbreadths above the collarbone. He made the incision and it bled briskly. Next, to my amazement, he moved quickly into the area between the windpipe and the major artery and vein that supplied blood to the brain. This was really tiger country, I thought.

I began to feel uneasy. This spot is where the Marines are taught to bayonet the enemy. I was amazed how close these vital structures are to the skin. The neck cut open looked like a bamboo pole. The high spots were the discs, he explained, and the flat part was the vertebral body. He took an x-ray to make certain that he was going to operate on the right disc. He told me that some cases had been reported where the wrong disc had been operated on. Again, I was horrified. The x-ray checked out fine, and he continued.

Using a knife, he cut directly into the disc and removed the front part of it. Then, using a special pincher, he took out the fragments of the damaged disc one by one. He warned me: "You have to be careful not to go too far back because you can push on the spinal cord or one of the cervical nerves." This admonishment didn't make me feel any better.

After the entire damaged disc had been removed, he moved down to the front of the pelvic area. He made an incision down to the part of the pelvis where you "hang your pants." He removed a lifesaver-sized piece of bone, which he used to fill the disc space above. It was carefully tailored to fit and then tapped gently into place with a mallet. I remember being impressed with how tightly it fit and how stable the neck area appeared after insertion of this bone graft. What a beautiful operation! What began as a terrifying experience ended up as a surgical triumph from which the patient reaped the benefit.

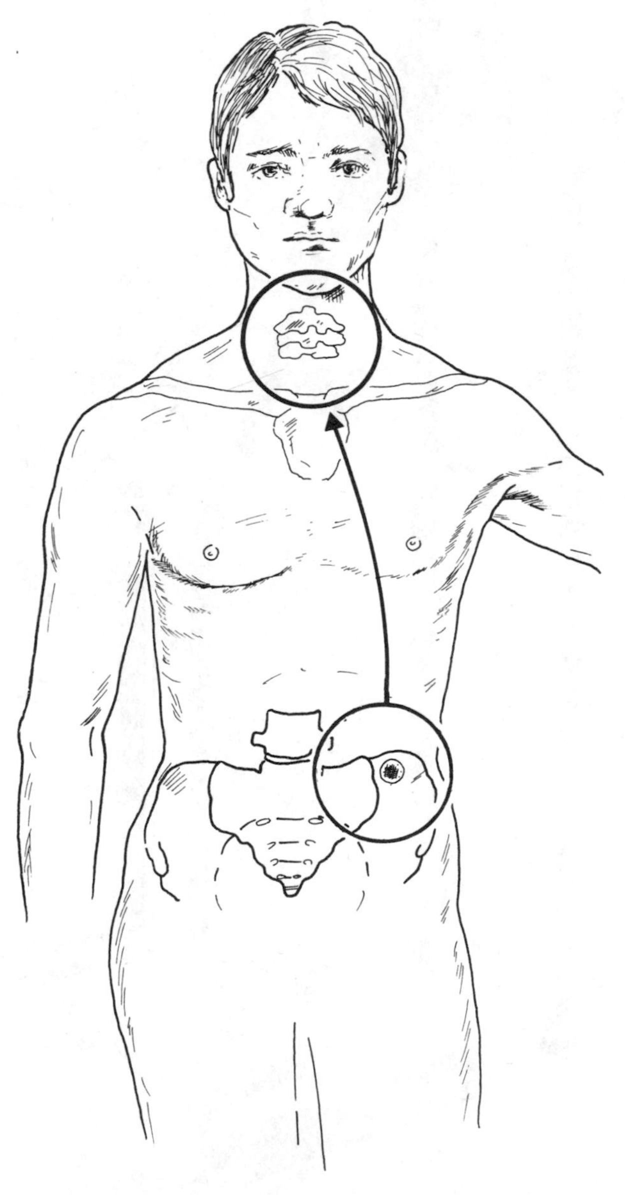

Cervical Disc Surgery

Nautilus Four-Way Neck Machine

exceeds the strength of the bones and ligaments in the cervical spine. The spinal cord is severed when its vertebral elements are forced out of alignment. Ironically, it becomes injured by the very bone designed to protect it.

DIAGNOSIS AND TREATMENT

If an athlete does not move either arms or legs within the first ten to fifteen seconds after an injury, the possibility of a broken neck flashes through my mind.

My first step is to make sure the breathing passage is clear. I feel for a pulse at the wrist to make certain that the heart is beating effectively. I ask the athlete if he can pick up his arms and then wiggle his foot. If there does not appear to be any motion or power in the arms or legs, then this patient must be treated as if the spinal cord has been injured.

Early treatment and movement of spinal-cord injuries has been well described in most first-aid manuals. The key: Use a long spinal board. This is slipped beneath the patient, and his head and neck are kept perfectly straight and aligned with the rest of the spinal column. Because the nerve injury can be made more severe by movement, sandbags should be placed next to each ear so that the head does not turn. Do not remove the football helmet or any other clothing. Never remove the helmet until x-rays are taken to determine the site and extent of the damage. Also, the neck should never be moved. This is crucial.

Unfortunately, if a fracture or dislocation of the neck has occurred and the spinal cord is severely injured, there is very little chance for recovery. If there is no improvement in the first forty-eight hours, seldom is there significant improvement after that. It is not unheard of for people to regain some useful nerve and muscle function after spinal cord injury, but this is rare after complete quadriplegia.

How can these injuries be prevented? A strong neck seldom breaks. Therefore, I strongly encourage athletes who play contact sports to build their neck muscles. This is not difficult. The three ways to strengthen your neck are isometric exercises, a head-halter weight-lifting apparatus, and a Nautilus neck machine. These are all equally effective and the results can be spectacular.

The Lumbar Spine

Lumbar Strain

CAUSES

The spinal column in the lumbar area is rugged and strong. The muscles supporting it are large and powerful, but despite their size and strength, they can rip or tear. In medical terms, these are called *strains.*

When heavy force is applied to the muscles or the muscles are suddenly called upon to fire before they are properly warmed up, they rip or strain. Warming up brings an extra blood supply to the muscle, thus making the muscles flexible.

If the muscle itself rips, the blood vessels within it rip, also. This causes an outpouring of blood into the rip; pain and swelling follow. This will usually lead to muscle spasm—which is painful in itself.

People who most commonly get lumbar strains are weight lifters, figure skaters, dancers, and baseball, basketball, and football players. Anyone who has to perform quick rotations which put great demands on the trunk is a strong candidate for a tear of these muscles. You know when one of the lumbar (called *paraspinal,* meaning "around the spine") muscles has been strained. The moment of injury is dramatic. However, it takes two to three hours before sufficient bleeding and irritation set in to produce disabling pain. Often, you will be able to finish out a contest or an exercise session only to suffer later.

"Because hockey players arch their backs in an unnatural position when they skate, they contribute to their own back problems," points out Jim Kausek, sports therapist of the Boston Bruins.

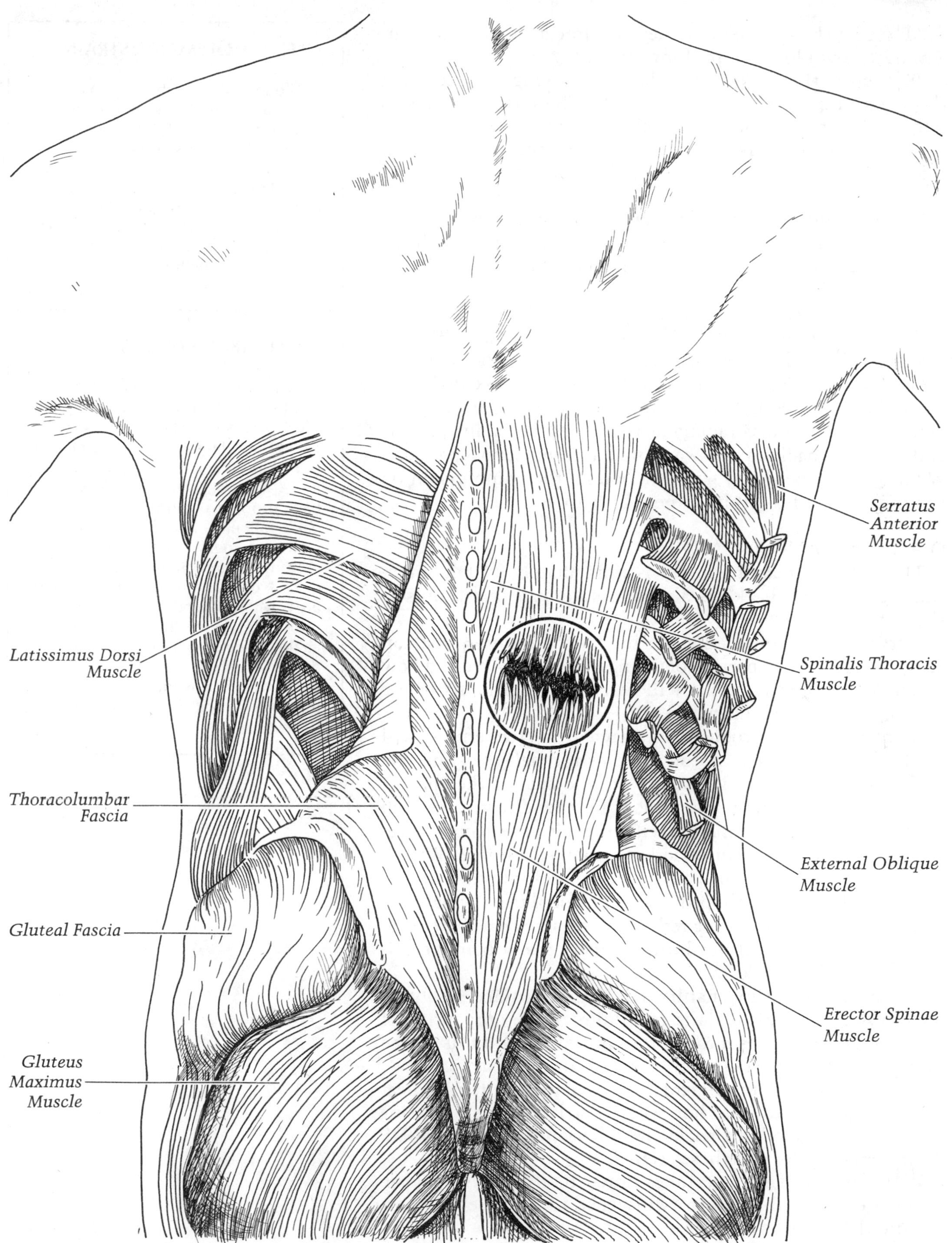

Lumbar Strain

"The best skaters look like they are almost sitting in a chair. They are down low and pushing off hard.

"Secondly, they injure their backs throwing and receiving those darn hip checks onto the boards. Those types of checks put a lot of pressure on the muscles and ligaments.

"Lastly, tight hamstring muscles—half the Bruins are in that category—predisposes them to low back pain."

Offensive linemen in football are noted for having low back problems. When they get off the line, they are underneath the defender and driving upward on him. In this position, their backs are arched.

"In my experience, the back is the weakest part of the body," says trainer John Lally, of the Washington Bullets. "Most basketball players will tell you that in one period of their lives, they grew from six feet to six feet six inches or from six feet three to six feet nine. The muscles in the back don't develop quite as rapidly as the bones. Secondly, basketball players' backs are always under tremendous pressure. When they come down with a rebound, the compression on their discs is at least three times the body weight."

DIAGNOSIS AND TREATMENT

The time to treat the injury is immediately. "I felt something give in my back" is almost a sure sign that a muscle strain has occurred.

Start the ice immediately. Within thirty minutes, you can get a good idea what portion of the muscle is involved and how extensive the tear is, but it will be twenty-four hours before you can get an exact diagnosis.

A muscle strain produces localized pain, tenderness, and swelling. When I push on the muscles through the skin, you will recoil. The strain is usually on one side of the spine—seldom both. Moving your back in the direction which stretches the injured muscle will intensify the pain. Movement in the opposite direction is usually pain-free. With a lumbar muscle strain, the pain never radiates into the buttock or the leg. If it were a slipped disc, you would usually have leg pain. I always check the leg reflexes to be sure that there is no nerve pressure. It will not hurt to push on any bony prominence in the low back.

Given sufficient time, nature will heal this injury. Apply ice for forty-eight hours. This is to minimize the outpouring of blood and tissue juices into the torn muscle area. If this can be minimized, the job of cleanup is easier for the body.

After the initial forty-eight hours of ice, switch to heat treatments. Hot towels, hot water bottles, electric heating pads, and hydrocollator packs are all useful. Diathermy (DP treatments) is ideal for these strains.

If the tear is small, it takes three to five days to heal. Moderate tears take one to two weeks to heal. Severe tears require three weeks or more.

GLENN HOFFMAN'S STRAIN

Glenn Hoffman, the Red Sox's third baseman, was taking grounders in a warm-up drill. He bent down and fielded the ball smoothly and then quickly turned and threw to second base. He felt something "pop" on the right side of his low back. I examined him; he had a lumbar sprain. Ice and heat treatments were started, and he was at full strength in ten days.

MY LUMBAR STRAINS

Two years ago I picked up an eighty-pound barbell to do an upright row exercise. I started to raise the weight and literally felt a muscle tear in my low back. "It's old age setting in," I thought, since I had never had back problems before. That night, my back stiffened. I could feel the localized tenderness in my injured muscles with my fingertips. If it was a slipped disc, I couldn't have felt the tender area. My wife iced my back for twenty minutes twice that night. I iced it again the next night. After the first two days, I used an electric heating pad twice daily for twenty minutes. Within a week I felt great. Then I started back on situps.

> **BACK TO BACK**
>
> In the 1960s, I was a consultant to the New York Rangers. At the time, four Ranger players—Jean Ratelle, Gil Gilbert, Harry Howe, and Camille Henry—had had discs removed from their backs and had back fusions. My job was to develop a back program for them, which I did.
>
> I gave them exercises to strengthen their abdominal muscles. I was careful not to have them overstrengthen their belly muscles because I didn't want the abdominal muscles overpowering the weakened back muscles—thus creating a new problem. My second goal was to assign them certain exercises to keep their backs flexible, especially at the site of the fusion.
>
> All these superstar hockey players were willing patients. They all did their exercise—and their successful careers continued.
>
> Earl Hoerner, M.D., clinical professor of rehabilitation and coauthor of *Sports Injuries—The Unthwarted Epidemic.*

Once sufficient scar tissue has formed across the muscle tear, returning to sports is safe. This is a day-to-day decision. Reinjury is possible. I insist that you be able to do at least thirty situps without discomfort before I will let you return to action. This tests the strength of the abdominal muscles as well as the lumbar muscle strain. Situps give me a good idea of where you are in the healing cycle. Lumbar strains take six weeks to heal completely and you should expect some stiffness and mild discomfort during that time.

Lumbar Slipped Disc

CAUSES

Millions of Americans suffer at one time or another with a slipped disc in the lumbar spine. That's because the lowest two discs in the bottom of your back are subject to huge forces, especially during sports. With the arms and upper part of the spine acting as levers on the bottom of your back, the force on your discs can equal 2000 to 3000 pounds per square inch. Also, the lower back muscles are tremendously powerful and can produce huge compressive forces on the discs.

Ninety percent of disc trouble in the low back is at the two lowest discs. It is unusual to see a slipped disc in athletes and exercisers under twenty years of age. But, after this age, under stress, discs can shatter or fragment. This is, in part, because discs tend to dehydrate with age, which makes fragmentation easier. If one of the fragments shifts outside the confines of the normal disc or applies pressure to the origins of the sciatic nerve which travels down into the buttock and leg, severe pain is caused.

The pain often radiates into the buttock and sometimes into the back of the leg. In severe attacks, the pain goes all the

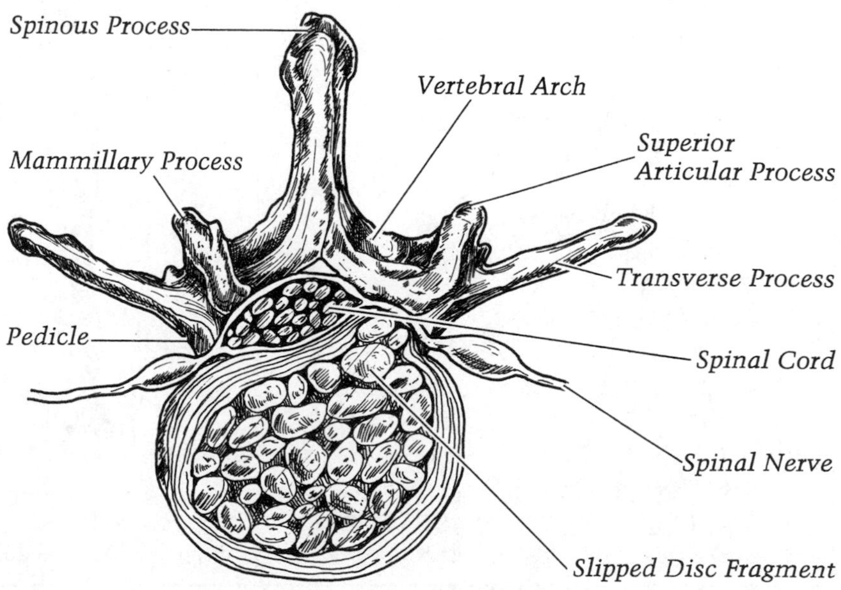

Fragmented Disc

way into the foot. Sometimes the pain is accompanied by a pins-and-needles sensation or numbness. High degrees of nerve pressure will also produce muscle weakness and a feeling that you cannot control your leg. We call this sciatica.

DIAGNOSIS AND TREATMENT

For most people, a slipped disc starts with some casual motion such as bending over to tie your shoe laces or to dry your leg with a towel. Even a sneeze can push a disc fragment into a new and painful position. Once the fragment shifts to

Level of Herniation	Pain	Numbness	Weakness
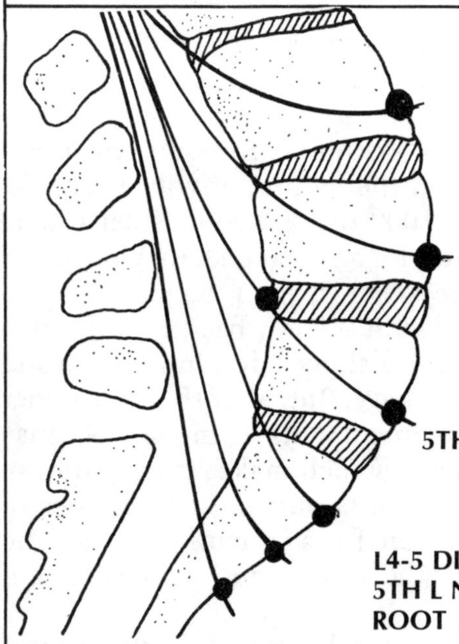 5TH L **L4-5 DISC: 5TH L NERVE ROOT**	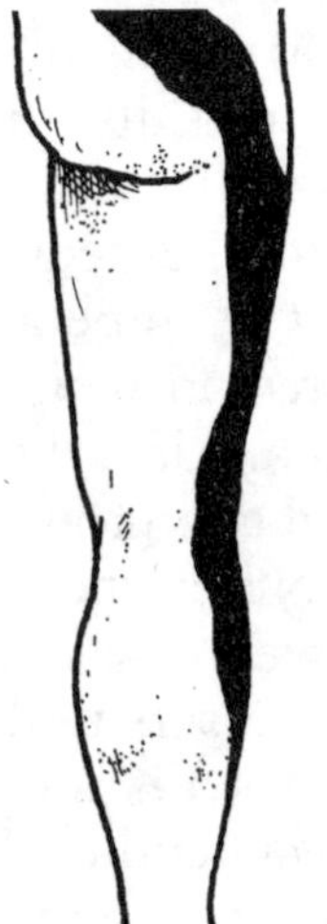 Over sacroiliac joint, hip, lateral thigh, and leg	 Lateral leg, web of great toe	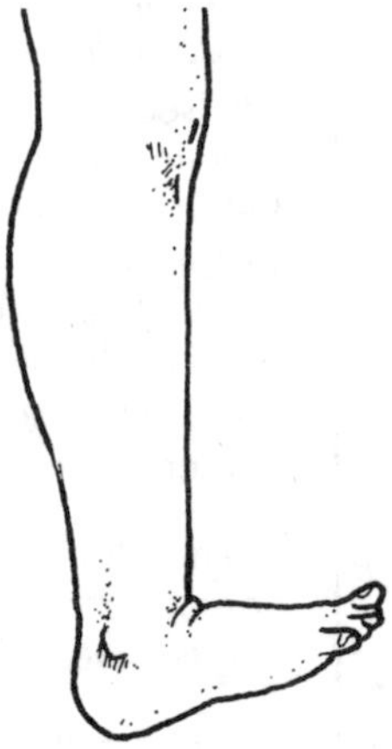 Weakness of dorsiflexion of great toe and foot; difficulty walking on heels; foot drop may occur
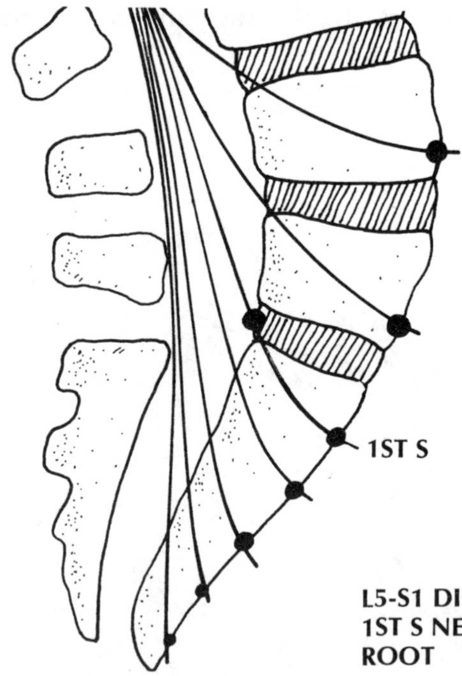 1ST S **L5-S1 DISC 1ST S NERVE ROOT**	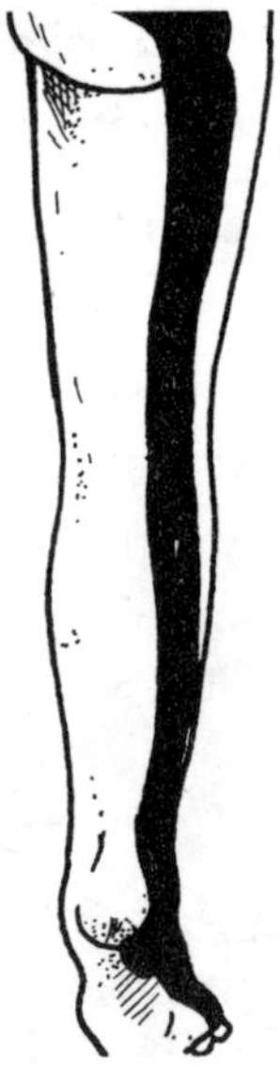 Over sacroiliac joint, hip, posterolateral thigh, and leg to heel	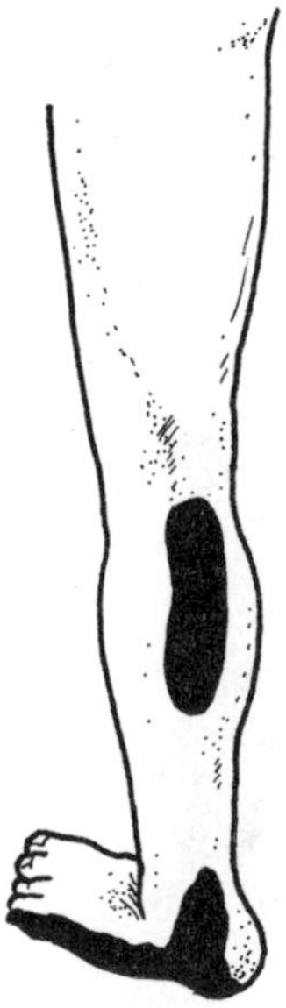 Back of calf; lateral heel, foot, and toe	Plantar flexion of foot and great toe may be affected; difficulty walking on toes

Atrophy	Reflexes
Minor	Changes uncommon (absent or diminished post tibial reflex)
Gastrocnemius and soleus	Ankle jerk diminished or absent

put pressure on the nerve, pain begins.

Within six hours of the shifting of the fragment, you will have excruciating pain. You will not be able to bend over to take off your shoes and socks. It will be hard even getting to the bathroom. Within forty-eight hours, most people are in a physician's office.

The first step in the examination of a disc problem is to ask you to stand as straight as you can. Usually, the muscles on the side of the back where the disc fragment has slipped go into spasm. I feel the muscle with my fingertips; it is as hard as a rock. The corresponding muscle on the other side of the spine is as soft as butter. Often, the sciatic nerve is tender when you push on it in the buttock area.

Next, while you sit down, I check the reflexes of your knees and ankles. I do this by tapping on the knee and ankle tendons with a reflex hammer. With severe nerve pressure, your reflexes will be sluggish or absent. Next, I check for a muscle weakness. I pay special attention to the muscles that allow you to cock your toes up toward your shin. I compare the two sides. With the tips of my fingers, I rub your leg to check for sensation. I am looking for patches of numbness. If I have any doubts, I will prick with a safety pin.

The last step is to put you on your back. I ask you to keep your leg straight, and then I elevate the entire leg off the table. You must keep your knee straight. This is called *straight leg raising.* If the sciatic nerve is under pressure from a slipped disc fragment, it will be very painful for you to bring your leg up ninety degrees from the table. Usually, the pain is on the side where the trouble lies. You can elevate your good leg far beyond the painful one.

Because I want to know if the spine is normal mechanically and if you have the right number of elements in the spine, I always take x-rays. In that way, I rule out the possibility of a tumor or an infection.

I cannot see the discs on the x-ray—only an empty space between the blocks of bone (called *vertebral bodies*) which form the spinal column. If you have had long-standing disc problems, the disc usually loses its normal height and the blocks of bone move closer together. In medical terms, this is called the *narrowing of the disc space.* If this happens, compensatory narrowing of the facet joints behind these vertebral bodies must occur as well. This throws these joints out of whack, and they develop spurs and become arthritic. Besides the pressure on the sciatic nerve, the spurs and the arthritis can be a source of pain.

Those with acute disc problems, I put to bed for three days. I recommend using an electric heating pad or a hydrocollator pack to help relieve the muscle spasm in your back. Use a firm mattress.

A disc does not heal quickly, and you must expect up to three weeks of discomfort. I always see patients back in my office one week after acute onset.

WILLIAMS FLEXION EXERCISES

1. Lie on your back with knees bent and hands clasped behind neck. Feet flat on the floor. Take a deep breath and relax. Press the small of your back against the floor and tighten your stomach and buttock muscles. This should cause the lower end of the pelvis to rotate forward and flatten your back against the floor. Hold for five seconds. Relax.

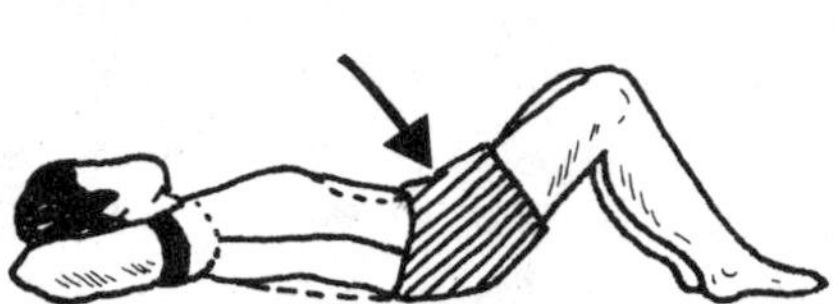

2. Lie on your back with knees bent. Feet flat on the floor. Take a deep breath and relax. Grasp one knee with both hands and pull as close to your chest as possible. Return to starting position. Straighten leg. Return to starting position. Repeat with alternate leg.

3. Lie on your back with knees bent. Feet on the floor. Take a deep breath and relax. Grasp *both* knees and pull them as close to your chest as possible. Hold for three seconds, then return to starting position. Straighten legs and relax.

4. Lie on your back with knees bent. Feet flat on the floor. Take a deep breath and relax. Draw one knee to chest. Then point leg upward as far as possible. Return to starting position. Relax. Repeat with alternate leg.

Note: This exercise is useful in stretching tight hamstring muscles, but is not recommended for patients with sciatic pain associated with a herniated disc.

5. a. Lie on your stomach with hands clasped behind back. Pull shoulders back and down by pushing hands downward toward feet, pinching shoulder blades together, and lift head from floor. Take a deep breath. Hold for two seconds. Relax.

b. Stand erect. With one hand grasp the thumb of other hand behind the back, then pull downward toward the floor; stand on toes and look at the ceiling while exerting the downward pull. Hold momentarily, then relax. Repeat ten times at intervals of two hours during the working day. Take an exercise break instead of a coffee break!

6. Stand with your back against doorway. Place heels four inches away from frame. Take a deep breath and relax. Press the small of your back against doorway. Tighten your stomach and buttock muscles, allowing your knees to bend slightly. This should cause the lower end of the pelvis to rotate forward (as in exercise 1). Press your neck up against doorway. Press both hands against opposite side of doorway and straighten both knees. Hold for two seconds. Relax.

The following exercises (7, 8, and 9) should not be started until you are free of pain and the other exercises have been done for several weeks.

7. Lie on your back with your legs straight out, knees unbent and arms at your sides. Take a deep breath and relax. Raise legs one at a time as high as is comfortable and lower to floor as slowly as possible. Repeat five times for each leg.

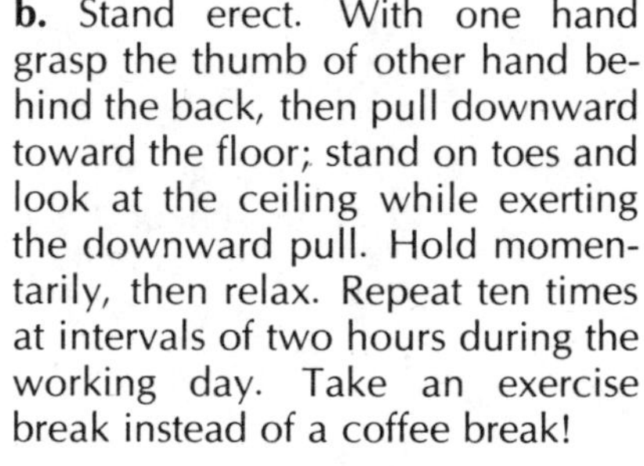

8. May be done holding onto a chair or table. After squatting, flex head forward, bounce up and down two or three times, then assume erect position.

9. Lie on your back with knees bent. Feet flat on floor. Take a deep breath and relax. Pull up to a sitting position keeping knees bent. Return to starting position. Relax. Having someone hold your feet down facilitates this exercise.

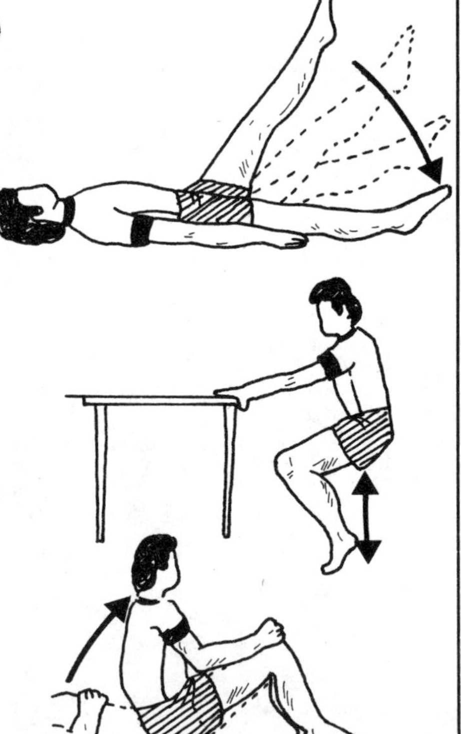

BACK RULES

1. Never bend from the waist only; bend the hips and the knees.
2. Never lift a heavy object higher than your waist.
3. Always turn and face the object you wish to lift.
4. Avoid carrying unbalanced loads; hold heavy objects close to your body.
5. Never carry anything heavier than you can manage with ease.
6. Never lift or move heavy furniture. Wait for someone to do it who knows the principles of leverage.
7. Avoid sudden movements, sudden "overloading" of muscles. Learn to move deliberately, swinging the legs from the hips.
8. Learn to keep the head in line with the spine when standing, sitting, or lying in bed.
9. Put soft chairs and deep couches on your "don't sit" list. During prolonged sitting, cross your legs to rest your back.
10. Your doctor is the only one who can determine when low back pain is due to faulty posture. He or she is the best judge of when you may do general exercises for physical fitness. When you do, omit any exercise which arches or overstrains the lower back: backward bends, or forward bends, touching the toes with the knees straight.
11. Wear shoes with moderate heels, all about the same height. Avoid changing from high to low heels.
12. Put a footrail under the desk and a footrest under the crib.
13. Diaper the baby sitting next to him or her on the bed.
14. Don't stoop and stretch to hang the wash; raise the clothes basket and lower the wash line.
15. Beg or buy a rocking chair. Rocking rests the back by changing the muscle groups used.
16. Train yourself vigorously to use your abdominal muscles to flatten your lower abdomen. In time, this muscle contraction will become habitual, making you the envied possessor of a youthful body profile!
17. Don't strain to open windows or doors.
18. For good posture, concentrate on strengthening "nature's corset"—the abdominal and buttock muscles. The pelvic roll exercise is especially recommended to correct the postural relation between the pelvis and the spine.

SOURCE: Reprinted with permission of the Schering Corporation, Kenilworth, New Jersey 07033.

I usually prescribe analgesic medication to kill the pain and a muscle relaxant. The pain itself produces muscle spasm. Once the muscles around the injured disc go into spasm, they cannot be properly nourished by their blood vessels. Once they are robbed of normal nutrition, they, too, produce pain. Therefore, attacking both the pain and the muscle spasm are an important part of the treatment. If I can keep the back free of pain for seventy-two hours, the muscle will be out of spasm.

If the muscle spasm decreases and you gain better motion and use of your spine, I begin a careful rehabilitation program. This takes the form of ultrasound treatments in the muscle area as well as starting gentle situp exercises. As with any other injured joint, the goals of physical therapy are to regain motion in the area as well as strength. If you had hurt your knee, your knee would be stiff and your leg muscles would be weak. It is no different in the back area. If your back has been injured, you need to regain motion in the lumbar spine and to restrengthen the muscles about the spine and in the abdominal area.

I use the Williams flexion exercise program over the next two weeks. You cannot return to athletic competition until three weeks after the initial attack. When you have 90 percent of normal motion in the back area and good strength in the muscle group, you can return to sports.

In 90 percent of disc cases, time effects a cure. However, because of a marginal blood supply in this area, it takes six to eight weeks for complete resolution of the symptoms. How the process happens is not known, because surgeons never operate on people who get well.

The other characteristic of lumbar disc difficulty is that it comes in repeated attacks. Often, these are two to three years apart, but all have the same type of pain and location.

What about the 10 percent who do not improve despite the two weeks of restricted activity? I worry a great deal about these patients since what I am looking for is steady improvement of the condition. These people I admit to the hospital for pelvic traction. This involves placing a harness around your hip area and hanging two fifteen-pound weights from this harness. I give the patients plenty of painkillers and muscle relaxants. The purpose of the traction is to pull on the spine and open up the disc spaces so that the slipped fragment can retreat to its normal location. This is effective in about 50 percent of cases. The patient spends seven to ten days in the hospital. Following successful traction therapy, I begin the same Williams flexion exercise program. These people seldom return to active sports participation in less than eight weeks from the time of the initial episode.

How do I know they are getting better in traction? They have less pain and, most importantly, begin to lose the pain that radiates down the leg. Usually, the back pain will disappear first and the leg pain (called *sciatica*) will be the last thing to ease off. Again, I am looking for steady, daily improvement of the symptoms.

For those patients that do not improve at all despite two weeks of rest at home and a seven- to ten-day in-hospital traction program, I do a *lumbar myelogram,* an x-ray that reveals pressure on the spinal nerves. The spinal cord and the spinal nerves live inside a fluid-filled tube. If you have a large slipped disc, it will show up as a dent on the myelogram. It looks like someone has punched the tube. This tells me exactly which disc is causing the trouble and how big the disc fragment is that is pushing against your nerve. It also allows me to make sure that no abnormality such as a tumor or an infection exists. If the piece of disc causing the trouble is large, surgery is usually inevitable. If it is quite small, another seven to ten days of bedrest is worthwhile. During that time, I usually give oral cortisone medicine to try to shrink the swelling around the pressured nerve. I never use oral cortisone for more than one week.

What does the surgery involve? It is straightforward. If you have a loose stone in your shoe, you would remove the stone to obtain relief from pain. It's the same in the lumbar disc situation. The object of the surgery is to remove the loose disc fragments that cause the pain. A small entrance is made into the area of the fluid-filled tube. A small bit of bone has to be removed to gain entrance into the spinal canal. This bone is called the *lamina.* Therefore, removal of the bone is called a *laminectomy.* After the laminectomy is performed, I carefully draw the nerves to one side so that I can see the fragmented

JEAN RATELLE'S BACK FUSION

In the late sixties, because of a herniated disc, Jean Ratelle, the high scoring center of the Boston Bruins hockey team, had a back fusion. But his back still causes him pain from time to time.

"The best treatment for chronic low back pain is traction," explains Jim Kausek, sports therapist for the Boston Bruins. "I used to use a True-Ease traction table, until I got an inverted chair. It is a great device.

"A player with back problems, like Ratelle, straps himself into the chair. The chair has an axis, and it rotates. Thus, the chair turns Ratelle backwards and upside down. Basically he hangs by his thighs, but his pelvis is held firmly into the chair.

"After seventy seconds, the vertebrae separates and relieves the pressure on the disc. After Ratelle got used to the inverted chair, he could sit in the upside down position for three or four minutes at a time.

"Besides the traction of the inverted chair, I apply heat or some galvanic muscle stimulation to his back."

disc which I remove. I approach the main portion of the disc using a special instrument. The area of the disc is approximately a half-inch in width. These instruments go into this area between the vertebral bodies, and other loose pieces of disc material are removed. This part of the operation is called a *discectomy.* I never remove the whole disc because the whole disc is not injured. Usually, only the back half of the disc is bad. Also, the main blood tubes, the *aorta* and the *vena cava,* lie just in front of the discs in this area. If you try to remove the whole disc, you could injure one of these main blood conduits. Once all the loose disc material is removed, the nerves are allowed to return gently to their normal position. The muscles of the low back which had been elevated away from the bone early in the operation are replaced. The skin incision is stitched and the patient remains at bedrest for forty-eight hours.

I then like to get my patients up and walking with a corset back support. I have them stay in the hospital for five days and then they are sent home to rest. During the next three weeks, I still like them to spend approximately half their time lying in bed. Walking is good for disc patients. Sitting is the hardest position on a disc; therefore, I like them to minimize that. I let them sit only for bathroom privileges and for meals. One month after surgery, they can begin to sit more and usually can return to their work if it is sedentary. Six weeks after surgery, they begin the Williams flexion exercise program and can begin activities such as bike riding and swimming. Three months after surgery, and if they have regained full strength and motion in the spine, they can return to full athletic competition. Three months is a safe time. Even if you break your thighbone, in three months it will be solidly healed.

What about the people who do not have an attack severe enough to warrant hospital traction or surgery? I see these people every eighteen months or two years or when their backs "slip out." They have an attack of low back pain, often accompanied by leg pain. These people are encouraged to treat the attack seriously.

Go to bed for forty-eight hours. Take painkiller pills and also start the muscle relaxant pills. If you jump on these attacks quickly and early, you can minimize their effect. Take it easy for a week, and usually, by that time, the attack will have blown over. Do not return to sports for two weeks. If these attacks recur more than every few years, this usually indicates a problem of the facet joints—those knuckle-size joints that lie behind the disc. This, then, becomes more the management of an arthritic joint rather than a slipped disc. I have you take anti-inflammatory medication for a month following an acute attack. I have you use heat on your low back for the first forty-eight hours and then immediately begin an exercise program using the Williams flexion exercises. When you return to sports three weeks after the incident, I ask you to wear a lumbosacral corset.

These attacks of facet joint pain can be disabling and often come on every two or three months. If they do not respond to my program, I inject cortisone into joints. This is done in the hospital, in the x-ray area. I use an x-ray machine to make certain that I can introduce a small needle directly into the joint itself. I then inject cortisone directly into the joint. This is called a facet joint cortisone block. Many patients will get six to nine months or even several years of relief from back pain out of this facet joint injection. This is a safe, simple way to relieve facet joint pain.

If you are a competitive or recreational athlete, a disc problem is a very threatening situation. Questions such as "Will I make my back worse if I continue with sports?" or "Will I have to quit forever, Doc?" are very common. I have not found an easy way to answer these questions with certainty. However, in general, you are better off staying active!

Spondylolysis and Spondylolisthesis

CAUSES

In Greek, *spondylo* means "vertebra," and *lysis* means a "loosening" or "release." *Spondylolysis* is a loosened vertebra. *Listhesis* means slipping of one spinal element on the other. *Spondylolisthesis* is a slipped vertebral element. These Greek words describe a process in the low back which is common in athletes.

In 90 percent of cases, spondylolysis or spondylolisthesis involves the lowest block of bone in the spinal column—the fifth lumbar vertebral element.

Spondylolysis starts as a crack in the arch portion of the vertebral element. Most medical scientists generally agree

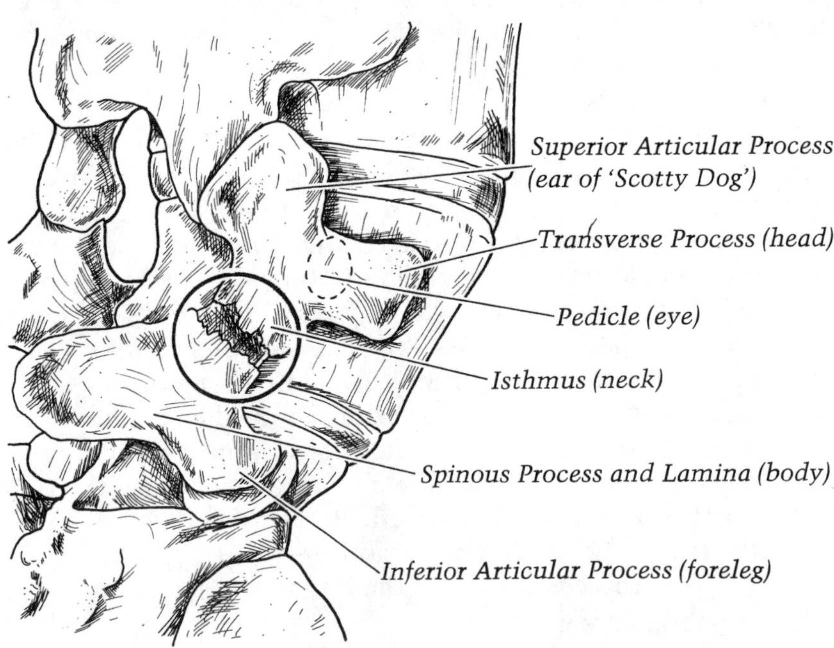

Spondylolysis

MY SPONDYLOLISTHESIS

When I was playing intramural basketball in high school in Fort Worth, Texas, I had my first attack of spondylolisthesis. The pain was so bad that I couldn't even walk up stairs. I went to an orthopedist in Fort Worth, who diagnosed my problem correctly. He put me on some exercises to strengthen my abdominal muscles and sent me to a therapist who administered ultrasound and heat treatments.

My back has bothered me for years. I control the pain with aspirin. Last year, I started a running program and got down to 180 pounds, but the running aggravated my back.

Stan Levine, M.D., the team physician of the Washington Bullets, x-rayed my back. He gave me a new set of exercises, called the Johnston Back exercises. They seemed to help.

The lifestyle I lead is hard on my back. I travel all the time. It is hard for me to get comfortable in an airline seat. Secondly, because I am the smallest member of the team, I usually end up sitting in the bulkhead seats on planes which means that I don't have much room.

Trainer John Lally,
Washington Bullets

that the crack is not present at birth. Rather, it comes on with increasing age and activity. I see most spondylolysis patients mainly after the age of twelve years. In most instances, the crack is present on both sides of the arch. In medical parlance, it is *bilateral*.

Does sports participation increase your chances of developing spondylolysis? The answer is yes. Dr. J. H. McMaster at the University of Pittsburgh did a study of interior linemen on their varsity football team. He x-rayed all of their backs and found that many had cracks in the lowest bone of their spine. This rate of spondylolysis was several times greater than that experienced generally. Dr. McMaster surmised that perhaps the four-point stance in football linemen put tremendous stress on the lower part of the back and caused the cracks to happen. Many great athletes are affected by this ailment: Olga Korbut, Carlton Fisk, and George Scott, former Red Sox first baseman.

A second study of lumbar spondylolysis of football players was made by Drs. R. L. Semon and D. M. Spengler at the University of Washington. They reviewed the medical records of 506 football players who participated over a seven-year period. Of these, 135 or 26.7 percent complained of back pain at some point in their football career. Forty-seven of these athletes had sufficient back pain to warrant spinal x-rays; 25.5 percent showed a bone defect in the low back—spondylolysis. Of note was the fact that these players with bone defects did not lose any more time from practice than did those players who complained of back pain but did not have the bone defect present. The conclusion was "that the presence of lumbar spondylolysis is of minimal clinical significance over a short (four-year) period"—the duration of a college football career.

"I have noticed the increasing incidence of spondylolysis in ballet dancers and gymnasts," points out Arthur M. Pappas,

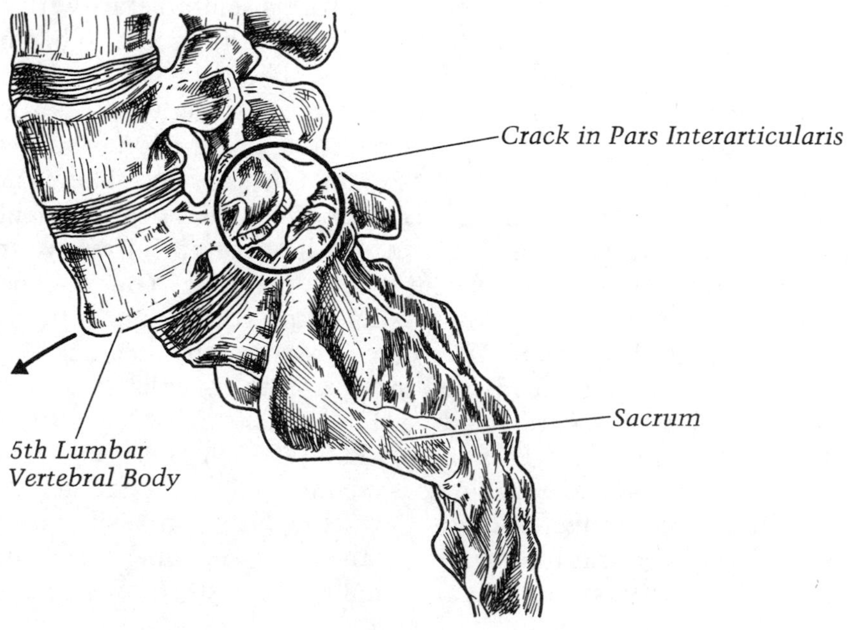

Spondylolisthesis

nationally known specialist on children's sports medicine. "One reason for the rise is the dramatic increase in the number of young people now participating in these activities. Secondly, children are starting at an earlier age. Lastly, there is tremendous pressure on the bones of the spine when a child dismounts from uneven bars or swings on rings. The stress on the back can be five to ten times body weight. Nobody knows for sure."

I think that some people are born with thin bone in this area. The stress and force applied to the already thin area in sports leads to a complete cracking through of the bone. If people with thin bone in the low back did not play sports, they would not develop the spondylolysis so early in life.

When the arch cracks through the vertebral body, it is no longer linked to the rest of the vertebral element. The vertebral body now is free to slide forward on the element just below it. This slipping forward is called *spondylolisthesis*.

Arbitrarily, physicians divide this slipping into four categories. These categories reflect the amount that one block of bone slips forward on its mate below. If it slips forward one-fourth of the way, it is called grade I spondylolisthesis. Grade II is a halfway slip. Grade III is a three-fourths slip, and grade IV is a complete slip.

Spondylolisthesis was first recognized in 1857 when a young woman died in childbirth. An autopsy revealed that she had a grade III spondylolisthesis which had blocked her birth canal and made it impossible for the child to leave the womb. This is the first time that medical science knew that this slipping phenomenon could take place. Once roentgenography (x-ray) was invented, the diagnosis became widely recognized.

DIAGNOSIS AND TREATMENT

All patients who come into my office with spondylolysis or spondylolisthesis complain of backache. When playing sports, the ache becomes worse. These conditions seldom start with a dramatic event or a specific injury. Rather, they seem to begin insidiously—over four to five months. The pain is located in the lower part of the back and tends to go straight across the back itself. Unlike a disc problem, the pain of spondylolysis and spondylolisthesis seldom radiates into the legs. The back hurts when the spinal column is mechanically loaded. You feel pain when you walk, sit, or run. With these activities, your spinal column is in a vertical-loaded position. You will be relieved of pain by lying down and putting your spinal column in a horizontal position. That takes pressure off the cracks.

Most people have restricted motion in the low back area to forward and backward bending. Often, they have very tight hamstring muscles and can thus barely reach the floor with their fingertips. If the slipping is advanced (i.e., second degree or more), you can feel a stepoff in the low back. This is because the slipped forward segment is ahead of the one below it. Sometimes, the slipping is responsible for a stretching of the

OLGA KORBUT'S BACK

"I met Olga Korbut in 1972 after she had won three gold medals in the Olympics," says Earl Hoerner, M.D., clinical professor of rehabilitation at Tufts University and coauthor of *Sports Injuries—The Unthwarted Epidemic*. "I noticed that she wore a special back brace all the time. It was made out of orthoplast, a very light, forming-fitting back support.

"I asked her coaches and doctors about the brace. They told me that she had a 25 percent slippage in her spinal column. In medical parlance, this is called a '25° spondylolisthesis.'

"In 1975, I met her again when visiting Russia and I talked with her doctors about putting her in a back strengthening program to control her symptoms. The Russians were anxious to try anything. I gave them my recommendations for exercises and I suspect that they used them. In 1976, I saw Olga again at the Montreal Olympics and she was no longer using her back support."

nerves in the area, which can make the leg reflexes sluggish or absent.

To discuss intelligently the diagnosis and treatment, I think it is helpful to divide this entire group of patients up into three major sections:

- Spondylolysis—without any slipping, the presence of the crack in the bone only.
- Mild spondylolisthesis—a mild slipping of one vertebral body on another.
- Severe spondylolisthesis—a marked degree of slipping of one vertebral body on another.

There are different concerns and treatments necessary for each of these groups.

Treatment of Spondylolysis without Any Slipping

Once the diagnosis is made, stop sports until the pain abates. I prescribe mild painkillers and the application of heat to the low back. Often, this reduces the pain. Usually within a week, the pain disappears.

Some orthopedists contend that if you have spondylolysis, you should not play contact sports. Their concern is that the stress of sports will develop into slipping or spondylolisthesis. There is no evidence in the sports medicine literature that sports participation will lead to gradual slipping in the spine of a person with spondylolysis. Therefore, I let athletes continue in the sports of their choice. I take them out of sports activities only if the stress of that sport causes recurrent painful episodes, which leads to a lot of missed time in that activity. Two former Boston Red Sox players—Rick Burleson (now with the California Angels) and Carlton Fisk (now with the Chicago White Sox)—have spondylolysis. Occasionally, they will miss a few days because of backache but play without any back complaints the rest of the time. In my opinion, it doesn't make sense for us to tell those two outstanding baseball players that their careers are over because they have small cracks in one of the vertebral elements at the bottom of their spines. I would use the same approach for all athletes that I treat with spondylolysis.

Treatment of First- and Second-Degree (Mild) Spondylolisthesis

"First-degree spondylolisthesis" means that the vertebral body in question slips forward one-fourth of the distance of the vertebral body below it. Second-degree means that it slips forward one-half the width of the body below it. Athletes with this condition have discomfort in their back more frequently than do those who have spondylolysis alone. When the pain comes, it is much more severe. It takes two to three weeks for

WHY GEORGE SCOTT LOST FORTY POUNDS

George "Boomer" Scott, the long-time first baseman of the Boston Red Sox, has a grade II spondylolisthesis—a 50 percent displacement in the normal alignment of the spine.

"During the 1978 season, Boomer complained of back pain," recalls Dr. Arthur Pappas.

"I'm having a problem bending over, Doc," Boomer said, pointing to his back.

"When I examined him, I found significant muscle spasm in his lower back," says Dr. Pappas.

"The x-rays revealed a grade II spondylolisthesis." According to Dr. Pappas and other physicians, a spine deformity of this magnitude usually prevents athletes from participating in competitive sports.

"I told Boomer about the back weakness," Dr. Pappas remembers.

"You mean that I have a disc problem?" Boomer asked.

"No, no, Boomer, you have two bones in your spinal column that are not lined up," Dr. Pappas said. "Your protruding abdomen is part of the problem."

"You mean that I'm going to have to lose weight?"

"That's right."

Between the initial examination and the start of the 1979 season, George Scott lost forty pounds. He also did stomach-strengthening exercises which helped hold his back straight.

"It was the most successful weight reduction I have ever recommended. Boomer has never been bothered by the back problems again," Dr. Pappas concludes.

it to go away. This shows that the spine simply is not as strong as normal. I worry a great deal more about these athletes, because I am concerned that the slipping may be an ongoing process. Therefore, I keep close track by x-raying them every three months during their first year under treatment. I do this to be sure that the bones are not slipping quickly one on the other.

I encourage these athletes to continue in noncontact sports. I think that a first- or second-degree slip is an unstable situation in the spine and that contact sports could make it worse. However, I see no harm in allowing them to continue in noncontact sports such as racquetball and tennis.

If the pain becomes unremitting and the slippage increases, you have only two choices. Either you stop sports activities altogether or give strong consideration to surgery. By quitting sports activities, you remove stress from your spine, and it may well be that you can live on a day-to-day basis with this degree of spondylolisthesis. Many athletes, however, do not want to quit sports. They would much rather have me

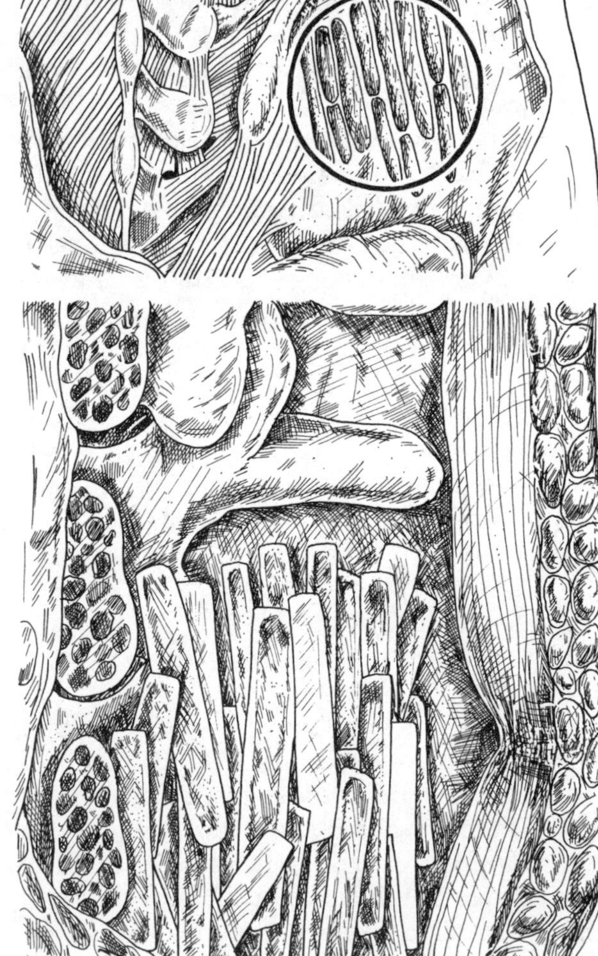

Bone graft material obtained from lateral aspect of ilium and packed over decorticated areas bilaterally

"HE JUMPS AS HIGH AS HE CAN"

Phil Chenier, a former Washington Bullet, has been plagued by back problems most of his career. His back condition is aggravated by his style of play.

"Instead of jumping four inches for a jump shot, Phil jumps to his maximum every time," says trainer John Lally, of the Washington Bullets. "Even if he is wide open, Phil jumps as high as he can. This puts a tremendous stress on his lower back."

strengthen their spine surgically and allow them to return to full sports activities.

The surgery is called a *spinal fusion.* Picture a piece of wood with a crack in it. If you were a carpenter, you would nail a second piece of wood across the crack to reinforce it. This is exactly what I do with a spinal fusion. I borrow bone from the back of the pelvis close to the spine. I surgically expose the weak area in the spine. I lay this bone across the weak area for reinforcement. Unlike the carpenter, however, we must wait six months for the bone to fully solidify. Most of my patients remain in the hospital for one week. I prescribe a rigid back brace for six months. This limits the mobility of the spine and allows the bone to heal completely. Once you have a solid spinal fusion, your back is as strong or stronger than normal. A period of an additional three months is necessary to do muscle strengthening with the Williams flexion exercises. This allows you to regain excellent strength in the fusion area. It also gives you enough time to regain stability in your spine. The spinal fusion only reduces your overall spinal mobility by about 10 percent. Most athletes can live with this.

Treatment of Third- and Fourth-Degree Spondylolisthesis

Here the spine is very unstable. Sports activity can lead only to predictable pain. You also run the risk of suddenly increasing the degree of slippage. Often, this degree of spondylolisthesis has put a lot of pressure and traction on the nerves of the spine. Therefore, you get leg pain. In this case, a spinal fusion is mandatory. Even for casual activities, it is important to make the athlete realize that his or her spine is very weak and unstable. Spinal fusion again is safe and gives predictably good results. The post-operative course for this fusion is the same as for the first- and second-degree spondylolisthesis patients.

Mechanical Low Back Pain

If you were to design an ideal body, you would make it symmetrical. You would want both halves to look the same. If you were to build a house, you would probably want to make both halves of the house symmetrical. You would want the first floor to be level so that the beams that went up to support the second story would have a stable platform from which to rise. The design of the human body follows these same physical principles. You would like your legs to be of equal length. Picture the legs standing like two beams underneath the pelvis. The pelvis is similar to the first floor of your home. It must supply a stable, level platform from which the spinal column arises. Then the spinal column itself should ideally be perfectly straight—like supporting beams for your house. Another example is a level deck on a ship from which arises the main mast for the sails.

In theory, this all sounds great. In practice, it just ain't so. Nature does make small errors. The two halves of the body are almost never exactly symmetrical. People have one eye a shade different color than the other, one ear larger than the other, one foot a larger size than the other, one leg a bit different in length from the other. Let's look at some of these inequalities and discuss their ramifications.

UNEQUAL LEG LENGTHS

If there is more than a half-inch difference in the length of your legs, this is a significant mechanical problem. It is like having unequal columns holding up the first floor of your house. Unequal leg lengths mean that the pelvis is tilted. When the spinal column rises from a tilted platform, it must somehow compensate. It does so by curving away from the short leg. If this did not happen and the spine remained straight, your head would be directly over the hip of the short leg. So if you have unequal leg lengths, the spine will have to curve away constantly from the short leg. This puts unequal stress on the spinal column with the joints, ligaments, and discs on a concave side of the curvature taking much more stress than those of the convex side. This leads to wear-and-tear changes and pain. Furthermore, people tend to stand on the short leg. If they stood on the long leg, the short leg would be off the ground. Once again, they are putting unequal stress on spinal column and body parts.

If you are a runner with unequal leg lengths of at least one-half inch, it is as if you were running on a side-hill slope. This puts a great deal of stress on the knee and hip on the short-leg side. Therefore, when I see athletes with back, hip, or knee pain and unequal leg lengths, I immediately give them corrective lifts in their shoes. Usually, a one-fourth to three-eighths-inch cork-type heel elevation is all that is neceessary. This slips inside the running shoe and in walking shoes. If more than one-half-inch correction is needed, you must also build up the front or the whole sole of the shoe. However, it is well worth it to equalize your leg lengths. I see many youngsters who are complaining of leg pain and back pain who simply need a correction of their leg-length discrepancy.

UNEVEN PELVIS

I see patients who stand with one hip lower than the other. However, when I measure their leg lengths, they are the same. X-rays reveal that one side of the pelvis is higher than the other. As far as the spinal column is concerned, this is the same problem as unequal leg lengths. It gives it a tipped platform. The spinal column has to curve to compensate. Once again, it is a simple matter to put a heel lift of one-fourth to three-eighths inch inside the athletic shoes. This will allow the pelvis to level out and give the back relief from its compensatory curving posture.

Nature's Lumbar Fusion

Nature is not perfect in the formation of the blocks of bone or vertebral elements that form your spinal column. They vary in size and shape. Some people only have four mobile elements in their low back instead of five. The fifth block of bone is fused to the pelvis by extra bone that has been present since birth. We call this *sacralization* of the fifth lumbar vertebral body. In other words, it is a solid part of the pelvis, and not part of the mobile low back. This puts tremendous stress on the other four remaining mobile elements and the discs that join them. I have seen at least 100 athletes who have had back pain from the disc and facet joints directly above a fused fifth lumbar vertebra.

In some instances, one side of the fifth lumbar vertebra will be fused to the pelvis while the other side is not. This is called *hemisacralization.* This leads to abnormal and asymmetrical forces across this area. It overloads the mobile side because the fused side cannot participate in absorbing any shock or force. This can lead to pain.

In some individuals, the bone structures are "programmed" so that the bones are shaped in such a way that the spine is curved. We call this *structural scoliosis.* It means that the curvature is built right into the structure of the bones. The curvature of the low back that occurs to compensate for uneven leg lengths is a functional scoliosis.

How do I tell the difference? I examine the patient's back in the sitting position. This takes the leg-length difference out of the picture. If the back straightens out when the patient is sitting, it is a functional scoliosis. If it stays curved in the sitting position, it is a structural or fixed scoliosis. Lumbar structural scoliosis is much more common in females than males. If it is severe, a corrective brace must be worn, and occasionally even surgery is necessary to eliminate pain.

What do all these mechanical problems have in common? They all have mechanical pain. By this, I mean that the back hurts when you use it. Rest eliminates the pain; use brings it on. The harder you use it, the more it hurts. This is mechanical back pain. The pain does not radiate down the leg as it does with a slipped disc and sciatica. The pain is localized in the low back region and is usually in the middle of the back. Remember, disc trouble tends to be on one side or the other. Physical examination of people with mechanical low back pain shows no evidence of nerve pressure. The reflexes are fine, the muscles are all strong, and the sensation in the legs is normal. However, there is usually some spasm in the muscles of the low back and a tender area can often be found by pushing fairly deeply into the low back area with your thumb. This says to me that it is a structural and/or mechanical problem, rather than a disc problem. Motion of the low back is usually restricted because of the pain of the structural abnormality. X-rays are essential. This tells me whether I am

DENNIS ECKERSLEY'S BACK

In 1979, Dennis Eckersley, the ace starter for the Boston Red Sox, had an excellent season, winning seventeen games and losing only ten and he finished third in ERA (earned run average) in the American League.

"Everyone was expecting a big year out of Eck," says Dr. Arthur Pappas. "We were all surprised when he pitched so poorly early in 1980.

"He began complaining of back discomfort. It was obvious that he was not pitching the ball, but just delivering it with his arm. It was a classic example of the importance of the back in the pitching motion.

"Dr. Southmayd and I checked his back and could feel muscle spasm. X-rays revealed a minor bone abnormality in the lower spine, a congenital malformation. The abnormality was making the lower back unstable and was the source of his discomfort. Why it flared up in the 1980 season is a mystery.

"It was obvious to us that he could not pitch for at least a month to overcome the back spasm. We hospitalized him. He was placed in pelvic traction for a few days and we gave him anti-inflammatory medication.

"Luckily, the spasms started to decrease. We do not know if it was the bedrest, the traction, or the medication which relieved the spasms. I suspect that it was a little of each.

"With prescribed exercises, Eck worked himself back into the starting pitching rotation in six weeks.

"Why this spinal abnormality caused problems in 1980 is difficult to explain. He did not have problems before 1980. Let's hope that he doesn't again."

dealing with a congenital abnormality, such as hemisacralization, a structural lumbar curvature, or a case of unequal leg lengths and normal low back anatomy.

Treatment is directed at the underlying disorder. Unequal

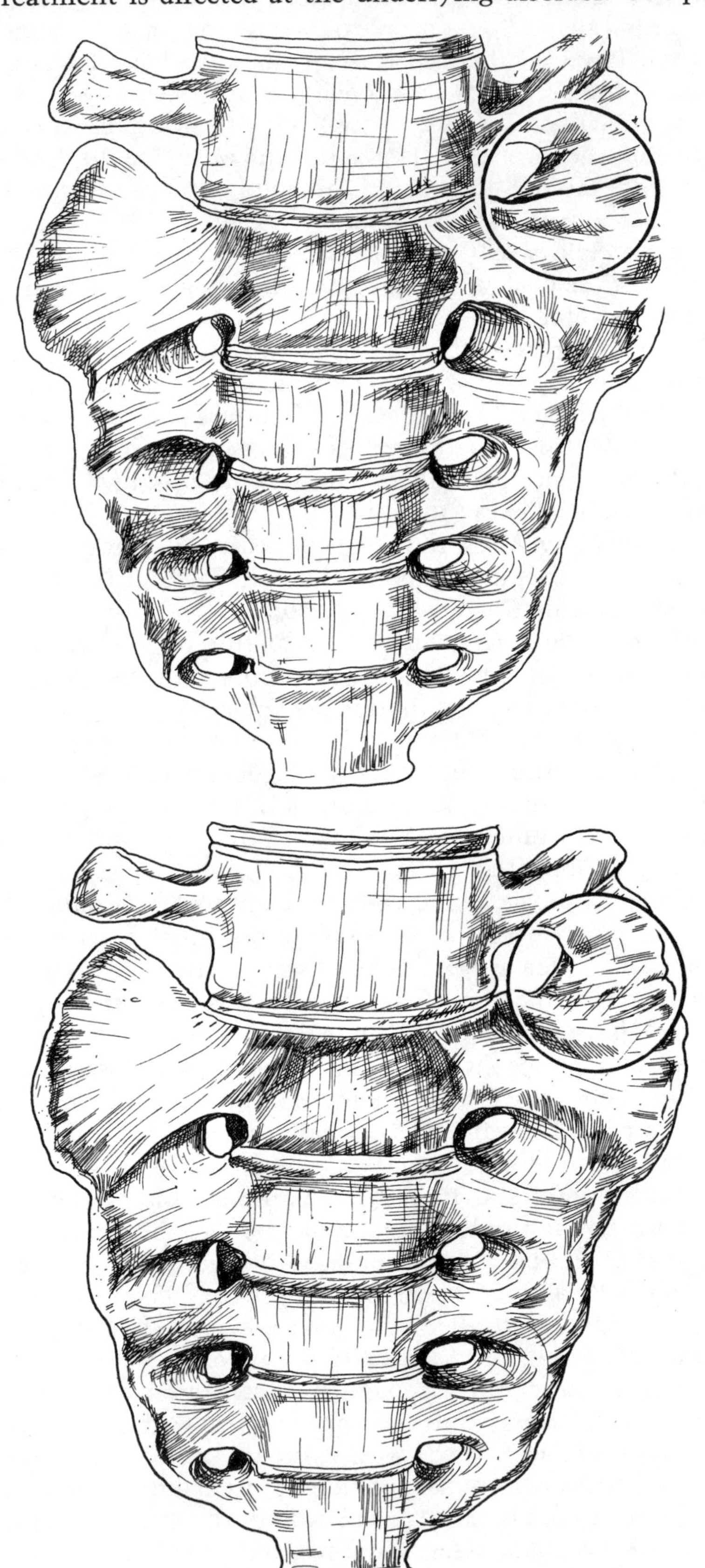

Nature's Lumbar Fusion (Sacralization of 5th Lumbar Vertebra)

leg lengths and a short-sided pelvic bone are corrected with shoe inserts. In all cases, I prescribe ten to fourteen days of restricted activity, and I like to use heat in the low back area to stimulate the circulation to come in and heal the irritation. Abdominal strengthening exercises are essential. I ask all patients to do at least thirty situps three times a week with their knees bent. Strong abdominal muscles are very important for the use of the low back in people with structural abnormalities. Medication stronger than plain aspirin or Bufferin is seldom necessary or effective. Finally, the athletes' competitive expectations must be carefully scrutinized. If long-distance running makes the back hurt despite our best effort, another sport must be chosen. Once the leg-length inequality is corrected, athletes can return to full function in all sports. If a structural bony abnormality is present in the low back, a change of athletic activity is necessary. In general, short-duration, high-intensity sports are easier on athletes suffering from mechanical low back pain. I sometimes recommend the use of an elastic or a semi-rigid lumbosacral corset to provide external support to the low back in these athletes.

Finally, what about the middle-aged athlete whose back is just getting arthritic? "Arthritis" in this sense means wear and tear in the area around the disc and the facet joints. These joints undergo a gradual wearing down of their surface material—similar to a loss of tread off a tire. This comes with age and use. Also, spurs build up along these joints. We call this *facet joint arthropathy.* This just reflects a mild structural abnormality of the facet joint (*arthro* means "joint" in Greek). People with this problem have mechanical low back pain, perhaps every six to nine months. The back will feel stiff the morning after sports activity. Again, the pain may radiate into the buttock, but never down the leg. In three to five days, it feels better and activity can begin again.

I encourage these people not to panic. The pain pattern is predictable and after a while, they come to understand it. I think it is important to be diagnosed correctly, and this can be done by a sports medicine or orthopedic doctor. Once you have an exact diagnosis and know what is causing the pain, you too can get in touch with your physician. I tell my patients that just because they have low back pain does not mean they are causing damage by sports activity. The pain is just the body's way of protesting the use of joints that are somewhat abnormal. I think it is very much to their advantage to continue athletics. This tends to maintain mobility of the muscular and ligamentous elements of the low back, which mechanically works in favor of spinal function. Furthermore, it maintains strength in the muscular elements of the spine; again this is an advantage. Therefore, if you have some wear-and-tear arthritis in your spine, play sports to your fullest capacity. I am absolutely convinced that it is entirely to your advantage, provided that you back off on those few occasions when you develop back pain and stiffness.

MITCH KUPCHAK'S BACK

Forward Mitch Kupchak of the Washington Bullets has had chronic back problems throughout his career. He has had surgery twice. He has been treated with traction, painkillers, muscle relaxants, exercises, heat, cold and many types of physical therapy. He built a Jacuzzi in his home to help relieve back pain. On the bench, during games, he uses a hydrocollator—and he has one at home, too.

In the 1979-80 season, Mitch tried to come back too quickly from back surgery. He explained: "Our team was playing so badly, I wanted to get back in the lineup."

I asked Mitch if anyone put pressure on him to return to the lineup. He said no. He's a great ball player—and he made the decision to return to the game himself.

Trainer John Lally,
Washington Bullets

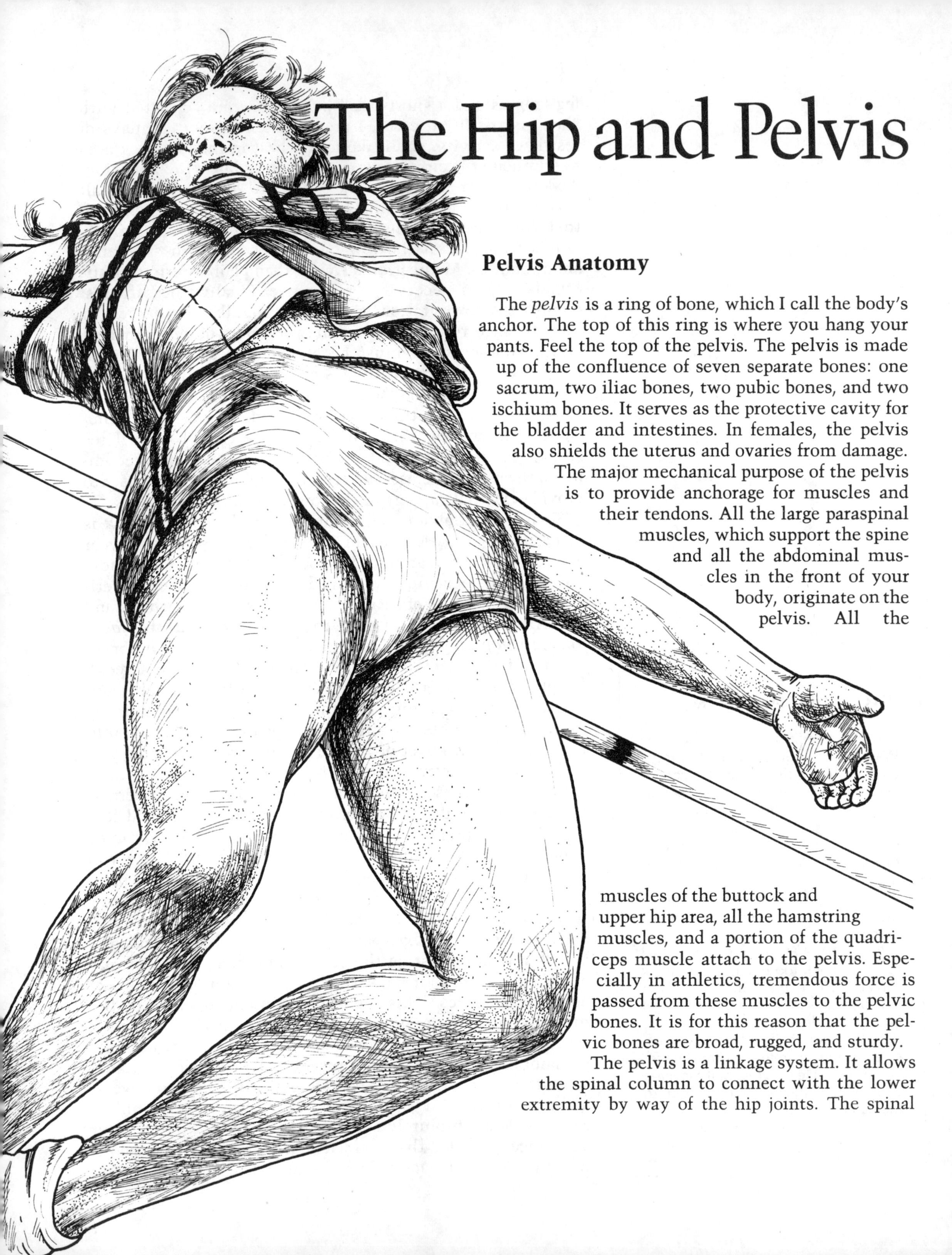

The Hip and Pelvis

Pelvis Anatomy

The *pelvis* is a ring of bone, which I call the body's anchor. The top of this ring is where you hang your pants. Feel the top of the pelvis. The pelvis is made up of the confluence of seven separate bones: one sacrum, two iliac bones, two pubic bones, and two ischium bones. It serves as the protective cavity for the bladder and intestines. In females, the pelvis also shields the uterus and ovaries from damage.

The major mechanical purpose of the pelvis is to provide anchorage for muscles and their tendons. All the large paraspinal muscles, which support the spine and all the abdominal muscles in the front of your body, originate on the pelvis. All the muscles of the buttock and upper hip area, all the hamstring muscles, and a portion of the quadriceps muscle attach to the pelvis. Especially in athletics, tremendous force is passed from these muscles to the pelvic bones. It is for this reason that the pelvic bones are broad, rugged, and sturdy.

The pelvis is a linkage system. It allows the spinal column to connect with the lower extremity by way of the hip joints. The spinal column attaches to the pelvis in the back of the ring.

The *sacrum* is a triangular, solid piece of bone which makes up one-third of the pelvis. The sacrum attaches to the last lumbar vertebral element of the spinal column, called the *lumbosacral junction*. On either side of the sacrum are the sacroiliac joints, which connect the sacrum to the iliac bones. The top of the iliac bone is the brim of the pelvis, which holds up your pants. On the outer side of each iliac bone is a socket-shaped hip joint. The front of the pelvis has two wishbone-shaped structures called the *pubis*, which meet to complete the pelvic ring. The junction of the two pubic bones is called the *pubic symphysis*. This is a flexible joint, and allows a certain amount of flexibility in the pelvic ring. The bottom of the pelvis is formed by two sturdy, stout extension bones, called the *ischium*. This is the bone that you sit on. If you feel with your hand deeply into your buttock, you can feel the ischium.

In growing athletes, many parts of the pelvis have prebones—soft bone. These are mechanically weaker than adult bones and therefore are vulnerable areas for injury.

Because portions of the pelvis are quite prominent and close to the skin, direct injuries are very common, especially in contact sports.

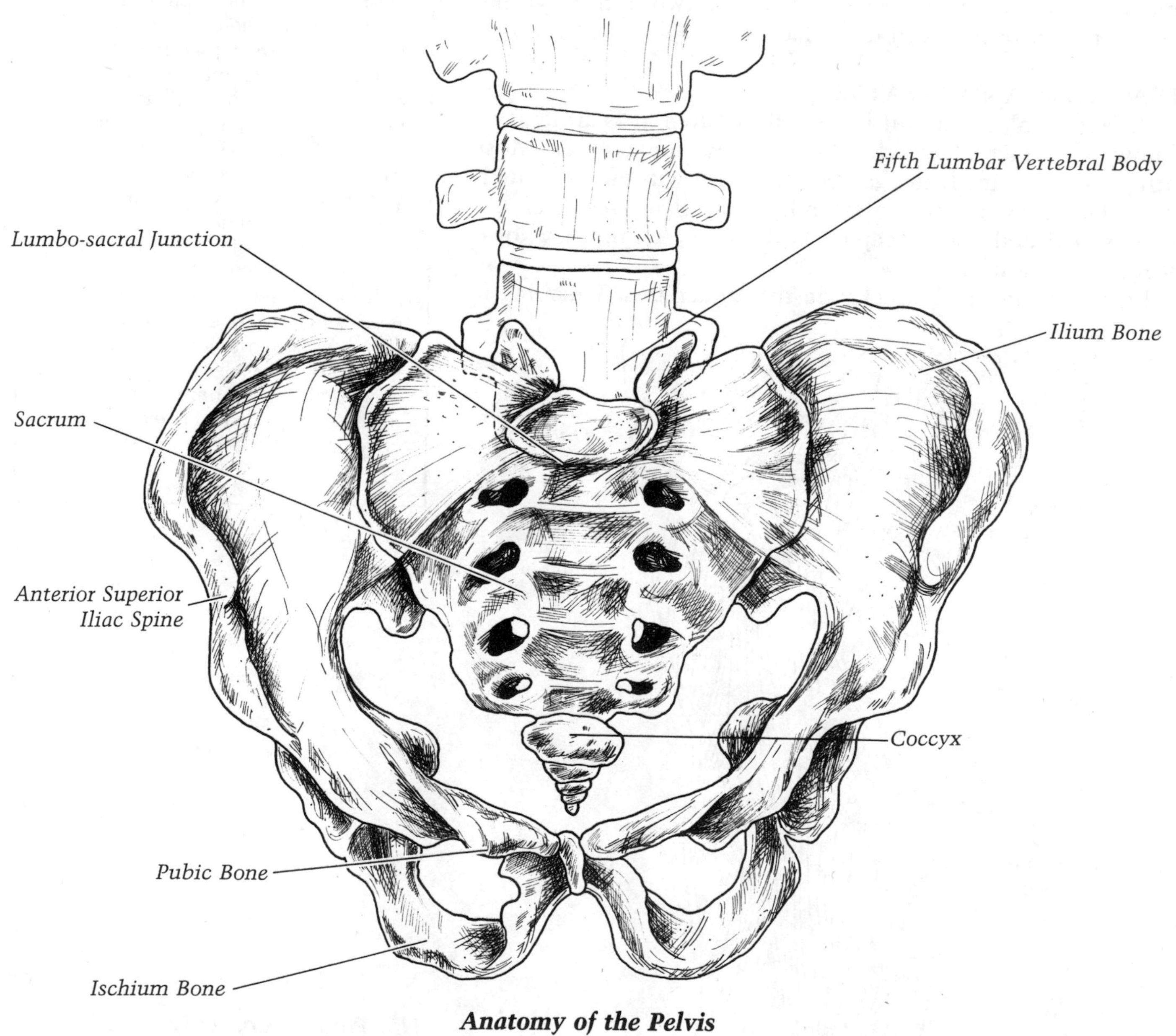

Anatomy of the Pelvis

Hip Pointers

CAUSES

The bone where you hang your pants is the *iliac crest*—the top part of the iliac bone. All the muscles of the buttocks, abdomen, and outer hip attach to this bony ridge. The muscles overlap one another along the bone and are elaborately connected to the bone by a broad tendon.

A *hip pointer* is the ripping away of the tendon which attaches to the iliac crest. In a sense, it is analogous to tennis elbow. That's why I call it the "tennis elbow of the hip."

Because it sticks away from the body, the iliac crest is vulnerable to injury—a blow from a helmet or a hockey stick—or to a fall. Most athletes in contact sports wear hip pads, but sometimes these pads are ill-fitting or too small. This can cause injuries. Once ripped, the tendon tries to heal itself. Its blood supply is not spectacular; it will be two to three weeks before proper healing can take place.

DIAGNOSIS AND TREATMENT

Because athletes are hit in the hip so often, it is difficult to know the exact blow which caused the hip pointer. Most athletes often finish the game or exercise without too much pain. However, in two to six hours, the iliac crest area becomes stiff and sore. Sleeping is difficult. A limp develops. Running is painful.

I can put my hand directly on the tender area. Touching it

HIP POINTER IN RUNNERS

Hip pointers are not a major problem in adolescent runners, according to a study of 310 track athletes by Dr. William G. Clancy, who heads the sports medicine section at the University of Wisconsin.

However, Dr. Clancy reported that of the 310 athletes in the study, which covered a period of thirty months, twenty-one (or 7 percent) sustained injuries to the iliac crest area.

All were undergoing growth, and the growth area of the iliac crest is highly vulnerable to injury. Symptoms were similar. All athletes had localized tenderness over the iliac crest and experienced pain when they tried to move their legs directly out from the centers of their bodies.

This activated muscles that were attached to the injured area. Initial treatment for all patients was three to four weeks of rest. Four of the patients were given cortisone injections. Within four to six weeks, all runners were able to resume training.

Dr. Clancy's diagnosis: the twenty injuries were traceable to the high mileage logged by the distance runners. All had suffered in an "apophysitis," or an irritation of a growing portion of the skeleton. It was further suggested that running 80 to 100 miles a week may be excessive for athletes whose bones are still growing.

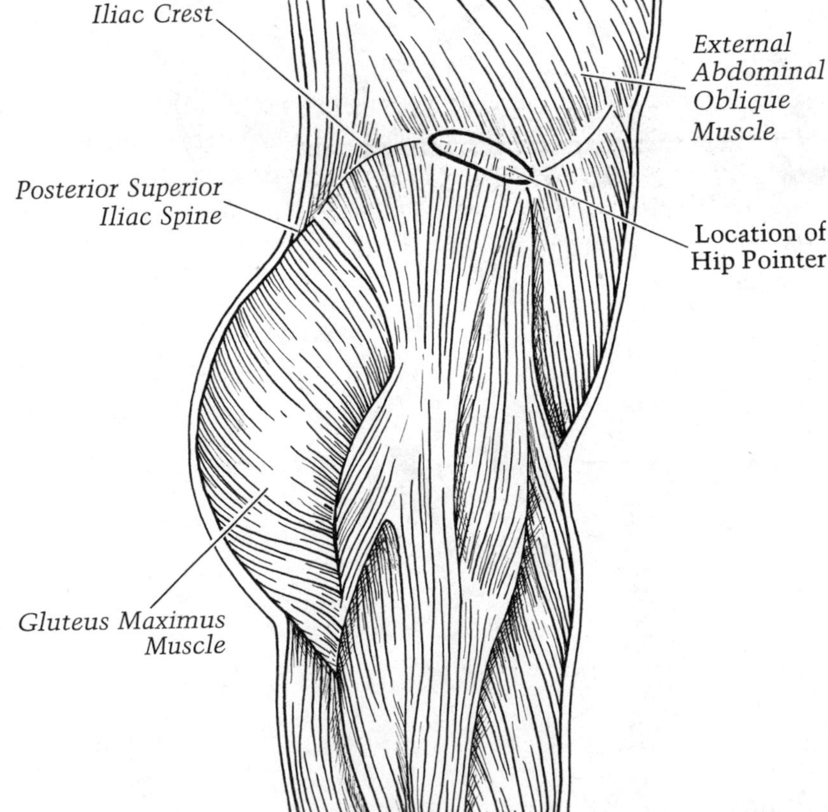

Hip Pointer Anatomy

causes pain. It lies just beneath the skin. It usually lies just at the side of the pelvis, about midway between the front and back of the body. Often, there will be some swelling.

If I ask you to bend your trunk away from the injured side, this will produce discomfort, but there is no limitation of hip joint motion. Seldom is a black-and-blue area seen.

I x-ray the injured part. I am looking for a fleck of bone, which sometimes tears off with its muscle-tendon attachment. A pelvic fracture in this area is rare in athletes. It is far more common in automobile or motorcycle accidents.

The best treatment is rest. That means no sports for a week. Ice the tender area twice a day for twenty minutes on two successive days. Then apply heat to the area three times daily for twenty minutes. This stimulates circulation which helps healing. Hot towels, an electric heating pad, hydrocollator packs, or diathermy (a more deeply penetrating heat) all work. If you have much less tenderness after a week, I can start you jogging.

Contact should be avoided for at least another week. If you go back into action too quickly and the injury is aggravated, then you may lose several months from competition because the injury was not allowed to heal properly. I seldom find anti-inflammatory medication necessary, and I do not give pain-killers for hip pointers. Nature heals this injury.

PAUL GUZZI'S HIP POINTER

Paul Guzzi, the former Secretary of State of Massachusetts, was a high school and college football teammate of mine.

In his junior year at Harvard, Paul, a defensive back, was hit with a downfield block in the right hip area. He finished the game without pain.

"My hip is killing me," Paul complained later that night.

"Oh, it's only a hip bruise," I said. "You'll be okay tomorrow."

The next morning, Paul could hardly walk. I helped him over to the Harvard fieldhouse to see Dr. Thomas Quigley. It was the first time I had ever heard of a hip pointer. Dr. Quigley explained to Paul the nature of the injury and he told my friend, "Hip pointers take several weeks to heal."

Paul was determined and tough. On the way back to the dorm, he said, "I can overcome this in two to three days." But the three days came and went and he still hobbled.

Paul's injury was a great lesson for me. No matter how tough you are, some injuries cannot be played through.

I remember the frustration in Paul's eyes as he waited for this injury to heal. He returned at the end of the season. What Paul and I thought was a "simple bruise" turned out to be an injury that cost him four weeks of the football season. It was six weeks before his discomfort was gone.

Bone Avulsion Injuries in the Pelvis Area

CAUSES

A *bone avulsion* exists when a tendon separates from the main trunk of the bone. I see bone avulsion injuries mainly in growing athletes. That is because their tendons are attached to

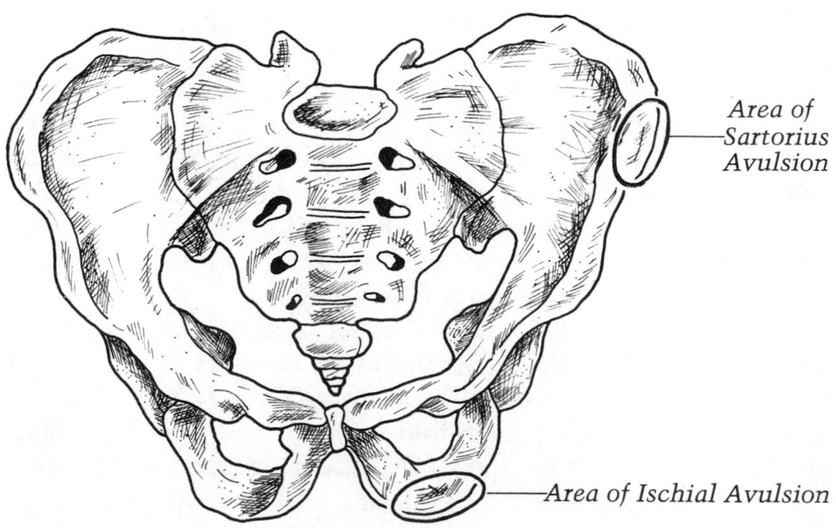

Pelvic Avulsion Injuries

prebone structures or growth areas which in turn attach to the trunk of the bone. The growth areas are much weaker than the mature bone.

Besides having prebone attachments, athletes in their early teens have growing and powerful muscles. Their muscles sometimes overpower the growth area. It is easy to see how a quick, forceful contraction of a muscle could actually pull the prebone away from the main trunk of the bone.

Eighty percent of the bone avulsion injuries I treat are in the hip area. The two main prebone areas that sustain them are the tendon attachments of the sartorius muscle and the ischium bone, the bone you sit on. The sartorius starts out on the front of the iliac crest and runs down across the front of the thigh. It becomes one of the tendons of the pes anserinus, which crosses the knee joint and attaches to the lower leg bone (tibia). If the prebone for the sartorius muscles is disrupted, the athlete will feel tenderness directly over the front of the iliac crest. This is the corner where your belt hangs.

The second area is the ischium where all of the major hamstring tendons attach. You can feel the bone by pushing deeply into your buttocks. The prebone area in teenagers is large and when it is disrupted causes discomfort and disability.

Bone avulsion comes with startling suddenness. About 80 percent of the time, you will be sprinting with your leg muscles contracting as hard as they can.

The contractions put tremendous force on the prebone area. You feel something let go in the pelvis area and you will not be able to run. The location of the sensation depends on which prebone disrupts. It happens because your muscles are just too strong for your bones.

DIAGNOSIS AND TREATMENT

In both cases, I start my diagnosis by carefully listening to your story of the injury. Secondly, I touch for tenderness directly over the disrupted area. I feel for swelling and a sense of fullness. You might have a black-and-blue area from the blood vessels which were ripped during this injury. They pour blood into the soft tissues. If I move the hip joints, which stretch the tendons involved, your pain increases dramatically.

X-rays are essential in diagnosing this disorder. I can compare the two sides of the pelvis. The injured side shows a large piece of prebone pulled away from the main portion of the pelvis. The amount of distraction can be measured with a ruler as compared to the normal side. If the injury is more than two weeks old, the x-rays show healing between the pelvis itself and the bone piece that has pulled away.

Although there was some dispute about bone avulsions among surgeons, I have never found it necessary to operate on these injuries. New bone will form quickly between the piece pulled away and the main bone. Once this is solidly reattached, the pain will disappear. As growth continues, this area will be remodeled and the old injury will be hard to detect.

Nature is able to cope with this injury. My job is to help nature along. I do this by placing you on crutches for seven to ten days. I have you use a partial weight-bearing gait. By this I mean you can put some weight on the injured leg, but every time you have that foot on the ground, I want you to push on your crutches.

This minimizes the amount of muscle contraction necessary to walk, and therefore creates much less tugging on the injured area by the muscle tendons. I have found that heat treatments and medication do not help.

Three weeks after the onset, I take more x-rays of the bone avulsion. If you have very little tenderness in the injured area, then you may come off your crutches. This injury takes at least six weeks to heal completely. During this time, you are not allowed to exercise. Following the six-week period, you can begin weight lifting to strengthen the muscles about the thigh and hip area. With six weeks of disuse, these muscles become much smaller. You should not return to sports until these muscles rebuild, which usually takes three to four weeks. During this time, begin to run straight, but not at full speed, and do agility drills. Ten weeks later, you are able to return to full sports activities. This generally means that you miss an entire sports season after such an injury.

Hip Anatomy

BONES AND JOINTS

The *hip* is the ball-and-socket joint which allows you to swivel your leg and lift your thigh. With a frozen hip joint, you would have a difficult time climbing stairs and stepping into a bathtub. The hip, which allows you to draw your knee up to your chest, has a total arc of motion approximately 110 degrees. This compares to 180 degrees for the shoulders.

Unlike the shoulder, the hip is very stable and sturdy. It needs to be because there are tremendous forces generated by the strong muscles that surround the hip, as well as the leverage effect of the long thighbone (femur) acting as a crowbar on the hip joint itself. It is practically impossible to dislocate the hip joint without first fracturing the socket.

The ball portion of the hip is, in fact, the top of the thighbone (the *femur*) which is called the *femoral head*. Directly beneath the femoral head is the *femoral neck*. The neck, besides supporting the head, connects to the main shaft of the femur bone. The normal femoral neck joins the shaft of the femur at a 135-degree angle. During the growing years, the ball portion of the hip is a prebone. It is connected to the femoral neck through a growth plate and is thus a mechanically weak area.

The hip socket lies on the outer side of the pelvis. As in the shoulder, the hip socket is deepened by a fibrocartilaginous area called the *labrum*, or lip. This makes the joint even more stable. As in all of the joints in the body, the ball and socket are held together by a sleeve-like structure, made of gristle-like ligaments, called the *capsule*.

Because it is a ball-and-socket joint, it can move in all three planes: front to back (flexion/extension), side to side (adduction/abduction), and rotation in and out (externally/internally). *Adduction* of the hip means moving the leg toward the center of the body. *Abduction* of the hip means moving the leg away from the center of the body or out to the side.

MUSCLES AND NERVES

The hip joint would be useless without the muscles surrounding it. These provide the motor power to make the joint move powerfully. The main muscle that allows you to bend your hip up toward your chest is the *iliopsoas muscle*. This attaches at the lumbar spine area and crosses the pelvis to attach to the hip bone. The muscle that opposes the iliopsoas is the *gluteus maximus muscle*. It straightens out the hip joint. This is the muscle which gives the buttocks its bulk and form. It is extremely powerful in athletes.

The abductor muscles, which run from the pelvis down to the lower part of the thighbone, bring the hip and the leg into the center of the body. These become very powerful in horseback riders, who are constantly bringing their knees together to control the horse. There are three components to the abductor muscles. The tensor fasciae latae muscle, located on the outer part of the iliac crest, runs down to the firm sheet of

MY BROKEN HIP

"I broke my hip and shoulder in one play in 1974," recalled Rudy May, the American League's 1980 ERA leader. "Cookie Rojas of the Kansas City Royals hit the ball slowly. I knew I had a play on him, so I chased down the ball at full speed. I bent over to pick the ball up and throw it with one motion. I bent over a little bit too far and flipped. I landed first on my shoulder, and then my hip came down hard on the astroturf. I dislocated my shoulder and split my hip bone."

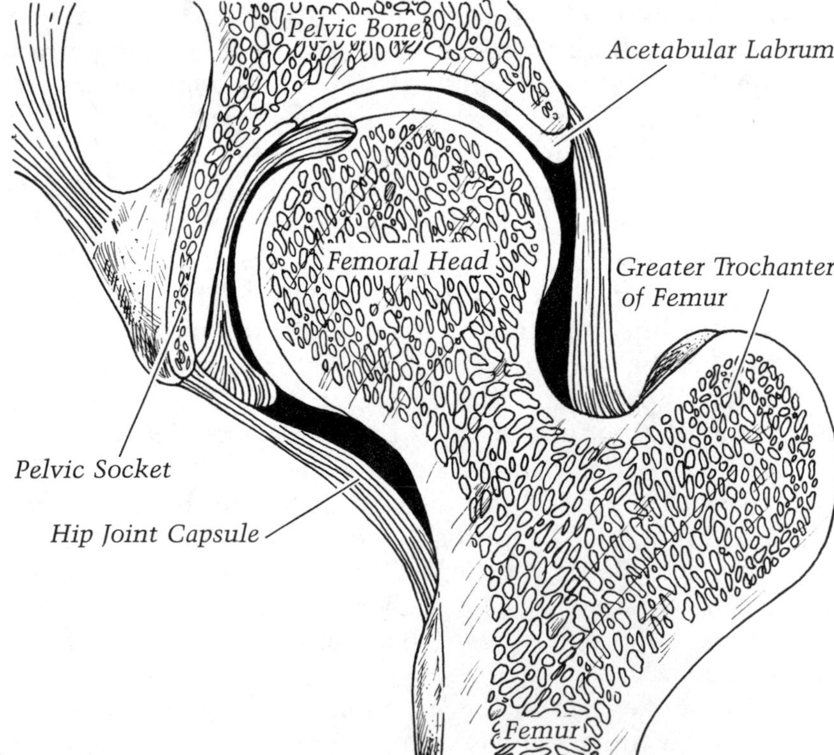

Hip Joint Anatomy

fascia which covers the outer aspect of the hip and thigh. By putting your hand on your thigh, you can feel the fasciae latae muscle tense when you walk or stand.

The other two abductor muscles, the gluteus medius and gluteus minimus muscles, give you the ability to stretch out or suspend your legs for long periods. Place your hand just below your iliac crest and move your leg out to the side. Feel these muscles contract. Finally, smaller muscles aid the larger muscles in producing rotational motion in the hip.

The two nerves that deliver the messages to these muscles are the *sciatic nerve* and the *femoral nerve*. The sciatic nerve is formed from five nerve roots which originate in the lumbar spinal area. They join together at the lumbar area and form the largest nerve in the body. It runs down the pelvis into the back of the buttocks. As it comes out of the pelvis, it supplies the muscles on the back and outer aspect of the hip (the extensor and abductor muscles). It then continues down the back of the thigh between the hamstring muscles. The muscles on the front and inner aspect of the hip (the flexor and adductor muscles) are supplied by the femoral nerve. This is a large nerve which also comes from roots starting in the lumbar spinal area. It comes out of the front of the pelvis and continues down into the front of the thigh. Pressure on either one of these nerves from their origin can produce muscle weakness.

As in all joints, you must rebuild strength and motion back into the hip area after an injury or ailment. I have included the exercises that I find effective.

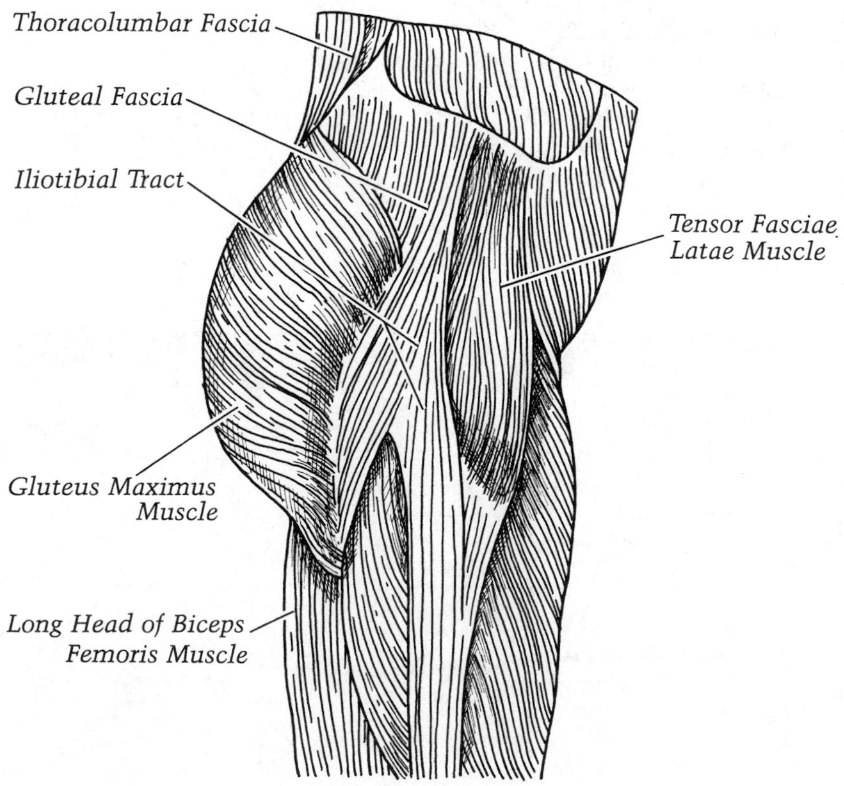

Muscles of the Hip Area

HIP EXERCISES

Do only those exercises checked for you. Do them *slowly!!* Do each exercise: 10 times twice daily.

1. *Heel Sliding.* Lie on back. Slowly bend the knee, sliding the heel up to the buttocks as far as possible without pain.

2. *Straight Leg Raising.* Lie on back. Keep the leg straight and lift it off the bed. Tighten thigh muscle before starting movement. *This exercise may be dangerous with certain conditions. Never use it unless instructed to do so!*

3. *Gluteal Setting.* Lie on stomach. Pinch the buttocks together.

4. *Hip Extension with Knee Bent.* Lie on stomach, head resting on hands. Turn face away from side being exercised. Bend heel to buttocks. Lift thigh.

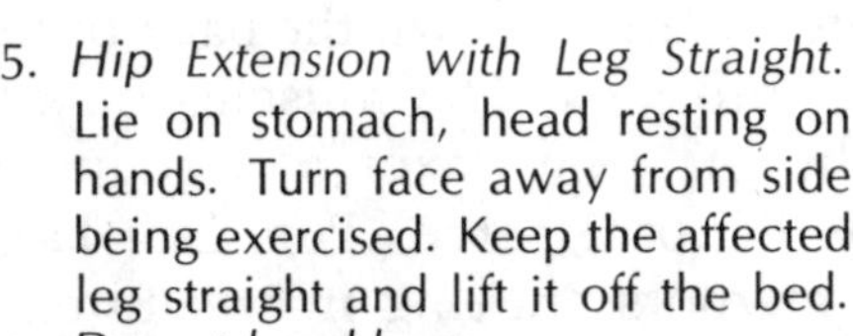

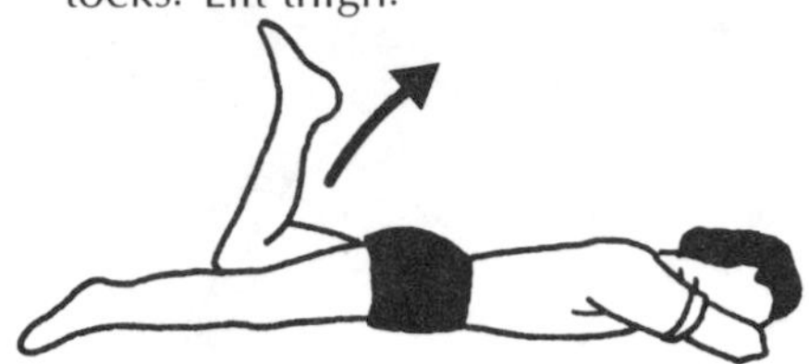

5. *Hip Extension with Leg Straight.* Lie on stomach, head resting on hands. Turn face away from side being exercised. Keep the affected leg straight and lift it off the bed. *Do not bend knee.*
   Resistive Do the exercise with three-pound weight around ankle.

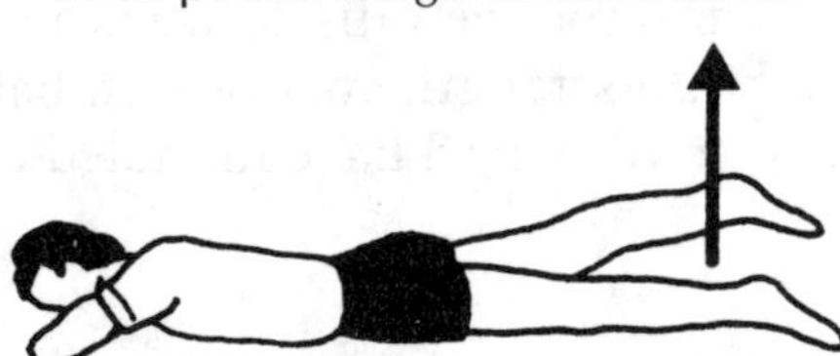

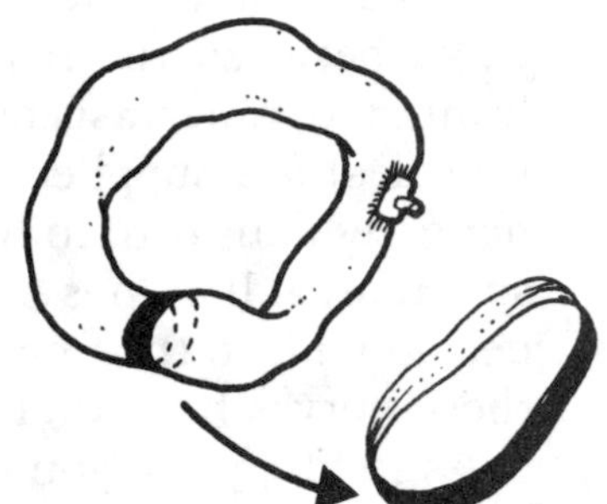

RUBBER BAND
FROM INNER TUBE

6. *Abduction with Band.* Lie on back. Place a rubber band cut from truck innertube around ankles. Keeping knees straight, spread legs apart and hold to the count of 3. The wider the band, the more resistance available.

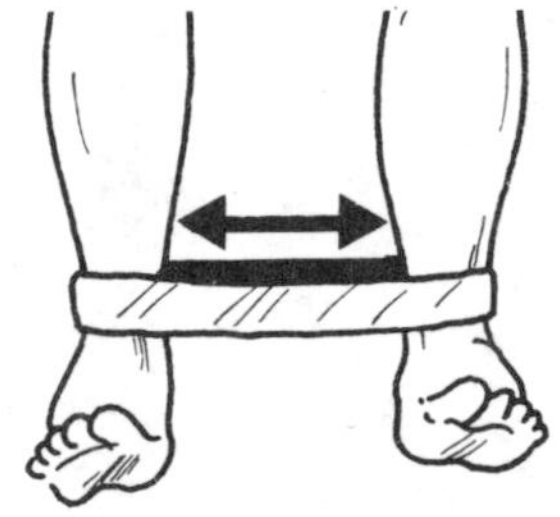

7. *Adduction.* Standing. Place a rubber band cut from truck innertube around your ankle and the leg of a table or other heavy object. Keeping knee straight, pull leg away from table crossing leg to opposite side of the body. Hold to the count of 3. The wider the band, the more resistance available.

8. *Internal Rotation.* Sitting. Place rubber band cut from truck innertube around ankles. Keeping knees together, spread ankles apart. Hold to the count of 3. The wider the band, the more resistance available.

9. *External Rotation.* Sitting. Place rubber band cut from truck innertube around ankles and chair leg. Keep your knee pointing forward and roll ankle and lower leg toward opposite foot. Hold to the count of 3. The wider the band, the more resistance available.

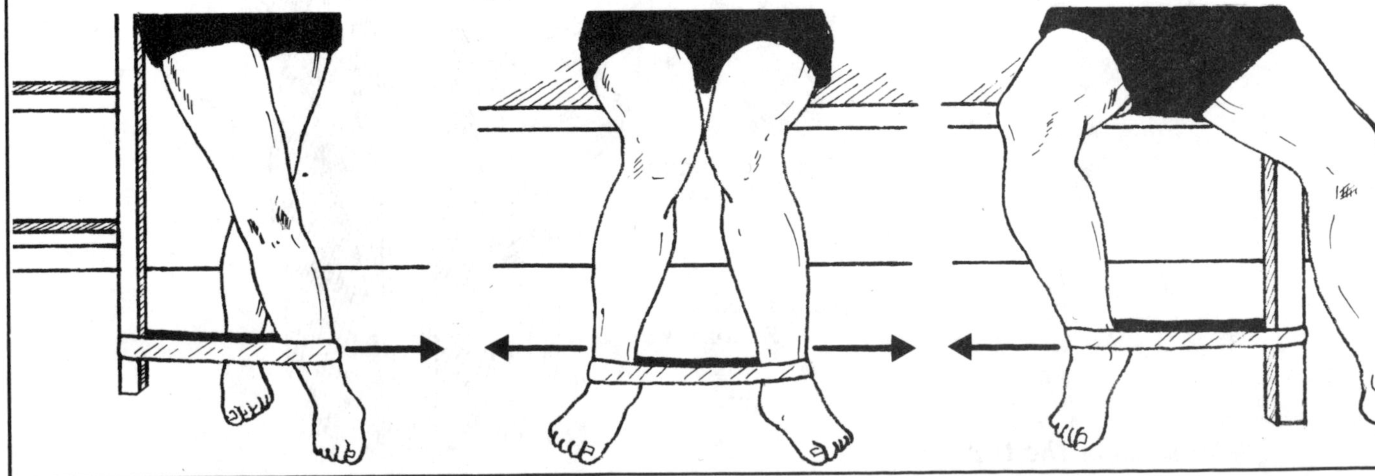

Trochanteric Bursitis

CAUSES

Trochanteric bursitis is an inflammation of the bursa sac that lies between the greater trochanter, the outer knob above the thighbone, and the iliotibial band—a strong flat sheet of fascia which provides a restraining layer for the large thigh muscles.

Often, a bout of this bursitis starts with severe exertion. You may be playing two to three extra sets of tennis or running an extra three or four miles on a given occasion. I often treat racquetball players who slam up against the wall of the racquetball court, causing direct injury to the bursa sac. Football players get trochanteric bursitis from being speared in the hip area with a helmet.

When the bursa becomes irritated, it starts to produce extra fluid, fills up, and becomes very painful. Once the inflammatory process starts, it tends to be self-perpetuating. This is why trochanteric bursitis seldom gets better without specific medical treatment.

DIAGNOSIS AND TREATMENT

Trochanteric bursitis sneaks up on you and becomes steadily more disabling.

As the bursa sac fills with fluid, the pain builds up over a four- to five-day period. Soon, you start to limp because of the discomfort. It will be painful to stand unsupported on the leg that has trochanteric bursitis. Everyday activities—climbing stairs or crossing your affected leg—will be difficult. The discomfort will be right on the outer edge of your hip, localized in an area about the size of your hand. In severe cases, the pain actually radiates down the outer aspect of the leg to the knee

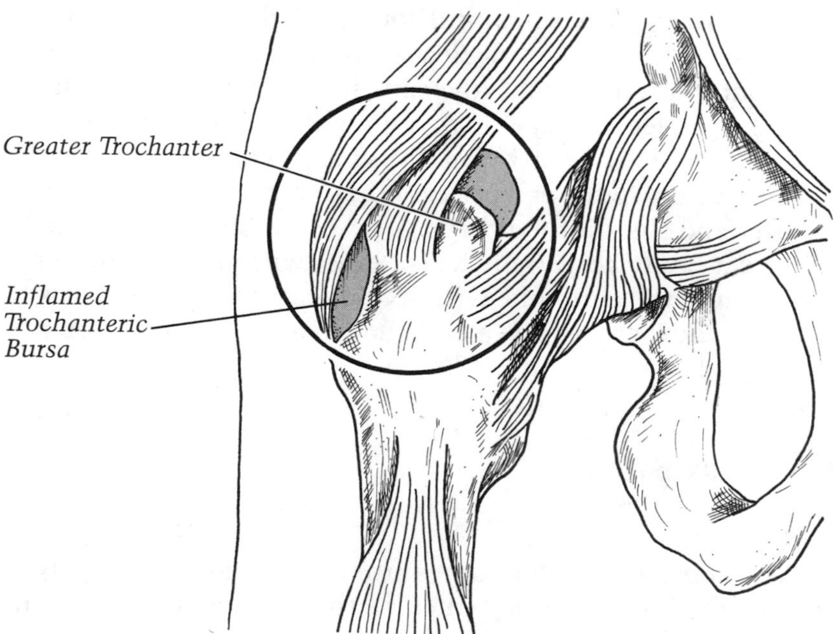

Trochanteric Bursitis

level. However, the pain never extends below the knee.

When I examine the patient who I suspect has this problem, I lightly probe the bony prominence. When I push on this area, I produce a deep agonizing ache. Next, I ask you to lie with the affected hip facing the ceiling. I ask you to raise your leg sideways, which fires the abductor muscles. This will squeeze down on the affected bursa and, if the trochanteric bursa is inflamed, it increases the pain. I also resist this motion with my hand to make it a bit more difficult and make the test more accurate diagnostically. Finally, I place you on your back. I move the hip in all directions to make certain that your hip joint is trouble-free. In trochanteric bursitis, the hip joint will be smooth and full. The chief goal here is to make sure which structure is producing more pain.

Finally, I take x-rays of the hip. It is important to x-ray both hips so that one can be compared against the other. It is also standard to take two views of the hip. The x-ray is front-to-back with the patient's legs out straight. For the second view, the x-ray technician asks you to move your hip up and out so that you assume the position of a frog. What I look for in trochanteric bursitis is a fleck of calcium in the area of the bursa sac. This is directly analogous to the calcium deposits in shoulder bursitis.

If you are experiencing pain, but do not limp, then I start with a twenty-minute application of ice three times a day. Ice diminishes the swelling of the bursa sac and produces a local anesthetic effect. I also start you on anti-inflammatory medicine orally—one with each meal. If this regimen is going to be effective, you should feel much better in seventy-two hours.

After three days of treatment, if you are in pain and cannot walk without a limp, I suggest cortisone injection into the trochanteric bursa. You lie with the painful hip toward the ceiling. I place a pillow between your knees so that the tension is taken off the iliotibial band. This makes the injection easier to administer and less painful for you. I then mark the most tender spot with a pen. I cleanse the area with alcohol and then put cortisone and Novocain into the bursa sac with a needle. I can feel the needle go through the iliotibial band, and I know then that I am in the bursa sac. The bursa sac is the only structure present between the bone of the greater trochanter and the iliotibial band. The Novocain takes immediate effect, and the patient feels better within five minutes. I remind him the cortisone will take at least forty-eight hours to work. In most cases, it is a one-shot treatment and recurrence is unusual.

PATTI LYONS CATALANO'S TROCHANTERIC BURSITIS

Patti Lyons Catalano was named by *Running Times Magazine* the best woman runner in 1980. One reason for her success is her husband, Joe Catalano, who is also her coach and an excellent marathon runner himself.

Before they were married, they were training together. Patti was trying to keep up with Joe, not only in distance but in stride. They were running over 100 miles a week.

Rob Roy McGregor of Sports Medicine Resource, who treated her for her hip bursitis says, "Patti has a classic female pelvis—a wide-faced pelvis. When she was running with Joe, she was inclined to increase her stride. The longer stride put pressure on her gluteus medius, the tensor fasciae latae and the greater trochanter.

"My suggestion was twofold; one, I asked her to shorten her stride. Secondly, I got her to exercise on the Nautilus equipment at Sports Conditioning Center. This helped build up her muscle mass. She took my advice. She is fine now."

Iliopsoas Bursitis

CAUSES

The *iliopsoas muscle* comes from the inside of the pelvis, crosses the groin, and attaches to the thighbone (the femur) just below the hip joint. It allows you to bend your knee up

> **NATE ARCHIBALD'S HIP BURSITIS**
>
> In 1980, Nate Archibald, the Boston Celtics all-star guard, inflamed his hip bursa when he dove to the floor to save a ball.
>
> "Nate was very sore after the game," says trainer Ray Melchiorre. "The bursa sac in the hip was obviously swollen. I treated Nate with RICE and galvanic nerve stimulation. It took about five weeks before Tiny was able to play without pain. But, because it was at the end of the season, he did play. I think it was a matter of personal pride, we had a championship at stake. I rigged a football-type pad to protect the inflamed bursa sac."

onto your chest in a forceful manner. Between the tendon of this muscle and the hip joint lies a bursa sac, about the size of a flattened out walnut.

This is one of the few things that can cause pain in the front of your hip. The other causes are a hernia and arthritis of the hip. It usually requires a doctor to sort this out.

The bursa swells up because it is irritated. It can be irritated by direct pressure, as in gymnastics. It is also irritated by heavy use of the iliopsoas muscle, as in wrestling and running up hills. The pain which is in the front of the hip is due to a filling up of the bursa sac with excess fluid from the lining.

Iliopsoas bursitis follows the same course as bursitis in other parts of the body. It starts as a dull aching pain which peaks in four to five days. Sometimes, it becomes so severe that you will have to quit exercising. You will feel an aching sensation over the front of the hip. It will become more painful with and after exercise.

If you have iliopsoas bursitis, I can induce the discomfort by pressing firmly on the front of the hip joint. Secondly, I ask you to bend your knee up toward your chest. I push down on the knee and you resist the motion, which forces the iliopsoas muscle to work harder and increases the pain of the bursitis. I always check for a hernia because that condition produces pain in almost the same location. Your hip joint motion will not be limited, as it is in arthritis.

Treatment is mainly based on anti-inflammatory medicine. Ice is also used. The best treatment is rest. Stop exercising for ten days. Ice your hip twice a day for twenty minutes. Take six to eight aspirins a day with meals. There is very little else I can do for you. This bursa is very difficult to inject with a needle. So, a cortisone injection is out. Fortunately, the rest, ice, and aspirin usually do the trick. I have never operated on iliopsoas bursitis.

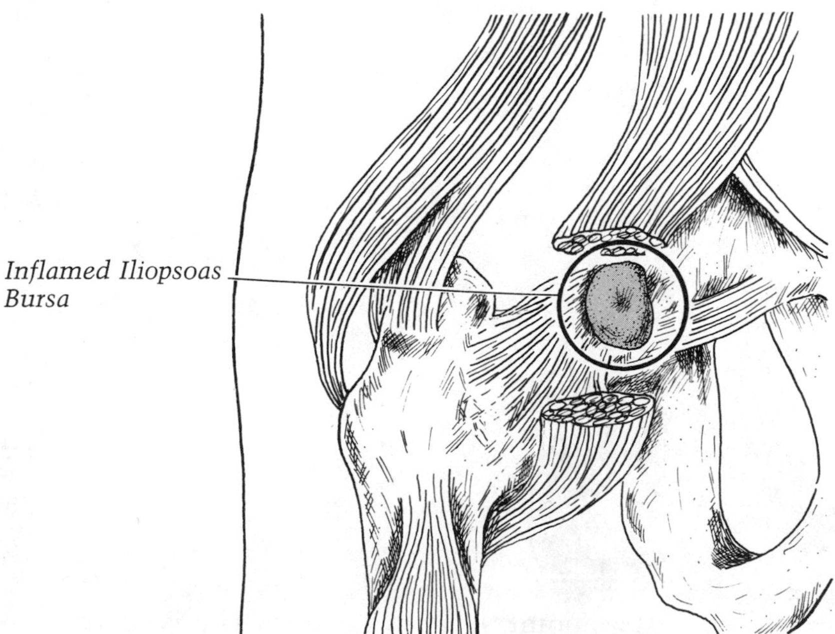

Iliopsoas Bursitis

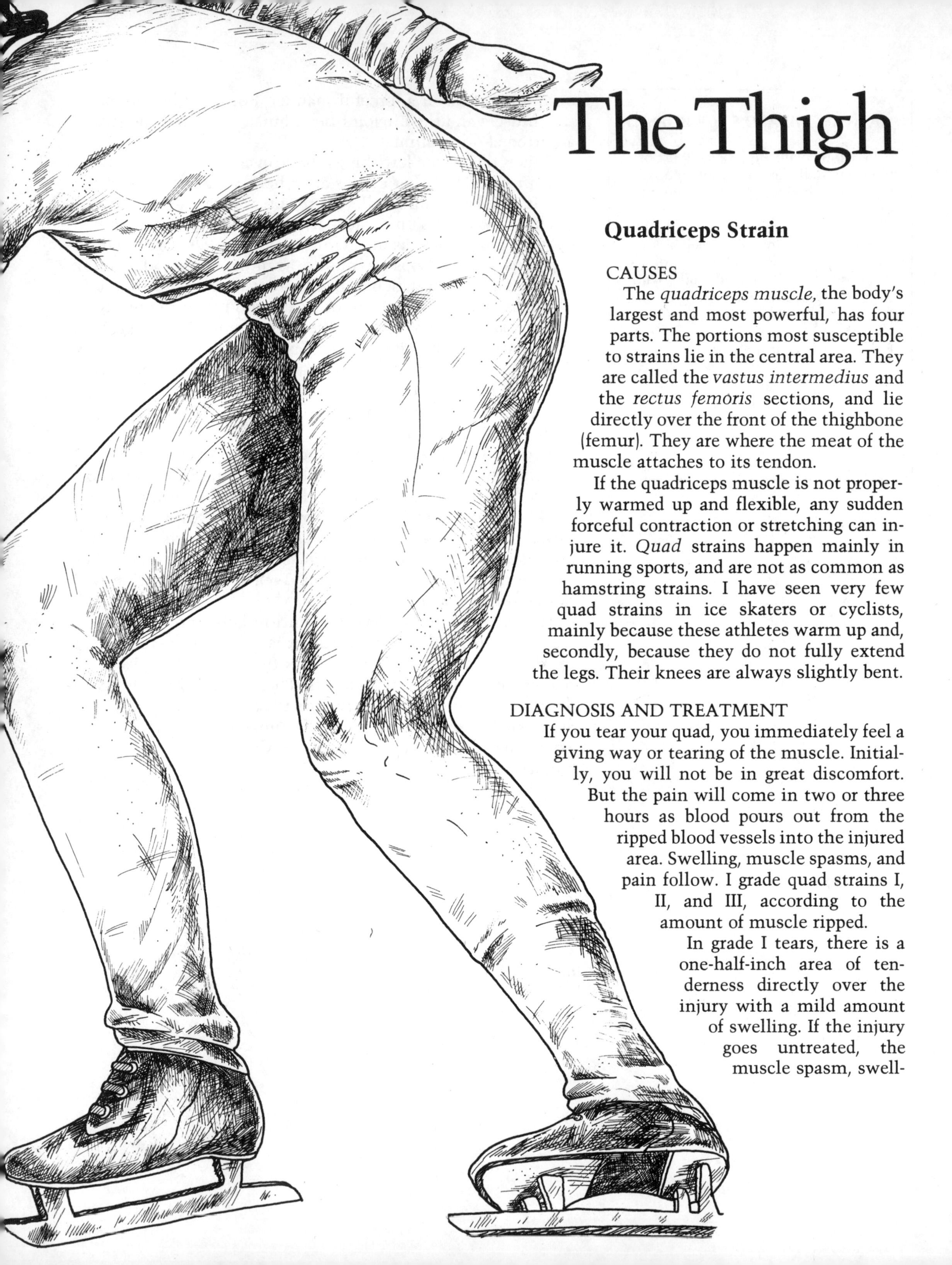

The Thigh

Quadriceps Strain

CAUSES

The *quadriceps muscle,* the body's largest and most powerful, has four parts. The portions most susceptible to strains lie in the central area. They are called the *vastus intermedius* and the *rectus femoris* sections, and lie directly over the front of the thighbone (femur). They are where the meat of the muscle attaches to its tendon.

If the quadriceps muscle is not properly warmed up and flexible, any sudden forceful contraction or stretching can injure it. *Quad* strains happen mainly in running sports, and are not as common as hamstring strains. I have seen very few quad strains in ice skaters or cyclists, mainly because these athletes warm up and, secondly, because they do not fully extend the legs. Their knees are always slightly bent.

DIAGNOSIS AND TREATMENT

If you tear your quad, you immediately feel a giving way or tearing of the muscle. Initially, you will not be in great discomfort. But the pain will come in two or three hours as blood pours out from the ripped blood vessels into the injured area. Swelling, muscle spasms, and pain follow. I grade quad strains I, II, and III, according to the amount of muscle ripped.

In grade I tears, there is a one-half-inch area of tenderness directly over the injury with a mild amount of swelling. If the injury goes untreated, the muscle spasm, swell-

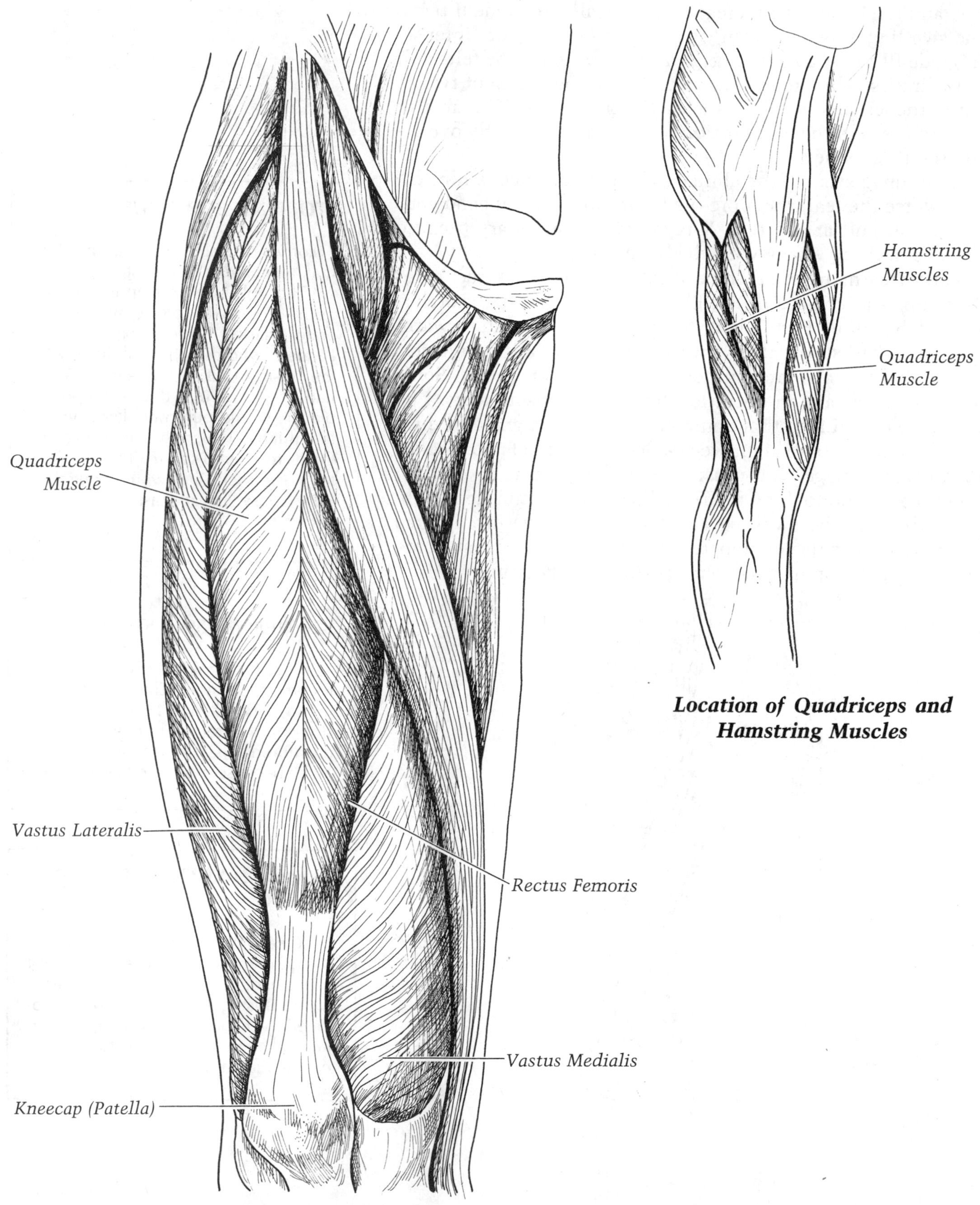

Location of Quadriceps and Hamstring Muscles

Anatomy of Quadriceps Muscle

ing, and tenderness all increase dramatically. In grade II tears, the swelling, muscle spasm, and tenderness are more intense. In grade III tears, an actual gap in the muscle can be felt with your fingers. This means that a cross-sectional area of two to three inches of muscle has actually given way. The area of tenderness is in the mid-portion of the muscle, directly overlying the thighbone.

In all quad strains, the ability to bend your knee is limited. The worse the tear, the more the limitation. As you bend your knee, you pull at the tear. X-rays are not necessary because there is nothing wrong with the bone.

Early treatment of all three grades involves rest, ice, compression, and elevation (RICE). In the severe tears, crutches should be used. Forty-eight hours after injury, start warm whirlpool baths or soak in a warm tub. Do it twenty minutes twice daily. I also recommend hydrocollator packs. Some surgeons advocate an operation for severe muscle tears. I do not. However, I find that complete healing of a grade III quad strain requires six to eight weeks. Because of the large amount of muscle damage, you must allow enough time to pass for solid scar tissue to draw the ripped ends of the muscle together. Only after the swelling and tenderness have disappeared can you begin muscle-strengthening exercises. If you start too soon, further ripping can occur, prolonging the injury.

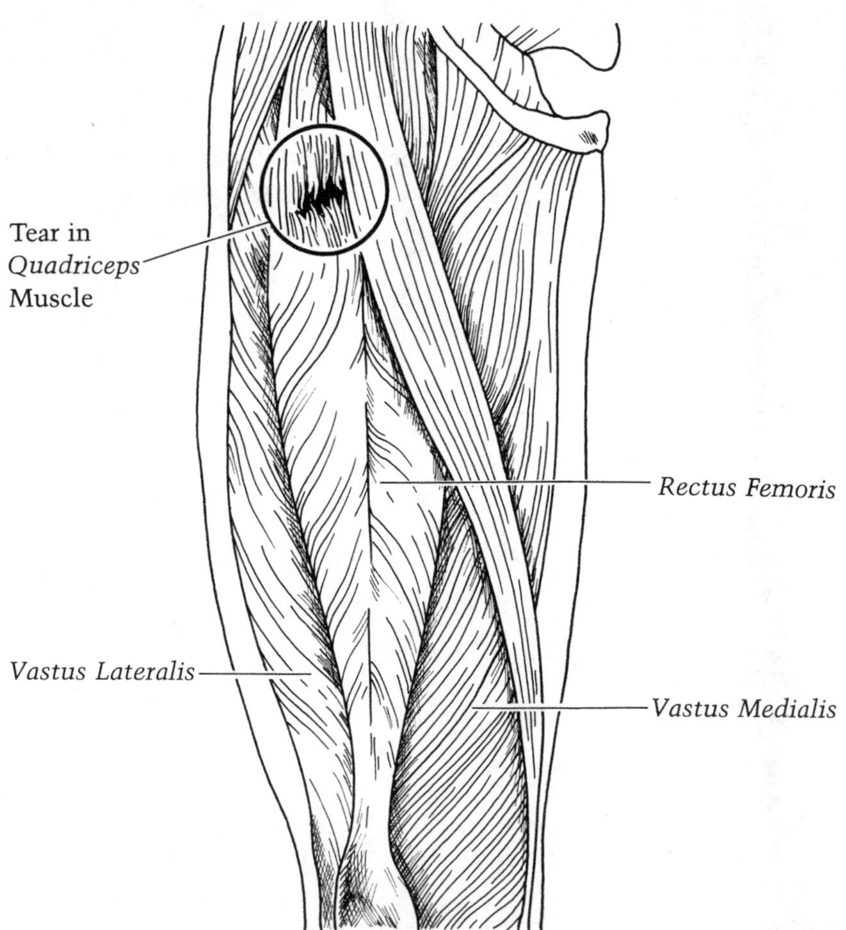

Quadriceps Strain

STOPPING BOB COUSY

For years, Bob Cousy was the playmaking guard for the Boston Celtics.

"The only way to stop Bob Cousy from scoring was to give him a deliberate charley horse," explains Jack Fadden, who did physical therapy for the Boston Celtics for years. "Cousy had very muscular thighs. Opposing players would purposely knee him in the thighs, which would cause his thighs to go into spasm. The old charley horse trick. This would neutralize Bob Cousy for at least a couple of games."

$200,000 MUSCLE STRAIN

In the finals of the 1981 $200,000 Richmond Tennis Classic, in Richmond, Virginia, Ivan Lendl of Czechoslovakia was forced to retire from the championship match with Yannick Noah because of a pulled leg muscle.

REHABILITATING THE QUAD MUSCLE

"One way I improve the strength of the quad muscle is to exercise against resistance," says Joe Bourdon, trainer of the Boston Teamen. "I use an isokinetic version of the stationary bicycle and an isokinetic fitron to strengthen the quad muscles."

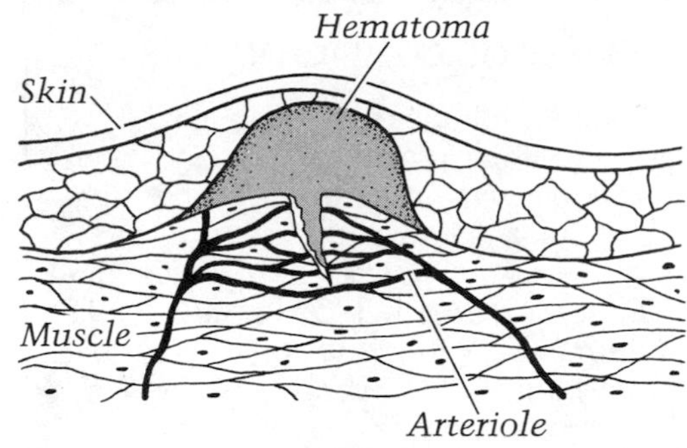

Contusion

Thigh and Arm Contusions (Charley Horse)

CAUSES

Contusions are bruises to the skin, to the fat beneath the skin, and to the muscles and the bones themselves. Contusions are very common in contact sports, especially football. They usually happen when a tackler drives his helmet into you. The force transfers directly to the skin and into the deeper layers of the leg.

The force of the blow ruptures the blood vessels in the skin, in the fat beneath the skin, and in the muscles themselves. When these blood vessels are broken, blood oozes out. In the skin, a contusion shows up as a black-and-blue spot (an *ecchymosis).* In the muscles, the force of the blow causes a pool of blood, or a *hematoma,* to form. A contusion can rupture the surface of bones, which is called the *periosteum. Peri* means "around" and *osteum* means "bone." When this layer is disrupted, live bone cells, called *osteoblasts,* leak out. These bone cells can form new bone.

The most common areas of contusion are the upper arms and thighs. The degree of disruption of the soft parts depends on the force and duration of the blow.

> **LARRY BIRD'S THIGH CONTUSION**
>
> In the middle of the 1980–81 season, superstar Larry Bird, the Celtic forward, was accidentally kicked in the thigh by Philadelphia's Darryl Dawkins. The contusion to Bird's thigh was massive and caused him to limp slightly.
>
> "There was a guy running around wearing uniform number 33," commented Boston coach Bill Fitch. "But that wasn't really Larry Bird. His timing was off and he favored his injured side."
>
> The toughest part about this injury for Bird, a tough competitor, was that he couldn't speed up the healing process. He played through the injury. But, as his statistics show, the injury took its toll. Until he fully recovered, he shot less than 35 percent from the floor. Bird usually is almost a 50 percent marksman.

DIAGNOSIS AND TREATMENT

How do you know if you have sustained a contusion? It is not apparent immediately, but within six hours you will notice pain and stiffness at the site of the blow. Untreated, this pain and swelling continues as the blood oozes out into the muscle. As the blood pools, the hematoma enlarges. It will restrict your ability to straighten your elbow or bend your knee. The contusion inhibits the muscle from stretching.

The problem with a contusion is that you may be unaware you have one. Most players continue their game.

If you started RICE immediately—the moment you got hit—you would reduce the swelling. But it will be six hours later before the contusion is evident.

In my examination, I always ask, "Do you remember the blow?" Invariably my patients do, and can describe it in detail. But the treatment is starting six hours late.

Ice the contusion for the first forty-eight hours. Then start warm whirlpool baths or hot-pack therapy. The warmth stimulates circulation, which carries away the old blood in the hematoma. It also warms the muscle and allows gradual stretching. A whirlpool bath should be taken twice daily for twenty minutes. A warm bath will do if you do not have access to a whirlpool. Continue for seven to ten days.

Ultrasound, a form of deep heat, seems to speed the breaking up of the hematoma. Although there is no hard and fast scientific evidence to support ultrasound treatment, I find it works.

In severe contusions of the leg, crutches should be used to take pressure off the injured extremity. A study on this subject done at West Point showed that the cadets fared better if they were taken completely off their feet and admitted to the infirmary when they had severe thigh contusions. Initially, x-rays of contusions are negative.

If you have a skin contusion, you will be out of action for two to three days. If the contusion extends deeply into the muscle tissue itself, expect to miss two weeks of sports. If you return to sports too quickly, the contusion is often reactivated, and more blood seeps into the contusion site. This compounds the injury and slows up your ultimate return to sports. Eventually, the hematoma turns into pain-free scar tissue.

The most complicated contusions are those that disrupt the bone surface. These occur almost exclusively in the biceps and quadriceps muscles on the front of the upper arm and thigh. When the bone covering is disturbed, the bone-producing cells leak out into the hematoma. They start to reproduce quickly and make new bone in the pool of blood. This is exactly like a fracture callus that forms between the ends of a broken bone. The blood clot organizes, or becomes "sticky," and then new bone is laid down in the muscle hematoma. This is called *myositis ossificans*—literally meaning bone in the muscle. Athletes with this problem have a great deal more swelling, and will begin to develop a firmness in the contused area within ten days. The contusion area produces heat which sometimes is taken for infection. The heat is due to the large amount of new blood pouring into the area, which is nourishing the new bone.

In myositis ossificans, x-rays taken after the first ten days show a light, hazy density, representing the new bone formation in the muscle. Over time, this myositis bone becomes more and more mature. It matures from the edges toward the center. This is an important pathologic fact. Fast-growing, healing bone looks very much like a bone tumor under the microscope. However, a bone tumor has its most mature elements in the center and its most primitive or wild-looking elements at the edge. The myositis organization is just the reverse.

How do I manage these myositis patients? The initial week of treatment is exactly the same as a non-bone-forming contusion. However, once the bone formation is documented by x-rays it means a three-month rest and recovery period. This is because myositis bone formations irritate the muscle. Many growths are like cactus leaves. Their sharp spikes dig painfully into the muscle and restrict its normal function. These patients cannot participate in sports activities, and rehabilitation must be very carefully conducted. Only rarely are bone formations smooth and not painful.

For six weeks, I allow only routine activities. There is no need to have a leg or arm immobilized by cast or sling. I take x-rays again at the end of this period. These usually show that the bone is consolidating and becoming more dense. As it does, it starts to shrink.

MYOSITIS OSSIFICANS

Jeff Southmayd, my nephew and a top football player at Wayland High School, was struck on the upper right arm in a game early in the season. Although he applied lots of ice, his upper arm swelled to almost twice its normal size in eight hours. The next morning, he could not straighten his elbow.

Within a week, his upper arm became very firm. It felt as if a baseball were tucked under his biceps muscle. The area became warm to the touch. I supported his arm with a sling. Ten days later, his x-rays showed a slight, hazy density just in front of the upper arm bone (humerus) in the biceps muscle. This is a sign that the bone surface was damaged and bone cells were spilling into the muscle. This condition is called myositis ossificans.

Three weeks after the injury, the x-rays showed an increased density in the mass of bone—a sure sign that myositis ossificans was forming. I removed his arm from the sling and worked his elbow up and down in a whirlpool bath.

Three months after the injury, the surface of the bone had sealed itself and matured. It was smooth and even. The ossificans was shrinking in size week by week. Jeff now was able to straighten his elbow. He was anxious to return to sports. I explained to him that the ossificans needs more time to heal.

"What do I have to lose?" he asked.

"If you get hit again, the whole process could start over," I warned. "This would be like restarting the biologic clock all over again."

Jeff decided it was worth the risk and returned without further injury. X-rays were taken one year later and showed only a very small (one-and one-half-inch) flat area where the myositis had been. Four years later, there was no sign of the ossificans.

STEP-UP EXERCISE

Robert Reese, the trainer for the Buffalo Bills, developed an exercise which he calls step-ups to strengthen the quadriceps muscle. It is beautiful in its simplicity.

Place the injured leg on a four- to six-inch block—a stair step will do. The good leg remains on the floor about four inches away from the block and set back about two inches behind the bad leg.

Lift your body with the injured leg. Do not push off with the good leg. Repeat until exhaustion. Do the step-up three times a day.

As you get stronger, increase the height of the block and move your good foot farther and farther away from the block.

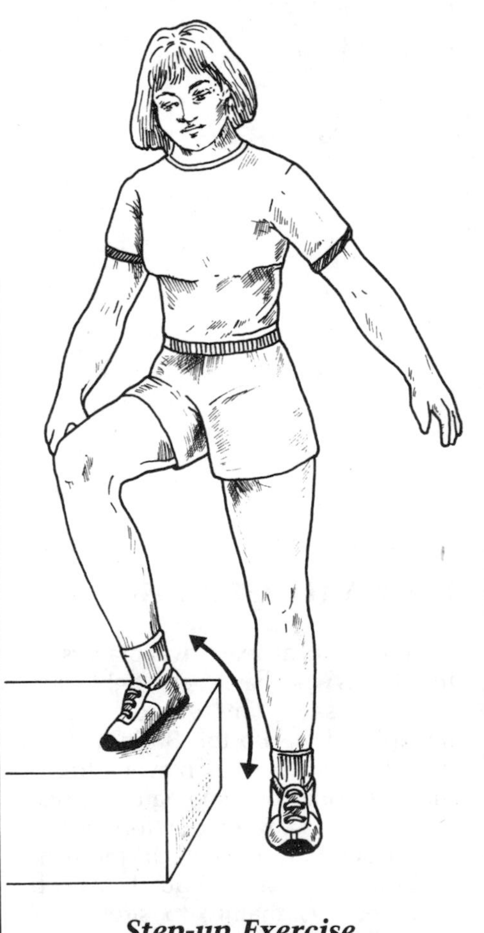

Step-up Exercise

After the six-week period, I start physical therapy. You can do gentle weight lifting to increase the strength of the muscle involved with the myositis. Step-up exercises also work well. It is important to work the joint below the area of involvement. If vigorous activity is started too soon, more bleeding will occur, and this in turn produces still more bone cells.

When you do return to sports, I recommend wearing a protective pad over the injured area for the next six months. The material that I like best is orthoplast, which cuts easily with scissors and can be molded after heating in warm water. Orthoplast is too rigid to apply directly to the skin, and therefore sponge rubber should be glued to the reverse side. The pad can then be taped in place using sticky adhesive spray on the skin and elastoplast tape.

Is surgery ever necessary for a myositis ossificans? Rarely. The body is very efficient in shrinking the bone mass. X-rays taken one year after a myositis attack will still show a small amount of residual bone. This may be present for life, but will never bother you. I have never operated on myositis ossificans, and I see patients with this problem frequently. If surgery is done, the immature bone can be mistaken for a bone tumor under a microscope. This could lead to entirely inappropriate treatment.

Hamstring Strains (Hamstring Pulls)

CAUSES

Fred Lynn, star outfielder, now with the California Angels, pulled his hamstring muscle early in 1979 and played the whole year in pain. Houston McTear, a world-class sprinter, has been bothered by hamstring pulls his entire career. *Hamstring pulls,* a rip or tear of a hamstring muscle or its tendon, are common ailments in sports. They happen mainly to sprinters. I have also treated these pulls in tennis, soccer, basketball, and baseball players. It is rarely a problem in athletes and fitness buffs who do not fully extend their knee—cyclists, snow and water skiers, hockey players, and ice skaters.

The *hamstring muscle,* located in the back of the thigh, pulls the knee down. The muscle originates on the bottom of the pelvis, a ring-shaped structure at the base of your spine, protected by the buttocks. The hamstring comes off the pelvic bone as a thick white tendon. It develops into a large fleshy muscle. You can feel the meat of the muscle by grasping the back or your thigh. As the hamstring runs to the knee, the muscle splits into two distinct tendon groups. Two-thirds of the hamstring tendons attach to the inner aspects of the knee and wrap around the front of the lower leg bone (tibia). The remaining one-third forms a single stout tendon which attaches to the top of the small outer lower leg bone (fibula). These *knee flexors* bend your knee powerfully downward.

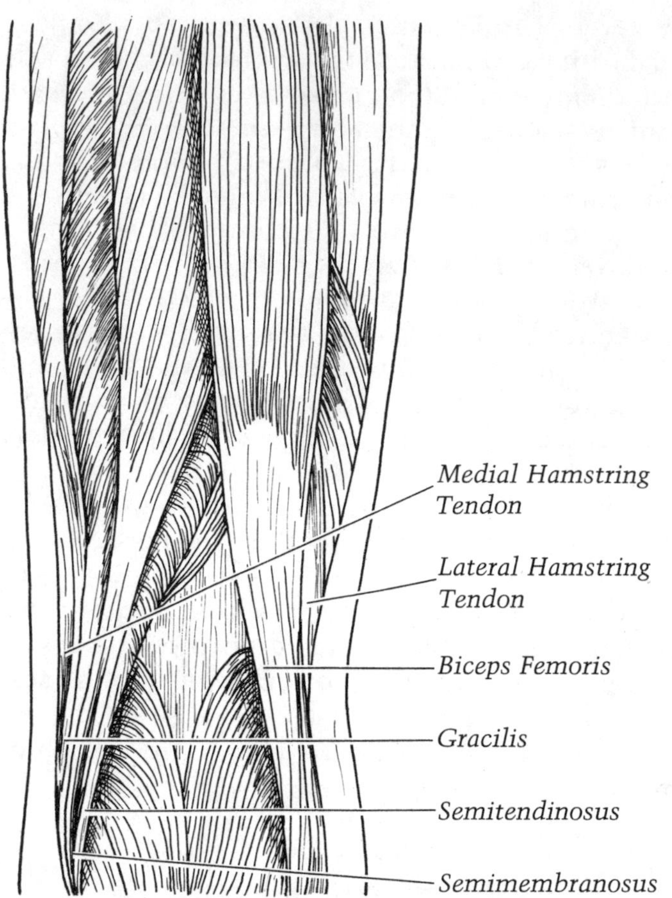

Hamstring Anatomy

Hamstring Muscles

Quadriceps Muscle

Hamstring Strain

The hamstring muscle is opposed by the *quad*, a massive muscle in the front of the thigh which raises the knee. When the hamstring moves in one direction, the quad moves in the opposite direction.

Lee Burkitt of Arizona State University, using a gauge to measure muscle strength, found that most athletes in running sports have quads that are one and a half times stronger than their hamstrings. Those athletes who had a higher strength ratio between those two muscles are the ones most likely to pull their hamstrings.

The hamstring rips with the muscle fully stretched out. Imagine the sprinter firing out the starting blocks. As he picks up speed, the quadriceps muscles powerfully contract to achieve straightening of the knee with each stride. The hamstring muscle tries to resist the quad. The pull happens when the quad overpowers the hamstring. Pulls can also happen if you are not properly warmed up or are especially tight.

In most cases, the meat of the muscle rips where the muscle attaches to its tendon, approximately one-third of the way down the thigh. Infrequently, the tendon will rip away from the bone of the pelvis, or at the knee level. The severity of the injury depends on the size and completeness of the tear.

HOCKEY HAMSTRINGS

"We have at least five players on the Bruins who have very tight hamstrings," says Jim Kausek, sports therapist of the Boston Bruins. "Even though all our hockey players stretch, the hamstrings become naturally tight in skating; many of our top players will show more strength in the hamstring muscle than the quad. My job, as trainer, is mainly to stretch the hamstrings and strengthen the quad."

DIAGNOSIS AND TREATMENT

When you tear your hamstring, you know it. You feel a definite ripping or tearing in the back of your leg. It is not painful at first, but you get a sense of uneasiness in the leg. Often, you can continue your athletic event until the pain intensifies. In a mild or moderate tear, the pain intensifies in three to six hours. In severe tears, the pain intensifies within thirty minutes.

If you think you have a hamstring pull, stop exercising or competing. The earlier you start treatment, the less disability you are going to have. The best treatment for a hamstring pull is RICE. Apply an ice bag immediately to your hamstring. I usually strap an ice bag on with an elastic wrap. This supplies compression to the injury and minimizes the bleeding and tissue fluid which can accumulate in the rip. The less fluid that accumulates, the faster the injury can heal. Ice for one hour three times daily. After one day of icing and the other RICE steps, start heat treatments twice daily. Hydrocollator packs, hot water bottles, electric heating pads, hot towels, or whirlpool baths are all about equally effective. Heat treatments stimulate the circulation and speed healing.

The most accurate time to evaluate the extent of the injury is twenty-four hours after it happens. That's when I like to grade the strain as mild, moderate, or severe: grades I, II, and III.

Seventy percent of the hamstring strains I examine are in the muscle belly itself. I can touch the tear with my fingertips. Mild strains are only one-half inch wide with very little swelling and no bruising. You will be able to contract the hamstring muscle completely with some discomfort. The healing time is three to five days.

In a moderate strain, the damage is larger—two to three inches. Thus, there is more soft tissue swelling and pain. It will be difficult to contract your hamstring muscle. You will heal in seven to ten days.

In severe strains, I can actually feel a gap between the ripped ends of the muscle. Also, a black-and-blue area, a sure sign of extensive tearing, develops in the first twenty-four to forty-eight hours. In the next two weeks, the black-and-blue area (ecchymosis) will gravitate down the thigh to the knee area. Your doctor should warn you about this movement. It sometimes causes anxieties.

With a severe hamstring strain, you will be unable to fully contract the muscle. I test this by having you lie on your stomach with your knee bent ninety degrees up on the examining table. I put my hand on the back of the ankle and start to pull the leg out straight. I ask you to fight me. This activates the hamstring muscles. By testing you in this position, I can actually watch the hamstring muscle contract and see the quality of that contraction. A severe hamstring pull takes at least three weeks to heal.

If the hamstring unit is injured at its bony attachment to the

LYNN'S HAMSTRING PULL

In May 1980, the Boston Red Sox were playing in Chicago against the White Sox.

"I remember the game well. It was a grim, overcast, early spring day. Because it had rained the day before, the outfield was especially sloppy," recalled Red Sox trainer Charlie Moss.

"A day or two after leaving Chicago, Fred Lynn, our all-star center fielder, told me that his hamstring muscle was bothering him. I assumed that he had pulled it in Chicago. The hamstring didn't seem very sore. I considered it a minor hamstring pull. I iced and compressed the muscle. Then, I wrapped the upper leg.

"I thought about the injury. Lynn is not a complainer. He does not miss games because of small injuries. The management and the players expect him to play. Secondly, Fred is one of the best-conditioned athletes I have ever known. He's big on stretching, and uses a baseball stretching chart developed by Bob Anderson. When I was taping his thigh, I recalled that Lynn was extremely flexible. He can touch his palms on the floor, so I don't think the hamstring pull happened because he was out of shape.

"I built him a special neoprene rubber sleeve which compressed the thigh and kept it warm. Fred liked it. After every game, I iced the hamstring muscle. I used ultrasound on it.

"In the third week in June, he came into the training room and said that his hamstring was bothering him again. It was a shock to me because he was playing so well. Dr. Arthur Pappas, the medical director for the Red Sox, recommended that Lynn be rested for a week. That was just before the break.

"Fred played in the all-star game and hit a homer.

"The second half of the season, Fred played with less pain, and wasn't significantly bothered by the hamstring pull since he won the American League batting title."

pelvis, the tenderness will permeate the crotch area. I can put my fingers deeply into the buttock and feel the *ischium* of the pelvis. This is the bone you sit down on. It is important to distinguish if the tear is off the bone in this area or whether it is in the muscle belly. It takes much longer to heal from bone injury—maybe twice as long.

When can you return to exercising or competition? First, the tenderness has to disappear. You have to be able to make a firm hamstring contraction without discomfort. At this point, you can begin to run. But, full participation is a day-to-day decision. Too early a return will lead to more injury to the ripped area.

When you return to sports, I recommend wearing external support such as elastoplast tape wrapped around the thigh directly at the level of the injury. I shave much of the upper thigh. The second step is to spray adhesive on the skin. I apply a prewrap. The elastoplast can then be wrapped over the prewrap to apply compression at the level of the tear. This wrap minimizes your discomfort.

I also like a neoprene rubber sleeve, which can be slipped up to the thigh. The sleeve is easy to apply. Sprinkle a little powder into it. This allows the rubber sleeve to come comfortably to the level of the injury. We use these a lot with the Red Sox players.

If the hamstring tears at the pelvis attachment, you must wear a groin-type wrapping around the leg and around the waist. It must go around the waist or else it will slip down and provide inadequate support. You must figure an additional one to two weeks of recuperation when the hamstring pull is in the groin area.

Occasionally, the hamstring pull remains sensitive for many months. It happens when you attempt to stride out and run at full speed. This is usually due to an irritation to the scar tissue in the previous muscle rip. I find that ultrasound treatments help eliminate this sensitivity. A ten-day program is usually sufficient.

How can hamstring pulls be avoided? Warm up thoroughly. Warming up brings blood to the muscle and makes it more pliable and less likely to tear. Most college and professional athletes spend at least an hour warming up. You should stretch the muscle before running or exercising. A tight hamstring is like a tight violin string. The tighter it is, the greater the chance it will tear when used.

Certain sports make people more susceptible to hamstring pulls. For example, baseball, football, and soccer contain much sprinting activity. Baseball players sprint out of the batter's box to first base. Football backs sprint through the line of scrimmage. Soccer players sprint on offense and defense. These sudden sprinting activities put tremendous stress on the hamstring muscles. That's why hamstring pulls are very common in running sports. These athletes must stretch much more than, say, the average fitness buff. Continued stretching during the game also helps ward off injury.

HAMSTRING STRENGTHENERS

Your hamstring muscles lie on the back of your thigh. If you grab the back of your thigh with your hand, you are holding the "meat" of the hamstring muscles. They give you the muscle power to bend your knee with force. We use the same progressive resistance technique as for knee crankers.

1. To strengthen your hamstring muscles, you must lie face down on the floor or on a firm table. Strap a weight boot to your street shoe or use an ankle weight. Begin with two pounds. Your hamstrings are not as strong as your quadriceps muscle on the front of your thigh. Starting with your leg out straight, bend your knee to ninety degrees. This makes your lower leg perpendicular to the floor. Take three full seconds to get there: 1000-1, 1000-2, 1000-3. There is no need to hold this position more than just a moment. Then let your leg down

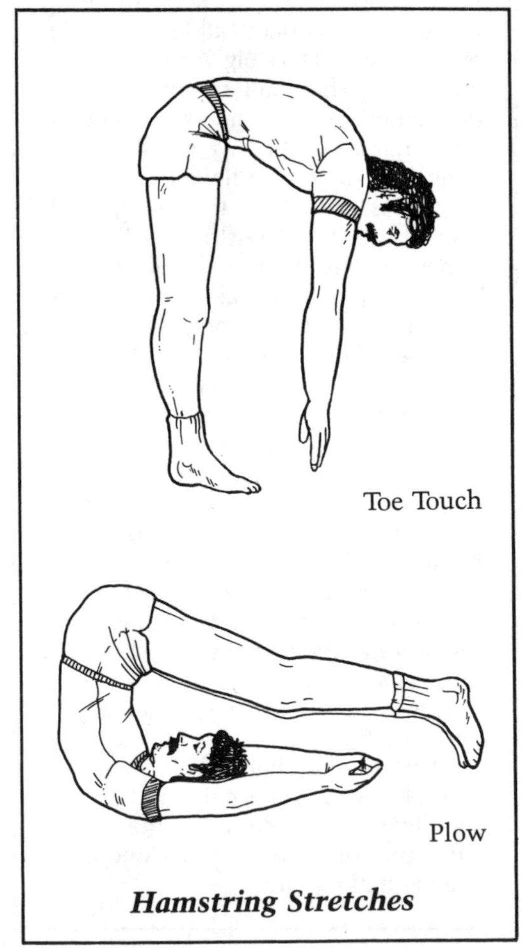

Hamstring Stretches

to the straight position again. Do this to a three-second count. Find a weight that you can lift this way ten times in a row. This is your maximum weight.
2. Now you are ready for a full progressive resistance program. First do ten reps (repetitions) of this exercise with one-half maximum weight. Rest five minutes. Next, do ten reps with three-quarters maximum weight. Rest five minutes. Finally, do your last set of ten reps with maximum weight.
3. If you are trying to gain strength, you should do these exercises three times a week. If you are coming off an injury or surgery, you should do them daily until the injured leg is as strong as the normal leg.
4. Add two pounds every five to seven days. Never use weights that are so heavy that you cannot finish every exercise.
5. In general, we like to keep a 3:2 ratio between the quadriceps and hamstring muscles.

THE DUGOUT STRETCH

Put your leg out straight on a table. (The Boston Red Sox do this exercise on the dugout.) Reach out slowly with both hands, touch the toes of the elevated leg. Hold for three seconds. Repeat at least ten times.

HAMSTRING EXERCISES

To Strengthen the Upper Hamstrings

1. Bolt an innertube firmly to a post or a door molding so that the bottom of the tube is six inches above the top of your knee when you are seated in a straight chair.
2. Put the leg closest to the tube through it so that the bottom rests under your thigh.
3. Forcefully lower your foot to the ground, then relax. Repeat rapidly ten times. Rest thirty seconds. Repeat procedure five times. Repeat with other leg.

To Strengthen the Lower Hamstrings

1. Fasten one end of a bicycle innertube to a stationary object such as a doorknob.
2. Sit down and put one leg through the other end, as pictured in the diagram. The innertube should be tightly stretched.
3. Bend the knee backward rapidly ten times, increasing tension on the tube. Rest for thirty seconds. Repeat the procedure five times. Repeat with the other leg.

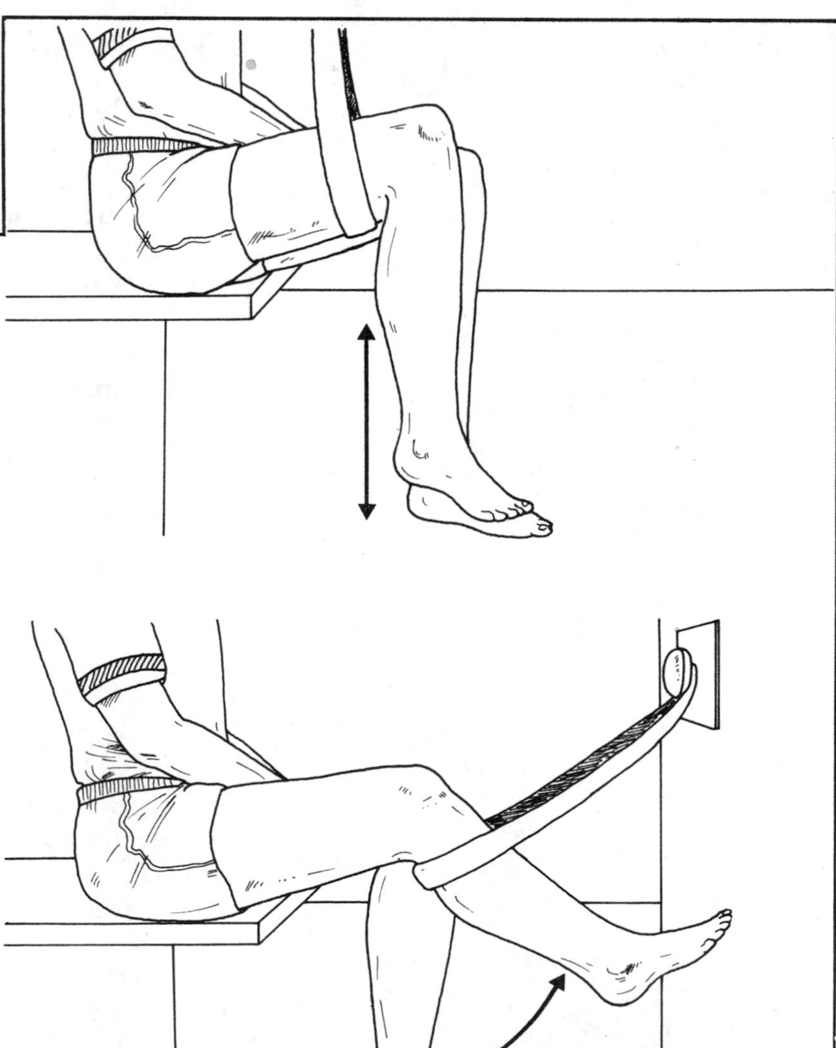

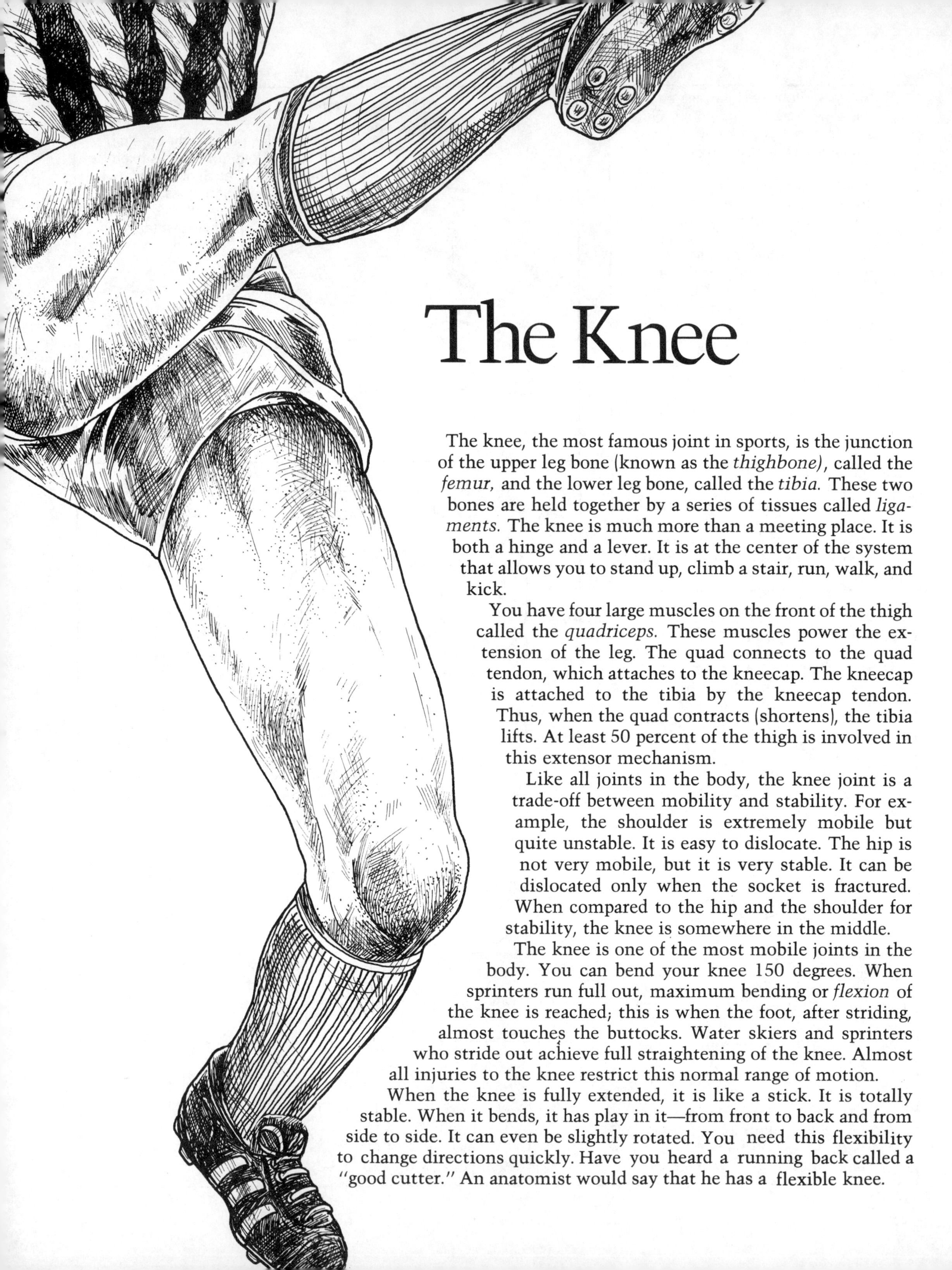

The Knee

The knee, the most famous joint in sports, is the junction of the upper leg bone (known as the *thighbone)*, called the *femur,* and the lower leg bone, called the *tibia.* These two bones are held together by a series of tissues called *ligaments.* The knee is much more than a meeting place. It is both a hinge and a lever. It is at the center of the system that allows you to stand up, climb a stair, run, walk, and kick.

You have four large muscles on the front of the thigh called the *quadriceps.* These muscles power the extension of the leg. The quad connects to the quad tendon, which attaches to the kneecap. The kneecap is attached to the tibia by the kneecap tendon. Thus, when the quad contracts (shortens), the tibia lifts. At least 50 percent of the thigh is involved in this extensor mechanism.

Like all joints in the body, the knee joint is a trade-off between mobility and stability. For example, the shoulder is extremely mobile but quite unstable. It is easy to dislocate. The hip is not very mobile, but it is very stable. It can be dislocated only when the socket is fractured. When compared to the hip and the shoulder for stability, the knee is somewhere in the middle.

The knee is one of the most mobile joints in the body. You can bend your knee 150 degrees. When sprinters run full out, maximum bending or *flexion* of the knee is reached; this is when the foot, after striding, almost touches the buttocks. Water skiers and sprinters who stride out achieve full straightening of the knee. Almost all injuries to the knee restrict this normal range of motion.

When the knee is fully extended, it is like a stick. It is totally stable. When it bends, it has play in it—from front to back and from side to side. It can even be slightly rotated. You need this flexibility to change directions quickly. Have you heard a running back called a "good cutter." An anatomist would say that he has a flexible knee.

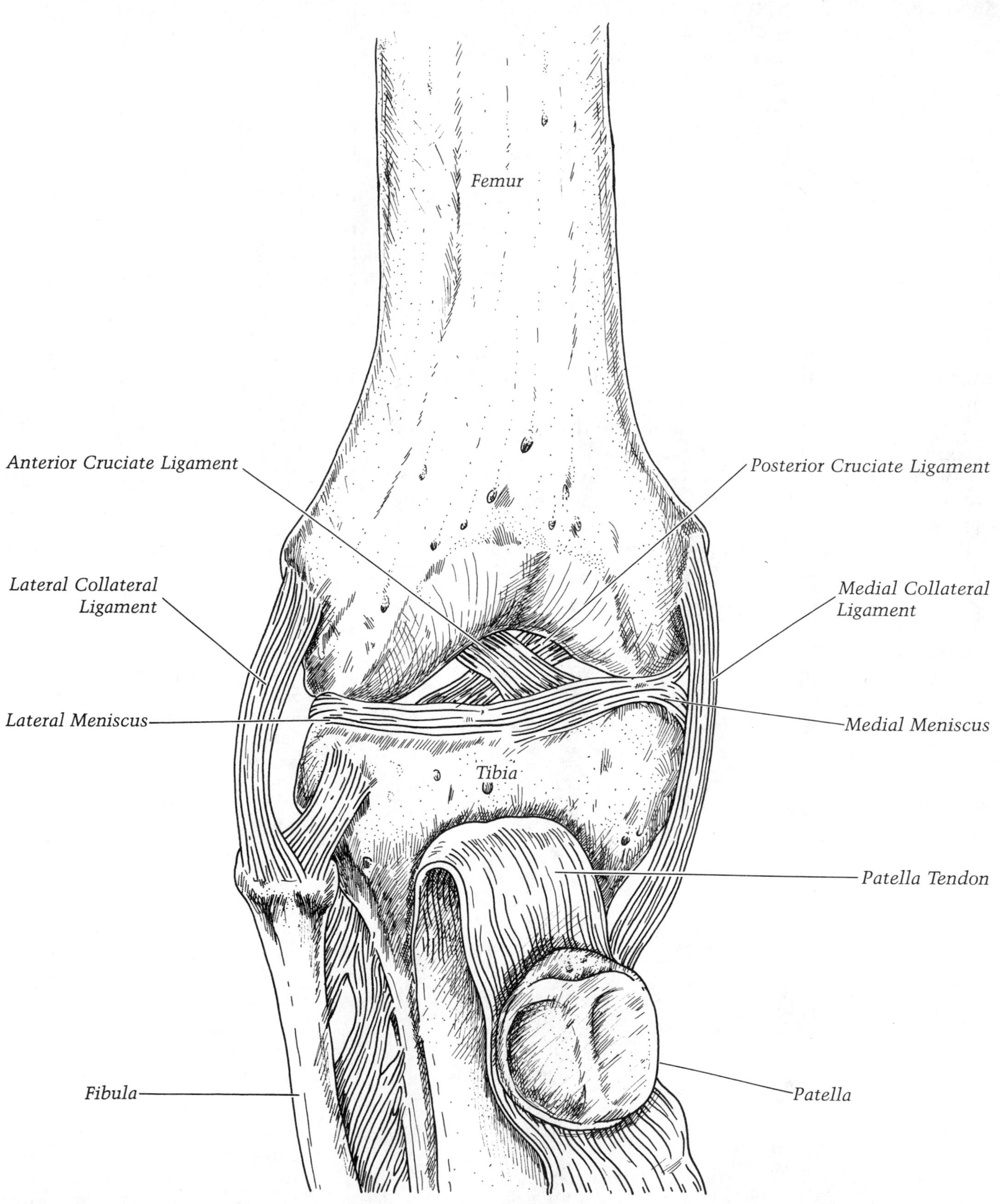

Anatomy of the Knee Joint

The bones of the knee are held together by ligaments. These are the gristle-like structures that join the femur and the tibia. They make the knee sturdy and keep it from wobbling.

The knee has two major ligament systems. One forms the sleeve-like connection between the two bones. This is called the *knee capsule.* Certain parts of this capsule are thicker than others. These thickened portions—and there are five of them—are distinct ligaments. They are like stripes sewn on the sleeve of your shirt.

Besides the five capsule ligaments, the knee has two additional ligaments that occupy the middle of the knee joint. Because they cross, they are called *crossed* or *cruciate* ligaments. The knee ligaments are what I term the "static knee stabilizers." The ligaments are static because they are girder-like.

The muscles and their tendons are the dynamic stabilizers of the knee. They are dynamic because they can quickly spring into action.

The geometric shape of the femur and the tibia add to knee stability. The end of the femur has two knobs, called *condyles.* They are round and smooth. They fit into the two shallow sockets in the top of the tibia. The ball and socket with the accompanying joint cartilages are rarely mentioned as a knee stabilizer, but they give the knee 10 percent of its stability.

Stability and mobility of the knee are achieved by its design. The diagram on this page gives you an idea of the complexities of the knee.

Because of its design, there are at least five places where the knee can be injured: the ligaments, the cartilages (called *menisci*), the muscles around the knees, the kneecap (a bone in front of the joint), and the tendons.

Here are some of the injuries to the knee and the extensor

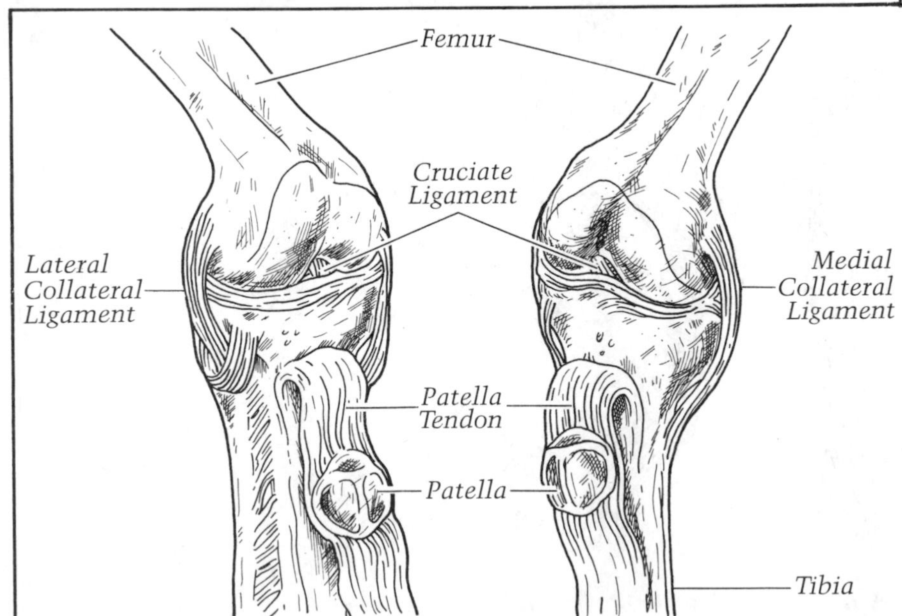

THE SEVEN LIGAMENTS OF THE KNEE

The capsule ligaments are:

1. The medial collateral (M-C) ligament. A broad, thick ligament running on the inside of the knee. This ligament runs from the thighbone to the top of the lower leg bone (i.e., femur to tibia). It is the strongest ligament in the knee. Yet, it is the one most frequently injured in sports because it is very common to be hit on the outer side of your knee. The force of the strike compresses the ligament on the outside of the knee and stretches the ligament on the inside—the medial collateral ligament. Ninety percent of knee sprains involve this ligament.
2. The lateral collateral (L-C) ligament. This ligament is much smaller than the M-C ligament and is on the outer side of the knee. It runs from the femur to the small outer lower leg bone (i.e., femur to fibula). The L-C ligament is rarely injured, because it is very unusual to be struck on the inside of the knee—the type of blow which would stretch this ligament. When this ligament is torn, there is often injury to the peroneal nerve, which lies next to it. About 5 percent of knee sprains involve the L-C ligament.
3. The posterior oblique (P-O) ligament. This ligament is a thickening in the capsule sleeve just behind the M-C ligament. I describe it as the strong part of the capsule sleeve on the inside back corner of the knee. It is almost never injured by itself. But in 50 percent of the cases of M-C sprains, the P-O ligament is also injured.
4. The arcuate ligament. This is a thickening of the capsule sleeve on the back outside corner of the knee. Three percent of ligament sprains involve it, and it can be stretched only if the athlete is struck on the front of the knee, driving it straight backward and forcing the bone apart in the back of the knee.

5. The oblique popliteal ligament. This is a thickening of the capsule sleeve on the very back of the knee. Its job is to keep the knee from being pried apart when force is applied to the front of the knee with the leg fully extended. Some stretching of this ligament is not uncommon, but complete tearing of it is rare.

In addition to these five capsule ligaments, the knee has two additional ligaments that sit right in the middle of the knee joint. Because they cross each other, they are called the *crossed* or *cruciate* ligaments. They lie side by side.

The front ligament—the anterior cruciate ligament—is a very common knee injury in sports. In fact, 60 percent of serious knee ligament injuries involve this ligament in some way. This is because this ligament, which is about the size of your fifth finger, lies in a position to be easily stretched by the kind of force that is applied to the knee in athletics. But many athletes seem to preserve satisfactory knee function even after this ligament is damaged.

The back cruciate ligament (called the *posterior cruciate ligament)* is the size of your thumb. It is a very important stabilizer of the knee. Injury to this ligament is quite unusual in sports.

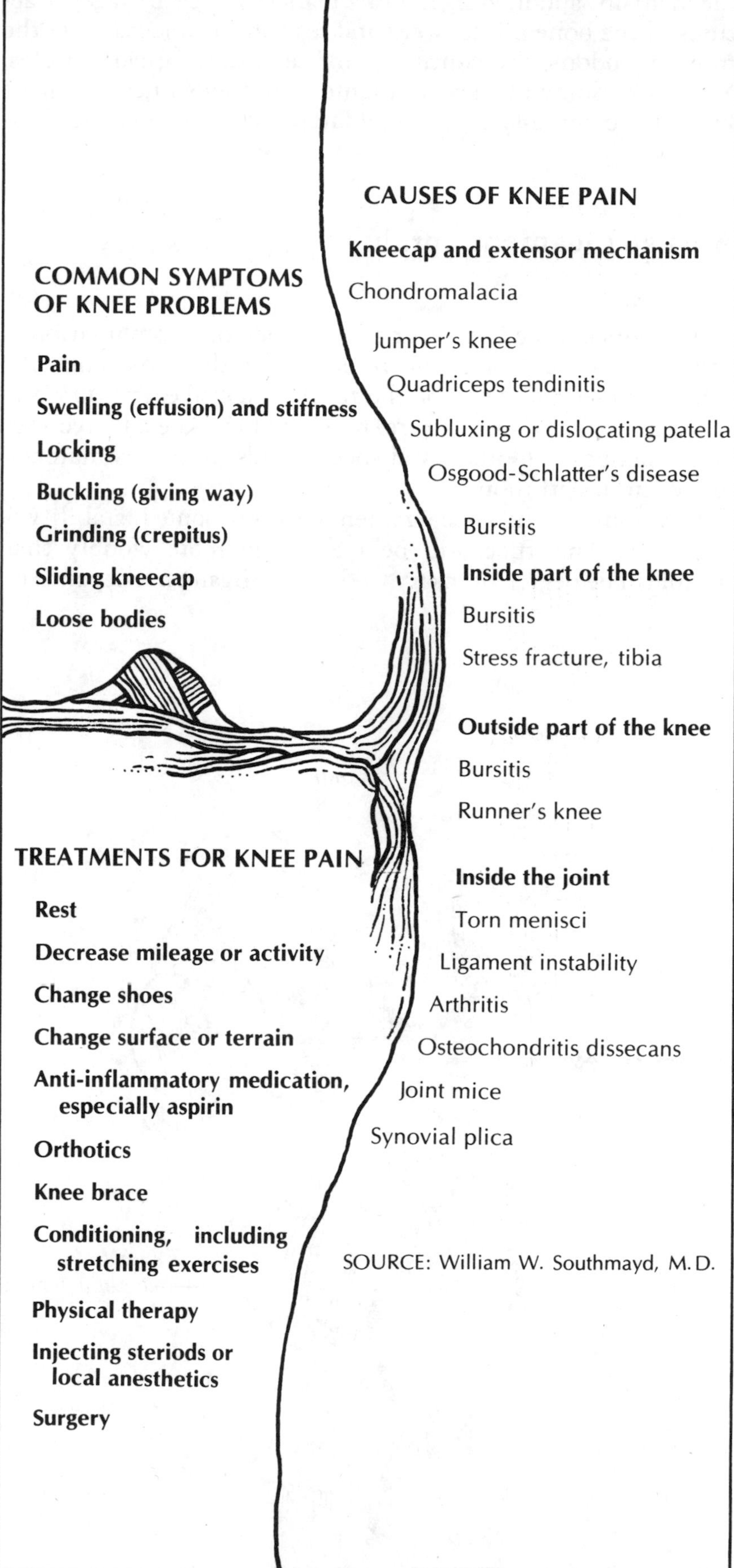

mechanism: sprained ligaments, tears of knee cartilages, fractures of the bone joints, wear and tear on the kneecap and the muscle tendons, and contusion and tears of the quad muscles. You know some of these ailments as runner's knee, jumper's knee, knee sprains, Osgood-Schlatter's disease, and the charley horse.

A Knee Ligament Sprain

CAUSES

A sprained knee is a rip or a tear of one or a combination of the seven knee ligaments. It is one of the most common injuries in sports. It can be the result of a strike, or *trauma* in medical parlance, or an overextension of the knee. I have seen knee sprains on football and soccer fields, in tennis matches, and even in bathrooms.

Any injury to the knee ligaments leads to some instability of the knee. The knee will be looser and more wobbly than normal. The degree of injury to the knee ligaments depends on

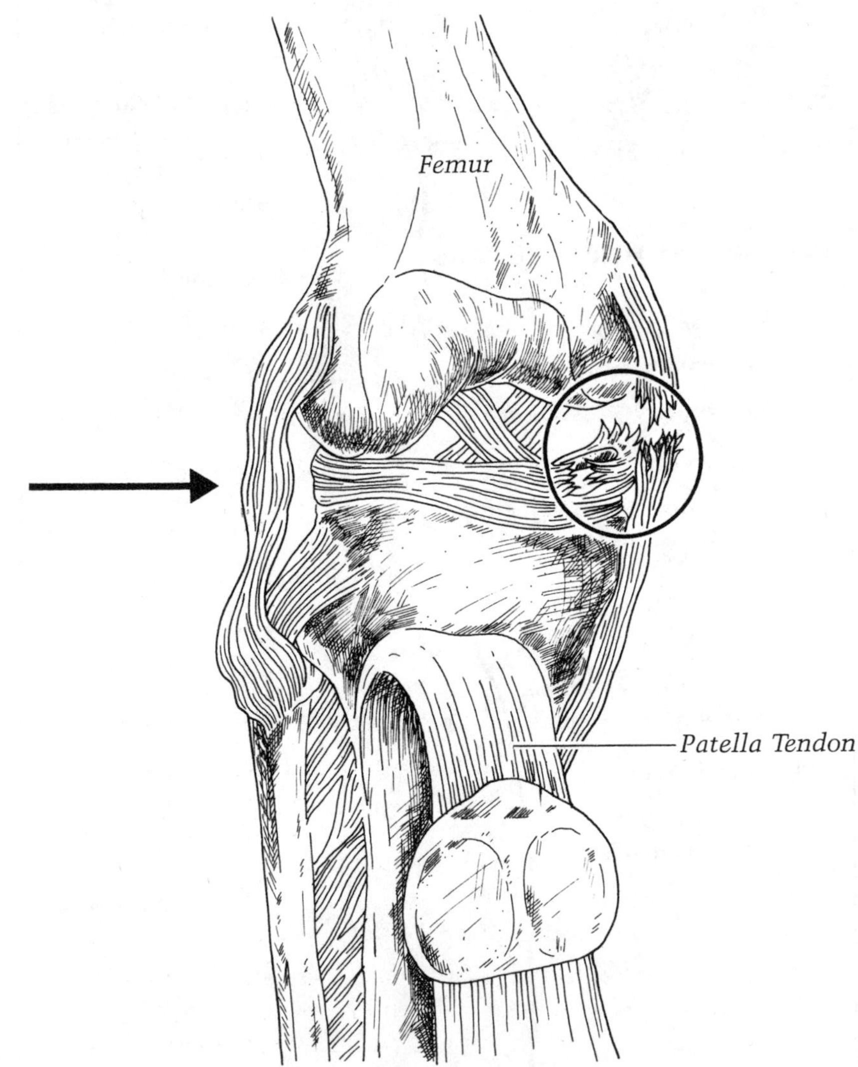

A Torn Medial Collateral Ligament

HOW TO GUARD AGAINST KNEE INJURIES

If you participate in contact sports, there are six good ways to help protect yourself against knee injuries:

1. Use cleatless shoes, as many pro teams do. Cleats anchor your foot to the ground.
2. Run with short, choppy steps. The less time the foot is on the ground, the less likelihood of knee injury. If you anticipate being hit, shorten your stride.
3. If possible, protect your knees from a hit, especially from the back and side. That's where the knee is most susceptible to injury and is one reason why the National Football League outlawed crack-back blocks.
4. Exercise the muscles around the knee. There is evidence that these exercises thicken ligaments and make them even more resistant to injury.
5. Don't do full squats with heavy weights.

the amount of force applied to them at the time of injury and how long the force was applied. The greater the force—and the longer it was endured—the greater the injury.

All ligament sprains can be separated into three degrees of severity. Hold a tissue with both hands. Begin to stretch the tissue apart. As you begin to pull, note the weakening of the center of the tissue as the fibers start to separate. This is a grade I sprain. If you pull harder, the tissue will separate in some other areas. This is a grade II sprain. When the tissue rips in half, this is a grade III sprain.

DIAGNOSIS AND TREATMENT

When you sprain your knee, you will feel pain. The degree depends on the severity of the rip. The immediate treatment is RICE—rest, ice, compression, and elevation.

When a knee ligament injury happens, you have a twenty- to thirty-minute grace period before the swelling starts. If I examine a knee sprain immediately, I get an excellent idea as to whether it is a grade I, grade II, or grade III injury. This is because when the sprain first happens, there is very little bleeding and no muscle spasm. Therefore, I can put stress on the injured ligament and determine whether it is just mildly stretched, partially torn, or completely torn. This is extremely important information.

A grade I sprain is relatively trivial and usually costs the athlete or fitness buff about seven days of idleness. A grade II injury requires two to three weeks to heal. A grade III injury must be operated on to stitch the ligament back together. Without an operation, your knee will never work right.

If you do not get examined for six to twelve hours after the injury, it is often difficult to be sure what grade of ligament injury has been sustained. As in shoulder separations, the muscle around the knee goes into spasm. Swelling occurs around the ligament. The knee joint often fills with blood, thus making it very painful to move and test the knee. Often, I remove the blood from the joint. This makes the examination more comfortable and more accurate. I used to insert Novocain inside the joint when I removed the blood. My hope was that the anesthesia would relax the muscle, but I have found this method very disappointing. People who exercise have very strong muscles, and once these muscles go into spasm, they are difficult to relax.

What I worry about is not recognizing a grade III injury—a complete disruption of the ligament. A strong muscle in spasm will make the joint appear more stable than it really is. Therefore, I do not hesitate to give injured people general or spinal anesthesia to completely relax their muscles. If I do not determine whether the ligament has been completely severed and fail to stitch the ligament back together, I have left the patient with a permanently weakened knee and future disability. Thus, when your doctor asks to examine your knee under anesthesia, do not protest. It is in your best interest.

A SIMPLE KNEE SPRAIN (First-Degree)

In 1979 I was in spring training with the Red Sox at Winter Haven, Florida. Fred Lynn was involved in a rundown drill. Trying to evade the tag, he darted back and forth between the first and second basemen. He was running toward second, stopped quickly, put all his weight on his right leg, and then turned quickly to run back to first. He felt a sharp pain in the inside of his right knee. He finished the drill, but within an hour he was limping. I examined his knee and found tenderness along the inside knee ligament (called the *medial collateral ligament*). I started him on RICE therapy. He iced his knee the rest of that day and all night. He perched the knee on two pillows, and I wrapped it with elastic bandages.

I reexamined his knee the next day. It was mildly swollen. There was some tenderness over the upper portion of the tibia, where the medial collateral ligament attaches to bone.

I examined his knee for stability. I did this by placing one hand on the femur and then pushing the tibia away from the femur. When I put my hand on the outer side of his femur and then moved his tibia away from the center of his body, I stressed the inside knee ligament. He squirmed; however, there was no sign that the ligament had been stretched or completely torn through. X-rays showed no evidence of any bone abnormality but did show swelling within the knee joint. My diagnosis was a first-degree sprain of the medial collateral ligament.

Fred remained in a compression dressing for seventy-two hours. He continued the ice for forty-eight hours. I had him use crutches for three days. Then, I started him on whirlpool baths and range-of-motion exercises for his knee. Within seven days, he was able to jog with elastic support on his knee. In two weeks, he recovered. When the season opened, Lynn was in center field.

A SECOND-DEGREE KNEE SPRAIN

During my junior year at Harvard, one of our toughest games was against undefeated Dartmouth. In the first half, the game was going well for Harvard. The score was tied. In the second half, Harvard kicked off. I ran down under the ball and was preparing to make the tackle on the ball carrier. My eyes were fixed on him. Suddenly, I felt a searing pain on the inner part of my right knee. My next memory was lying on the ground with one of the Dartmouth players across my knee. It was a legal block, but a hard one. Unfortunately, my foot was fixed in the turf while I was preparing to tackle the ball carrier. When the Dartmouth player's shoulder hit me at knee level going all out, something had to give. It was my medial collateral ligament. I was carried off the field and taken into the medical area. X-rays showed no bone fracture. Dr. Thomas Quigley, the team physician, carefully examined my knee. It was tender on the inside. The outside did not hurt. I remember him saying that it might be either an injury to the inside knee cartilage or a bad sprain. He applied ice and installed a compression dressing. I watched the end of the game on television. After the game, he gave me crutches and I hobbled back to my room. As I lay with my throbbing leg elevated in the air, I thought this was the worst day of my life. I limped back to the fieldhouse the next day. Dr. Quigley examined my knee again. He told me that I had a second-degree sprain of the medial collateral ligament. He explained that a grade II injury was a partial tearing of the knee ligament. I had difficulty picturing my injury. All I knew was that my knee was killing me and I felt like a wounded moose.

Dr. Quigley kept my knee bandaged for five days. I was on crutches for seven days. The bulky bandage was comfortable because it kept my knee from wriggling. Without the knee wrap, I would have been unable to sleep.

Five days after the injury, I was started on whirlpool treatments. Because my leg was buoyant, I could move my knee, which had become quite stiff. Jack Fadden, Harvard's trainer, used ultrasound on the inner part of my knee to stimulate the blood supply and break up the inflammation. At that time, it was medically acceptable to inject cortisone into the ligament. My ligament was injected every two days for a week. The only way I agreed to the shots was that I was told that it was an aphrodisiac. Finally, two weeks after the injury, I started to run with my knee heavily taped. Jack Fadden applied a Duke-Simpson tape support, which gave excellent stability to the inner aspect of the knee and was very comfortable. It made my knee feel secure, but I was unable to cut because there was too much pain. I missed the next two games. With the continued whirlpool baths and ultrasound treatments, I was able to return to football with the tape support in three weeks. Two months after the injury, I went duck hunting. When I turned quickly on the muddy ground, my knee still hurt. I was convinced that it would never heal. Later, in medical school, I learned that pain two to three months after a second-degree sprain is not uncommon.

Dr. Quigley's treatment was unusual for grade II sprains. If I had been treated for this injury someplace else, I am sure that I would have been placed in a cast for four to six weeks. The purpose of a cast is to let the ligament completely rest so that it can heal correctly. In my case, that course of treatment would have meant the loss of the whole season. Because of Dr. Quigley, I was able to play in the last two games of my junior year.

Years later, Dr. Quigley, who became my mentor and medical colleague, explained why he did not use a cast on me. He said that over the years at Harvard, he had found that casts were unnecessary. The ligament healed just as well if early motion was started. He emphasized that careful rehabilitation should begin as soon as the swelling stopped and the tenderness permitted. It was his successful formula. It must have been right.

AN UNUSUAL KNEE SPRAIN

Wayne Cashman, the captain of the Boston Bruins, has had a history of mild sprains to the medial collateral ligament. But with a two- or three-week rest, Cashman overcame the problem and returned to the ice.

In December 1979, Cashman's knee was hit in a game. The knee became tender.

"He had tenderness just over the medial collateral ligament," recalls Jim Kausek, sports therapist for the Boston Bruins. "The knee didn't respond to treatment or rest.

"Cashman was rested until late in December. We put him in a Lenox-Hill brace and he started playing, but the pads of the brace irritated his knee.

"I had a double-hinged polycentric brace custom-made for him. That seemed to do the trick until a game in Winnipeg when Cashman got hit into the boards. The knee started to give him the same problem.

"This time he had his knee arthrogramed and arthroscoped. The test revealed a partial tear of the medial collateral ligament. The ligament had a hook tear which flapped between the bone and acted like a torn meniscus. Apparently some granulation tissue formed and the ligament couldn't adhere to the bone even with rest.

"At the end of the season, Cashman had an operation. The ligament was reattached to the bone. He is fine now."

When I was a resident at Massachusetts General Hospital in 1973, I x-rayed my knee. It showed no abnormality such as bone spurs or arthritis.

Eighteen years later, my knee functions normally, but there is a bit of looseness on the inner side. I can feel the looseness when I put it in certain positions quickly. Occasionally, after heavy use or in cold, damp weather, my knee will ache. Otherwise, my knee functions perfectly.

A COMPLETE TEAR OF THE KNEE LIGAMENT (Third-Degree Tear)

Royce Terrell, an eighteen-year-old high school defensive tackle, was playing football in a traditional Thanksgiving Day game. He was running to his left when one of the opposing players blocked him from the side, crashing into his right knee. Royce fell like a tall tree. He was carried from the field on a stretcher. The team physician, Dr. Charles Thompson, examined his knee and diagnosed a complete medial collateral ligament tear. He asked if I would take a look at Royce. I agreed with Dr. Thompson's analysis. Because there was no ligament to restrain extreme movement of the separated bones, a complete tear without muscle spasm or swelling was easy to diagnose. I had no choice but to operate. Royce was admitted to the hospital and placed in a half-cast. His leg was elevated and ice bags were applied to the front of the knee. Surgery was carried out the next day. Royce's ligament was completely shredded and pulled away from the femur. Fortunately, his inside knee cartilage (called the *medial meniscus)* was not damaged. It had been separated from the edge of the joint, but I was able to stitch it back in place and to stitch the shredded ligament together. Next, I pulled the ligament up against the femur. To get a good tight repair of a third-degree sprain, I stretched the ligament up as much as possible. I used two metal staples to anchor it in place on the femur. These hold the ligament in place until it heals. I closed the incision and placed Royce in a front-and-back plaster splint. These are held together with Ace bandages and allow for swelling in the knee, which is inevitable after this major operation. Five days after surgery, I removed his plaster splints and applied a long-leg cast from the toes to the groin. The knee was kept at an eighty-degree bend. This angle of the knee takes the tension off the staples and the repaired ligament. He remained in the cast for six weeks. Following removal of the cast, he started physical therapy.

If Royce had not undergone surgery, he would have had a permanently loose and sloppy knee. Now, his knee functions normally. Although the inner aspect of his knee sometimes aches after hard use or damp weather, he can play sports and needs no brace or tape support. But if you have a third-degree tear of the medial collateral ligament of your knee, you will be out of action for six months.

Anterior Cruciate Ligament Injury

CAUSES

Sixty percent of all significant knee ligament injuries involve at least a partial injury to the *anterior cruciate ligament,* which is one of the two ligaments inside the knee joint itself. In 30 percent of cases of torn knee cartilage, there is also damage to the anterior cruciate ligament.

Why is this ligament injured so often? In my opinion, it stems from the anatomy of the knee itself. When the knee is completely straight, all the ligaments are tight and the leg becomes a straight stick. There is no side-to-side, front-to-back, or twisting motion available in the knee in this position. However, as soon as you bend the knee even slightly, play develops between the femur and the tibia. When the knee bends, the anterior cruciate ligament restrains the amount of front-to-back and twisting motion between the two bones. It

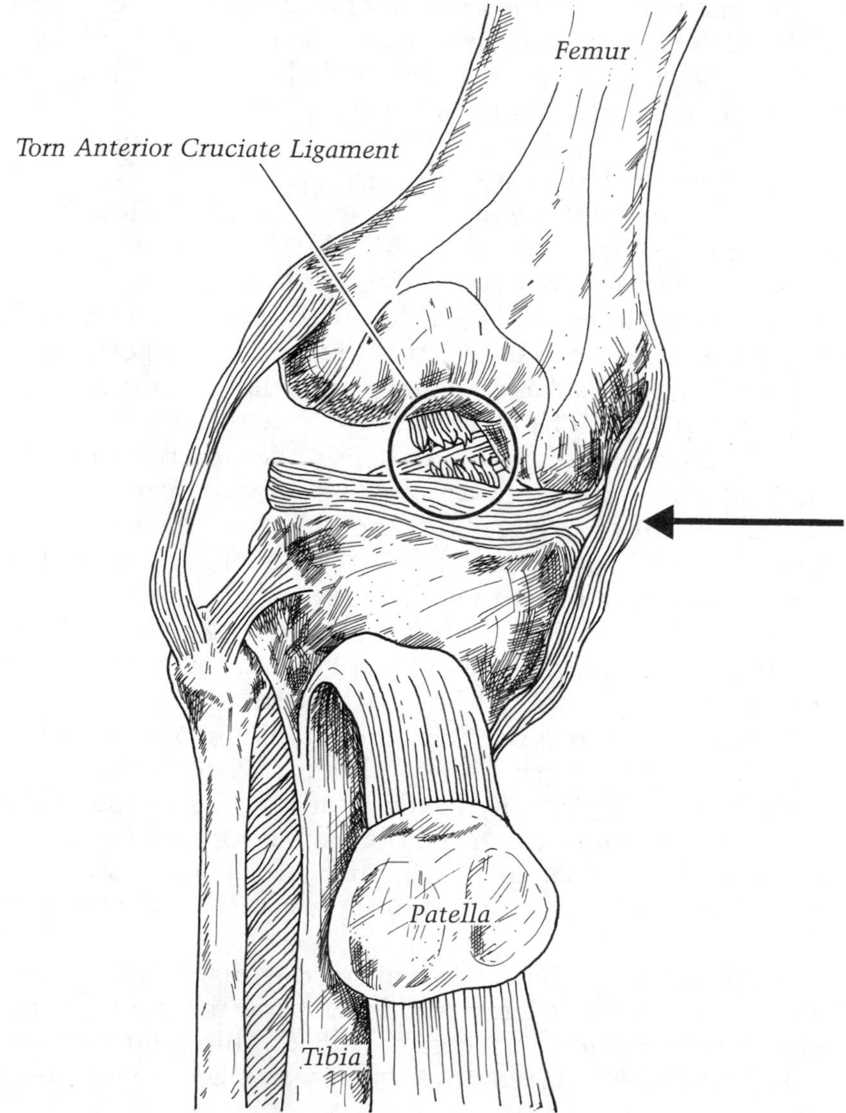

Tear of the Anterior Cruciate Ligament

functions as a restraining guidewire to keep the two bones in proper relationship to one another.

When either a large external force is applied to the knee in a bent position, or when tremendous muscular force is applied to the knee by the athlete himself, the anterior cruciate ligament is put to the supreme test. It tries desperately to hold the two bones together, but if the force applied to the two bones exceeds the strength of the ligament, it stretches or rips completely. The anterior cruciate ligament usually tears in one of three places. It can pull away from the femur or the tibia. Also, it can tear right in its middle. The middle is the most common place.

Most anterior cruciate ligament tears occur in contact sports—football, soccer, rugby, and lacrosse. They often are caused by a blow on the outer part of the knee. Because the medial collateral and the anterior cruciate work in tandem, they usually get injured together.

DIAGNOSIS AND TREATMENT

How can you tell if the anterior cruciate ligament has been damaged? When it tears, the ruptured blood vessels in it bleed and the knee joint quickly fills with blood. This happens within an hour of the injury. Almost all ligament tears fill the knee joint with blood. When the ligament tears without you being hit, you feel a pop as the ligament tears. You will get that "just not quite right" feeling. Because of the pain, you will not be able to continue in the game.

Swelling sets in rapidly. If the anterior cruciate ligament is the only injury, there will be no localized area of tenderness. It is impossible to touch the ligament because of its position. If another ligament is damaged along with the anterior cruciate ligament, there will be tenderness over the ligament.

The easiest way for me to tell if the anterior cruciate ligament is damaged is to perform the *anterior drawer test.* I put the patient on a table with the knee over the edge of it, with the tibia hanging perpendicular to the floor. I grasp the upper end of the tibia with both hands and pull forward. This is the motion that the anterior cruciate ligament usually restricts. If the tibia comes forward farther on this injured knee than on the normal knee, I suspect that the anterior cruciate ligament is torn. This anterior gliding motion of the tibia on the femur can also be checked with the knee out straight. Again, I place my hands behind the upper end of the tibia and pull it forward. If the tibia glides forward farther on the injured side than the uninjured side, I know that the anterior cruciate ligament is torn. This is called the *Lachman test.*

Occasionally, x-rays are helpful in this diagnosis, but not often. This is because ligaments do not show up on x-rays. In a few cases, if a bone chip comes with the ligament end, the x-rays will show it. I recommend that x-rays be taken in every case.

It is very important to make an accurate diagnosis on torn

ACUTE KNEE INJURIES WITH BLOOD IN THE JOINT

Dr. Kenneth DeHaven, head of the section of athletic medicine at the University of Rochester (New York) School of Medicine, examined 113 athletes who had acute knee injuries. All had blood in their knee joint (called *hemarthrosis).* Ninety percent of these athletes had serious problems. The anterior cruciate ligament was found torn in 72 percent of the athletes. Of these, 67 percent had torn cartilages, 6 percent had bone chips, and 3 percent had a tear of the posterior cruciate ligament. Only 4 percent had no cartilage or ligament injury. Dr. DeHaven concluded that acute knee injuries with early formation of blood in the joint is a definite indication for arthroscopic examination of the joint to accurately pinpoint the injury.

GREER STEVENS' KNEE

Greer Stevens, a professional tennis player from South Africa, was injured in a mixed doubles match in 1978, while playing for the Boston Lobsters. She went for a ball hit by Ille Nastasi and planted her left foot on the ground. In doing so, she twisted her left knee, tearing her anterior cruciate ligament, and both menisci, the washer-type cartilages which separate both knee bones.

"She had severe pain and immediate swelling of the knee," says Dr. Bert Zarins, co-director of the Sports Medicine Clinic at the Massachusetts General Hospital. "I found Greer's knee unstable. An arthroscopic examination revealed that her anterior cruciate ligament was shredded in such a way that it could not be repaired. I recommended surgery for Greer. She agreed.

"I replaced the anterior cruciate ligament with a tendon from another part of the body, and I removed the torn menisci. One week after her operation, I built a special hinged cast for her. This allowed her to begin moving her knee, and avoid knee stiffness. She used the cast and crutches for about six weeks. With a physical therapist, she gradually gained motion in her knee and started to rebuild her atrophied quad muscles. It took her about nine months to fully recover. But she has done very well. She even won a major tournament, beating Chris Evert. In the following year, she won the Wimbledon mixed doubles championship and is rated as one of the top ten tennis players in the world."

anterior cruciate ligaments. In numerous medical studies in the last five years, it has been shown that the anterior cruciate is vital to the knee joint. Without it, the knee has a 50 percent chance of becoming permanently unstable. I need to know if the anterior cruciate ligament is injured, what part of the ligament is injured, and if it can be repaired.

Some doctors use a test called an *arthrogram,* which is an x-ray of the knee after the injection of a radiopaque dye, to diagnose the problem. I do not think that an arthrogram is reliable for this type of injury because technical difficulties arise during the test. The blood in the knee mixes with the dye; thus the x-rays are sometimes unclear. I prefer an arthroscope test, which can sometimes be done under local anesthesia. In an acutely injured knee, I use general or spinal anesthesia. Because of the muscle spasms, it would be impossible to get an accurate diagnosis otherwise.

The arthroscope is inserted on the outer part of the knee, beside the kneecap tendon. I wash out the joint thoroughly with large amounts of saline solution. I systematically check both inside and outside knee cartilages. I examine the lining of the joint. I check the bone ends to make sure that no fractures have occurred. I then look into the central part of the knee where the anterior cruciate ligament lies. It is easy to see. If it is torn in its middle, I see the two torn ends, which look like paintbrushes. There is generally a clot of blood between these ends. If it is torn at the femur end or the tibia end, this can be seen directly. If a piece of bone is torn off, I can see that, too.

If the ligament is stretched, I do nothing. If it is partially torn, I also do nothing. But if the ligament is torn off the bone or severed, it can be reattached surgically.

If the ligament is torn away from the bone but very close to its bony attachment, I stitch the ligament back to the bone itself. I drill into the bone and thread the anterior cruciate ligament through it. The knee seems to be able to heal the reattachment nicely, and the ligament itself remains alive. That is because the blood supply in the middle of the ligament has not been injured. After the surgery, the knee is restored almost completely to normal. Patients may have discomfort with extremely hard use of the knee. Some patients do, however, run marathons after this operation.

Unfortunately, 50 percent of the time, the ligament is torn in the middle. I find that it is fruitless to try to stitch the ends back together. The nourishing blood supply to the ligament is interrupted, and the repaired ligament does not survive. It shrinks back into a small nubbin of scar tissue. If the ligament cannot survive, then why operate to stitch it back together at all? If I have to operate on the knee because other knee structures are damaged and need surgery, I will put a couple of stitches across this ligament to keep it from flopping into the joint and interfering with normal function.

Studies show that 90 percent of patients can get along without the anterior cruciate ligament, which means that they can return to sports.

Anterior Cruciate Ligament Reconstruction

What happens if you are one of the 10 percent who has knee difficulty because the anterior cruciate ligament is torn or ripped? You should suspect that this has happened when your knee feels unstable. It gives out unexpectedly. This happens mainly while you try to cut. It can also happen as you slow down from a hard run or sprint. When going down stairs, your knee will feel "funny." After an episode of buckling, your knee will swell and hurt. Because of the buckling, you will lose confidence in your knee and will be unable to play sports effectively. You will be intimidated by your knee.

How do I make this diagnosis? I am looking for the "funny knee" symptoms. If you tell me that your knee buckles, it tips me off that there is an anterior cruciate problem. In my physical examination, I try the "drawer" and Lachman tests.

Finally, I do a *pivot shift* test. This is done by having you lie on your side with the bad knee up. With the foot resting gently on the table, I can push and pull on the thighbone and lower leg bone in opposite directions. I bend and straighten the knee as I am doing this. If the tibia glides forward on the femur as the knee comes toward the fully straight position, you have a positive pivot shift test. I can feel a definite shifting of the two bones, one on the other in an abnormal fashion. If the anterior cruciate ligament were not torn or stretched, it would ordinarily prevent this erratic motion between the two bones. I always x-ray a knee for this type of evaluation. However, x-rays are usually negative.

The simplest treatment is a Lenox-Hill brace. This is a brace invented by Dr. James Nicholas and his staff at Lenox Hill

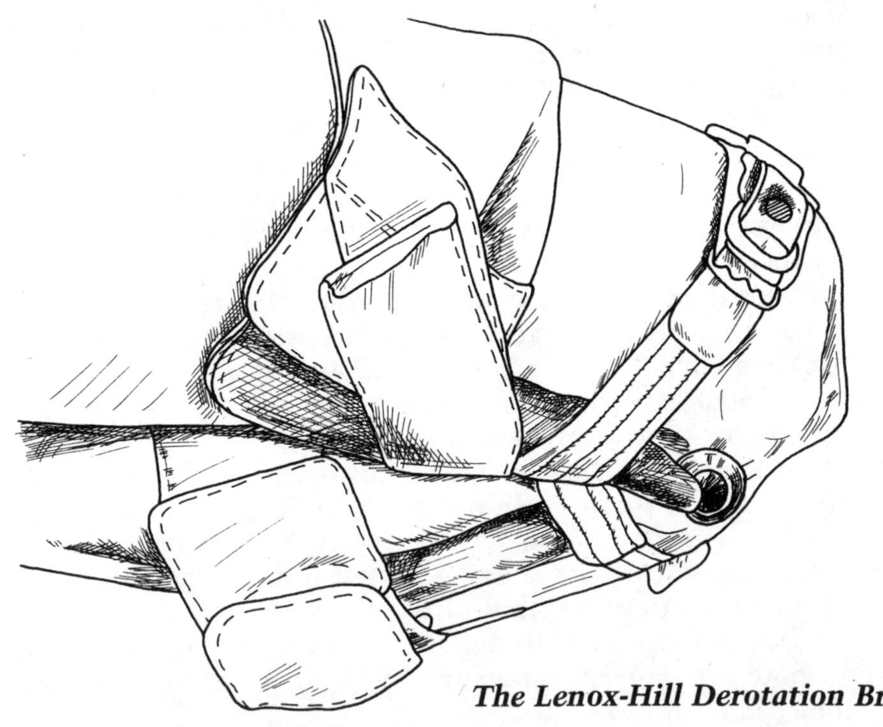

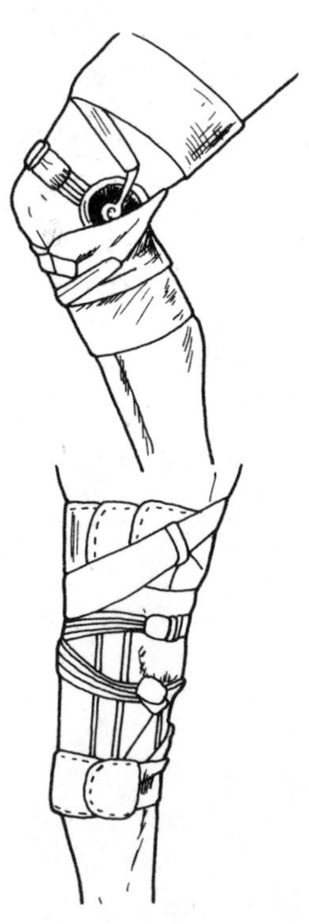

The Lenox-Hill Derotation Brace

Hospital in New York. In my opinion, it is the best athletic brace available for unstable knees; it is custom-fitted and totally protects the knee. It is made from a plaster mold of the patient's knee. My patients have the mold taken of their leg, and I ship it to New York, where the brace is made. The brace is then returned and fitted. This brace is perfectly legal in most sports except rugby. (It is not allowed in rugby because the brace has some exposed metal, which might injure a rugby player.) The brace can be worn not only for competitive athletics, but also recreational athletics. I find that about 50 percent of patients with anterior cruciate ligament instability can function using the brace. They are not bothered by the brace routinely—and it controls the knee when they are playing sports. The brace costs about $350 and can be obtained only with a physician's prescription.

In some patients, even the brace is inadequate. They still have a sensation that the knee is giving way and that the bones are shifting inside the brace. These same athletes tend to have difficulty with their knees even in routine activities. They have difficulty descending stairs, getting out of their car and turning corners quickly. They lose confidence in their knees. At this point, most surgeons recommend surgery.

The purpose of the surgery is to substitute some structure for the absent anterior cruciate ligament. Among orthopedists, there are two schools of thought. One group recommends replacing the anterior cruciate ligament in the same location where it once lived. Usually, one of the hamstring tendons is used for the substitution and rerooted through the knee joint. Other surgeons use a piece of the kneecap tendon to replace the anterior cruciate ligament. In both cases, my criticism is that the rerooted tendon does not have a blood supply of its own. Rather, it is now bathed in joint fluid which cannot nourish the tendon adequately. The final verdict is not in. It is unclear whether these tendons can predictably function in the place of the missing anterior cruciate ligament.

Because of this biologic uncertainty, I favor the alternate approach. This involves reconstructing the anterior cruciate ligament outside the knee. I like the Ellison procedure, developed by Arthur Ellison, a noted sports medicine orthopedist from Williamstown, Massachusetts. In this operation, a strip of the tough fascia on the outer side of the knee (called the *iliotibial band)* is used. This strip is then rerooted under the lateral collateral ligament so that it runs a course parallel to that of a normal anterior cruciate ligament. The rerooted course lies in an area with an excellent blood supply. Therefore, biologically, it makes sense to me that this reconstruction has a better chance of surviving than it does within the joint (in medical parlance, *intra-articular)* anterior cruciate ligament substitution. The Ellison procedure is an extra-articular anterior cruciate substitution.

The operation takes approximately two and a half hours. A long-leg cast is then worn for six weeks with the knee bent at about seventy degrees. The cast is removed, and it takes at least another six weeks of physical therapy before the knee starts functioning normally. It will be six to nine months before the patient is back to sports. On return, I recommend wearing a Lenox-Hill brace for one year. A tape support around the knee, if it is professionally applied, is also effective. In my opinion, an athlete should always wear an external support of some kind throughout his or her sports career. For recreational athletics, this type of external support is not necessary. No operation can make the knee completely stable. But the operation improves stability tremendously—from 60 to 90 percent function.

Posterior Cruciate Ligament

CAUSES

The posterior cruciate ligament sits directly behind the anterior cruciate ligament inside the knee joint. It has been called the "keystone" of the knee. It is very strong and seldom injured. The anterior cruciate ligament is injured 200 times more frequently. To stretch or rip this ligament, force has to be applied to the front of the lower leg bone with the knee bent approximately ninety degrees and the foot off the ground. A blow like this allows the whole tibia to come back on the femur. This puts the posterior cruciate ligament under great tension and occasionally the force is great enough to sever it.

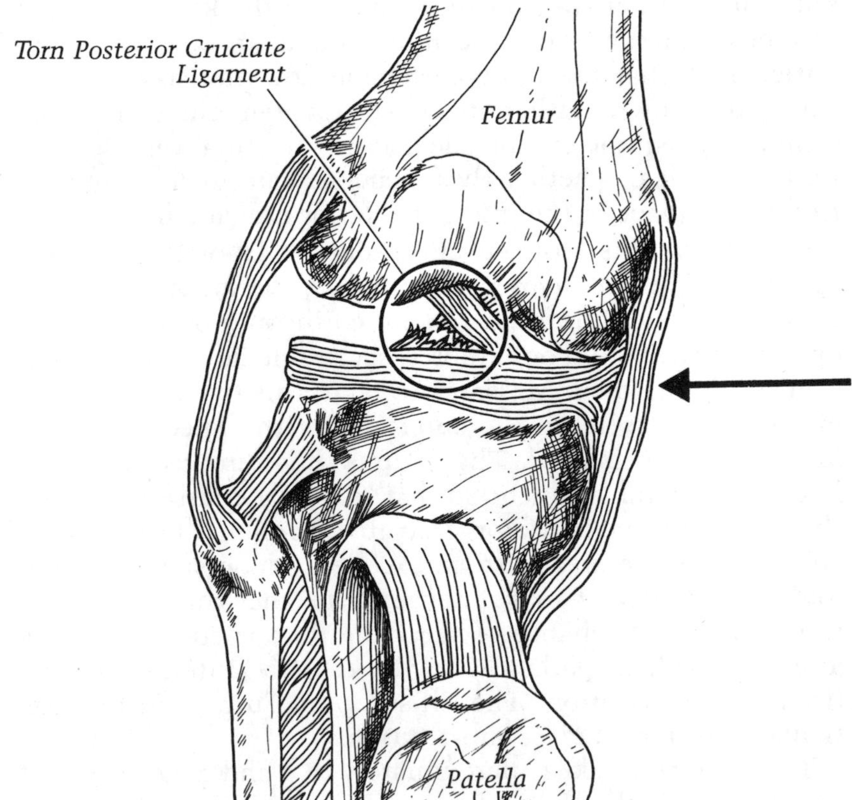

Tear of the Posterior Cruciate Ligament

DIAGNOSIS AND TREATMENT

The same general principle for the posterior cruciate ligament tear applies as for the anterior cruciate ligament tear. The diagnosis must be made as early as possible. The ligament should be repaired if it is pulled away from either of the bones. If it is torn in the middle, stitching it back together does not work. When it is torn in the middle, my choice has been to substitute or reconstruct the ligament. I have used a stout, nearby tendon (called *semimembranosus*) as a substitute for the damaged posterior cruciate ligament. Patients return to full sports activity after this type of repair.

Knee Cartilage Injury (Torn Meniscus)

CAUSES

The knee joint cartilages, called *menisci,* sit between the ends of the bones that make up the joint. They act like knee shock-absorber pads. Sometimes, I think of them as a washer in a faucet. They deepen the sockets into which the knobs or condyles of the femur fit. In this way, they increase the stability of the joint. They are identical to the socket-deepening effect of the glenoid labrum in the shoulder. In fact, both of these structures are made out of the same biologic material, called *fibrocartilage.*

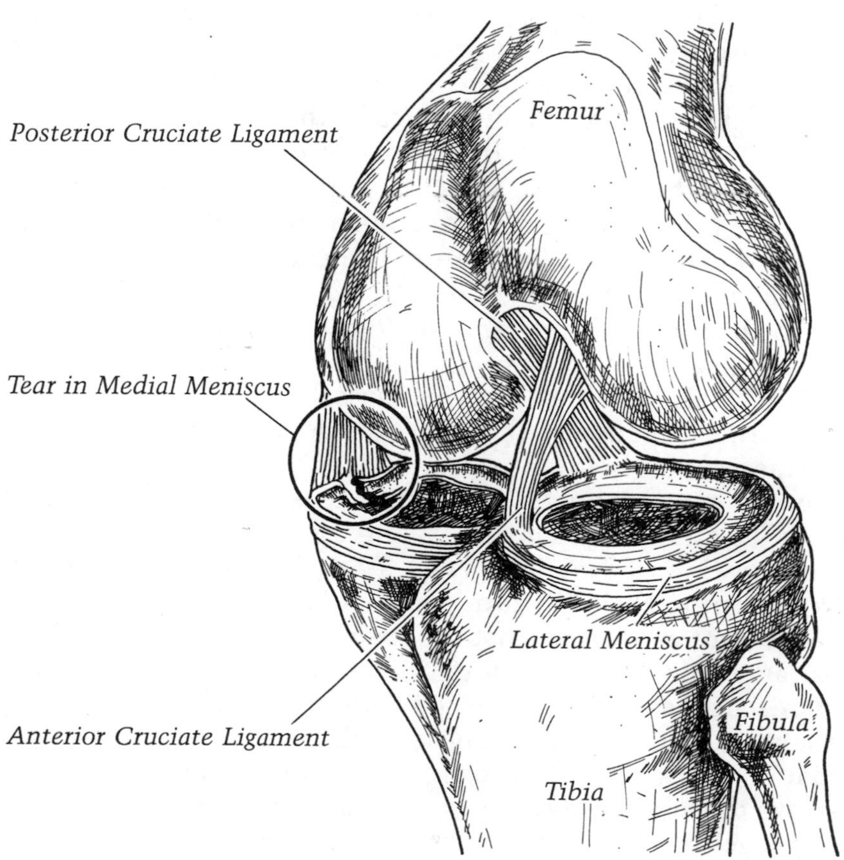

Torn Medial Meniscus

In addition to their stabilizing influence, the knee joint cartilage also helps expand the surface area of the femur and the tibia.

They are triangular in shape and are wedged between the bone ends. Because of the shape, they increase the contact area between the two bones. For example, when you walk, there is less wear and tear on the bones that form at the knee because the cartilage distributes the weight over a wide surface.

The covering surface of the ends of the bone, the joint surface cartilage, is much like a tread on a tire. This tread will wear down more quickly if there is a great deal of force applied to it per square inch. The knee cartilage seems to prevent this type of wear and tear. Loss of any tread on the surface of the joint is called *arthritis.*

Knee cartilages have no blood supply of their own; they are unique in the human body. Because of this biologic fact, a tear or a rip cannot heal itself. That is why so many athletes end up needing surgical treatment for a torn knee cartilage.

A torn knee cartilage is the third most common injury to the knee. Nationally, it is estimated that 52,000 knee cartilages are removed each year from football injuries alone. At least twice that number are removed from other sporting injuries.

How does cartilage get torn? It is very much like tearing the washer in your faucet. If you twist down hard enough on a washer, it will rip.

It is practically impossible to tear your knee cartilage by just running or cutting under your own power. The rip usually stems from a direct contact in collision sports or by falling awkwardly when rebounding in basketball or running into an outfield wall in baseball. In these injuries, the femur acts as a grinder on the cartilage. The tibia is fixed on the ground, the cartilage is fixed to the tibia, and the body weight comes down on the knee through the femur. The cartilage is firmly compressed between the two bones. If you start to turn the femur on the compressed cartilage, the forces exceed the strength of the cartilage. At this point, the cartilage fails. It rips or tears.

The inside knee cartilage (the medial meniscus) is torn four times more frequently than the outside knee cartilage (called the *lateral meniscus).*

If you get a tear in your cartilage, it is torn for good. A tear may be quite small initially, but then with further sports participation, it extends. When a cartilage tears completely, from front to back, the torn portion often flips into the center of the joint and lies against the cruciate ligaments. In medical parlance, this tear is called a "buckethandle" tear of the meniscus, because it is shaped very much like a handle of a bucket. The reason that surgery is almost always necessary is that the torn meniscus tears up the knee joint itself. It destroys the cartilage on the ends of the bones (the joint surface cartilage). This leads to arthritis.

ANDRE THORNTON'S CARTILAGE

Andre Thornton, the heavy-hitting infielder of the Cleveland Indians, tore his knee cartilage in 1980 spring training. He bent down to field a grounder, his spikes caught and his knee twisted.

"I heard a very loud pop," says Cleveland trainer Jim Warfield. "I think that everyone on the field knew that something dramatically had gone wrong with Andre's knee."

Andre's knee was operated on twice for tears of the medial meniscus—the cartilage on the inner side of the knee.

DIAGNOSIS AND TREATMENT

When the meniscus rips, you have a sensation that something is giving way in the knee. Gradually, over the next six hours, the knee fills with synovial fluid. This is extra joint fluid produced by the lining of the joint. I describe this as the knee "crying over what has happened inside." With this excess fluid, your knee becomes stiff and impossible to bend fully. The pain is usually on the side of the knee where the cartilage has been injured. If it is the medial meniscus, the inside of your knee hurts. If it is the lateral meniscus, the outside of your knee hurts.

If the cartilage tear is large enough, the torn piece of cartilage slips into an abnormal position in the knee. This acts like a marble slipping into a gear work. In this situation, you are unable to fully straighten your knee and must walk with a limp. This is called a "locked knee."

If there is a very small tear in the cartilage, and the knee does not lock, you will be able to recover in three to five days and return to normal activities. Remember, however, that the cartilage cannot heal. Therefore, the torn knee joint cartilage will continue to plague you. You will have good and bad days. Sometimes, your knee will feel absolutely normal. However, at

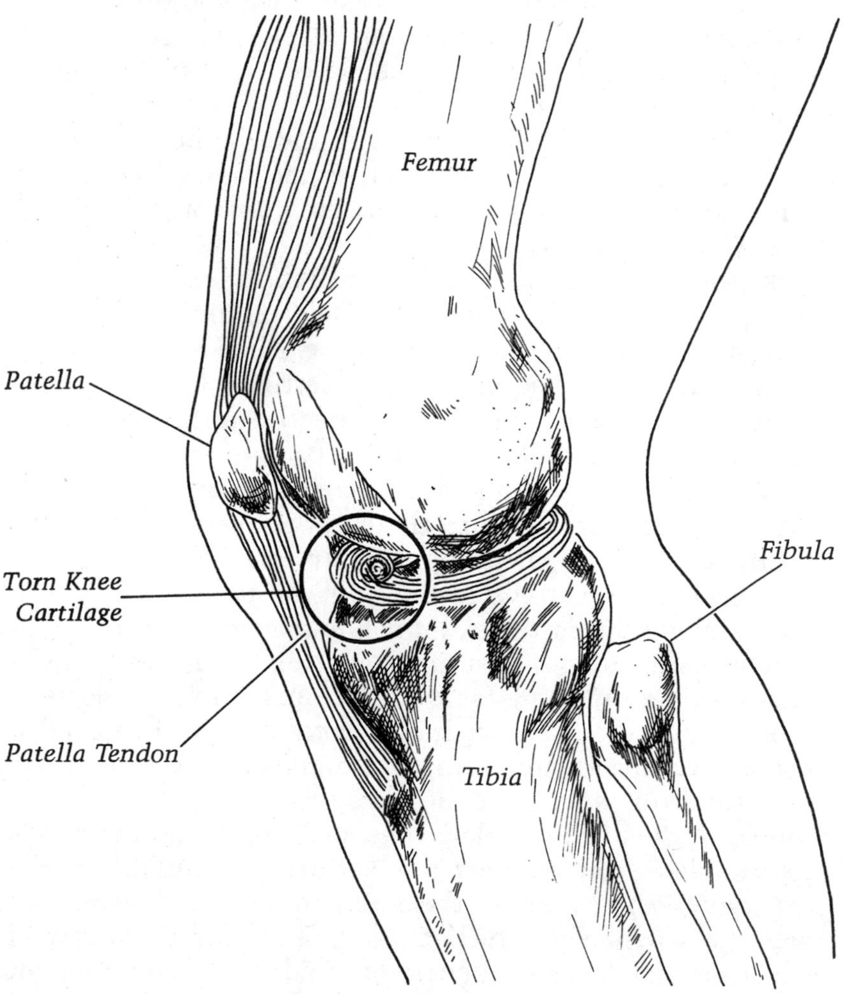

Locked Knee

other times your knee will "buckle," or give way in an unexpected fashion. You will lose confidence in your knee function because you will never know when it is going to give out. You may experience momentary episodes of locking when your knee cannot straighten. This happens especially when you sit down. With your knee bent, the torn piece of cartilage sometimes slips into the joint. When you get up, you cannot straighten your knee. Over time, you learn a little dance to do to readjust the piece of cartilage within the joint. After one of these episodes of giving way, or locking, the knee will swell again. It will be stiff for three to five days and then gradually recover. You will continue to have an aching sensation on the inside or outside of the knee depending on which cartilage is torn. You will find it difficult to perform cutting motions because the torn cartilage will be pulled when you turn on your knee. Unfortunately, no amount of taping can change this sensation because it is inside the joint.

Often, the type of cutting motion reveals which cartilage is injured. If you have a torn inside cartilage on your right knee, it hurts to run bases in a counterclockwise fashion (this is an inside cut). However, it would not hurt you to run the bases in the opposite direction (this is an outside cut).

If you have a torn knee cartilage, you will feel uncomfortable not only in athletic activities, but also when you turn over in bed at night or try to get out of your car. Whenever you put a rotational force (called a *torque* force) on your knee, you are likely to experience pain.

The physical examination to determine whether or not you have a torn cartilage is extremely important. Nowhere in diagnostic orthopedics is there more emphasis on "dead reckoning" of the anatomy.

I like to have the patient sit first with the knee over the side of the examining table. The knee is bent at ninety degrees. I feel all around the contours of the knee. I like to feel directly over the cartilages. The inside knee cartilage lies just above the top of the tibia. I put my finger directly on the cartilage in the front, in the middle, and in the back of the knee. Any tenderness along this "joint line" suggests a cartilage injury.

Next, I run my fingers directly over the outside knee cartilage. Again, it lies a half-inch directly above the top of the bone. I feel for tenderness in the front, on the outside and in the back along the cartilage. Joint line tenderness is an excellent diagnostic clue for torn cartilage. Next, I have the patient lie flat on the examining table, facing up. I hold both heels in my hands and see if the knees are both straight. I can tell if there is swelling in the joint by simply looking at the two knees from a distance of about three feet.

I look for the knee dimple on the inside of the normal knee. This dimple is immediately lost if fluid accumulates in the knee joint. When I move the kneecap back and forth, the patient gets a splashing feeling. Next, I check for stability of the ligaments. I move the knee side to side with it out straight

KNEE CRANKERS

These are designed to strengthen the large muscle on the front of your thigh—your quadriceps muscle. We call them *progressive resistance* exercises because you progressively increase the amount of weight that you lift during the exercise period. These were developed during World War II by Boston orthopedist Dr. Thomas DeLorme.

1. Sit on a firm surface such as your kitchen table. Have your knee bent at a right angle (ninety degrees) over the edge of the table (diagram).
2. Use a weight boot strapped on to your street shoe or an ankle weight. Start with three pounds. Bring your knee out straight. Take three seconds to get it straight. Count 1000-1, 1000-2, 1000-3. Now, hold it straight for three full seconds. See the front thigh muscle enlarge as it contracts. Count again, 1000-1, 1000-2, 1000-3. Now let your knees return to the right angle start position. Repeat this ten times. We call this ten "reps" (repetitions).
3. Keep adding weight until you find the most weight that you can lift ten times. This is your maximum.
4. Now, your progressive resistance exercise program begins (PRE). You start each session by doing ten repetitions with one-half maxi-

mum weight. Rest for five minutes. Next, do ten repetitions with three-quarters maximum weight. Rest five minutes. Finally, do ten repetitions with maximum weight. It is most important to do the exercise properly and slowly. Forget the amount of weight—you are not going for the world's record.

5. Do this three times a week if you are just trying to get stronger—that is, for strength training. If you are coming off an injury or a surgery, do it daily. Have a trainer, a physical therapist, or your doctor supervise the program.
6. Add two pounds to your maximum every five to seven days. Keep adding weight until you have rebuilt the weak leg to the same strength as the opposite, normal leg.

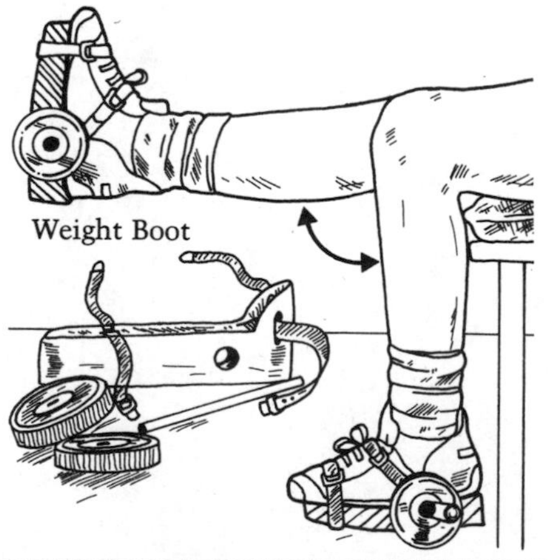

and bent at thirty degrees. This tests the collateral or side ligaments of the knee. I then bend the knee to ninety degrees and push the tibia back and forth on the femur. This tests the cruciate ligaments.

I then do the McMurray test. This involves bending the knee as far as possible and then turning the tibia on the femur. With the knee in maximum bend, the torn cartilage can slip into the knee joint. With the lower leg bone held and rotated on the upper leg bone, the knee is quickly straightened. If there is a torn cartilage, often the torn portion will become trapped between the bones in the bent position and then snap back into place as the knee is straightened. It feels like a piece of gristle snapping in your fingers. This clunking sensation is called a *positive McMurray test* and is a sure indication of a torn cartilage. But only 60 percent of patients get the clunking sensation.

Regular x-rays of the knee do not show the knee joint cartilage. However, a knee x-ray must be taken for a complete evaluation. I once examined the knee of a college quarterback. Many elements of his story and the physical examination suggested a torn cartilage.

Just to be on the safe side, I took x-rays of his knee. They revealed a baseball-size cyst in the upper end of the tibia. If regular x-rays had not been taken, the cyst would have been completely missed.

If I cannot see the cartilage on regular x-rays, how can I know for sure that it is torn? I have two other diagnostic tools: the arthrogram and the arthroscope, mentioned earlier (the word *arthro* means "joint" in Greek). Dye is injected into the knee so that the cartilages can be coated with a dense fluid material. This dye fluid has heavy molecules in it that give it almost the same density as bone. Therefore, it shows up as a white line on an x-ray film. After Novocain has been used to numb the skin, the dye is injected. The knee is filled with air. This improves the accuracy of the dye test. If there is a tear in the cartilage, the dye will run down into the tear, much like fluid flowing into a canyon. I can see the dye in the tear. This test is 95 to 98 percent accurate for the inside knee cartilage. However, because of anatomic and technical problems, I find it highly inaccurate for the outside knee cartilage.

The arthrogram is done for outpatients. An x-ray physician (known as a *radiologist)* performs the test. It takes about one hour to do and is not painful. It is uncomfortable. The knee feels full and squishy for the next two or three days. You may get crackling noise in the knee because of the air that is injected.

The other test that I have available to make the diagnosis of joint cartilage pathology is the arthroscope. This is like a small telescope—no bigger than a ball-point pen—that can be inserted into the knee joint. In my opinion, the arthroscope is a highly accurate, safe, straightforward way to obtain precise information about the status of the knee joint. It can be introduced into the knee through a tiny skin incision. I frequently do this test under local anesthesia. The skin and the inside of the joint are numbed with Novocain. The arthroscope is inserted into the joint, and the joint is filled with a saline (salt) solution. It is inserted on the outer part of the front of the knee. When the examination is finished, a simple skin stitch is placed at the insertion site. Then you are able to go home and rest the remainder of that day. If everything is normal, you can return to daily activities the next day. In five days, you can restart full sports activities.

By positioning the knee in various degrees of bend, the complete knee joint can be visualized. The front seven-eighths of the inside knee cartilage can be seen easily and clearly. The remainder is difficult to see because of the shape of the inside knob on the femur (the medial femoral condyle). However, this is the portion of the knee which is best seen with the dye test (the arthrogram). Combining these two tests, I feel that the diagnosis on the inside knee cartilage is 99 percent accurate. The arthroscope is incredibly accurate for the outside knee cartilage. The entire cartilage can be seen, and if there is any question about the diagnosis, small probes can be introduced into the knee to push on the cartilage. While the arthroscope is inside the knee, the other structures and the joint lining can be checked as well. The back surface of the kneecap is visualized. The anterior cruciate ligament can be seen without difficulty. I always check for loose pieces in the knee.

Why is it important to know the status of the knee cartilage as soon as possible? I have had bad experiences twice with athletes continuing to play where there was a suspicion of knee joint cartilage injury. These athletes sustained dramatic malfunctions of the knee. When surgery was eventually performed, they not only had large tears in their knee cartilage, but also had suffered knee ligament injury (anterior cruciate ligament injury). My suspicion is that if the cartilage problem had been dealt with quickly, their knee ligaments would have been preserved. There is a big difference in the long-term function of the knee if it is only a cartilage problem versus a cartilage and ligament combination.

Once I know the status of the meniscus, I can treat the patient intelligently. Because the menisci have no blood supply of their own, they cannot heal. Once torn, always torn. The symptoms never disappear. Sometimes the knee will feel absolutely normal, but at other times it will lock or buckle. It will continue to swell intermittently. The knee will not perform normally in either the long or the short run. The patient's athletic performance will be diminished as a result.

Even if the short-term symptoms could be overcome, there are long-term consequences of leaving a torn cartilage in the knee joint. The ends of the bone and their covering material—the knee joint surface cartilage—rub against the damaged meniscus. This acts like a file on the joint surface. Eventually, the surface material wears down in an uneven and irregular

Operation Using Arthroscope

way. This is arthritis. Thus, if you leave a torn meniscus in your knee at an earlier age, you will end up with an arthritic knee. For sixty years, surgeons have been removing torn menisci.

How do I remove a torn meniscus? The traditional approach has been to put the patient under general anesthesia. An incision is made on the side of the knee where the cartilage is damaged. The joint is entered, and with special small knives, the cartilage is cut out of the joint. I prefer to leave a small rim of cartilage right at the edge of the joint to make certain that the surgery does not damage any of the ligaments. The operation takes only sixty to ninety minutes. The patient remains in the hospital for two to three days. I keep my patients from bending their knees with an immobilization splint. Two weeks after surgery, I remove the stitches from the skin incision. Generally the patients are on crutches for three weeks. I then begin them on daily physical therapy sessions. Whirlpool baths, range-of-motion exercises, and weight-lifting exercises are all begun at the same time. Six weeks after surgery, most patients can jog short distances on soft surfaces. Swimming is also good. In three months, you can return to full sports activity.

I remove many torn menisci with the arthroscope. Let me explain the method. The patient is given general anesthesia in the operating room. I introduce the arthroscope into the knee, and the knee is filled with a saline solution. Then, a mini-television camera is attached to the arthroscope. The image of the inside of the knee joint is projected onto a television monitor. The monitor is set in front of me, so I can see exactly what I am doing inside the knee joint on the TV screen. Through a very small incision—usually less than one inch—I introduce small instruments into the knee joint. These are specially made instruments—none larger than a knitting needle—that allow me to remove the damaged cartilage bit by bit through the skin incision. These instruments include a small probe for feeling the surface of the cartilage, a small pair of scissors, a small cutting knife and a grasping forcep, and a steam shovel-like biting instrument.

Removing a torn meniscus with an arthroscope is an oper-

BRAD PARK'S KNEES

Brad Park, an all-star defenseman for the Boston Bruins, has had a series of knee operations. He has had menisci removed from both knees.

"Because Brad is skating on articular cartilage now," says Jim Kausek, the sports therapist for the Boston Bruins, "he has had some degenerative changes in the knee.

"Last January, Park had osteophyte formations and bone spurs removed from the knee. The operation was performed by Bert Zarins, one of the team physicians and an orthopedic surgeon at Massachusetts General Hospital. The operation was successful and Park is again skating."

ation that is a bit longer and technically more difficult than the traditional procedure. However, it pays rich dividends. Hospital time is cut to two days. The patient uses crutches for only two to three days instead of two to three weeks. Once the swelling in the knee subsides and the stitches are removed, the patient can begin exercising and even run. This occurs within two weeks. In six to eight weeks after surgery, the patient can return to full sports activities. In the next five years, much of knee cartilage surgery will be done with the arthroscope.

What can you expect of the knee once the meniscus has been removed? The knee is not 100 percent normal. After all, it has absorbed an injury of sufficient magnitude to cause the torn meniscus in the first place. Next, a surgical procedure has to be performed. Considering both of these facts, it is unrealistic to think that your knee would emerge as good as new.

Do not despair. Four players on the Red Sox have had their meniscus removed. Their knees all function normally, and they have never missed any time from playing as a result of this surgery. However, at the end of a doubleheader, especially on artificial turf, they will feel some knee discomfort. The next morning, their knees will be stiff and painful on the side of the operation. However, the knee still functions normally. As long as you have no knee ligament damage, there is no need for taping or a knee brace for support.

What about later in life? If you have to have your cartilage removed, won't you get arthritis? There is no question that you are more prone to arthritis. It would have been nice if you had not been injured at all. A study from the Mayo Clinic found that people who have had their meniscus removed do develop wear-and-tear arthritis at an earlier age than normal knees. This means in your fifties and sixties, you will develop a bit more pain and stiffness in the operated knee.

I also see patients who do not have meniscus surgery and live with the torn cartilage in their knee into later life. These people end up with a complete knee replacement. In other words, the surface of a joint is so battered from rubbing against the torn cartilage that it wears out. The surface material grinds down to bare bone. It is like a tire that loses its tread. The only way to solve this problem is to do a total knee replacement. This involves a resurfacing of the joint with implanted metal and plastic components. These components are cemented into the bone and can bring about dramatic relief from pain. However, after a knee replacement, you cannot return to sports.

In summary, you are far better off to have a damaged cartilage removed from your knee. In a short time, your knee regains normal function and you can return to full sports participation. If you decide against an operation, the torn meniscus will wear out the bone cartilage in the knee joint. Eventually, this will cause arthritis. There is a strong possibility that you will need a total knee replacement—and this is a major operation.

TUCKER FREDERICKSON'S REHABILITATION

In 1964, Auburn's Tucker Frederickson was one of the top college football players in the country. He was drafted by the New York Giants. He had so much potential that his coaches had a difficult time deciding whether he should play fullback or middle linebacker.

In his first training camp in 1965, Tucker injured his knee so badly that he needed knee surgery. In 1966, he injured the other knee, and sat out the season.

"It was in 1966 that I started to work with Frederickson," says Dr. Earl Hoeiner, clinical professor of rehabilitation medicine at Tufts University School of Medicine. "In the mid sixties, the way to rehabilitate a knee was to strengthen the quadriceps muscle. There was no thought of improving the flexor mechanism. The state of the art was relatively primitive. We did not have Cybex units to measure different strength between muscle groups nor did we have the lightweight knee braces.

"I devised a new knee strengthening program for Tucker which I called 'The Seven Positions.' My program was designed to improve the internal and external rotation of both his knee and hip; to strengthen both the flexor and the extensor mechanisms of the knee; and to build up the ligaments in the knee.

The Extensor Mechanism

Have you ever thought about the anatomy of your leg? What allows you to generate tremendous force and endurance in the upper part of your leg—to kick footballs, hurdle, run marathons, and cycle long distances? In medicine, we call the anatomic system which allows you to thrust your leg powerfully the *extensor mechanism.* It involves the combined function of the quadriceps muscle, the kneecap, and the tibia. Because of the tremendous size and strength of this muscular apparatus, you can generate forces four to five times your body weight with it.

Think of the extensor mechanism as a steam shovel. The quad muscles are the *control cab* and the *winch.* When the foot and the tibia are placed firmly on the ground and the knee is straight, the extensor mechanism or the winch is fully wound up. When you sit in a chair or squat down, the quad muscle, the kneecap and the kneecap tendon stretch and control the squat. Once you are in a full squat position, the quad muscle continues to work to keep you balanced. Now,

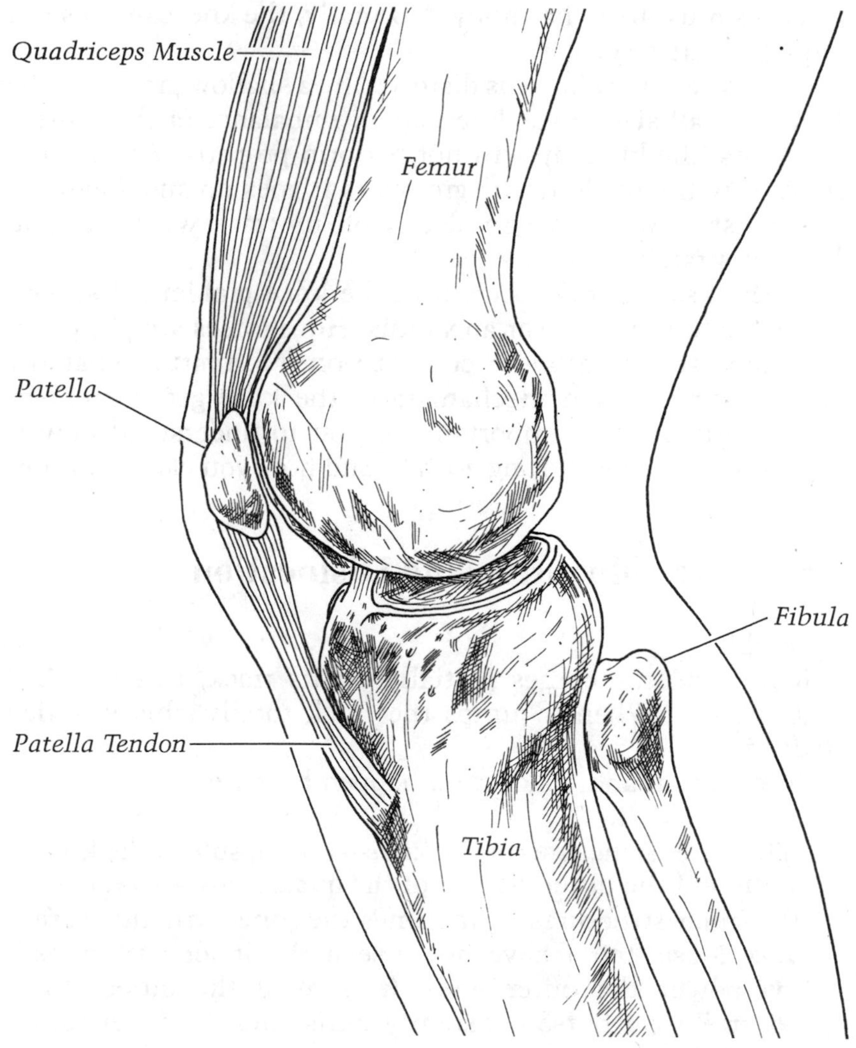

Extensor Mechanism of the Knee

stand up. The steam shovel works in reverse. The muscle-bone-tendon unit shortens and brings you back up into a standing position.

When you look at the extensor apparatus from the side, it is clear that the knee acts as a pulley. The ropes of the pulley are the kneecap tendon and the quad tendon. The kneecap, about one inch thick, sustains forces of 1000 to 2000 pounds. The larger your kneecap, the more mechanical advantage it offers. In other words, it is a larger pulley. If you do not have a kneecap, your extensor apparatus is weakened by at least 20 percent.

The kneecap travels like a train on a track. It fits in a V-shaped groove in the end of the femur. This cradles the kneecap as the kneecap slides up and down. It is also stabilized in the track by the capsule of the knee joint and the quad tendon. These are passive restraints on side-to-side motion of the kneecap. The dynamic stabilizers of the kneecap are the muscular heads of the quad muscle itself. The muscular element acts like reins on a horse. The outer rein is the outer quarter of the quadriceps (called the *vastus lateralis muscle).* The inner rein is the inner quarter of the quadriceps muscle (called the *vastus medialis muscle).* To slide properly, these muscles must be in harmony. Normally, the kneecap does not wobble in its track.

If the kneecap balance is distorted by a shallow groove in the femur, small size of the kneecap, or imbalance of the muscle elements, the kneecap will not perform properly. At the very least, it will wobble in the groove as it goes up and down. At the worst, it will slip completely off its track whenever you bend or straighten your knee.

I would say that 50 percent of the knee problems I see are traceable to the extensor apparatus. Huge forces are placed on this apparatus in athletic competition. No matter what the sport, your extensor mechanism of the knee gets a tremendous workout. It is important for you to understand how it works, what can go wrong with it, and how you can sustain it.

Kneecap Subluxation and Dislocation

CAUSES

If the kneecap comes partially off its track, this is called *subluxation.* When it jumps the track totally, this is called *dislocation.*

Subluxation and dislocation happen because:

1. The outer structures of the knee—the capsule of the knee joint and the outer quarter of the quadriceps—overpower the inner structures. Sometimes the inner structures are very loose, but I have never seen the inner structures overpower the outer ones. It is as if the outer rein controlling a horse constantly pulls, and the inner rein

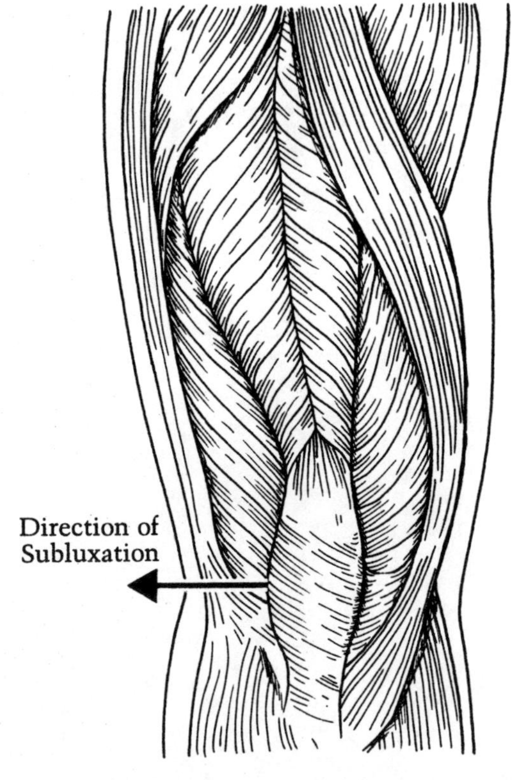

Patella Subluxation

was not used at all. Thus the horse, or in this case, the kneecap always veers toward the outside.

2. A kneecap rides high. This looks as if the kneecaps are too far up the leg. Usually, tall, thin people have high-riding kneecaps.
3. Kneecaps are too small. Only 1 percent of people have these.
4. The groove that the kneecap sits in is too shallow. This is another genetic abnormality. You inherit the shape of your bones.

The normal kneecap glides up and down in its track and has no tendency to jump laterally to the outside of the knee. In the most abnormal case, the kneecap jumps off the track with every bending and straightening action of the knee. Most people who have difficulty with this mechanism fall somewhere in between these extremes.

Many athletes have a mild tendency toward outer instability of their kneecaps. This is basically an accident waiting to happen. If the inner side of the kneecap is struck by an opponent during sports, the kneecap is forcefully dislocated. This converts a mild abnormality to a major abnormality. The inner structures, which are weak to begin with, are weakened still further by a dislocation and the outer structures become more overpowering.

DIAGNOSIS AND TREATMENT

How can you tell if you have a loose kneecap? Try this. Sit with your knee bent over the edge of a table. Straighten your knee out. Grasp the kneecap with your thumb and index finger. Flex your lower leg. Hold the kneecap as it glides up and down. Does it travel in a straight course? If you have kneecap looseness, the cap will veer off to the side during the last ten to fifteen degrees of straightening of your knee. We call this a "J sign" because the veering off looks like the letter J. Now, place your leg flat on the table. Relax your muscles. With your thumb and index finger, wiggle your kneecap back and forth side to side. If it moves more than a half-inch in either direction, you have a loose kneecap. If you can push your kneecap more than a half-inch to the outer side of your leg, you undoubtedly have subluxation of the kneecap during function. In other words, as you use your knee vigorously and contract your thigh muscle, your kneecap comes off its track. If you can push your kneecap completely off the side of your leg (some people can), you are experiencing dislocation of the kneecap. If the knee tightens up when you push, this is a positive apprehension test; it means that you have loose kneecaps.

To make the diagnosis of kneecap instability more scientific, the Q-angle test was devised. The kneecap becomes the center point for an angle formed by the long axis of the quadriceps muscle and the kneecap tendon. The angle between the quad muscle and the kneecap tendon is normally less than ten degrees in men and fifteen degrees in women. Any higher measurement indicates that there is a stronger tendency for your thigh muscle to pull your kneecap to the outer side of your knee. This produces subluxation or dislocation.

Kneecap problems run in families. To a great degree, the structural problems—the shape of the bone, ligaments, capsule, and muscles—are determined genetically. Of these elements, only the muscles can be strengthened to change kneecap instability.

What happens when you have kneecap instability? Usually, running straight ahead is not a problem. However, cutting or quick changes of direction are difficult, painful, and put the knee under stress. On sudden changes of direction the thigh muscles have to gain quick control of the kneecap. If the kneecap is imbalanced, the thigh muscles lose control and the kneecap slides out.

If the kneecap comes partially off the track, it usually will snap back into place and you can continue playing. After it happens a few times, you can anticipate the subluxation. Inevitably, it will disturb your concentration. It is hard to think about your knee, the game, and the opponent all at the same time.

In mild knee subluxation cases, external support of the kneecap is enough to stabilize the kneecap. Many prefabricated knee braces help dramatically. The primary requirement is that braces support the outer aspect of the kneecap. This applies pressure to the kneecap to keep it on track.

Two distinct types of knee supports are equally effective. One is a neoprene rubber sleeve with a support built for the outer part of the kneecap. It has a hole over the kneecap itself, so no direct pressure is applied there. In mild cases, an external support is sufficient to restore balance to the extensor mechanism and works effectively for the athlete. Its chief advantage is that it is easy to use. You do not need a trainer to apply it.

In some cases of subluxation and dislocation of the kneecap, supports give insufficient control. Surgery is the only answer in such cases. There are many operative procedures, but all have the same goals. First, I free up the tight structures on the outer aspect of the kneecap. This is done by making an incision through all the layers on the outer side of the knee directly down into the joint. I carry this release or incision right up to the level of the vastus lateralis muscle. The purpose is to defeat the right outer rein, thus causing the kneecap to pull toward the outside. The second part of the operation involves strengthening or tightening the structures on the inner aspect of the kneecap—the inner rein. The operation that I favor is called a modified Goldthwaite procedure. In this operation, I split the kneecap tendon in half. I move the inner half over approximately one inch and stitch it down to its new location in the top of the tibia. Thus, instead of one kneecap tendon, you now have two! Next, I enter the knee joint on the inner side. I examine the back surface of the kneecap for wear and tear. If it is rough, I smooth it out with a sharp blade. I remove the roughened surface of the kneecap. If the cartilage is worn down to the bone, I drill the bone with a small drill. This draws blood into the area so that a clot will form which hopefully will convert to heavy and pain-free scar tissue. In sewing up the joint, I take a tuck in the capsule on the inside to strengthen the inside structure still more. Finally, I mobilize the vastus medialis muscle. By using a careful surgical technique, the meat of the muscle and the tendon itself can be separated from the rest of the quadriceps group. It looks like a large tongue. I then advance this down on top of the inner side of the kneecap. I stitch it in place with permanent sutures. This gives me a dynamic stabilizer on the inner aspect of the kneecap.

In all, I have three strengthening parts on the inner side of the kneecap: the inside half of the kneecap tendon, a tuck in the inside joint capsule, and an advancement of the vastus medialis muscle. I have weakened the outer rein of the knee joint, enabling the kneecap to travel a straight track. I have rebalanced the kneecap.

After the operation, I keep my patients in a soft splint for the first five days. Then I convert to a plastic cylinder cast, which runs from mid-thigh to mid-calf. I keep the patient in this cast for six weeks to allow the other structures to heal. Then I start the patient on physical therapy in the form of whirlpool baths, range-of-motion exercises, and strengthening exercises for the quadriceps and hamstring muscles. It takes four to six months after surgery to return to full sports. Once the patient is fully rehabilitated, however, external supports are not required. With rebalancing of the extensor mechanism and straight tracking of the kneecap, there is no reason for further wear and tear on the back surface of the kneecap. Therefore, the surgery is an insurance policy against further instability of the kneecap.

Increasingly, there is interest in doing just the outside or lateral weakening portion of the operation. This can be done "through the arthroscope." We call it a *lateral release* of the kneecap. The advantages are that of a short hospital stay, two or three days, as well as only seven to ten days on crutches. A sponge rubber pad is worn over the outer part of the knee for one week to stop bleeding. My opinion is that this works well for the milder degrees of kneecap instability, but the regular operation will still be necessary for more severe forms of involvement.

CAUSES OF PATELLA CHONDROMALACIA

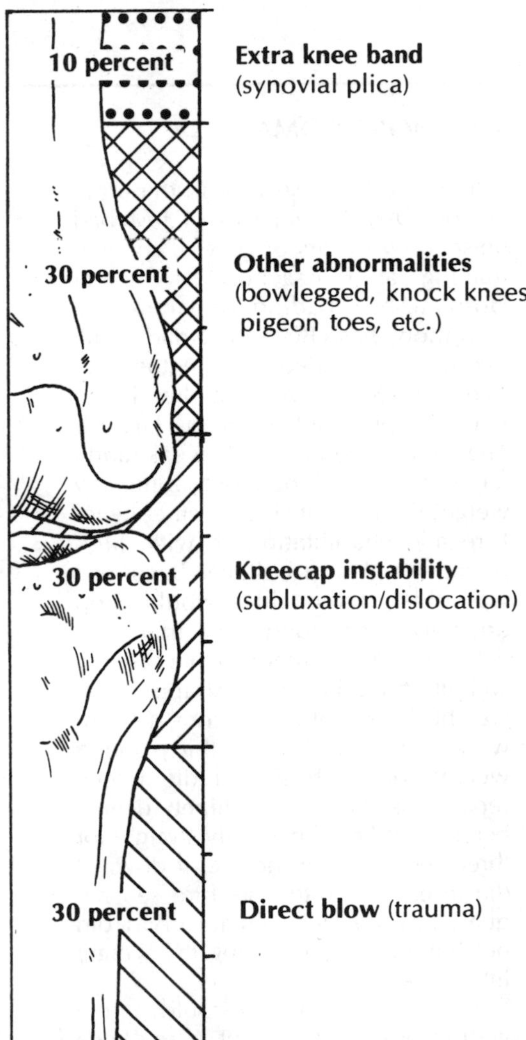

SOURCE: Statistics compiled by Sports Medicine Resource, Inc.

Kneecap Wear and Tear (Patella Chondromalacia)

In medical terms, wear and tear on the back surface of the kneecap is called *patella* (kneecap) *chondromalacia*. In Greek, *chondro* means "cartilage"; *malacia* means "softness." In

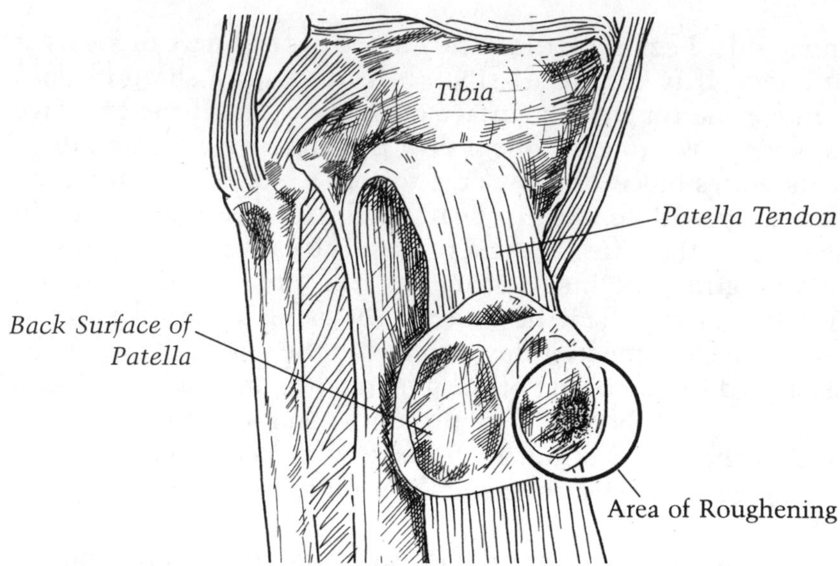

Patella Chondromalacia

fact, patella chondromalacia is a roughening or softening of the joint surface cartilage behind the kneecap.

The pain of chondromalacia is directly behind the kneecap. But why the back of the kneecap becomes painful is a mystery to medical scientists, because it has no nerves.

CAUSES

"Chondromalacia" is a term which describes only the symptoms. The problem is due to any one of these conditions:

- Kneecap instability. The kneecap moves out of its groove.
- A direct blow to the front of the kneecap. This is called *traumatic chondromalacia.*
- A genetic abnormality causing growth of an extra band in the knee joint. This is called *synovial plica.*
- Unknown. Some people get chondromalacia for no apparent reason. This is called *idiopathic chondromalacia.*

At birth, the ends of the bones that form the joints and the back surface of the kneecap are all coated with white, glossy joint surface cartilage. It's like new tread on a tire. Over time, with use, the tread wears down on the back surface of the kneecap. If it wears down early in life, it's called *chondromalacia.* If it wears down when you are in your fifties or sixties, it's called *osteoarthritis.*

I have taken a specimen from the back surface of the kneecap of a young person with chondromalacia and compared it to a specimen from an older person with kneecap arthritis. They look exactly alike under the microscope. If I use special strains and chemical tests, I find that the specimens are also similar in their chemical makeup. Medical scientists don't know why the surface material wears down over the years.

How do I know the degree of chondromalacia? If you look at the surface of the kneecap, you can see the different degrees of

MY CHONDROMALACIA

In 1970, I was serving in the army at Fort Dix, New Jersey. I just had finished two years of general surgical training at Massachusetts General Hospital, but I had no formal training in orthopedics. Because I had some free time, I decided to strengthen my right leg. Seven years earlier I had injured it playing football at Harvard. The leg was feeling a little weak, and I felt guilty that I had not done any weight lifting for it since I left college. I recall rehabilitating it with fifty-pound weights in college. Because I was older, I thought I would drop down to thirty pounds for starters.

I used our kitchen table as my weight-lifting bench. I strapped the weight boot onto my construction-worker boots. I put thirty-pound weights on each side. I did fifteen repetitions starting at a ninety-degree bend. I held my knee fully straight for three seconds and then let it down. I did three sets. After the third set, my quariceps muscles felt quivery. I did not have any pain during the weight lifting itself.

The next morning, I could hardly get out of bed. The front of my knee was killing me.

"What have I done to myself?" I

roughening. In the 1940s, Drs. Bennett and Bower, two physicians at Massachusetts General Hospital, discussed the grades of chondromalacia; they took kneecap specimens from dead patients of various ages. What they found was that the back surface of the kneecap is the first part of the knee joint to wear out. They called early changes grade I chondromalacia. The surface looks as if it had been sandpapered. As you advance into your twenties, thirties, and forties, the wear-and-tear changes become more severe. In addition to the roughening, the surface material starts to crack. This is grade II chondromalacia. Despite the cracking, no real loss of the thickness of the tread occurs. In grade III chondromalacia, the surface material looks like crab meat and the surface starts to wear thin. In grade IV chondromalacia, there is a complete loss of surface material. The bare bone of the back surface of the kneecap is exposed and is very painful.

It is impossible, I think, to predict how severe a patient's pain will be with any particular degree of chondromalacia. I've seen some people with grade III chondromalacia function quite well. On the other hand, some athletes with grade I changes experience constant pain. First, the pain is a reflection of how hard you use your knee. Second, it is related to your pain threshold. A high pain threshold allows some people to function without discomfort. From a medical point of view, these are important factors in assessing your problem.

asked my wife.

I bent my knee to put on my shoes. That was painful. Sliding into my car hurt. I could barely climb the stairs in the hospital. Because I was just starting in orthopedics, I really did not know what was wrong. I asked George Unis, one of my buddies, to check my knee. George had just finished his orthopedic training at St. Luke's Hospital in New York. As I bent and straightened my knee, George felt some grinding on the back of my kneecap. This was painful. He did the quadriceps inhibition test, and I nearly went through the roof. I described to him what I had done. He laughed, "You learned a very important lesson. Never start a patient off with too much weight."

George explained that I had challenged my kneecap so brutally that I had irritated its back surface. "You have chondromalacia." I had never heard the term before, but I knew it was disabling. He prescribed six aspirin tablets a day for one week. The pain gradually subsided over the next ten days. It has never returned. Any weight lifting I have done with my knee since that time has been very gradual. It was an important lesson for me.

DIAGNOSIS AND TREATMENT

If you get chondromalacia, how do you know it? Pain! You will have pain in the front of the knee. In describing the pain, you will usually put your hand directly over the kneecap itself. The pain intensifies with prolonged periods of sitting with the knee bent. Often, the knee hurts after you sit through a movie; thus doctors call this the "positive theater" syndrome. When sitting, you will want to shift your knee position from a ninety-degree bend to a straight position. When you rise from a sitting position, you will need to walk around a bit to loosen up your knee. Because climbing requires powerful activation of the quad muscles, going up and down stairs produces kneecap pain. Squatting or kneeling directly on the kneecap are also painful. In all these cases, the pain is caused by the rough kneecap rubbing against the femur. Running straight ahead at low speeds seldom causes discomfort. However, sprinting or running up hills does. If you have severe chondromalacia, only sleeping relieves the pain.

In my physical examination, you sit on the edge of the table with your knee bent ninety degrees. I ask you to flex your knee back and forth. I place my hand on the kneecap and feel for a grinding sensation. This grinding is called *crepitus*.

As you bend and straighten your knee, the normal kneecap slides smoothly over the thighbone. If the kneecap is rough, it will grind as it moves in its groove against the femur. Only chondromalacia produces this grinding sensation. Because the extensor mechanism lifts only the weight of the tibia, lifting your leg while sitting on the examining table does not produce pain.

Next, I ask you to tighten your knee up with your leg fully extended. I place my thumb and index finger at the top of the kneecap. In this way, I can control the kneecap. Then, I ask you to push down as hard as possible. This activates the quadriceps muscle and draws the kneecap past my grasping fingers. If the kneecap comes to a screeching halt, it is a sure sign that you have chondromalacia. With a normal kneecap, this movement does not produce pain on the back of the kneecap. You should be able to forcefully contract the quad muscle and pull the kneecap from my grasp. This is called the *quadriceps inhibition test*. If you have chondromalacia, the pressure and the quadriceps contraction together produce great pain. Only chondromalacia produces this reaction.

Finally, I ask you to step up and down at least fifteen times on a stepping stool. When you step up, you have to lift your whole body weight through your knee joint. If your kneecap is irritated, it will be a painful exercise. Often, in severe stages of chondromalacia, you will look to balance or support yourself while attempting to do this test.

X-rays, which show only bone, are always negative. Since the surface material is only rough, an arthrogram does not help. The arthroscope (the body telescope) can confirm the problem. With it, I can see the degree and extent of roughening on the back surface of the knee. Usually, it is the size of a quarter or a fifty-cent piece.

How do I treat patella chondromalacia? Rest. You must quit sports for six weeks. I prescribe anti-inflammatory medication by mouth for two weeks—one with each meal. I also begin you on straight leg raising exercises to strengthen the quad muscles. Usually with chondromalacia the quadriceps muscle weakens. This is easy to spot. I measure the circumference of the leg six inches above the top of the kneecap. If the circumference is smaller on one leg than the other, weakness has set in.

MOUNTAIN CHONDROMALACIA

"For our ascent of Mt. Himal Chuli, our expedition used Sherpas, high altitude porters," says Richard A. St. Onge, a world-class mountain climber and physician at Sports Medicine Resource, Inc. "Sherpa means people from the East—the Tibetan ethnic group who came across mountains 600 years ago. Pound for pound, they are probably the strongest people in the world. A typical porter will carry sixty pounds in oxygen-thin altitudes. That's tremendous, considering that they normally weigh between 90 and 115 pounds.

"Almost all our porters had some form of chondromalacia. I suspect that it was caused by climbing up and down mountains.

"When the porters found out that I was a doctor, I became very popular. 'Sab, my knee hurts. What can you do for it?'

"I examined all their knees. Almost every knee clicked and grinded, a sure sign of creptius. I put them all on aspirin, which is not all that different from what I would do back in the States. This would alleviate some of the pain. They seemed to have more trouble with their knees when going downhill. I don't entirely understand the mechanics, but downhill days were far worse for them than uphill days."

THE FOUR TYPES OF CHONDROMALACIA

The following paragraphs give a detailed explanation of the four types of chondromalacia.

I: CHONDROMALACIA DUE TO KNEECAP INSTABILITY

CAUSES

If your kneecap subluxates or dislocates, you are a candidate for early chondromalacia. Medical scientists are not sure why a loose kneecap leads to chondromalacia. In my opinion, it produces "shear" force. This shearing of the surface tends to be a destructive side-to-side force. When the kneecap tracks normally, the force against it is strictly a compression force. The kneecap is well designed to accept compression force but poorly designed to accept shearing force. Patients with kneecap instability and chondromalacia will have two sets of symptoms. You have the instability feeling caused by the kneecap quickly sliding to the outer part of your knee. This occurs with vigorous athletic use. In addition, you will have chondromalacia symptoms—pain behind your kneecap when sitting for long periods, climbing stairs, squatting, or kneeling. All these movements cause pressure on the kneecap. Often both knees are involved.

DIAGNOSIS AND TREATMENT

No matter what the cause, the method of diagnosing chondromalacia is the same, and was explained earlier (See page 273). When an unstable kneecap causes chondromalacia, the kneecap must be restabilized to allow it to track properly. Once this is achieved, the shearing force on the back surface is stopped and symptoms of chondromalacia end.

In mild cases, correcting kneecap instability with quadriceps exercises is usually enough. The pain disappears. The grating or crepitus may continue but is usually not painful. If the chondromalacia is severe, surgery is needed to stabilize the kneecap. When the surgery is performed, the back surface of the kneecap is smoothed down with a special knife blade. This removes the shaggy chondromalacia surface and allows the patient a fresh start.

What happens if the chondromalacia continues to progress despite surgical stabilization of the kneecap? This used to be one of the most difficult problems in orthopedic surgery.

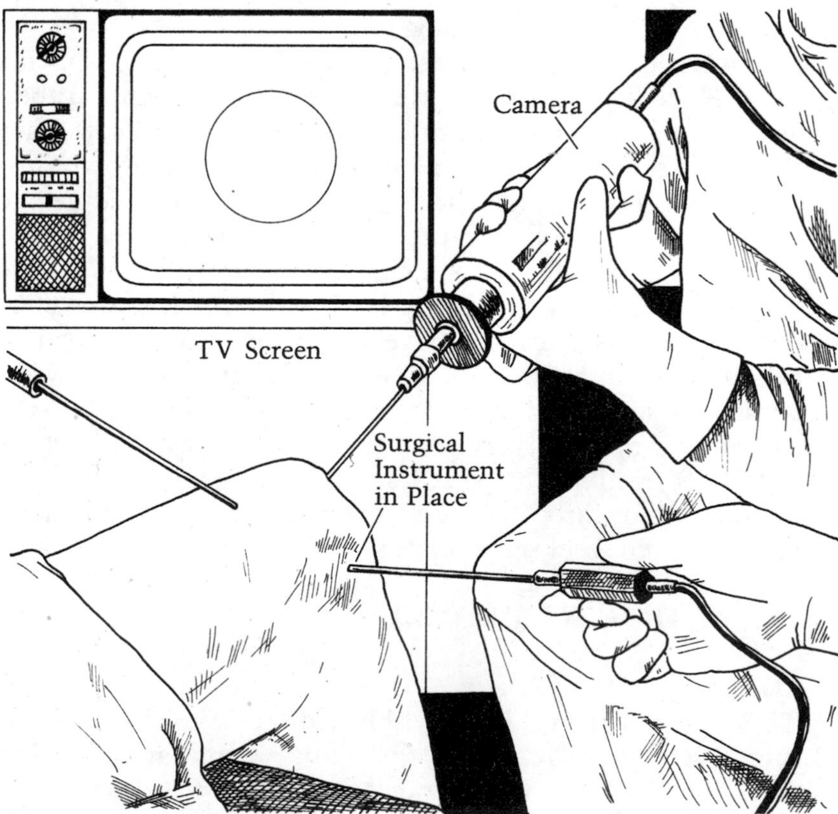

Knee Operation with Camera Attached to an Arthroscope

However, a new technical advance has helped. This involves shaving of the back surface of the kneecap "through the arthroscope."

The patient is given general anesthesia. The arthroscope, with a television camera attached, is introduced into the knee. The television camera allows me to see the patient's knee on a large video monitor. This frees my eyes and hands to use other instruments in the knee joint. I use a kneecap shaving device, smaller than a ball-point pen, to shave off the loose fragments. The device, a small rotatory cutter, looks and works like a lawnmower. While it cuts off the fragments, it sucks them into the shaver itself. They are removed through a long plastic tube into a suction bottle. As this shaving and removal process takes place, large amounts of fluid are forced through the knee joint. It takes approximately forty-five minutes to shave the back of a kneecap.

Biologically, it is unclear exactly what this procedure accomplishes. Because loose fragments contain harmful enzymes, they are apparently detrimental to the joint. Enzymes are large protein molecules which are designed to digest things in the body. For instance, the enzymes in your stomach help digest food. The enzyme in the knee joint, called *collagenase,* breaks down the collagen material which is the basic framework of the articular cartilage (joint surface cartilage). The more severe the chondromalacia, the greater the enzymes present. The enzymes are washed out by the high amount of fluid that I push through the knee during the procedure. The fluid also removes the loose fragments. With the shaving, it seems likely that a chemical change in the joint is taking place in addition to a physical change.

In any case, these patients improve. The kneecap shaving gives them a clean start. When they bend and straighten their knee, they have much less grinding behind the kneecap. They also experience a dramatic reduction in pain. They are discharged from the hospital the day following the procedure and are able to resume full activities two weeks later.

The long-term effect of shaving is not yet known. I have done the procedure for only the last two years. It seems clear to me that these patients have everything to gain by this procedure and nothing to lose. It is safe, effective and allows us to attack the chondromalacia without actually cutting into the knee joint. If open surgery were done to smooth the kneecap in a similar fashion, the patients would be out of sports for at least three months. With the new method, they are out of sports for three weeks at the most.

BARRY'S CHIPS

For years, Rick Barry, a professional basketball player, was bothered by swollen knees. He continued with knee injuries until 1980, when he fell and had a bone spur knocked loose in his knee and couldn't perform—because of severe swelling and limited flexion of the knee. X-rays showed degenerative changes and narrowing of the knee joint—a sign of his old knee injury.

"Rich had a swollen, painful knee when I first examined him," says Dr. Bert Zarins, Boston Bruins team physician and codirector of the Sports Medicine Clinic at the Massachusetts General Hospital. "Using the arthroscope, I located a large piece of bone which had fractured inside the joint. Rick was actually playing with this loose piece wedged in the back of his knee joint. With the aid of the arthroscope, I removed the loose body. This fairly simple procedure relieved most of the pain and allowed the knee to straighten."

Rick says, "Though I retired the next season, my knee hasn't felt so good in more than five years. I actually can sleep through the night and sit in a movie theater without pain or discomfort. I wish that I had gone to Bert Zarins years ago!"

TRAUMATIC CHONDROMALACIA

CAUSES

This is brought on by a hard blow directly to the kneecap, driving the back of it against the femur. Traumatic chondromalacia can develop when you fall, when you get kicked in athletics, or when your kneecap is driven against the dashboard in an automobile accident. This type of injury produces a bruising of the back surface of the kneecap itself. In many cases, the kneecap never recovers fully. Most often, it involves only one knee.

DIAGNOSIS AND TREATMENT

The knee usually swells with joint fluid (called *synovial fluid)* immediately after the injury. Often, the blow is hard enough to break the skin over the kneecap. If you rest your knee a few days, it seems to improve, as long as the kneecap is not actually broken.

Traumatic chondromalacia usually develops over a two- to three-month period. It starts with stiffness in the knee. You begin to have pain sitting in a bent-knee position. You are unable to squat and have difficulty climbing stairs. Running is painful. Whenever you activate your extensor mechanism, you produce discomfort.

SYNOVIAL PLICA (KNEE BAND)

CAUSES

Between the sixth and twelfth weeks of embryonic life, the knee joint is formed. It starts out as a solid mass of gristle-like material. Through a mysterious process, it develops, as if a sculptor were working inside the joint. Bit by bit, the solid material is removed and a joint cavity is formed. After twelve weeks, you have a knee with all its structures—the knee joint cartilages (menisci), the anterior and posterior cruciate ligament, and the lining of the joint (synovial membrane).

Some people have more than just these structures. Imagine being inside a pillowcase with one side sewn tighter than the

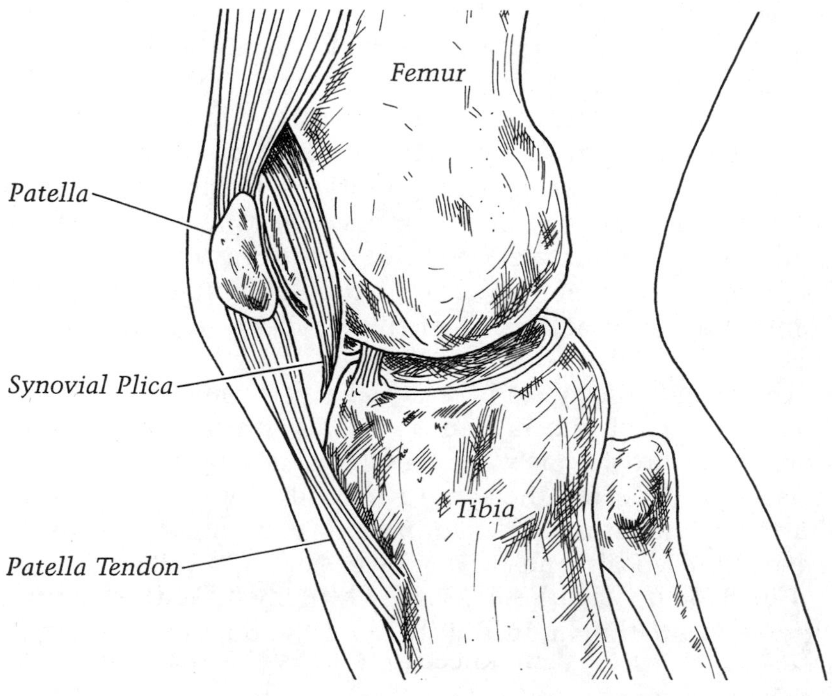

Extra Knee Band (Inner Shelf Plica)

other. From inside, this would look like a pleat or a ridge. This is exactly what the extra bands in the knee look like from inside. In medical jargon, these extra bands are *plica.*

The most common plica is a band running from side to side in the joint directly above the kneecap. It is called the *transverse plica.* This band is not troublesome. The band which causes the most symptoms lies on the area just beside the kneecap on the inner part of the knee. It's called the *inner-shelf plica.*

These bands are made out of scar tissue-like material. The problem it seems is that they cannot grow as quickly as the rest of the joint. Generally, the chondromalacia symptoms appear at twelve or thirteen years of age. (I have seen very few patients with this problem older than twenty years.) At this point in the growth cycle the bones and other structures of the knee have enlarged so much that the band has become too short. When you bend your knee, the band bowstrings across the thighbone and rubs on the bone itself. As you use your knee vigorously in sports, the band rubs back and forth and acts as a file on the surface of the joint. This filing action produces irritation and pain. The more you use your knee, the more irritated it gets. It can even become so irritated as to produce extra fluid.

DIAGNOSIS AND TREATMENT

How do I know that you have an extra band in your knee? First, there are the chondromalacia symptoms. You have trouble using your knee in a bent position. Sitting for prolonged periods of time with the knee bent causes discomfort. Climbing stairs, squatting down, and driving the car all produce pain. Second, the pain is on the front of the knee on the inside. I have never seen a plica on the outer side of the knee. If you bite the bullet and use the knee hard despite the pain, the knee will swell.

Diagnosis of synovial plica is tricky. It is the diagnosis of exclusion. By this, I mean that the tests that are usually positive for chondromalacia are not positive in these patients. I do not find crepitus or grinding on the back of the kneecap with bending and straightening motions. Rather, this type of activity produces pain on the inner aspect of the end of the femur. You do not have a positive quadriceps inhibition test. The remainder of the knee examination is completely normal. Sometimes I can actually feel this band and flip it with my fingers.

An accurate diagnosis is essential for treating this problem. The only sure diagnosis comes with an arthroscope examination. (An arthrogram is worthless in diagnosing the knee plica. It is impossible for the dye to individually coat the band.) During the arthroscope examination, you are wide awake. I use only Novocain and local anesthesia.

The band looks like a broad white shelf. With the knee extended out straight, it is quite loose and floppy. However, as the knee is bent in a ninety-degree position, the band tightens down over the end of the femur. There is often roughening of the femur in the area where the band has been rubbing. Sometimes, the back of the knee is rough. The remainder of the joint is normal.

I wash the knee out thoroughly. I remove the arthroscope and put a dissolvable stitch in the skin. Within five days, you can resume full activity. Once I have made the diagnosis, I recommend rest for one month. This is to break the inflammatory cycle and let the knee "cool off." During this period, take two aspirin tablets with each meal to reduce inflammation. Usually, the pain and swelling will diminish greatly over this period. However, only about 10 percent of patients are able to resolve the extra-band problem with rest; one month of restricted activity will usually do the trick if it is going to work. Then I return you to full athletic activity. If the knee immediately becomes painful or swollen again, I suggest surgery.

Removing a plica is different from most operations, because I am taking out a piece of the knee which does not belong there. Therefore, once it is removed, you are left with all the components that a normal knee needs to function. In contrast, the purpose of other types of knee surgery is to remove fundamental portions of the structure.

I start the operation with a two-inch incision on the inner aspect of the knee. The capsule of the knee joint is opened lengthwise and the lining tissue becomes exposed. I remove the plica, which is approximately one and a half inches in width and four inches in length. The removal extends to the area above the kneecap itself. After the operation, I immobilize the knee in a soft splint for two weeks and then begin physical therapy. In eight weeks from the time of surgery, you can return to full activity.

Lately, with the arthroscope and a television camera, I have been able to remove this band without actually opening the knee joint surgically. Through a small incision, a special biting-type instrument is inserted into the knee. Using the television to locate the exact area of the band, I remove it bit by bit. The advantage of this is that the patient does not need crutches after the operation. A splint for two to three weeks will do the work. Physical therapy can be started one week after surgery, and return to sports is faster. This is quite new, and the long-term results are not yet known. However, in the future, it is likely to become the most common surgical approach to removing synovial plica.

IDIOPATHIC CHONDROMALACIA

CAUSES

Some patients have chondromalacia for no apparent reason. They have had no direct injury to their kneecaps, they do not have small kneecaps, they do not have dislocating or subluxatating kneecaps, and they do not have arthritis; they do have chondromalacia. This is called *idiopathic chondromalacia. Idiopathic* is the medical word that means "we don't know."

DIAGNOSIS AND TREATMENT

Idiopathic chondromalacia is like any other chondromalacia. Sit with your knees bent, and you have pain; climb stairs, or squat down, and you have pain; you can't kneel. Sometimes the knee swells, but it doesn't become hot—like an arthritic knee.

My examination shows no evidence of abnormality of kneecap function. It tracks straight and true with no side-to-side hypermobility. There is no evidence to indicate direct injury. The leg and the foot are aligned. Because you have no bone damage, the knee x-rays are negative.

When you bend and straighten your knee over the edge of the examining table, I can hear the grinding. You feel discomfort. You have a positive quadriceps inhibition test. Idiopathic chondromalacia is one of the most discouraging conditions that I see. If it lasts for more than six months, it almost always requires surgery. I have never gotten a patient back to normal without it. The only alternative is a sedentary life.

I find it hard to treat because I don't understand it. There are no mechanical factors to correct. The kneecap has not been traumatized; it is not hypermobile or dislocating. There is no synovial plica to contend with. There is just wear and tear on the back of an otherwise normal kneecap.

I always explain my dilemma to patients with chronic idiopathic chondromalacia and first try out conservative treatment—rest and anti-inflammatory medications. The chance is one in a thousand that this will help. My patients appreciate any action short of surgery. In this way the patient and I come to the surgery with a common sense of commitment. Together, we feel we have done everything short of surgery to relieve the problem. If the patient wants to continue in athletic activity, surgery is the only answer.

Prior to 1978, the only surgical procedure was to cut directly into the knee and shave down the back of the kneecap. This was done with a flat scalpel blade which looks like a tiny sword. All the shaggy material was removed. If the surface material had worn down to the bone, then the bone area was drilled with a small drill, thus allowing the bone to hemorrhage. Hopefully, a blood clot from the hemorrhaging converts to heavy scar tissue and functions as surface material. After the operation, the patient needs a knee splint for two weeks. Then physical therapy starts, very slowly so that the irritation doesn't recur. Sometimes it's six months before the patient can return to competition. The results are unpredictable.

With the arthroscope I can shave the kneecap in about forty-five minutes. The next day, the patient is discharged and in two weeks, can start general physical therapy. Very little rehabilitation is necessary. I've found that the patient usually can return to full sports activities six to eight weeks after the procedure. The difference from the old routine is tremendous.

Lanny Johnson, M.D., the pioneer of this procedure, finds that occasionally he has to repeat it one to two years after the first shaving. Even so, it is a relatively simple method—and one hard to criticize.

Jumper's Knee (Upper Patella Tendinitis)

CAUSES

The final link in the extensor mechanism is the kneecap tendon. It runs from the bottom of the kneecap to the top (called the *tubercle*) of the tibia. It is about as thick as your thumb. It is approximately two inches in length and slides only a one-fourth inch. When you run, 1500 pounds of force pass across this tendon. When you jump, the force increases to 2500 pounds. The force comes from both the firing of your

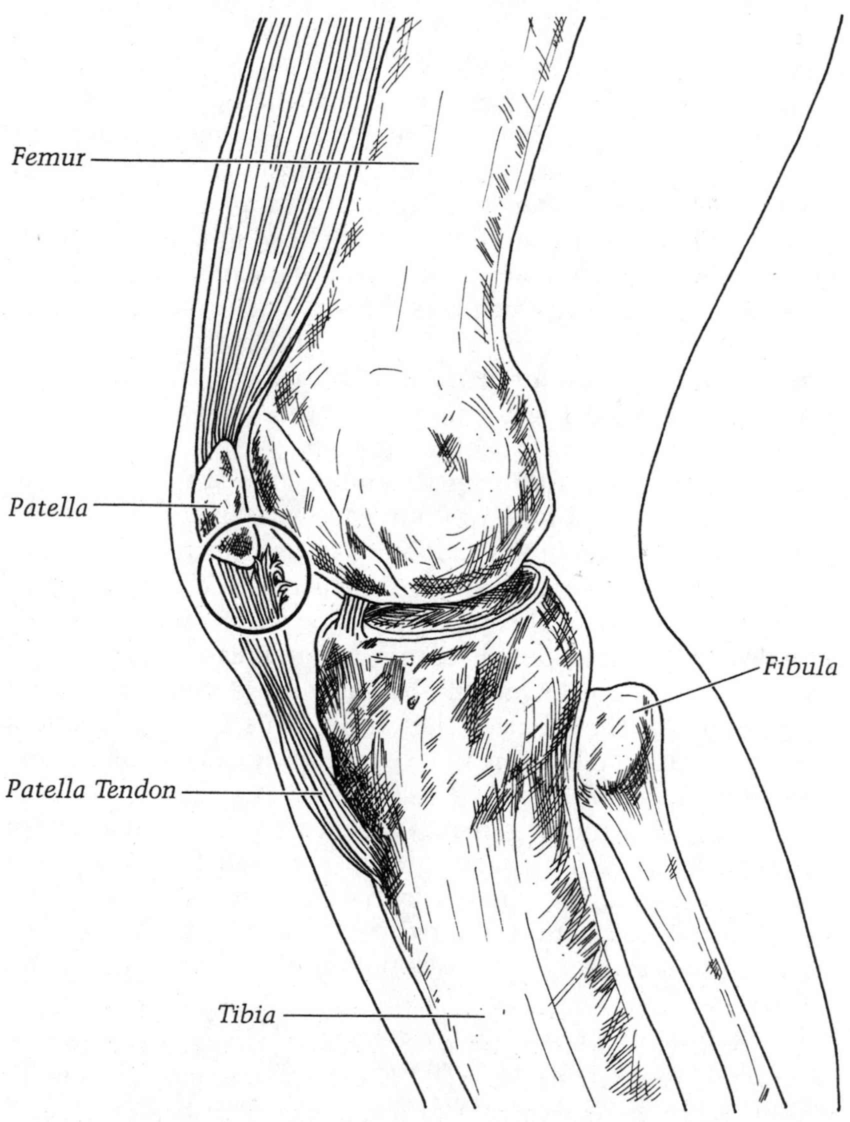

Jumper's Knee (Patella Tendinitis)

quad muscles and weight of landing. Therefore, the kneecap tendon is under tremendous stress, especially in jumping athletes—basketball and volleyball players and hurdlers. Because it is such a common ailment in leaping athletes, it is called "jumper's knee."

The force of jumping irritates the tendon near its attachment to the kneecap and eventually a tendinitis develops.

I think of jumper's knee as the tennis elbow of the knee. The kneecap tendon is similar to the muscle structure in the elbow. It has a poor blood supply and, thus, heals slowly. When it is inflamed, it produces fighter cells and pain.

Jumper's knee is not a predictable happening. Why some people get it and others do not is a medical mystery. At one time, I believed that people with high-riding kneecaps were more susceptible to jumper's knee, but I got too many patients with the opposite condition—loose kneecaps—who also were suffering from the ailment.

Jumper's knee is more common in male athletes than female athletes. It does not seem to occur in athletes over thirty years of age. It is not related to direct injury to the knee.

DIAGNOSIS AND TREATMENT

Like most tendinitis, jumper's knee develops slowly over a two- to three-week period. The pain centers at the point where the kneecap and the tendon meet.

The pain comes when you squat or jump. Sometimes the pain is present when you first get out of bed in the morning. It decreases as you use your knee during the day.

I have treated many athletes who play through pain with jumper's knee. They report mild discomfort during the sport. Afterward the pain intensifies. Often you will have some swelling and tenderness in the kneecap tendon area. Your knee will not give way, buckle, or swell.

The examination for jumper's knee is one of touch. You sit on an examining table with your knee bent at ninety degrees over the edge. I feel along the kneecap tendon itself. I start at the bottom of the tendon where it attaches to the tibia. I then walk my fingers up the tendon to the kneecap—the tender spot. You feel immediate discomfort from pressure. Next, I ask you to extend your knee straight out. I resist your movement by pushing on the front of the ankle. This usually produces pain. The remainder of the knee examination is normal. There is no skin discoloration.

The best treatment is rest. When I first see patients with jumper's knee, I ask them to stop sports for two weeks. I recommend icing the tendon for twenty minutes twice daily. The icing should continue for three days. I begin the patients on anti-inflammatory medication by mouth. I do not use pain pills.

I reexamine your knee two weeks later. If it has become less sensitive, I allow you to return to sports activities. You remain on the medication. In 50 percent of my cases, this treatment completely eliminates the problem.

If the tendinitis is resistant to the initial treatment, I consider a cortisone injection. I am against direct injection of cortisone into the tendon. This has been shown experimentally to weaken the tendon, and some world-class athletes have actually had ruptures of the kneecap tendon after cortisone injection. Therefore, when I inject cortisone into this area, I inject it behind the tendon so that the cortisone can bathe the irritated area. I feel that the patient has everything to gain and nothing to lose by cortisone injection when the tendon is resistant to conservative treatment. One cortisone injection is enough to cure an additional 25 percent of jumper's knee victims.

If the pain still persists, I apply a cylinder cast. This is made of plastic and runs from mid-thigh to mid-calf. It keeps the knee straight and prevents it from bending. If the knee cannot bend, you cannot put very much stress on the kneecap tendon. You wear the cast for one month.

Ordinarily, the cast completely resolves the tendinitis. After the casting, you must start physical therapy. I recommend whirlpool baths and knee crankers to regain motion in your knee. A second important exercise is straight leg raises. This is to regain power in your thighs. It takes one month of these exercises to get back into sports. Go slowly, so you don't reirritate. I consider casting a radical form of treatment, yet it is a way to manage difficult cases.

When all else fails, there is surgery. However, it is only necessary for one out of every 200 cases of jumper's knee. The operation I do was designed by Martin Blazina, an eminent California sports medicine physician. This involves an actual cutting away of the tendon from the lower aspect of the kneecap. A small bit of bone is removed from the bottom of the kneecap. This provides a new fresh bone bed for the tendon. Two holes are drilled in the kneecap. The tendon is brought through the holes. I tie them with a strong stitch. After the operation, the patient must wear the cylinder cast for six weeks. The rehabilitation of the knee requires at least another six weeks. This operation is successful in those few cases where all forms of conservative therapy have failed and the patient continued to have symptoms a year later.

Bursitis of the Knee

CAUSES

The knee has fourteen bursa sacs around it that help the joint, the tendons, the bones, and the skin function. Three cause problems. Think of bursa sacs as body ball bearings. For example, if you did not have a knee bursa on your kneecap, you would split the skin over the cap whenever you fell down.

There are two ways to irritate a bursa sac. One is to tear open the skin around the knee and cut the bursa sac. But a more common cause of bursa inflammation (called *bursitis)*, however, is repeated falls on the sac.

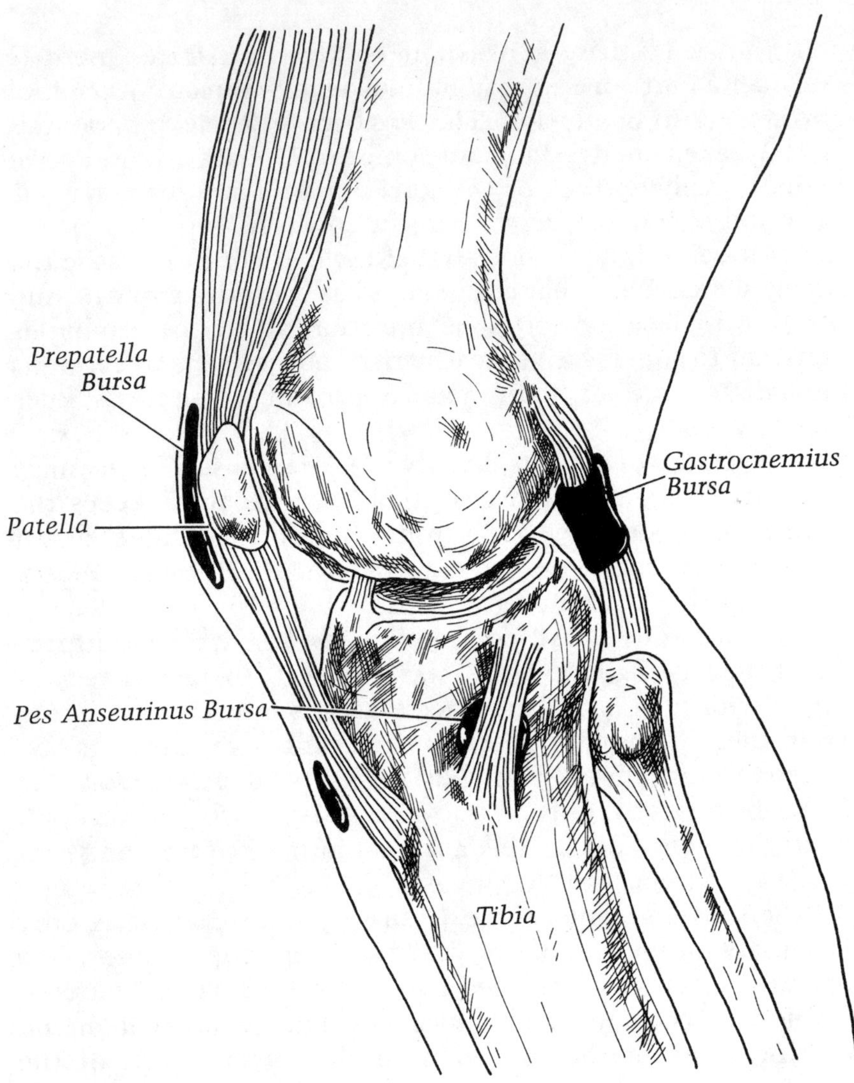

Bursa Sacs of the Knee

Prepatella Bursitis

DIAGNOSIS AND TREATMENT

Most collision sports allow padding over the knee to avoid the simplest form of this ailment, called *prepatella bursitis,* which is irritation to the bursa sac lying in front of the kneecap. For centuries, it has been called "housemaid's knee," because women who scrubbed floors on their hands and knees were its most common victims.

Prepatella bursitis develops over a forty-eight- to seventy-two-hour period. Gradually, you feel discomfort and pressure in the front of your kneecap. Then the swelling starts. If you do not ice it, it can grow as large as an egg. Synovial fluid is produced in the sac. You will be unable to kneel.

The initial treatment is ice three times a day for twenty minutes for two days. Take two aspirin tablets with each meal. This should produce a significant decrease in swelling and pain after three days.

If it does not, I inject cortisone into the sac. I freeze the skin over the bursa with Novocain. I insert a needle into the bursa sac and draw out all the fluid. The synovial fluid or bursa fluid, which is straw-colored, is the same fluid produced in all the joints in the body. The lining of the bursa sac is the same material that lines joints.

Leaving the needle in place, I inject cortisone into the bursa sac. I bandage the knee tightly to keep the fluid from coming back. The compression bandage squeezes the walls of the bursa sac together. I have my patients wear the bandage for forty-eight hours. Usually one cortisone injection is all that is necessary. Rarely does the problem recur. If the bursitis does come back, I reinject cortisone. If the problem becomes chronic, I favor surgical removal of the bursa sac. This can be done easily through an incision that looks like a smile on the front of the knee. The body remakes a bursa sac. You do not miss the surgically removed bursa sac. After the operation, you will not be able to bend your knee for five days. In three weeks, you can return to full activities.

Goose-Foot Bursitis (Pes Anserinus Bursitis)

The second most troublesome bursa sac around the knee lies under three tendons on the inner side of the knee itself. It is this spot where the hamstring muscles form tendons that attach to the upper part of the tibia. It is the hamstring muscle and the tendons which draw the lower leg back. The three tendons look like a goose's foot and are thus called the *pes* (foot) *anserinus* (goose). The bursa sac is positioned between the tibia and the tendons. It allows these tendons to slide up and down on the bone smoothly when you bend and straighten your knee joint. Why this bursa sac becomes inflamed is a mystery. It comes and goes almost without explanation. Occasionally, a direct blow to this area will produce bursitis.

With goose-foot bursitis, you feel an aching pain on the inner side of the knee and on the upper portion of the tibia. The pain intensifies with running, cutting, or jumping. In many cases, you can run through the pain on any given day, but after you stop exercises, the pain becomes more intense. Finally, the bursa becomes quite swollen and tense. It is impossible to exercise.

I diagnose goose-foot bursitis by dead reckoning. When I feel along the goose-foot tendons, I can feel a fullness. There will be tenderness when I touch the spot where the tendons attach to the tibia. When you stretch your hamstring tendons, it is painful. I ask you to bend your knee. I place my hand behind your ankle and resist this bending effort. This forces the hamstring muscles to work vigorously—to tug on the tendons. The tendon, in turn, pulls against the inflamed bursa and causes pain.

X-rays never show bursitis, but they are important because they rule out a tibial stress fracture, which can occur in the same location. Initial x-rays in a stress fracture will often be negative, but repeat x-rays two weeks after the onset of symptoms reveal new bone growth around the stress fracture. Thus, pain in this area is either goose-foot bursitis or a stress fracture of the tibia.

The treatment is the same as for prepatella bursitis. Ice and aspirin are the first lines of defense. If this is unsuccessful, a cortisone injection into the bursa sac is indicated. One injection is enough in most cases. Recurrent attacks of bursitis in this area are rare.

Baker's Cyst

The third bursa sac that can give difficulty in the knee lies directly behind the knee joint itself. As the knee bends and straightens, the tendons of the calf muscle, which attach above the knee, rub against the back of the femur. This is the reason for the strategically placed bursa sac. Irritation to this

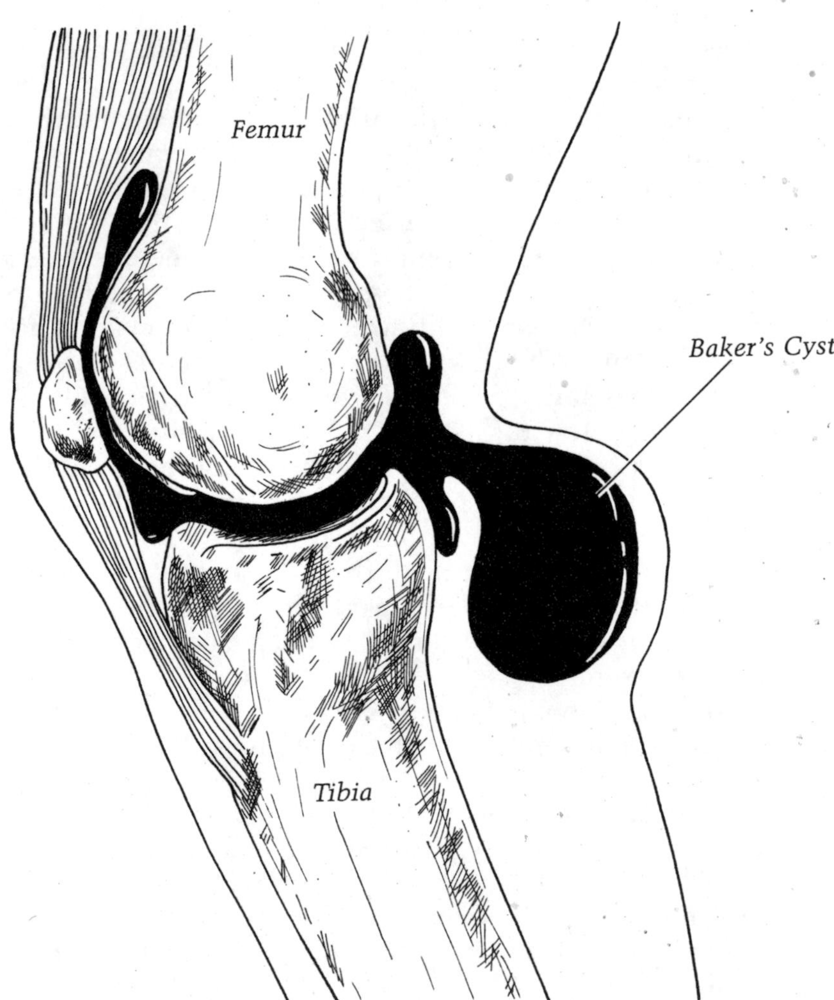

Baker's Cyst

sac should be called *gastrocnemius bursitis. Gastrocnemius* is the medical name for the calf muscle, but the inflammation is misnamed Baker's cyst, after the surgeon who first discovered it.

In adults, Baker's cyst almost always indicates that there is mischief inside the knee joint that leads to excessive joint-fluid production. This constant swelling leads to increased fluid pressure in the joint. The extra fluid is then forced into the bursa sac through a very small opening which connects the bursa sac and the knee joint.

Thus, both the knee joint and the bursa swell and become enlarged and painful. The most common cause of knee joint abnormality and Baker's cyst is a torn meniscus. Occasionally, loose bodies or arthritis can produce sufficient long-term swelling of the knee to blow up the bursa.

In children, Baker's cyst is the result of a congenital variation—a large opening between the joint and the bursa sac. Joint fluid from the knee is free to escape into this bursa sac. Interestingly, the opening is a one-way valve. That is, the joint fluid can get into the sac but cannot get out. This makes the bursa sac distend and become tense and uncomfortable. In children, no abnormality exists within the knee joint itself.

In both children and adults, the symptoms are the same. You have a heavy feeling in the back of your knee. You cannot bend your knee fully; you have tension when you try to straighten it all the way out. Full motion is possible. Occasionally, the bursa sac becomes so large that it actually pushes on the nerves that lie alongside it. This can produce radiating pain down into the calf and even onto the sole of the foot. This nerve pain can be very annoying and quite intense. But numbness as a result of nerve pressure is almost never present.

In Baker's cyst, I always notice a swelling over the back inside aspect of the knee joint. Sometimes, it can be as large as an egg. When there is a lot of fluid in it, it can be quite tender. I always compare with the back of the other knee.

In children, the balance of the knee examination, apart from the Baker's cyst, is completely normal. In adults, evidence for a torn cartilage is usually present, and swelling (an effusion) is found within the knee joint. Because there is nothing wrong with the bones, regular x-rays are seldom helpful. The Baker's cyst is strictly a soft-tissue problem. That is why I inject it with dye and it becomes obvious on x-rays. It looks like a party balloon hanging off the back of the joint.

I always listen to this area with a stethoscope because in patients over forty-five years of age, this type of swelling can occasionally represent an enlargement of the main blood vessel on the back of the knee (called an *aneurysm of the popliteal artery*).

If ice, aspirin, and rest don't work, I try the simplest surgical treatment, a cortisone shot. I numb the skin with Novocain directly over the cyst and draw out the straw-colored joint fluid with a large needle. I then inject cortisone into the cyst and apply an Ace bandage for compression. The treatment is the same for adults and children. In 75 percent of children's cases, one *aspiration* (emptying the cyst with a needle) and one cortisone shot are effective. If the cyst recurs, I give a second cortisone injection. If it recurs after that, I opt for surgery.

In adults, aspiration and cortisone injection do not get to the root of the problem. The trouble lies inside the joint. In 90 percent of the cases, this involves a torn knee meniscus. Appropriate diagnostic steps as described in the meniscus section should be taken. If you do not remove the abnormality in the joint, the joint will continue to swell and the cyst will be a chronic problem.

The surgery, in patients less than eighteen years old, is fairly straightforward. The leg is operated on from the back, and the inside portion of the calf muscle is retracted away from the bone itself. Directly underneath the calf muscle is the swollen Baker's cyst. The bursa sac itself is then removed and traced down to its connection with the back of the knee joint. The hole into the joint is stitched shut. The body will make a new bursa sac which will not have any connection with the joint. Problems after this type of surgery are unusual.

In adults, however, 90 percent will be cured of their bursitis if the joint abnormality—joint mice (loose cartilage in a joint), torn cartilage, and so on—is taken care of. Therefore, I remove the torn cartilage or the joint mice and do nothing about the bursa sac other than to drain it with a needle. By solving the knee joint problem, I eliminate the chronic fluid production and the sac is then no longer distended. In 10 percent of patients, however, the sac seems to have gained the ability to continue to produce extra fluid on its own. It must be removed surgically at a later date. I don't remove every Baker's cyst in an adult, when only one out of ten will act up later on.

Osteochondritis Dissecans

CAUSES

The bone surfaces that form the knee joints are covered with cartilage (joint surface cartilage) that has space-age qualities. The two joint surfaces are very slippery and slide as if on ice. The cartilage, nourished and lubricated by joint fluid, has considerable elasticity and can recover from denting remarkably well.

As with any body material, joint bone cartilage can be damaged. The most common way is as follows. In the normal tibia, there are two spines or peaks which sit at the very top of the bone near the spot where the cruciate ligament attaches. In some people, these spines are particularly tall. When the two bones torque one on the other during sports activities, one of these spines can hit the end of the femur. This injures the

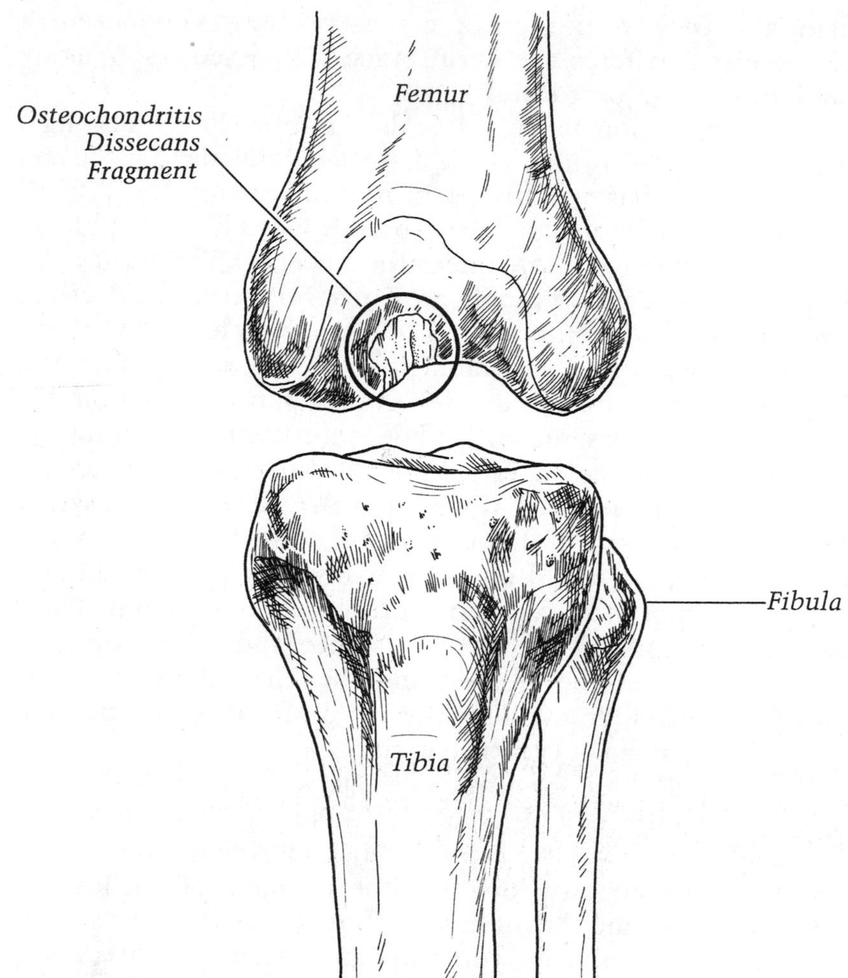

Osteochondritis Dissecans

surface of the femur. The injury looks very much like a divot dug with a golf club. In time, the piece loosens and eventually starts to cause pain. This process is called *osteochondritis dissecans.* The term *dissecans* refers to a dissection of a loose piece from the surface of the joint.

If the piece falls free, it acts as a marble in a gearbox and falls in the knee mechanism and locks it. When the piece falls free into the joint, it is called "joint mouse" because it is white and runs around the joints.

These two problems—dissecans and joint mice—are mainly seen in athletes between the ages of twelve and sixteen years—before they reach skeletal maturity, or full bone growth.

DIAGNOSIS AND TREATMENT

Pain comes when the cartilage piece is loosened from its bed. It is apparently this wiggling of the piece in its crater that causes discomfort. The pain is mechanical in nature. When the knee is used heavily, the pain starts. Light exercise will not bother the knee. Although most people find the pain on the inner half of the knee, the exact site is not easy to pinpoint. The pain gradually subsides two to three hours after using the knee. With the pain comes swelling. The looseness of the piece irritates the entire knee. The knee responds by producing extra joint fluid.

My examination usually reveals a normal knee. The ligaments are stable, as long as the dissecans remains in place and the knee joint has a full range of motion. Sometimes a small amount of fluid builds up in the joint. It looks like a dimple on the inner side of the knee near the kneecap. The fluid is an important indicator of problems. Thus, I try to produce a fluid wave in the knee joint, to find out how much fluid is there. Picture a balloon full of water. If you push on one side of a balloon, a ripple of fluid crosses the balloon and puffs out the other side. It's the same with a knee joint containing fluid. So, I lightly tap on the outer side of the knee and look on the inside to see if the fluid wave cross the joint. This is a very subtle but accurate test for excessive joint fluid. When you distend the knee with fluid, it does not bulge out in uniform fashion. Rather, it is like an inner tube with a bleb on it. There are some points of minimal resistance and other points of rigid resistance. The area around the kneecap is the most easily distended. You will see fluid in the front of the knee, but not in other areas of the knee joint.

If you bend your knee ninety degrees, I can push the skin against the surface of the femur. In this way, I can almost feel the tender area over the loose osteochondritis dissecans fragment. Bend your own knee to ninety degrees. Push just beyond the inner and the outer side of the kneecap. If the surface of your femur is tender, you may have osteochondritis dissecans.

If the loose piece has dropped from its crater, the knee can lock. By this, I mean that you will not be able to straighten it fully. You will feel resistance. You will be able to bend the knee, but not straighten it out. This is because the loose body is lying directly between the bones which form the knee and mechanically stop the knee from extending.

Plain x-rays should always be taken. I take four views of the knee to try to find the loose piece. If the piece is present, but not loose, I will see the crater. The piece of bone that is attached to the fragment is more dense than the normal bone, because it has lost its blood supply and is dead.

I cannot see the surface of the joint on regular x-rays—only the bone. To see the surface of the joint, I must put dye in it. This test is called an *arthrogram*, as described earlier. The dye outlines the fragment and the bone surface.

I can also determine the status of the fragment with the arthroscope, a body telescope which can peer into the joint through a narrow opening. If the piece is loose from the surface, I look for distinct fragments. They are usually the size of a quarter. If the dissecans has not separated, I will see a very smooth-looking end of the femur. This is very encouraging if the surface has retained its smoothness and evenness. This means that the fragment is not about to drop out. However, if I see a distinct fragment through the arthroscope, I know the fragment is getting ready to fall out. If the fragment has already dropped out, the arthrogram is not necessary since you can usually see the displaced fragment on the regular x-ray. The arthroscope should be used only to confirm this diagnosis.

When I introduce the arthroscope into the knee, I fill it with saline solution. I can see the displaced fragment sitting between the ends of the bone. Sometimes the fragment has moved to a remote part of the knee joint and the entire knee must be examined to determine the exact location of the joint mouse.

How do I treat these patients? If the fragment is present on the plain x-ray and the surface is intact, I just restrict their activities from sports, and the patient has an excellent opportunity to heal with rest only. Three to six months of no sports is usually necessary. Follow-up x-rays are taken every two months. I do not find that a cast is necessary to restrict the knee function during this time. This also eliminates the muscular atrophy (decay) that accompanies cast immobilization.

After six months, if the patient still experiences pain and the x-rays are unchanged, surgery must be considered.

In young people, whose bones are still growing, it makes sense biologically to try to stimulate healing of the fragment. This is done by introducing small pins into the fragment to hold it in place and to help bring a blood supply from the main portion of the femur to the fragment itself. These are called *Smillie pins,* after their developer, I.S. Smillie, a world-reknowned knee surgeon. They are introduced from the side of the femur down to the joint itself. This must be done in an operating room with x-rays to control the position of the pin. You remain in a hospital bed for forty-eight hours after surgery; then, you can walk on crutches. You need them for three weeks. Next, you must start physical therapy to strengthen thigh and hamstring muscles. In most cases, it takes six months after the surgery before you can return to full sports.

In adults with fully grown bones, I have found that it is almost biologically impossible for the bone to heal itself. If the symptoms are persistent, I recommend surgery.

With the patient under anesthesia, I check the knee with the arthroscope before surgically entering the knee joint. Ordinarily, I can see a wrinkled dissecans. Then I proceed to open the knee joint surgically and carve out the loose fragment and remove it. Next, I drill deeply into the bone at the base of the crater. Blood is allowed to flow into the crater and form a clot. Eventually, the blood clot converts to solid scar tissue. I call this "body wisdom." The body substitutes scar tissue in place of the missing surface material. After the operation, I immobilize the knee with a splint for about two weeks. I do not allow you to bend your knee during that period. Following removal of the stitches, I will start you on physical therapy—whirlpool baths, range-of-motion exercises, and weight-lifting exercises for the quadriceps and hamstring muscles. I like to keep my patients on crutches for about four weeks to allow proper healing of the crater. Three months after the surgery, you can resume full sports activities.

If the osteochondritis dissecans falls completely out of its crater, it becomes a loose body in the knee or a joint mouse. Often, the piece becomes lodged between the ends of the bones forming the joint. This is like dropping a marble into a gearworks. The joint locks suddenly, and you are unable to put any weight on the knee whatsoever. This happens suddenly and does not require any great force of external injury to produce it. I have had a lock knee happen to patients who were walking off the basketball court.

The problem is almost unmistakable. The knee swells and cannot be straightened. It usually locks fifteen degrees from full extension. Because you cannot feel the loose piece through the skin, you will not have any localized tenderness. The remainder of the knee examination is normal. Plain x-rays are an important diagnostic tool. The joint mouse, plus its crater, is clearly visible. In my mind, there is no question but that surgery is necessary once the x-ray and clinical picture are positive.

I put my patients under general anesthesia. I introduce the arthroscope into the knee. I fill the knee with saline solution. I look for the loose pieces; I also want to identify the crater or craters. As in osteochondritis dissecans, I surgically open the knee and expose the loose fragment. I remove it and then drill the base of the crater with a small drill. Once again, we are trying to introduce a blood clot into the crater which can convert to firm and durable scar tissue. The post-operative course is exactly the same as in the osteochondritis dissecans.

A few times, I have used the arthroscope to remove the joint mouse. I have introduced very small grasping-type instruments into the knee joint and grabbed the loose fragment. The problem is that this technique does not allow me to drill into the base of the crater. This would seem to be a theoretical drawback, and yet the advantage of removing it with an arthroscope is that the patients remain in the hospital only overnight and use crutches for only two to three days. They can begin physical therapy five days after surgery. To allow maximum healing inside the joint, I ask them to rest one month before returning to sports.

Osgood-Schlatter's Disease

CAUSES

Osgood-Schlatter's disease is not really a disease. It is a tendinitis of the kneecap tendon. The inflammation is at the junction of the kneecap tendons and the tibia.

In young athletes, the tendon is attached to prebone, which is weaker than the normal adult bone. With 800- to 1200-pound stresses on the tendon from running and jumping, it becomes irritated and a tendinitis begins.

Osgood-Schlatter's disease occurs only in growing people—usually between the ages of ten and sixteen years. I have never seen it in a patient older than twenty years. The tendinitis often shows up as a bump one and a half inches below the kneecap, on the front of the top of the tibia.

DIAGNOSIS AND TREATMENT

Osgood-Schlatter's disease begins with low-grade pain. It is located directly over the tendon attachment to the tibia. In New England, I see this tendinitis frequently in April and May. After young people have been inside all winter, they tend to play too hard in the spring. This puts tremendous stress on the tendon.

Initially, you have pain when you first get up in the morning. After you start using your leg, the pain intensifies. However, you should be able to walk and play sports. It usually takes two to three weeks for maximum symptoms to build up. Children seldom complain initially, and, therefore, I tend to see only the more advanced cases. The most common symptom is pain when kneeling. When it becomes quite severe, the child is unable to run at full speed; some children limp. All parents recognize these signs as trouble and bring the child to

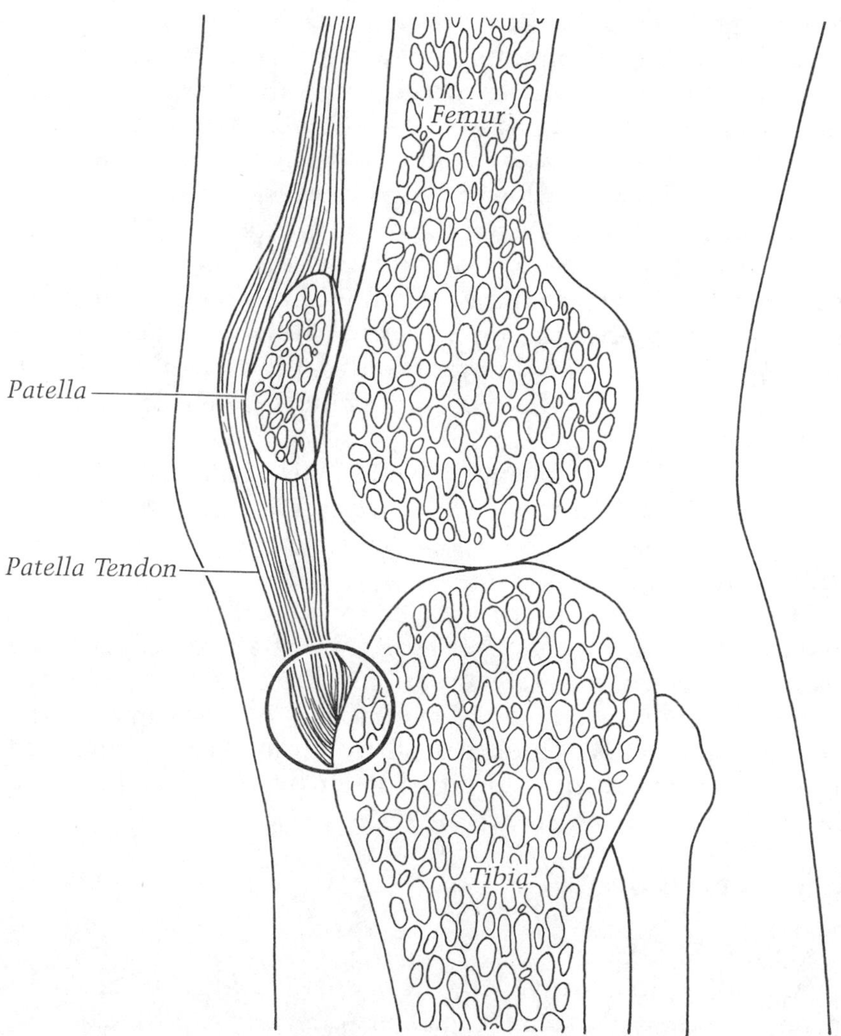

Osgood-Schlatter's Disease

a physician. The discomfort is made worse by squatting, climbing stairs, and walking up hills.

The physical examination shows tenderness and swelling directly over the tendon attachment. I can feel a bump. With your knee bent ninety degrees over the edge of the examining table, I ask you to kick out and straighten your knee. I hold down the front of the ankle joint. This puts tremendous stress on the irritated area and causes discomfort. I ask you to squat down, and this, too, causes pain.

The remainder of the knee examination is normal. There is no tenderness over the knee cartilages, and the knee ligaments are sturdy. The kneecap is seldom hypermobile.

How do I treat Osgood-Schlatter's disease? The best treatment is no treatment. Osgood-Schlatter's disease sounds and looks worse than it is. Even in the most severe case, the only remnant is a small knob on the front of the knee when the person reaches adulthood. Such a bump is sometimes cosmetically annoying or uncomfortable to kneel on. Aside from this, however, the knee is completely normal. Therefore, with this in mind, I allow my patients to continue full tilt in athletics. I make only one provision. Small amounts of pain from the tendinitis can easily be handled by some aspirin. It is impossible to expect athletes at this age to rest. If the pain is severe, more aggressive treatment is sometimes necessary.

In this setting, I immobilize the knee with a cylinder cast. This is a cast that runs from mid-thigh to mid-calf. It keeps the knee out straight. If you cannot bend your knee, you cannot put much stress on your tendon. You wear the cast for one month. In the following three weeks, you can perform only routine activities. After this period, I start you on straight leg raises against resistance. With this exercise, you do not have bending stresses on the tendon with the weight. Generally, two months after the application of the cast, you can return to full sports activities. I never had to apply a cast twice to the same patient. Once the tendinitis quiets down, it seems to stay away permanently.

I have only once operated on Osgood-Schlatter's disease. This was in my military days. It was on a recruit who had to kneel repeatedly on his Osgood-Schlatter's bump. The bump created tremendous pain.

His x-ray showed a separate piece of bone formed within the tendon. This meant that the prebone had fragmented, creating pain. This is a rather rare situation. I took the fragments out surgically, which cured his symptoms. At the time of the surgery, the tendon looked healthy around the fragmented bone. The tendinitis had cured itself but had left behind a bone element that required surgery.

Runner's Knee

CAUSES

According to a survey by *Runner's World Magazine,* 30 percent of the 15 million joggers in America are afflicted with a painful knee condition called "runner's knee." That makes runner's knee the number one ailment of the lower body with the exception of ankle sprains.

Runner's knee comes from the stress of running. It is an overuse syndrome, due to micro-trauma to the sleeve of the knee joint. The pain is mainly in the area of the iliotibial band, lying on the outside edge of the kneecap. The key point about runner's knee, in my opinion, is that it is a problem not inside the knee joint, but *around* it. In this way, it can be easily distinguished from internal knee-joint problems such as a torn cartilage or ligament or a loose body (a joint mouse).

In more than a decade of treating athletes, I have treated runner's knee only in runners, mainly distance runners. I have not seen it often in sprinters.

The reason became clear when I began to study the biomechanics of running. First, I learned that the average jogger makes 3000 foot strikes per mile. That is 60,000 foot impacts for every twenty miles, which is an average weekly distance for recreational runners. Second, each leg impact must bear the entire body weight by itself—because in running, you have only one foot on the ground at a time. In walking, 30 percent of the time, both feet are on the ground. The force of landing is about three times your weight. That means if you weigh 150 pounds, the force when you land is about 450 pounds.

In runners with normal feet, the force of running is dissipated by the foot. However, if you have a minor abnormality in your foot anatomy, like low or high arches, the force of the foot strike is passed to the knee area. Walking or performing

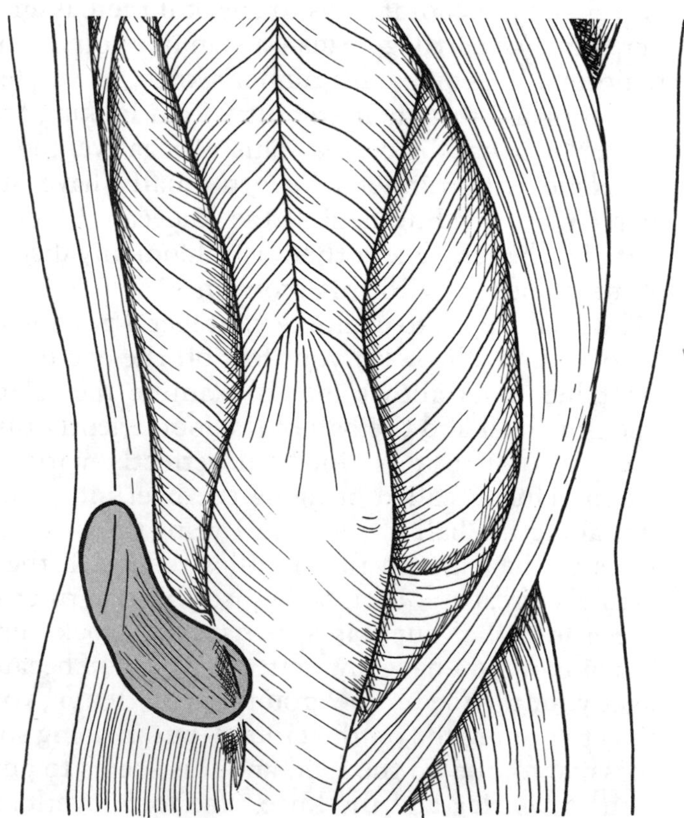

Location of Pain in Runner's Knee

daily activities does not cause knee pain. Thus, the knee often pays the price for an abnormal foot in runners.

Medical scientists have determined that the pain from runner's knee is caused by multiple areas of tearing in the knee sleeve. This produces pain that is not sharply localized. Rather, it seems to lie over a broad area on the outer part of the knee joint.

DIAGNOSIS AND TREATMENT

Runner's knee usually starts with minor discomfort, which becomes progressively worse.

Sometimes it begins after a single long run. "I felt particularly good, so I ran an extra four miles, and the pain developed after the run" is a common scenario. The pain is located almost always on the outer aspect of the knee.

The degree of discomfort runs from a dull aching sensation to sharp stabbing pain. The pain is not localized. Most sufferers of runner's knee cannot put their finger on one particular tender spot. They will usually use the flat of their hand to describe the location of the pain.

My examination of the athlete with this type of complaint begins at the knee itself. I want to be sure that nothing is deranged inside the knee. I start with the knee bent ninety degrees over the edge of the examining table. With my fingertips, I feel along the outer border of the kneecap and down under the outer knob of the femur. This is the area where tenderness is found. Because it is around the outer edge of the kneecap, physicians call this *peri-* (around) *patella* (kneecap) discomfort. There is usually no swelling or discoloration.

I feel along the joint lines to check if there is any tenderness over the knee cartilage pads—a sign of a torn cartilage. These are normal in people with runner's knee. Next, I feel along the ligaments and test their stability. I am looking for a sprained knee or one that dislocates. This is negative, too. Interestingly, people with runner's knee normally have stable knees. There is no grinding behind the kneecap. I feel down along their shin bones, where there is seldom any discomfort. Shin splints do not go with this problem.

The final step is to evaluate your feet. I put you face down—on your stomach—with your feet off the end of the examination table. This allows me to examine the relationship between the heel and the ball of the foot. I gently turn your heel outward so that it is lined up perfectly with the Achilles tendon. This is called bringing the heel into neutral. I put a ruler along the ball of the foot. I want the ruler to be at right angles to the calf and the Achilles tendon. If the area underneath the great toe is tipped up so that the ruler tips up, you have a foot that supinates or naturally cocks up. With this kind of foot, you have an ineffective pushoff because it is hard to get your big toe on the ground. By flattening your arch, you can get your big toe on the ground, but in doing so, you put an excessive torque on the foot, and this seems to put extra force on the lower leg and the knee. The force overloads the outer aspect of the knee joint and produces the micro-tearing—runner's knee.

If your foot supinates, your running shoes will wear out in a curious pattern. The outer edges of the heel and the shoe itself wear down quickly. The inner areas over the ball of the foot look brand new. That is why I ask my patients to bring in their running shoes.

Virtually all people with runner's knee have this type of foot alignment. I have had great success in curing runner's knee by using a foot orthotic, a custom-made arch support which fits into both athletic and regular shoes. Needing orthotics is like needing eyeglasses. You must have your own pair. It does not work when you use somebody else's.

The purpose of the orthotic is to fill the area between the ground and your cocked-up foot. In other words, the orthotic is a space filler. In running with orthotics, your foot lands on the orthotic, which then pushes on the ground. Thus, you do not need to contort your foot to run effectively. This changes your foot plant and eliminates the contorting force to the knee. In my opinion, an orthotic shoe correction is essential to the management of runner's knee.

For the immediate pain of runner's knee, I recommend using ice for twenty minutes twice daily for one week. This should ease the tenderness.

Take two aspirin tablets with each meal for seven days. The aspirin will ease the inflammation. If the aspirin is not effective, I prescribe oral anti-inflammatory medicine.

Stop running for at least a week. This gives the micro-tears time to heal.

After a week, you may start running again, but only at half the daily distance you ran prior to your knee problem. If you can run pain-free for three weeks, you can return to full distances. I practically never use cortisone shots to relieve the pain of runner's knee and surgery is never a consideration.

VARIOUS SPORTS AND THE DEMANDS THEY MAKE ON THE KNEE
James D. Key, M.D.

	Functions of the Knee							
Sports	Bends	Straightens	Angles from Side to Side	Slides	Rolls	Rotates	Subject to High External Force	**Total**
Football	5	5	5	4	4	5	4	**33**
Ice Hockey	4	4	5	5	5	5	5	**33**
Basketball	5	5	5	4	4	4	3	**30**
Soccer	5	5	5	4	4	4	3	**30**
Skiing	3	3	5	5	5	5	4	**30**
Wrestling	4	4	4	4	4	4	4	**28**
Karate	4	4	3	2	3	4	4	**27**
Baseball	3	3	3	3	4	4	2	**22**
Track (Sprinting)	5	5	3	2	2	2	2	**21**
Tennis	3	3	3	3	3	3	1	**19**
Swimming	3	3	3	2	2	2	1	**16**
Golf	2	2	3	2	2	3	1	**15**

Rating Scale: 5—Very Strong use, 4—Strong Use, 3—Medium Use, 2—Light use, 1—Little or No Use

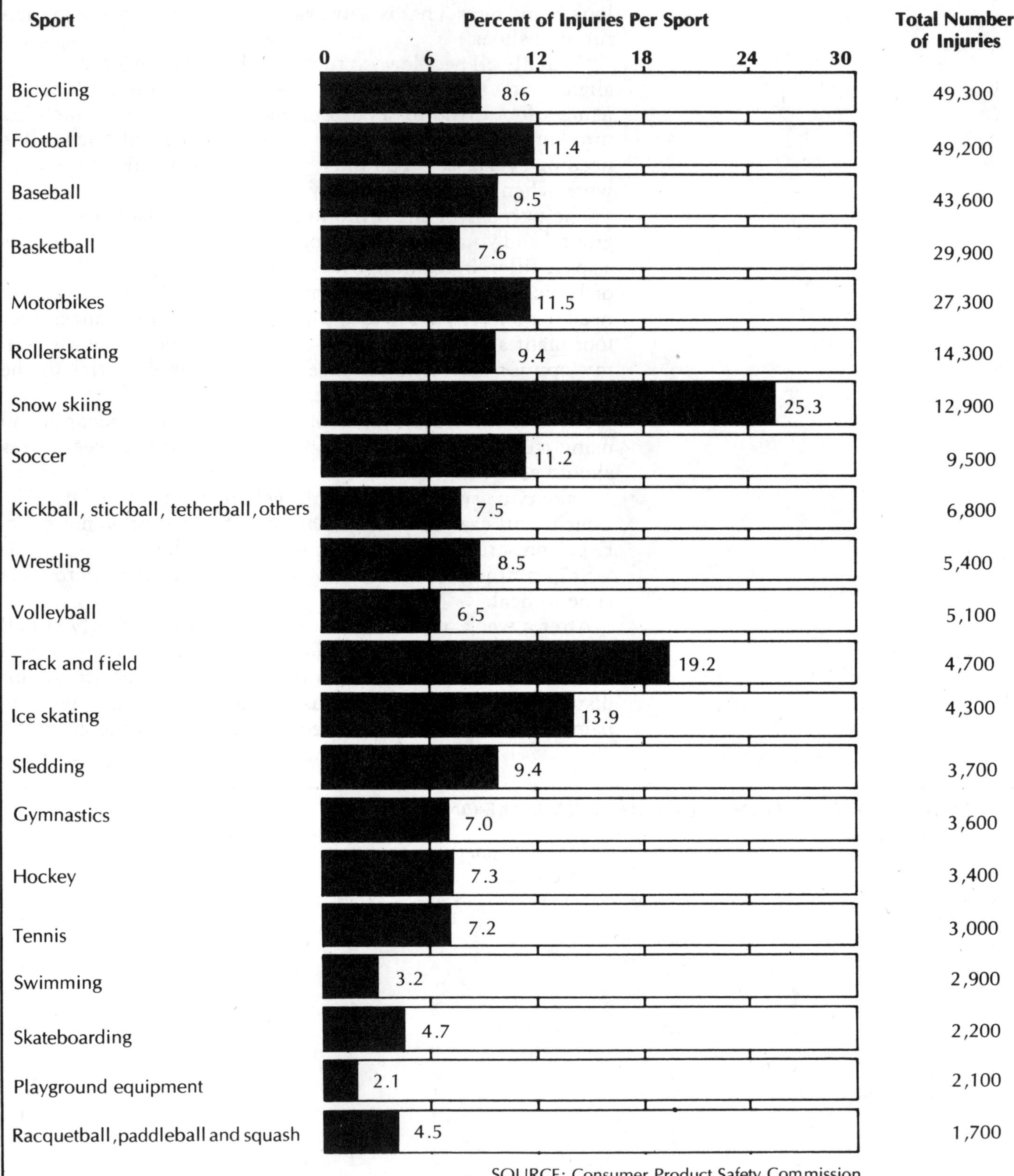
MORE THAN 298,000 PEOPLE SUSTAIN KNEE INJURIES EVERY YEAR
1979
Sport
Percent of Injuries Per Sport
Total Number of Injuries
0
6
12
18
24
30
Bicycling 8.6 49,300
Football 11.4 49,200
Baseball 9.5 43,600
Basketball 7.6 29,900
Motorbikes 11.5 27,300
Rollerskating 9.4 14,300
Snow skiing 25.3 12,900
Soccer 11.2 9,500
Kickball, stickball, tetherball, others 7.5 6,800
Wrestling 8.5 5,400
Volleyball 6.5 5,100
Track and field 19.2 4,700
Ice skating 13.9 4,300
Sledding 9.4 3,700
Gymnastics 7.0 3,600
Hockey 7.3 3,400
Tennis 7.2 3,000
Swimming 3.2 2,900
Skateboarding 4.7 2,200
Playground equipment 2.1 2,100
Racquetball, paddleball and squash 4.5 1,700
SOURCE: Consumer Product Safety Commission

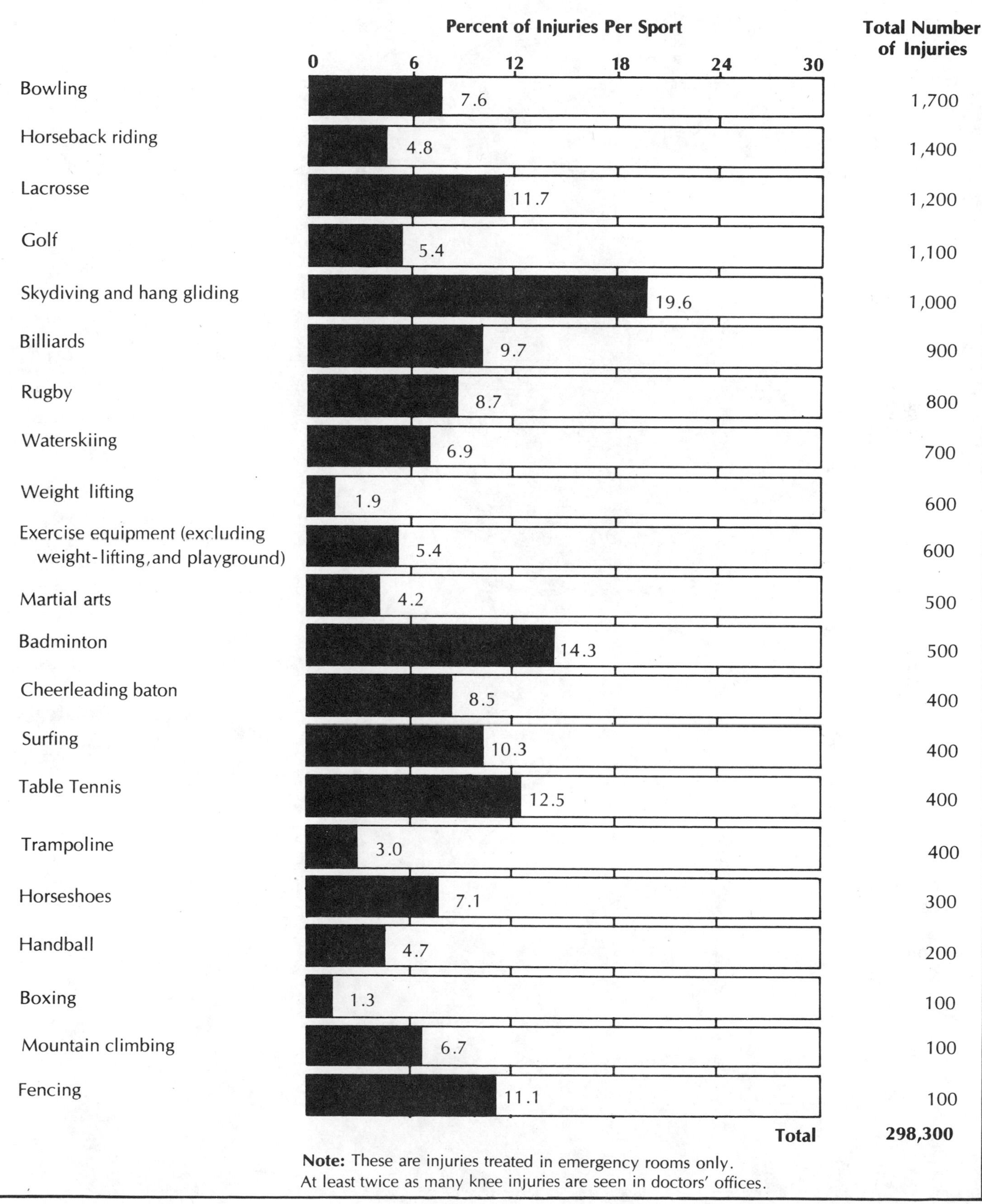

Note: These are injuries treated in emergency rooms only.
At least twice as many knee injuries are seen in doctors' offices.

The Lower Leg

Calf Muscle Strain

The calf muscle, the *gastrocnemius* in medical language, is a two-joint muscle. It starts out above the knee joint and ends up below the ankle joint.

It has two major portions, an inner portion called the medial head and an outer portion called the lateral head. These portions attach on the back of the thighbone and go down behind the knee joint. They form much of the bulk of the back of the lower leg.

About two-thirds down the lower leg, the calf muscle, which is extremely strong, stems off into the largest tendon in the body, the *Achilles tendon*. In general,

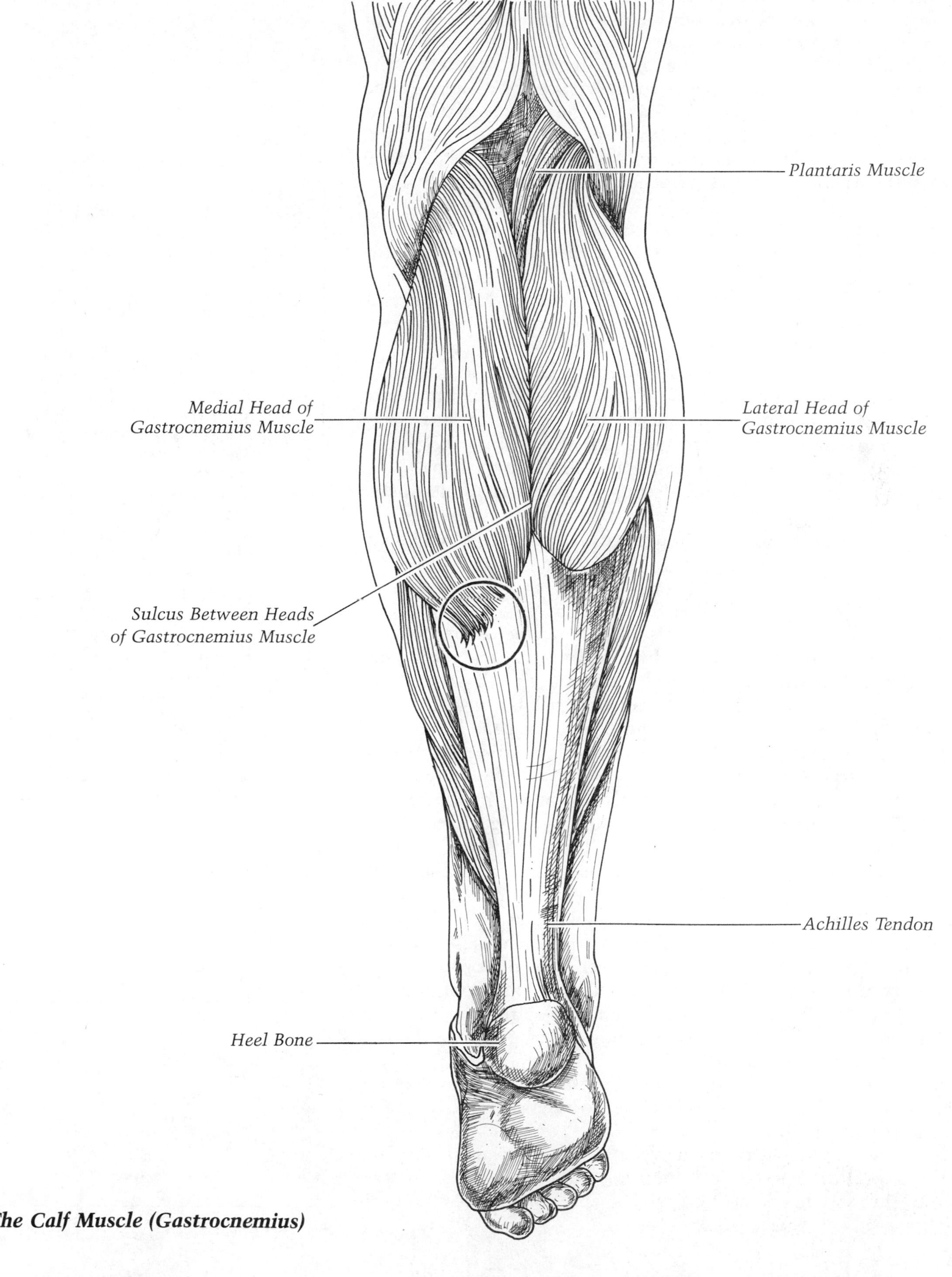

The Calf Muscle (Gastrocnemius)

the stronger the muscle, the larger the tendon it needs. Without the calf muscle-tendon unit, you could not rise on your toes, climb stairs, run, or walk.

The weakest portion of the muscle-tendon unit is where the meat of the muscle attaches to the tendon. We call this the musculoteninous junction. This has been demonstrated experimentally by hanging weights on calf muscles of rabbits; they tear first at the musculotendinous junction. The tendon itself will not tear unless it has been weakened by a scalpel cut.

CAUSES

A *calf strain* is a ripping away of the inner portion of the muscle from its tendon attachment. It is caused by putting your calf muscle under too much pressure. Most running athletes—players of baseball, football, basketball, and soccer—are the victims of calf strains.

The worst calf strain I have ever treated was in a twenty-eight-year-old tennis player. He started up to the net to return a drop shot. The quick forward acceleration and upward tilting of the foot caused him to put great pressure on his calf muscle. He felt it "go twang" and feared his calf had ripped apart.

About 80 percent of the calf muscle strains occur in athletes and fitness buffs who are more than thirty years old. The reason is not clear. But I do not see these strains in eighteen-year-olds.

DIAGNOSIS AND TREATMENT

There is no secret about this diagnosis. You feel immediately disabled and have an intuitive feeling that something has happened to your calf. For the first twenty to thirty minutes there is very little pain only because the calf is "in shock." It is numb. During the first three hours, swelling, pain, and muscle spasm all set in. Many large blood vessels are ripped and blood pours into the local area.

Stop exercising, or you will only aggravate the injury. Use RICE immediately. Do not wait for the swelling and pain. The ice, compression, and elevation will minimize the swelling and the healing period will be shortened. Untreated, the swelling will continue over the first twenty-four to forty-eight hours. A black-and-blue area often develops.

I have treated some athletes with gigantically swollen calves—sometimes two inches larger than the opposite calf. With this much swelling, I worry about clots forming in the leg veins. The clots, called thrombophlebitis, can travel to your lung and kill you. That is how Harry Agganis, the Red Sox's first baseman, died in 1955.

I ask you to use crutches for the first few days to take the pressure off the calf muscle. After forty-eight hours of RICE, start heat applications.

The body can heal this injury only so fast because it involves a large muscle mass. In 90 percent of the cases, it will be two to three weeks before you will be able to jog successfully. It

RUBBER TUBING EXERCISES

Here are exercises to strengthen the muscles on the front of your lower leg. These allow you to bring your foot up toward your shin.

1. Take the inner tube from a bike tire, or cut a circular strip from a car tire inner tube. Loop it under the leg of a heavy chair. Place the opposite end of it over the top and outer part of your foot.
2. Sit in another chair. Move the chair to the side until the big rubber band is tight against the outer border of your foot. Now, move your foot slowly up and down. Hold the up position for three full seconds. Do this 30 times. Take a five-minute break. Repeat three times. Do this twice daily.
3. As your muscles get stronger, you need more resistance. Simply pull the chair further back to make the rubber tighter. This will give the muscle a harder work out.

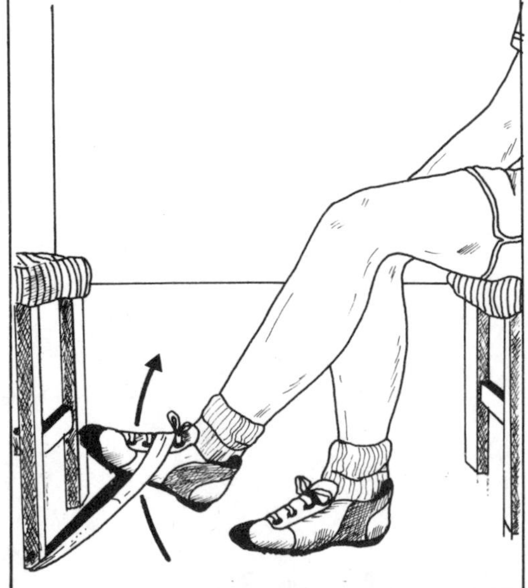

takes that long for a solid scar tissue to form. After you are able to jog pain-free, you should allow an additional week of conditioning before you run or exercise full out, including long-distance running, multiple sprints, and calf strengthening exercises. If you return suddenly to full activity, you may rip the calf again and be laid up for many more weeks. In total, it requires four to six weeks for complete healing of this injury.

Once the scar tissue heals the rip in the muscle, it is as strong as normal. However, scar tissue is not as elastic as normal muscle, so you must stretch out your calf carefully before sports. Use the same stretches as for the Achilles tendon—wall push-ups and the heel stretch board.

Sometimes, a calf muscle strain is mistaken for a plantaris tendon rupture. The plantaris muscle and its tendon is a small muscle in the back of the leg which is a remnant of early man. Twenty percent of us do not have one.

The plantaris tendon is approximately one one-hundredth the size of the Achilles tendon. Any popping or injury to its muscle or tendon could not possibly cause the tremendous swelling and pain caused by the calf muscle strain.

Shin Splints

A *shin splint* is a catch-all term used by coaches, trainers, and some physicians to describe an aching pain on the front of your lower leg. Shin splints occur most frequently to runners and running backs, but almost any running athlete is a candidate for this condition.

Shin splints are an early season phenomenon. They happen mainly when your legs are out of shape. When I was a student at Harvard in the early sixties, about 20 percent of the crew got shin splints. The reason: The coach had the oarmen run up and down the 110 stairs at Harvard Stadium.

There are four separate reasons why you develop pain in the foreleg.

1. A sprain or rip to the posterior tibial muscle which originates in the back of your lower leg bone and holds your arch up. About 75 percent of shin pain is due to overuse of this muscle.
2. An inflammation of the bone covering of the lower leg bone. This is called tibial periostitis. This is about one-tenth as common as posterior tibial problems.
3. Anterior compartment syndrome is a painful condition brought on by an interruption of the blood supply to the three muscles of the front of the lower leg.
4. Tibial stress fracture, which is a crack to the lower leg bone.

These four conditions induce pain with activity; they go away with rest. The pain is all centered within an area of an inch or two.

INJURY PREVENTION

"Many sports injuries can be prevented," says Earl Hoerner, clinical professor of rehabilitation at Tufts University and coauthor of *Sports Injuries—the Unthwarted Epidemic*. "About 60 percent of the runners and joggers that I see in my running clinic are in trouble because of lack of flexibility or imbalance of their muscle systems. It is really not a skeletal problem at all, but usually a soft tissue problem, which has never been analysed until the injury.

"Most injured athletes and fitness buffs reach a barrier where their bodies can no longer compensate for the lack of flexibility or the muscle imbalance; at that point the injury occurs. If they were properly assessed before they started their exercise program, many of these overuse injuries might be avoided."

Posterior Tibial Shin Splints

CAUSES

The *posterior tibial muscle* originates on the back of the lower leg bone (the tibia). Feel your shin bone, the bone in front of your lower leg. As you move your hand to the inner part of your leg, you can feel the point where the posterior tibial muscle originates. The meat of the muscle forms a strong tendon that goes down behind the inner ankle knob and attaches to the top of the arch of your foot (the tarsal navicular bone). This tendon holds up the arch of the foot.

The pain results from overuse—the stress of exercise. Every time you put your foot down, the posterior tibial muscle strains to hold up your arch. In running a mile, you are

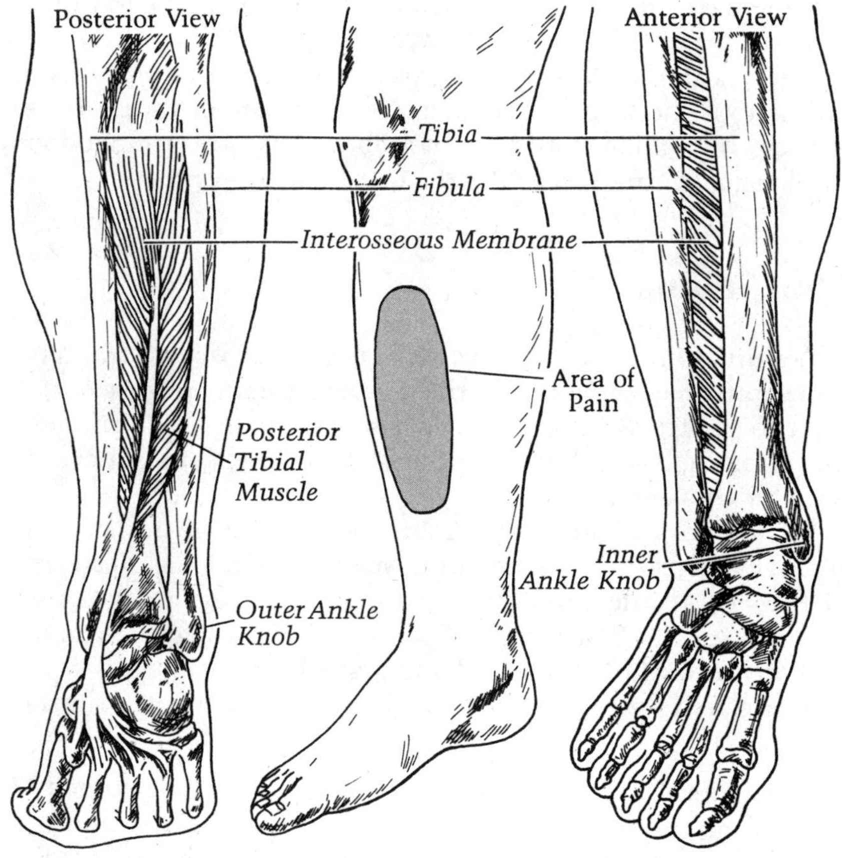

Posterior Tibial Shin Splints

stressing the muscle fifty to seventy times per minute for each foot, because that is how many times your foot is striking the surface.

If you have flat feet (excessive pronation), the posterior tibial muscle-tendon unit has to work much harder than with a well-developed foot arch. I tell patients that having a flat foot is like having a gnome down on your foot pulling constantly on your posterior tibial tendon as you run or exercise. The stress pulls directly up to the origin of the muscle on the back of the tibia. If you have a flat foot, the muscle attachment can tear away from the tibia bone. Ninety percent of the time, the separation from the bone occurs gradually, but the symptoms intensify over time.

DIAGNOSIS AND TREATMENT

In the first bout, the pain from a posterior tibial shin splint starts two to three hours after you stop exercising. It is a dull soreness, and results from the tearing of the muscle fiber. Subsequently, the pain starts as soon as you begin exercising.

The best treatment is rest. Stop running or exercising for a week.

Ice the point of tenderness twice a day for twenty minutes, for two days. Then, start heat treatments twice a day. Take six to eight aspirin each day to lessen the inflammation.

Start exercises to stretch and strengthen the posterior tibial muscles. To strengthen these muscles, do toe raises. To stretch them, either use an inclined plane or do wall push-ups. If the rest, ice, aspirin, and stretching do not give you relief in a week, see a physician.

I make the diagnosis by touch. First, I feel along the shin bone. The surface of the bone, which is directly beneath the skin, is not tender. As I bring my fingers around to the inner edge of the bone, however, I find a three- to four-inch segment of tenderness. This spot is the origin of the posterior tibial muscle.

Next, I examine the foot anatomy. I am looking for fallen arches, which is the key to the diagnosis in 95 percent of the cases.

Treatment involves the use of an orthotic device, a custom-made arch support which prevents the arch from flattening and eliminates the tugging on the muscle attachment at the bone. I favor a semi-rigid orthotic device. A rigid plastic arch support is uncomfortable. Wear the orthotic device in all your shoes.

Even with the shoe correction, it takes two to three weeks for the muscle attachment injury to heal. Stop all running for another week, and then run only half-speed and half-distance for the next ten days. By this time, healing is complete and you can return to full sports activities. With the proper arch support correction, you should never suffer this disorder again.

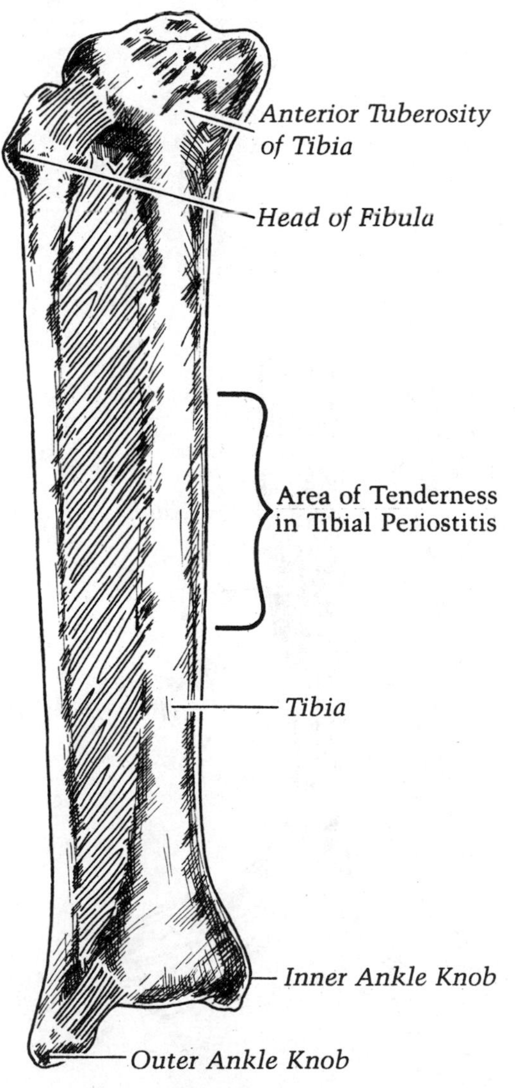

Tibial Periostitis

Tibial Periostitis

CAUSES

A second form of shin splint, *tibial periostitis*, is also common in running athletes and is due to an irritation of the covering of the lower leg bone (the periosteum). In certain individuals, the force of the foot hitting a very hard running surface transmits irritation to the front of the shin bones. This, in turn, irritates the covering of the bone and produces pain.

Although it has never been proven, it seems likely to me that there are small areas of bleeding under the covering itself. Tibial periostitis will probably never be totally understood because it is never necessary to biopsy the bone covering and examine it.

This type of shin splint comes on in the same way as the posterior tibial shin splint. It hurts in approximately the same location and is aggravated by running. Often, you can run through the pain for several weeks, but it will intensify. Finally, it becomes so intense you must curtail your sports activities.

Tibial periostitis is only one-tenth as common as the posterior tibial shin splint. I have never been able to link any disorder—flat feet, bowed legs, or an inverted heel bone—with the onset of this problem, but it seems more common in those runners who are doing long-distance events and running on hard surfaces such as asphalt.

DIAGNOSIS AND TREATMENT

The diagnosis is based on touch. The tender area lies directly beneath the skin on the front portion of the shin bone, the anterior surface of the tibia. The tenderness starts about three inches above ankle level and extends for two to three inches. Often, I can feel a slight swelling and fullness under the skin on top of the bone. This point will be sensitive to pressure. The key to diagnosis of an inflamed bone covering is that it is tender right on the bone as I push through the skin.

The best treatment is to stop running and other sports activities for seven days. This removes the pounding that has created the problem. Next, the irritated area should be iced for twenty minutes twice each day for two days. For the inflammation, take two aspirin with each meal for seven days. If the aspirin and ice do not relieve the pain, I prescribe stronger anti-inflammatory medication taken orally. There is no other treatment.

I like to take x-rays of both lower leg bones. For this condition, these x-rays are standard practice. I take them to be sure I am not dealing with a stress fracture (which I will discuss in the next section).

After the inflamed bone covering quiets down, you may return to sports. It is important to restart running or exercising on a soft surface, such as a grassy field or a composition track. For the first week, your workout should be one-fourth of your normal activities. You should then increase to half a workout for the next week and allow one month to return to full activities. Be sure to wear good athletic shoes with plenty of shock absorption in the heel. This means a generous heel wedge made of materials that will provide shock absorption.

Anterior Compartment Syndrome

CAUSES

A third form of shin splint, the *anterior compartment syndrome*, is a condition involving the three muscles that lie directly beside the lower leg bone on the front of your leg. These powerful muscles—the anterior tibial, the extensor

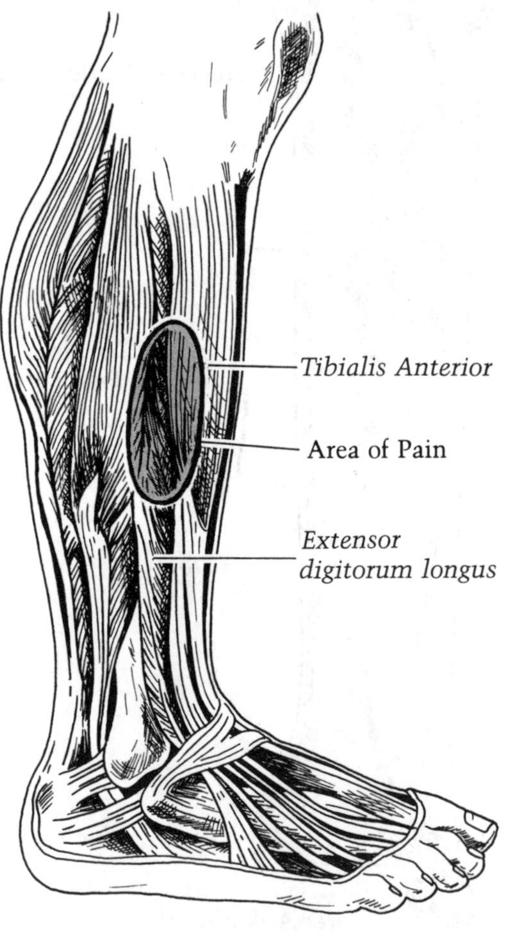

Anterior Compartment Syndrome

SURGERY RELIEVES COMPARTMENT SYNDROME

Mary Decker was a world-class runner at age fourteen, but was forced to quit competing in 1977 because of recurring pain in both legs. Her problem was diagnosed as a recurrent tibial compartmental syndrome by Glen Almquist, a Santa Ana, California, orthopedic surgeon, who recommends surgical release of the affected muscle compartments.

Decker decided against surgery because Almquist had never performed it on a runner. She had spent three years searching for relief when she met Dick Quax, the 1976 Olympic 5000-meter silver medalist. Quax had undergone surgery for compartmental syndrome in New Zealand in 1975, just eight months before the Olympics. He convinced Decker that it would solve her problem, and in July 1977 Almquist operated on the tibial compartments of both legs.

With Quax as her new coach Decker rebuilt her training schedule, and in August of 1979 she set a new American 1500-meter record.

SOURCE: *The Physician and Sportsmedicine.*

hallucis longus, and extensor digitorum longus—fill out the contour of the front of your lower leg and lie between the two lower leg bones. Their tendons cross the ankle joint and continue down onto the foot. They allow you to bring your foot up toward your shin and bring your toes up off the ground.

Anterior compartment syndrome stems from your anatomy. The muscles which they involve occupy a box formed by the lower leg bones with a fascial covering over them. As they are put to hard use in exercise, the muscles swell with blood. The amount of swelling differs among athletes. The problem arises when the box containing the muscles pushes back against the swelling. As the swelling increases, the pressure in the box increases and the blood flow to the muscle is restricted. This leads to a relative starvation of the muscles, which produces pain in the muscle itself. Because this pain occurs in the front of the lower leg, it is often diagnosed as a shin splint. This is the same kind of pain produced by a heart attack.

DIAGNOSIS AND TREATMENT

You know that you have anterior compartment syndrome by the pain located just beside the shin bone area. The discomfort increases with running. It comes on at about the same distance each time. But, it takes a physician to accurately diagnose the problem.

Milder degrees of anterior compartment syndromes can be "run through"; however, if you have severe pain, you should stop running. Allow ten minutes for the swelling to decrease and for the blood supply to reestablish itself in the muscle. Begin running again slowly. Over the next few weeks, the difficulty seems to pass away. Ice down the front of both lower legs after running for twenty minutes. The ice diminishes the swelling and makes your legs feel more comfortable.

Alteration of the running shoe or changing the running surface does not help this problem. Fortunately, the muscles seem to be able to adjust to the training over time.

In sports medicine centers throughout the country, actual pressure measurements can be made in the compartment to gauge tightness. It is done by injecting a special needle into the compartment. Normal pressure in the compartment is thirty-five millimeters of mercury; any reading above fifty to sixty millimeters of mercury is considered abnormal.

In a few athletes, the pain will not subside. These are the most severe cases and constitute only 1 percent of all anterior compartment syndrome conditions. In such patients, I do a fasciotomy (to cut into the fascia). This is nothing more than surgically opening the box in which the muscles live. Through a short incision, I use a long pair of scissors to make a slit in the fascia over the front of the compartment. This is just beneath the skin. A fasciotomy requires hospitalization only overnight and the rehabilitation is very quick. In three weeks, you can be back to full activity. This operation is extremely effective, safe, and gives predictably good results.

Tibial Stress Fractures

CAUSES

A *tibial stress fracture*, a small crack of the lower leg bone, occurs when a bone is subject to pounding over a short period of time. I see stress fractures two to three weeks into the training season. They occur when you come off a relatively inactive period and start going quickly into heavy training.

The most common place for a stress fracture to develop is at a junction on the top one-third and the middle one-third of the tibia, where the architecture of the bone changes quite suddenly. One part of the bone is funnel-shaped and the other is cylindrical. The difference leads to increased stress concentration and this is where stress fractures develop.

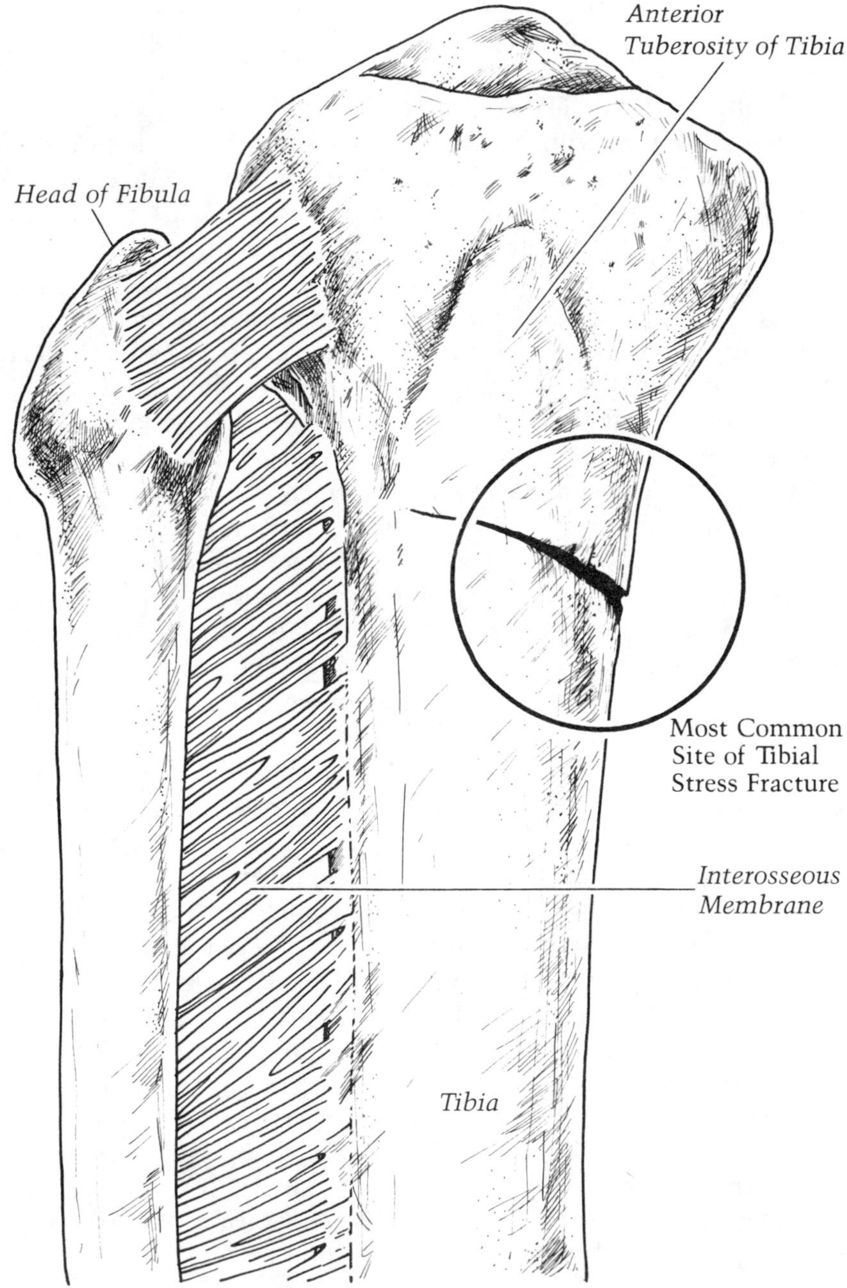

Tibial Stress Fracture

DIAGNOSIS AND TREATMENT

At first, the pain from the stress fracture will be dull and aching; the discomfort will be centered in one spot on the lower leg bone. Pain in other forms of shin splints is diffuse.

Initially, you will be able to run through the pain. But each time you do, your leg will become more painful. Finally, the pain becomes so severe that running is impossible. At this point, you usually check with your doctor.

The physical examination tells all. I run my fingers up the front of the shin bone. There is no tenderness until I get to the exact location of the stress fracture. When I touch the point of the crack, you will scream. The maximum area that hurts is no more than one inch. This is much smaller than the area of tenderness that comes from having tibial periostitis, posterior tibial shin splints, or the anterior compartment syndrome.

I always take x-rays. But, if they are taken within the first week of the onset of pain, they will often be negative. Given enough time, however, new healing bone will appear on the x-rays. Even if the initial x-ray is negative, I proceed with the provisional diagnosis of tibial stress fracture. I take more x-rays two weeks later, and this gives me the answer.

The patient must stop running. Rest is the only cure. Icing, taping, or a cast are not necessary. It requires four to six weeks of rest for complete healing to take place. One month after the diagnosis, you can start running short distances, for short durations. Gradually reestablish the conditioning in your legs. It usually takes three weeks to return to full sports activities.

I have never seen a tibial stress fracture come apart or become displaced. That is why a cast is not necessary. Once healed, it is unusual for this to happen again.

The Ankle

Ankle Joint Anatomy

The *ankle* is the most primitive joint in the body. It moves only up and down. Look down at your own ankle and foot. Bring your toes up toward your shin bone as far as they will go. In medical terms, this is called *dorsiflexion*. Now, push your foot completely down the other way. This is called *plantar flexion*.

The ankle cannot rotate side to side, nor can it tilt inward or outward. Because the ankle moves in one direction, it is very stable and relatively resistant to serious injury.

BONES AND JOINTS

There are three major bones and two major sets of ligaments that comprise the ankle joint. The two lower leg bones (tibia and fibula) join to form the top half of the ankle joint. Together, they are the receptacle into which the ankle joint fits. In medical terms, this receptacle is called a *mortice*, a French word that means a hole into which something fits. It is the ankle bone that occupies this hole. The ankle bone is called the *talus*, or astragalus. It is unusual in that no muscle or tendon attaches to it. Three-quarters of its surface is covered with joint surface cartilage (articular cartilage). The remainder of its surface is regular bone for ligament attachment.

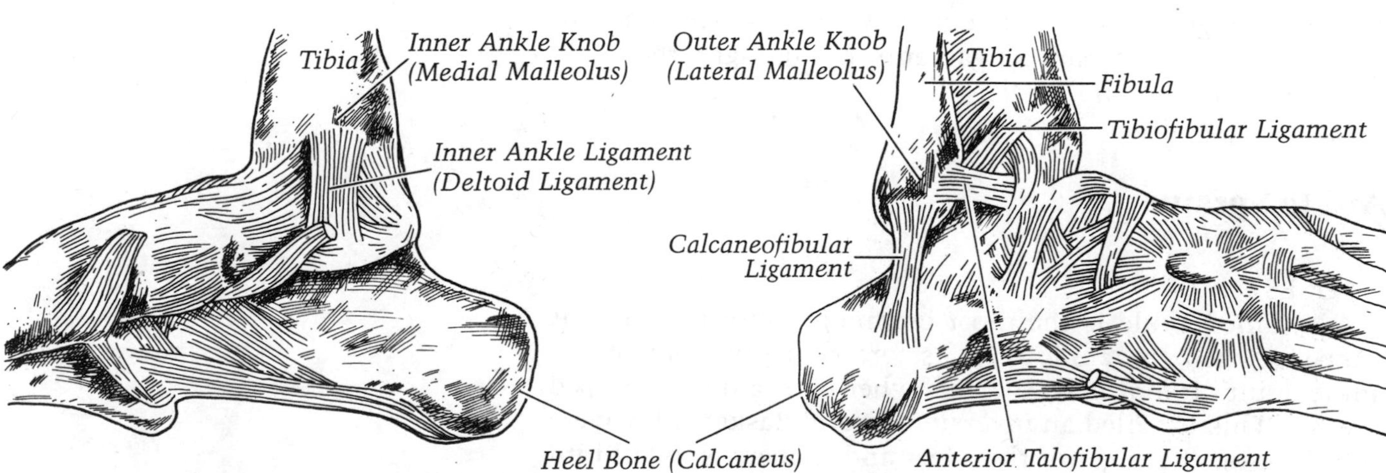

View of Inner Side of the Ankle **View of Outer Side of the Ankle**

Anatomy of the Ankle

The outer ankle knob is the lower end of the smaller of the lower leg bones, the *fibula*. The inner ankle knob is the lower end of the larger of the lower leg bones, the *tibia*. They are held tightly together by *ligaments* (the tibiofibular ligaments). The ends of these bones are also covered with bone cartilage (articular cartilage). These surfaces match those of the ankle bone.

The ankle bone is held in the mortice by ligaments. The large ligament on the inside of the ankle is called the *deltoid ligament*. It is triangular in shape—like the Greek letter 'delta.' The apex of the triangle attaches to the inner ankle knob. The broad part of the triangle attaches to the ankle bone itself.

The outside ligament system is a bit more complex. As you look at the outside ankle knob (the lateral malleolus of the fibula), you see that there are three directions in which ligaments can be directed. The front ligament that goes forward from the outer ankle knob is called the *anterior talofibular ligament*. This ligament runs from the lateral malleolus to the ankle bone. It is put to the test when your foot turns under. This may rip or sprain the ligament. Sprains of the front ligament are the most common injuries in all sports.

The second ligament goes from the tip of the lateral malleolus of the fibula straight down to the heel bone. This is called the *calcaneal fibular ligament*.This, too, is put under pressure when the foot is turned under. It is a second line of defense and only tears after the front ligament tears. The third ligament runs directly back from the outer knob of the ankle to the ankle bone. This is called the *posterior talofibular ligament*. This is never injured in sports.

The ankle is another joint that has a trade-off between mobility and stability. The ankle ligaments must be strong enough to keep the ankle bone from tipping out of the ankle mortice. If the ligaments can't hold the ankle firmly, then the athlete experiences multiple episodes of the ankle "giving out." This simply means that the ankle bone is too loose in its socket and is unstable. On the other hand, the ligaments must be long enough and allow enough play so the ankle bone can move properly. If the ankle ligaments are too tight, the ankle has restricted motion.

EIGHT PAIRS OF SIZE 14 FEET

One reason for the tremendous amount of sprains in professional basketball is that the big men tend to concentrate in a sixteen-foot area called the three-second lane.

When you get eight pair of size 14s and larger feet in the lane, it is pretty tough to come down on a piece of wood. I call this the lane sprain alley.

Trainer Ray Melchiorre,
Boston Celtics

Ankle Sprains

CAUSES

A sprained ankle is the most common injury in sports. It happens when the ankle bone is forced or pried out of the ankle joint. It normally occurs when the ankle is tipped inward. This is called an *inversion sprain*. Basketball players who land on the edges of their feet and runners who fall into pot holes are principal ankle sprainers.

In mild sprains, the ankle bone is forced slightly out of place for an instant. This causes only mild stretching of the liga-

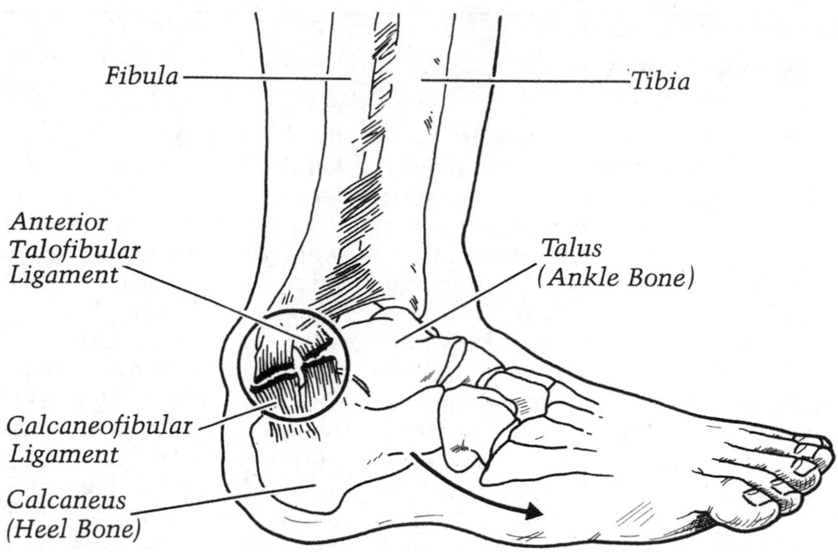

Ankle Sprain (Double Ligament Injury)

ments on the outer part of the ankle. In severe sprains, the ankle bone actually comes out of the ankle socket. This causes complete tearing of the ligaments.

When your ankle is suddenly tipped inward, the first ligament to restrain this motion and thus sprain is the one that runs from the outer ankle knob to the ankle bone itself. It is called the anterior talofibular ligament.

The second ligament on the outer side of the ankle commonly injured in sprains is the one which runs from the outer ankle knob down to the heel bone. It is called the calcaneal-fibular ligament. This is the second line of defense against inversion sprains. I've never seen the second ligament sprained by itself; usually the anterior talofibular ligament tears first.

Depending on how much force is applied to this ligament, it either stretches a small amount (grade I sprain), tears partially through (grade II sprain), or tears completely (grade III sprain). The degree of sprain depends entirely on the amount of force applied to the ankle ligaments and the duration of the force. The bigger the force and the longer it is applied to the ligament, the worse the sprain.

If only the front ligament is sprained, it's called a one-ligament sprain on the ankle. If both ligaments, the front and the bottom, are sprained, it's called a two-ligament sprain. Because the treatment is different, it is very important to make the distinction between one- and two-ligament sprains. A two-ligament sprain takes longer to heal.

DIAGNOSIS AND TREATMENT

When you sprain your ankle, you will know instantly. You will feel something tear on the outer part of your ankle. Occasionally, you may feel a popping sensation.

Immediately, the outer part of your ankle joint becomes painful. Usually, you can walk off the pain in the next five to ten minutes. Your ankle then goes into a period of pain-free

MY ANKLE SPRAIN

In my senior year at Harvard, near the end of the season, we played Brown University. Harvard was vying for the Ivy League championship that year; we had to win this game. By the third quarter, we led by three touchdowns.

When I was about to make a routine tackle, I suddenly felt a searing pain in my left ankle. A Brown player had crashed into my ankle, helmet first. Even though my ankles were heavily taped, I knew that my ankle was damaged. The pain was severe for the first couple of minutes, and then it gradually stopped. I was helped off the field by teammates. I was unable to put any weight on the ankle for two or three minutes.

Suddenly, it seemed to get better. I tried to "walk it off." The ankle was stiff and just didn't feel right. Jack Fadden, the Harvard trainer, removed my shoe and sock, and Dr. Quigley, the Harvard team physician, examined my ankle. "It's definitely the lateral ligament, but it is hard to know just how badly it's hurt," said Dr. Quigley. "Put the ice on it, Jack."

Jack Fadden put a huge plastic bag full of ice cubes directly on the outer aspect of my ankle. He told me to keep the ankle elevated. It began to feel a bit better. The ice numbed the ankle. I could hardly feel my ankle at all.

Within ten minutes, the game had turned around. Brown had scored twice and they were threatening to score a third time. I was dying to get back in the game.

"Isn't there any way I can go back in?" I asked Jack.

"If I strap your ankle again with tape, you'll be able to run on it."

"Strap it up, Jack."

I went back into the game. It was the fourth quarter. We were able to kill the Brown rally, and we won the game.

By the end of the game, my ankle was trying to burst through the tape. I limped off the field. I was delighted that we had won. But I knew my ankle was in big trouble.

In the dressing room, Jack removed the tape. The swelling was now rapidly increasing. It hurt all along the outer aspect of my ankle. Jack put the famous Harvard sponge rubber donut on my ankle and wrapped it with an Ace bandage. He gave me crutches. Because it hurt tremendously to put weight on my ankle, I was grateful for the crutches. I went out of the locker room like a three-legged dog to meet my fiancee, Sally, and my parents. I was really worried. I knew that my ankle was badly injured.

I checked in with Jack Fadden and Dr. Quigley on Sunday morning. "It is only a one-ligament sprain," said Dr. Quigley, "but it is torn all the way through."

The outside of my ankle had an egg-shaped swollen area on it. It was turning black and blue and it was painful to move. I couldn't put any weight on it. Next Saturday, Harvard played Yale. It was the most important game of the season.

Jack Fadden was very kind. I looked steadily at his eyes and asked him if he thought I'd be able to play in the Yale game. He explained that the human body is not a machine and that anything is possible. He never flinched. If it was humanly possible, I knew that they would get me to play.

The next day, Jack took off my elastic bandages. He sent me for a whirlpool treatment. Next he ran the ultrasound machine over the outer aspect of my ankle and massaged the edema fluid out of my ankle. The fluid which had accumulated from the injury was sitting in the tissue under my skin. By mechanically pushing this fluid up the leg, he helped the healing process. The warm whirlpool baths made my ankle feel good.

By Thursday, with the ankle firmly taped, I was able to run straight ahead, but not at full speed. I couldn't make cuts. I knew I was a doubtful starter for Saturday's game. I had almost resigned myself to that fact. The team went to New Haven on the train on Friday morning. We ate lunch on the train and then went to the Yale Bowl for our pregame workout.

It was November 23, 1963. When we first entered the Yale Bowl a television cameraman told us that President John F. Kennedy had been shot. Instantly, the football game paled in comparison to the national tragedy. The game was postponed. The team returned to Boston that afternoon. It was strange, I thought. This national disaster had given my ankle a week's reprieve.

The following week, I continued with whirlpools and tape support. The ankle felt better and better. By Wednesday, I was finally able to cut on it. Saturday, I was ready to play. In my mind, Jack Fadden and Dr. Quigley's physical therapy and rehabilitation program had saved me at least a week. It was an important medical lesson for me.

To this day, if I try to play squash without using high-top sneakers, the outer part of my left ankle aches. With this extra support, my ankle functions fine. It has been a permanent reminder to me that there is no such thing as a "simple ankle sprain." The body never forgets.

"shock." During this time, you have very little swelling, pain, or muscle spasm. You get a false feeling of security. Even severe ankle sprains feel good for the first thirty minutes. This sometimes makes it very difficult to determine whether one or both ligaments have been torn and also the degree of tearing. Even the doctor can get a false sense of security.

However, after the thirty-minute grace period, tenderness develops. At this point I can more easily determine whether both ligaments have been injured. I do this by touch. First, I begin to feel gently just in front of the ankle knob. This is the front ankle ligament. If there is tenderness here, then this ligament has at least been stretched. I feel down at the bottom of the tip of the outer ankle knob. If there is tenderness here, then you have a two-ligament injury. It is impossible to determine the degree of severity of the sprain during the first twelve hours. This is why I reexamine ankle sprains twenty-four hours after the injury. In football, the time to determine the degee of severity of an ankle sprain is not Saturday afternoon, but rather Sunday. Even with early treatment, swelling, tenderness, and muscle spasm will happen by the next day. The more swelling, tenderness, and muscle spasm, the worse the sprain.

Treatment of Single-Ligament Sprains

By examination, I have determined that you have a single-ligament sprain. Next, I grade the sprain. If there is only mild swelling and tenderness, it is a grade I sprain, and you will miss four to five days from competition. A grade II sprain has more swelling and more tenderness. You will have trouble moving your ankle up and down. You will miss seven to ten days of activity. A grade III sprain has severe swelling and bruising (ecchymosis). The swelling is so bad that you will not be able to move your ankle up and down. You won't be able to

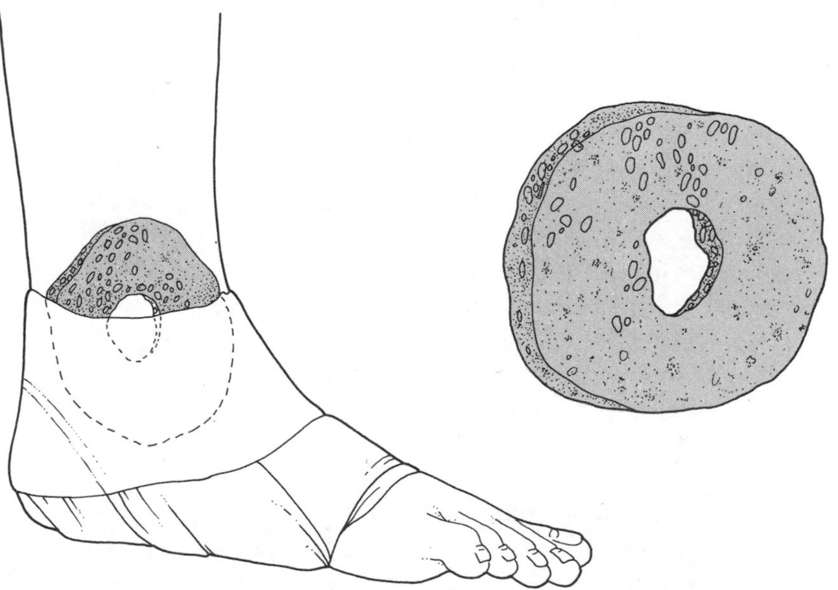

Harvard Sponge Rubber Donut

walk without limping; crutches are usually necessary. This will cost you two to three weeks of active competition.

All sprains, single- or double-ligament, should be immediately treated with RICE: rest, ice, compression, and elevation. This should be carried out over the first twenty-four hours.

A favorite compression device used in Harvard athletics is a sponge rubber donut. This can be placed with the hole directly over the outer ankle knob. The remainder of the sponge rubber compresses in the area of the sprain. This is attached to the ankle with an Ace bandage. I compress all sprains for the first twenty-four hours. It is important not to place the ankle in a rigid circular cast, because the swelling could exceed the size of the cast and cause severe pain. Therefore, I always go with a soft bandage in the first twenty-four hours. I always put you on crutches for one day. I don't want you putting any weight on the injured ankle. Then, I reevaluate the next day.

For grade I sprains, I continue the elastic compression for seventy-two hours and begin warm whirlpool baths. The athlete may return to action with this type sprain five days after injury, with tape support on the ankle. For grade II sprains, I use the elastic support, but keep it on for five days. Then, I begin you on warm baths or whirlpools. Within ten days, you can return to competition with tape support of the ankle. With grade III sprains, I use a gel cast support. This is a gauze roll which is impregnated with zinc oxide. It is rolled on like an Ace bandage, but congeals to form a semi-rigid cast. It gives excellent support of the ankle and keeps the injured area from wiggling and causing more irritation. However, the advantage is that you can wear your shoe and sock with this type of support. The gel cast is left on for one week, and then I reevaluate the ankle. If there is still a lot of hemorrhaging, swelling, and muscle spasm, the cast is reapplied for a second week. Usually, within ten days of the injury, you can start whirlpool treatments or warm baths. I recommend two daily treatments. Range of motion exercises of the ankle should be carried out while you are in the whirlpool bath. After a week of whirlpool treatments, start reverse stair climbers. Within three weeks, you should be able to return to full athletic competition with tape support.

After full motion has been regained, it is important to strengthen the peroneal muscles on the outer aspect of the ankle. These muscles originate on the outer part of the calf. Their tendons drop down behind the outer ankle knob. The outer ankle knob serves as a body pulley. The pulley increases the mechanical advantage of these tendons. They are the ones that enable you to lift up the outer border of your foot. They become weakened after a sprain.

Treatment of Two-Ligament Sprains

Two-ligament ankle injuries usually happen when the first ligament is completely torn (grade III sprain). When I make the

REVERSE STAIR CLIMBERS

After an ankle sprain, the hardest motion to regain is that one which brings your toes up toward your shin muscles. In orthopedic parlance, this is called *dorsiflexion*. When you sprain your ankle, the most comfortable position is to let your foot hang down a bit. It is this position which accommodates the ankle's swelling most comfortably. Therefore, if your ankle remains in this position for seven to fourteen days, it is difficult to get the ankle moving in the opposite direction. That's why I start range of motion exercises as soon as possible.

One way to avoid a stiff ankle is to climb up stairs backwards. In order to mount the stair successfully, you must drive your foot up toward your shin bone. It is the ideal exercise to regain dorsiflexion of the ankle. You should climb three flights of stairs daily. Always lead with the ankle that has been injured. Until full motion has returned, this exercise should be continued.

PERONEAL STRENGTHENING EXERCISING

Here are exercises to strengthen the peroneal muscles on the outer part of the lower leg.

1. Take the inner tube from a bike tire, or cut a circular strip from a car tire inner tube. Loop it under the leg of a chair. Place the opposite end of it over the top and outer part of your foot.
2. Sit in a chair. Move the chair to the side until the big rubber band is tight against the outer border of your foot. Now, move your foot slowly up and out as far as it will go. Hold for three seconds. Do this thirty times. Take a five-minute break. Repeat three times. Do this twice daily.
3. As your muscles get stronger, you need more resistance. Simply pull the chair further back to make the rubber tighter. This will give the muscle a harder workout.

diagnosis of a two-ligament injury, I place the patient in a walking cast twenty-four hours after the injury. Within that time, the maximum swelling has occurred. This cast runs from the toes to below the knee. The ankle is placed in a ninety-degree angle. This is the neutral position. I like to use a synthetic cast material—a plastic. It is lightweight, allows plenty of breathing of the skin, and it can become wet. Some of my patients have even water skied wearing the cast. It requires about two hours of drying time in the sun or thirty minutes of drying time with an electric hair dryer.

I leave this cast on for three weeks. The cast allows the ligaments to rest at their normal length. Then, the healing scar tissue can reestablish continuity of the ligament without disruption. This is important. If both ligaments are allowed to heal properly, the ankle becomes stable again. If they heal stretched out, the ankle will be permanently loose. After three weeks in a cast, I remove the cast, and you can start physical therapy.

I recommend two warm soaks a day and a series of motion exercises to regain ankle mobility. I also ask you to do reverse stair-climber and peroneal strengthening exercises. Generally, a two-ligament injury will put you out of competition for four to five weeks. Three weeks is spent in the cast, and then an additional ten days to two weeks is required for full rehabilitation. After this type of injury, tape support for sports should be continued for at least six months.

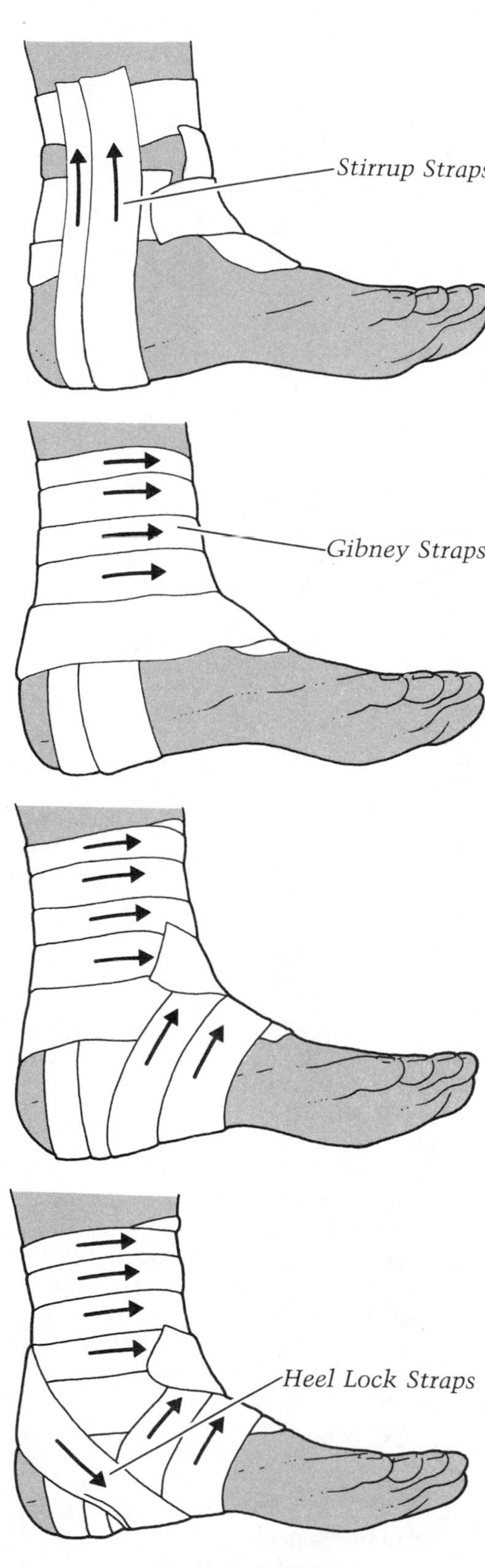

Ankle Taping

HOW YOU CAN PREVENT ANKLE SPRAINS

There have been numerous medical studies on preventing ankle sprains. The results: The more external support the ankle has, the smaller the chance of sprain. Especially for high school athletes, I recommend high-top athletic shoes. They come up over the ankle knobs and resist the forces of sprains more effectively than low-cut athletic shoes. It is for this reason that many high school basketball players are now wearing high-top sneakers.

A second excellent method of preventing sprains is to tape your ankles before competition. About one-third of the Cleveland Indians have their ankles taped says Jim Warfield, the team trainer. A first-class tape job runs from the mid-foot to mid-calf. Ankle wraps also are effective. They are easier to apply than tape, but are less effective. They can be worn over socks.

Lenny Moore of the Baltimore Colts was the first national athlete to tape his ankles over his athletic shoes. This looks flashy, but has been no more effective than applying the tape on the skin.

These are measures that can be used for players who have never had an ankle injury. If you have had an ankle injury, it is absolutely essential that some form of external support be used for a year following injury. Even if you haven't had an ankle injury, I would recommend high-top athletic shoes or ankle wraps.

Lateral Ligament Instability of the Ankle

CAUSES

If you repeatedly sprain the ligaments on the outer aspect of your ankle, they can become stretched out and loose. Eventually, they become unreliable in keeping the ankle bone in the ankle mortice. You will experience repeated episodes of "giving way." This can happen running straight ahead, cutting, or even just strolling down the street.

Each time the ankle falls out of its mortice, the ligaments are stretched more. The ankle, then, has all the characteristics of acute sprain. It is a vicious circle. The more it falls out, the more the ligaments are stretched. The more stretched out the ligaments become, the more frequently it falls out. The result is a permanent instability of the ankle.

All joints are designed to act in a stable fashion. If they are too loose or unstable, a small amount of damage is done to the joint every time it gives out. The surface of the joint is scraped and can gradually start to wear down. Eventually the joint becomes arthritic. As part of the wear-and-tear phenomenon, spurs build up on the edge of the joint. These are like small stalagmites in caves. It is not clear why the body produces spurs, but they are a sure sign of joint instability. They are bad because they misshape the joint and also dig into the soft parts around the joint, causing pain.

DIAGNOSIS AND TREATMENT

How can you avoid this problem? I feel the only way to prevent ligament instability is by treating acute sprains aggressively. That is why I always cast two-ligament sprains. If the acutely sprained ligament is immobilized in a cast, it will probably heal at its normal length. If you are allowed to walk on a bad sprain without support, the ligaments heal in a more stretched out fashion. It is for this reason that I feel there is no such thing as a "simple ankle sprain."

"The ankle gives way, Doc." This is a sure sign of ligament instability of the ankle. I have treated patients with as many as forty ankle "give ways." When the swelling and pain have abated, they have gone back to athletic competition.

Danger signals: Even with tape support, the ankle gives out. If an ankle is very unstable, tape is no match for the forces involved. Another danger sign is the increasing frequency with which the ankle gives way. If both conditions are present, then surgery may be the only solution.

Examination of an ankle with instability usually reveals some long-standing puffiness over the outer ankle ligaments. The ligaments have never fully healed. The ankle is too mobile. In examining your ankle, I test to see if the ankle bone (talus) can be shifted out of the ankle joint. I grasp the heel bone with one hand and control the lower leg bone with the other. First, I pull forward on the heel and push back on the lower leg bone (tibia). In loose ankles, you can feel the ankle

CASTINO'S ANKLES

John Castino, the 1979 rookie of the year, twisted his left ankle as he crossed first base in a game aginst the Red Sox in May 1980. He fell to the ground immediately and was carried to the Minnesota Twins dugout. Dr. Arthur Pappas, the medical director of the Red Sox, examined Castino minutes after the accident. He found tenderness over the outer part and the front part of the ankle joint. The swelling started immediately.

"Have you had many sprains?" Dr. Pappas asked.

"Sure, Doc. I have sprained my left ankle ten times and my right one eight times. I guess I have weak ankles," he said mournfully. "I ordinarily tape my ankles, but I have been lazy lately. I have just been using an elastic tape support instead of regular white athletic tape. I guess I am going to have to go back to the stronger tape."

Dr. Pappas strapped a full bag of ice on Castino's ankle and he was sent to a nearby hospital for x-rays. He returned forty-five minutes later to the Red Sox medical room. I have never seen anybody look sadder. "The x-ray doctor told me I have arthritis in my ankle, Doc," Castino said dejectedly. "Is that true?"

I examined his x-rays. He had small bony spurs throughout the ankle.

bone actually shift forward under this pressure. It is shifting out of the ankle joint. Doctors call this a *positive ankle drawer sign*. This is a good way to determine whether the front ankle ligament (anterior talofibular ligament) has been damaged. This is the ligament which is designed to keep such motion in check. It is very much similar to the anterior drawer sign of the knee.

Next, I put an inversion stress on the foot and ankle. I push the ankle underneath the lower leg bone in an inward direction. I hold the lower leg bone tightly while I am doing this. Once again, I can feel an excessive amount of shifting of the ankle bone with this stress. The normal ankle is always compared with the sprained one since some people are just loose-jointed and I could misinterpret this as instability. Do stress tests on both sides.

Special x-rays, called *stress views* of the ankle, should always be taken to confirm the diagnosis. Using an x-ray machine, I repeat the anterior drawer and the inversion stress test. In this way, if the ankle bone shifts out of the ankle joint, I can document it by x-rays. The good ankle is tested in a similar fashion and x-rayed. If the ankle bone pulls forward more than half a centimeter, and the drawer test or tips are out of the ankle joint more than ten degrees in the inversion stress test, then lateral ligament instability of the ankle is proven. Do stress x-rays on both sides.

How do I treat unstable ankles? Tape. Your ankle must be taped. This includes practices, games, and exercise. Simply wrapping your ankle with an Ace bandage or using an elastic ankle support will not do. These devices are useful for keeping the swelling out of the ankle, but they are not strong enough to support an ankle with ligamentous instability.

You must use white athletic tape applied over prewrap. A good tape job is often enough to tip the balance in your favor and prevent the ankle giving way.

Secondly, I recommend high-top athletic shoes for people with such ankles. This gives good external support and adds stability. I start you on exercises to strengthen the muscles on the outer aspect of the calf that help control the in and out (inversion and eversion) motion of the foot and ankle. These are called the *peroneal muscles* and are the second line of defense against sprains. Sometimes, tremendous strength in the peroneal muscle group is sufficient to overcome laxity of the ankle ligaments.

In 60 percent of my patients with unstable ankles, any one treatment—taping the ankles, high-top shoes, or muscle exercises—controls the problem. They can fully participate in sports and are not apprehensive about their ankles in competition. In 40 percent of the ankle instability cases, these measures do not permit normal participation in sports. It is decision time: Have an ankle operation or quit sports.

The operation I favor is the Watson-Jones Procedure. This is an elegant operation. The principle of the surgery is to substi-

Bone spurs are commonplace in athletes who have numerous ankle sprains. "Don't worry about arthritis," I said. "Your x-rays look like those of hundreds of other athletes who have had their ankle sprained in athletic competition. You have at least 10,000 more miles to go on these ankles."

"Whew," said John. "What a great relief, Doc. I thought my career was over."

He had not fractured his ankle. But by this time, his ankle had swelled, despite the icing. An egg-sized welt had developed over the outer side of the ankle, but the tenderness spread to the front of the ankle. This is a frequent pattern of ankle sprain for baseball players.

As we chatted, John revealed that his left ankle had suddenly given way a few times when walking. Medically, this indicates a ligament instability of the ankle.

It has always interested me how top-flight pro athletes, such as John Castino, can play in the professional ranks with ankle instability. But, a little athletic tape neutralizes the problem. If his ankle were giving way despite the tape support, I am sure that he would be a candidate for surgery to recreate his ankle ligaments.

tute a tendon for the stretched out and incompetent ligaments. The tendon that I choose is the peroneus brevis tendon. This is one of the peroneal tendons that lies behind the outer knob of the ankle. The operation involves detaching the tendon from its muscle belly. I drill a hole in the outer knob (the fibula). The tendon, which is attached to the outer aspect of the foot, is now brought through the drill hole. This portion of the tendon that runs from the foot to the outer ankle knob recreates the lower, outside ankle ligament (the calcaneal fibular ligament). I then bring the tendon through the ligaments that run between the ankle bone and the heel bone. This recreates the front outside ankle ligament (the anterior talofibular ligament). The patient remains in the hospital for approximately three days and then is put into a short-leg walking cast for six weeks. This allows the tendon to heal properly and gain sufficient strength to act in its new capacity. It takes another six weeks after the cast is removed to regain motion in the ankle and strength in the muscles. Your peroneal muscle does not lose strength because of the substitution. I attach the peroneus brevis muscle belly to its neighbor, the peroneus longus muscle.

After this operation, you must tape your ankles for one year. This allows complete biologic maturation of the surgical reconstruction. After the twelve months, you may play noncontact sports without any tape support.

I want to put ankle sprains in proper perspective. For every one hundred people who sprain an ankle, only twenty will have a two-ligament injury. Of these twenty, only five will subsequently develop lateral ligament instability. Three of those five will have their symptoms controlled by peroneal muscle strengthening, tape support, and high-top athletic shoes. The other two will require surgery. Therefore, only two out of every hundred people who sprain an ankle will ever face the option of surgery.

ANKLE BALANCE TEST

"One way I determine whether a player is ready to go back into the game after sustaining a sprained ankle is a balancing-type test," says Ray Melchiorre, NBA trainer. "I ask the player to stand with the injured ankle on a brick or block, something that is only two to three inches off the ground. If they can balance without falling off, I know the ankle is fully strengthened. If they fall off, its back to strengthening exercises.

"Often times, I will tape the ankle; that's just enough to stabilize the ankle."

Ankle Fractures

CAUSES

An *ankle fracture* occurs when the ankle knobs break. The reason ankles break frequently in sports is that the foot sometimes acts as a crowbar or lever to break the ankle joint. Picture the foot as a crowbar with its tip in the ankle joint. Any force that is applied to the end of the foot multiplies at least three times by a leveraging action.

Here are some examples, from my files, of how ankles break:

- A high school soccer player misses the ball with the inner side of his foot. Instead, his foot hits the ground, bringing it to a screeching halt, but the momentum carries it forward to the ankle joint. The foot and the

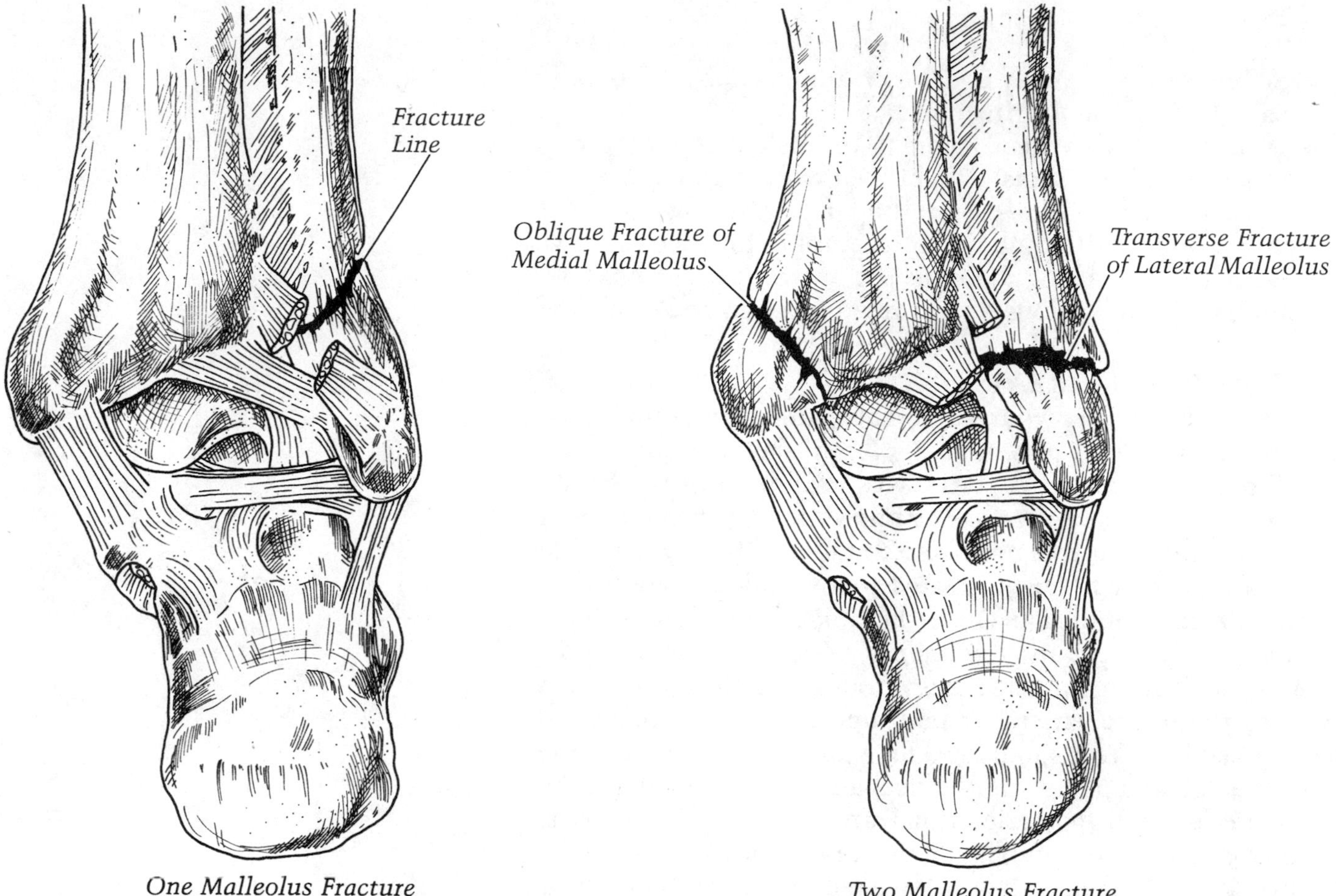

One Malleolus Fracture

Two Malleolus Fracture

Ankle Fractures

ankle bone are heading in opposite directions. The force is great enough for the ankle knobs to fracture. The player is out for the season.

- A baseball player hook-slides into second base. His foot stops on the base, but momentum carries his body forward. Again, the foot acts as a lever to break the ankle knobs.
- A female gymnast vaults over a horse. She lands awkwardly on the outer border of her right foot. The foot quickly snaps under her. The inner ankle knob breaks through as the foot and ankle bone slam against it. Surgery is necessary to repair the damage.

The ankle joint can fracture in three ways:

- A break of the outer ankle knob, called a *single malleolar* fracture of the fibula. Because the outer ankle knob is the first to come in contact with the crowbar effect of a fall, it is the most common fracture of the ankle bone. Often, ripping of the deltoid ligament on the inner aspect of the ankle accompanies this fracture.

- A break of the outer and inner ankle knobs. This is called a *bimalleolar* fracture.
- A fracture with displacement or dislocation of the ankle bone (the talus). When tremendous force is applied to the ankle—e.g., falling from a height—the ankle bone will come out of the ankle mortice along with the fracturing. This is the worst type of ankle injury. Combinations of all three types can be seen.

DIAGNOSIS AND TREATMENT

When you fracture your ankle the pain is awesome. One patient, a basketball player, said, "I can't believe how much this hurts." The swelling starts within an hour of the break. You will not be able to walk. Putting weight on the ankle will be too painful. You will need help to move. Often, your ankle is grossly distorted. I have treated many athletes whose feet were pointing southeast. Once the ankle knobs are broken, the foot can point in any direction.

Upon arrival in an emergency room, a fieldhouse, or a training room, the patient's blood supply into the foot should be checked first. You should feel for a pulse from the top of the foot (the dorsalis pedis artery pulse) and the pulse behind the inner ankle knob (posterior tibial artery pulse). Sometimes, with gross distortion of the foot and ankle, the arteries will be slightly kinked because they are twisted. In these situations, it is very important to get the ankle and foot properly aligned as soon as possible.

If you even suspect a fracture, a trained medical person should apply a supportive splint to your ankle. I prefer a pillow splint or an inflatable air splint. A wooden splint is not comfortable. The splint reduces the amount of wiggling of the broken bone ends—it limits soft-tissue damage.

An ice bag should be strapped to the ankle immediately with an Ace bandage. Elevate the leg.

X-rays are the main diagnostic tool. They show the fracture lines in the bones as well as the degree of dislocation of the ankle bone. After examining the x-rays, I plan the treatment. If the bone ends protrude through the skin (an open fracture), you must be taken to the operating room for proper cleansing of the wound and straightening of the fracture. (See page 90.)

If the skin has not been violated by the bone ends, then several decisions must be made. First, I must realign your bones back to their normal architecture. This is called reducing the fracture.

My second concern is maintaining that reduction. Sometimes, this can be done by applying a cast. Other times, surgery is necessary to keep the bones in their proper position. In surgery, screws, pins, and other metal

TOE RAISES

Here is a set of exercises to strengthen your calf muscles. The calf muscles allow you to rise up on your toes. Try it. Rise up on your toes and reach down and feel your calf. You can see the muscle bulge out as it contracts. Feel how firm it is compared to feeling your calf when you are sitting down. When you rise up on your toes, you are lifting up your body weight.

1. Stand arm's length from a wall so you won't lose your balance. This allows you to steady yourself if necessary.
2. Rise up on the toes of both feet. Take two seconds to get the top. Stay up for two seconds. Count it out, 1000-1, 1000-2 . . . then go back down.
3. Repeat the toe raise thirty times. Your calves should feel tight when you finish.
4. When you can do thirty toe raises with ease, add some weight to your body. Hold a heavy book in each hand. Hold a five- to ten-pound weight in each hand.
5. To get a harder workout, use a weight-lifting bar. Put a twenty-five-pound weight on each end of the bar. Now put the bar across your shoulders behind your neck. Do the toe raises as above.

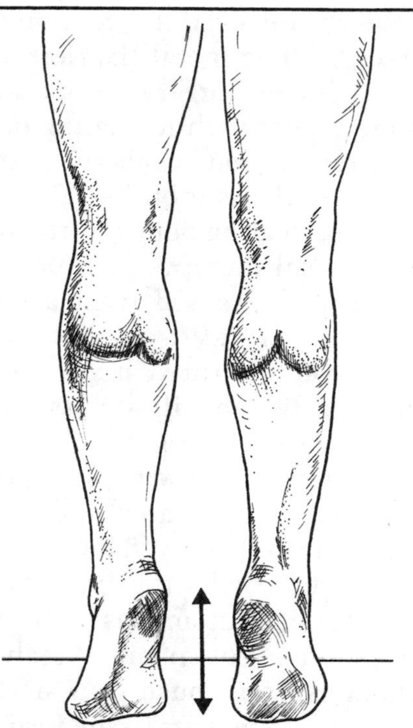

RUBBER TUBING EXERCISES

Here are exercises to strengthen the muscles on the back of your lower leg. These allow you to bring your foot down away from your shin.

1. Take the inner tube from a bike tire, or cut a circular strip from a car tire inner tube. Sit in a chair. Hold one end in your hand. Place the opposite end of it under the bottom of your foot.
2. Pull up with your hand until the big rubber band is tight against your foot. Now, move your foot slowly down. Hold the down position for three full seconds. Let your foot come back up. Do this thirty times. Take a five-minute break. Repeat three times. Do twice daily.
3. As your muscles get stronger, you need more resistance. Simply pull harder with your hand to make the rubber tighter. This will give the muscles a harder workout.

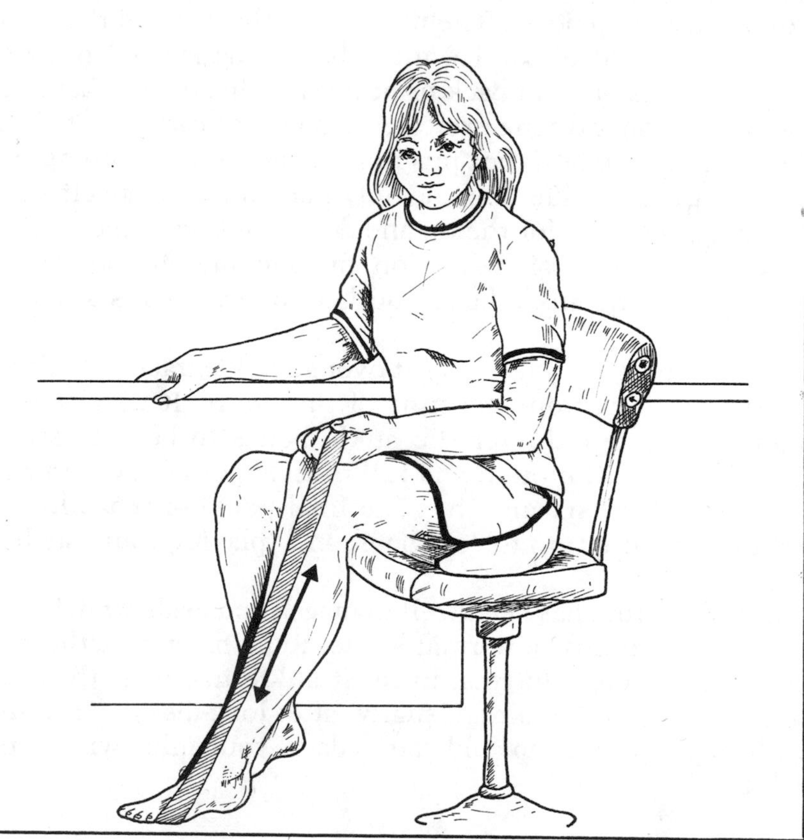

devices are used to guarantee that the bones will remain straight. The metal fixation devices implanted in ankle fractures during surgery do not speed healing; they just keep the bone straight while healing occurs.

After the bone heals the cast is removed. Often, you are in a cast for three months. The ankle bone is one of the five slowest healing bones in the body.

Physical therapy is a must. If you took a normal, uninjured ankle and encased it in a cast for three months, it would become very stiff. So, you can imagine what has happened with your fractured ankle. I am convinced that people cannot effectively rehabilitate fractured ankles on their own. You need professional help.

Therapy starts with whirlpool baths. The foot and ankle are buoyant in the water, and this makes it easier to move your ankle up and down. The residual swelling and stiffness can be massaged out by the trainer or therapist. Elastic support—an Ace bandage or an elastic ankle wrap—are important in keeping fluid out of an ankle fresh from a cast. Finally, a trainer or therapist must push your ankle up and down gently to stretch adhesions and scar tissue formed during immobilization. I like reverse stair climbing as an additional method of regaining ankle motion. (See page 316.) Until you regain full motion, it is impossible to run at full speed.

After the motion begins to return to the ankle, you can begin strengthening exercises. There are no muscles around the ankle, only the tendons of the muscles from the calf that run by the ankle joint. The muscles of the calf and those that lie on the front of the leg must be strengthened. Toe raises with weights on your back or heavy weights in your hands are helpful. Rubber tubing exercises can also be carried out. Finally, an orthotron or Cybex machine is unexcelled at strengthening a damaged ankle. It is the only machine that can effectively offer resistance for the up and down motion of the ankle as well as the inversion/eversion (in and out) motion at the junction of the ankle bone and heel bone. This is a distinct advantage.

How long does it take to get over an ankle fracture? Nine to twelve months from the time of the injury. It usually takes about three months for the ankle bones to become strong. Recovery of full motion and full muscle power takes an additional three to six months. The final phase of rehabilitation involves your return to your sport and "playing your way back into shape."

If the fracture has been well managed and heals straight, you can expect almost a normal ankle. Remember that the body never completely forgets. In most ankle fractures, the break goes right into the joint. Usually, after long-distance running, or heavy activities on cold, damp days, your ankle will protest a bit.

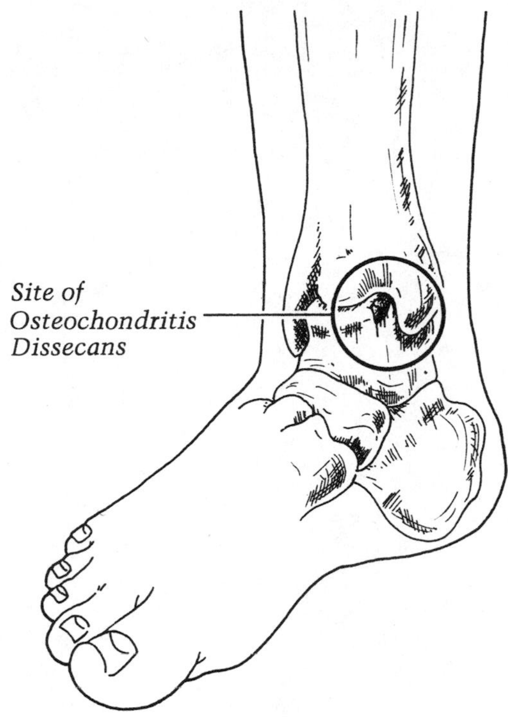

Osteochondritis of the Ankle

Osteochondritis Dissecans of the Ankle

CAUSES

When you walk, the weight of your body is distributed over the top surface of the ankle bone (the talus). When you play sports, that weight is dramatically increased. For example, when you rebound in basketball, the ankle bone comes under three times as much stress as walking generates.

The real problem arises when you land heavily with your foot turned in underneath you. This jams the inner part of the ankle bone against the lower leg bone. A tremendous amount of force is concentrated over a small area. Such repetition can damage the surface of the ankle bone. It is as if the ankle bone has been hit with a golf club and a divot has been taken out of the top of it. In medical terms, this is called *osteochondritis dissecans*.

Why some athletes develop osteochondritis dissecans and others don't is a puzzle to medical scientists. In my experience, the condition occurs in those sports where the ankle is used hardest. Ballet dancers develop this problem in a different location from running athletes because they use their ankles in the "on-toe" position. The osteochondritis dissecans produces pain when the divot-like piece becomes loose in its crater.

DIAGNOSIS AND TREATMENT

All the athletes with dissecans tell the same story. Pain in their ankle develops over a three- to six-month period. In use, the ankle hurts. Because swelling develops in the joint, the pain may intensify after a workout, and the ankle gets stiff. This is not a matter of instability of the ankle joint, so taping the ankle does not help. Only rest seems to relieve the symptoms.

My examination of the ankle does not give me much of a clue. The ankle generally has close to normal motion, and because the ligaments are unaffected it is not loose or unstable. It is impossible to push directly on top of the ankle bone through the skin to find a tender area.

The diagnosis for osteochondritis dissecans is totally dependent on an x-ray evaluation. Plain x-rays sometimes show the loose bone fragment. It is seen lying in a shallow crater. Because the loose piece of bone has lost its blood supply, it becomes more dense on the x-ray than the remainder of the ankle bone and it is easily identified. If I have any question, I always do tomograms of the ankle. This is an x-ray technique where the x-ray machine and the x-ray film are moved simultaneously as the x-ray is taken, which produces the effect of cutting through the bone.

I can see the front of the bone, the middle of the bone, and the back of the bone all separately. In the midst of x-raying the ankle bone, I can clearly see the osteochondritis dissecans fragment.

If the fragment causes constant pain, I have never known

conservative treatment to bring relief of the symptoms. Putting you into a cast or restricting the use of your ankle does not work. Biologically, the fragment does not have the ability to heal itself. The only answer is surgery.

The surgery starts with an incision over the front inside part of the ankle. The ankle joint is just beneath the skin in this area and the joint itself is opened. By putting the foot in a downward position, the ankle bone tips forward in the ankle joint. In this position, most of the ankle bone with its loose fragment can be easily seen. I carve out the dissecans. It is like taking a worm out of an apple. This leaves just the crater. With a small drill, I make many holes at the base of the crater in the main substance of the ankle bone. Blood flows through the holes, creating a new blood supply.

Eventually, a blood clot forms in the crater which, over time, hardens into firm scar tissue. In this instance, the hardened tissue acts as an excellent substitute for normal joint surface material.

After the surgery, I make a cast which I divide in half. This is called a bivalved cast. It allows you to remove the cast three times a day for motion exercises of the ankle, and while it is on, it assures that the ankle will be rested.

I do not allow you to put weight on your ankle for one month. That means crutches. Six weeks after the surgery, strengthening exercises are begun for the calf muscles and the muscles on the front of the leg. Three months after surgery, if you have a full range of motion of your ankle, you can begin running and exercising. Four to six months after surgery, full athletic activity is allowed. I have been very pleased with the results of this type of surgery. All of my patients have returned to their former athletic pursuits. Occasional pain and stiffness on damp, cold days or after extremely heavy use have been the only complaints.

Achilles Tendinitis

CAUSES

The *Achilles tendon* is the largest tendon in the body. It is the connecting fibrous link between the calf muscles—the gastrocnemius and soleus—and the heel bone. Think of the Achilles tendon as a large cable that transfers the power of the calf muscles to the heel bone. Without this tendon, you could not rise up on your toes or run; you could not walk up and down stairs.

As with any body tendon, the Achilles can become inflamed. This inflammation, sometimes chronic and painful, is called *Achilles tendinitis*. I call it the "tennis elbow" of the lower leg. Achilles tendinitis is common in athletes who participate in running sports—basketball, soccer, football, baseball, squash, and track.

The inflammation is caused by trauma to the tendon. The

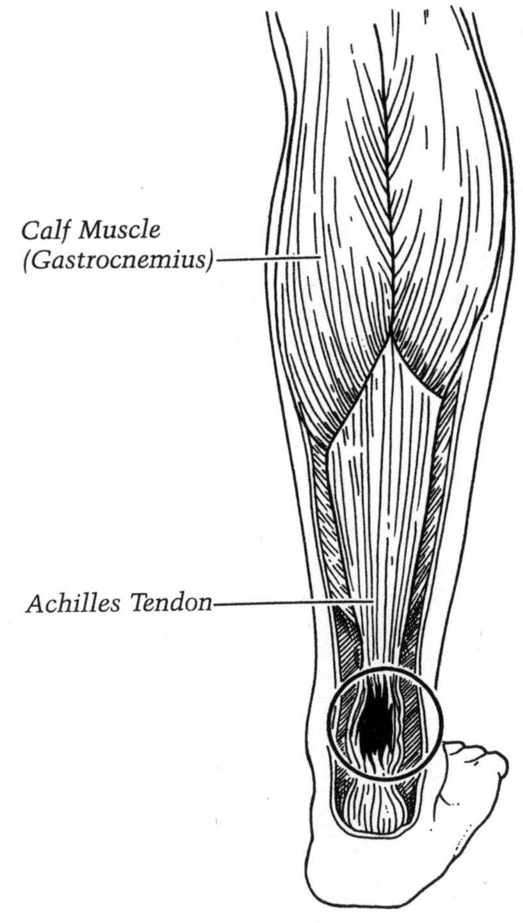

Achilles Tendinitis

THE MILITARY BATTLE AGAINST ACHILLES TENDINITIS

When I was stationed at Fort Dix, New Jersey, in 1968, we had an average of 600 new cases of Achilles tendinitis every basic training cycle. It was never clear to me why this malady was so common. Other basic training centers across the country had experienced the same kind of problem.

There are two possible explanations. One lies with the army boot itself. The few times that I wore army boots, I found that the heel area was much too large. My heels were swimming in my boots. I asked the supply sergeant for a new pair of boots.

"They all fit the same way," he said. "You are supposed to wear two or three pairs of socks with the boots." I took the supply sergeant's advice. The socks filled up the front of the boot, but the heel portion was still much too loose.

Some soldiers developed creases in the sides of their boots which pushed against heel counters, which, in turn, pushed against the Achilles tendon.

A second explanation of the tendinitis outbreak could be traced to problems of marching and running in the sandy soil of Fort Dix. I tried marching in the training areas. I found that my heels sank into the sandy soil, stretching my calves and Achilles tendons. After a five-mile march in boots, my Achilles tendons felt uncomfortable. I could not do much about the situation. Certainly, the drill sergeants did not have much time or interest in getting the recruits to stretch their calf muscles or Achilles tendons prior to training. I recommended putting the soldiers in sneakers, but my senior officer told me that "sneakers weren't military."

Ironically, one colleague who recently finished a military tour told me that his base was now using running shoes in the early stages of basic combat training to avoid the Achilles tendinitis problem.

Most cases at Fort Dix responded to five to seven days of sick call, ice, and aspirin. The more resistant cases were placed on oral anti-inflammatory medicine. Nothing was done about the boot problem.

trauma happens when the calf muscles contract hard or suddenly. Because the tendon has very little elasticity, it does not contract. The result of the calf shortening is tiny tears in the Achilles tendon. The tears occur when the muscles contract suddenly in sports requiring quick bursts of speed, or they can happen gradually from the wear and tear of endurance running. Physicians call this *micro-trauma*.

Lack of flexibility of the muscle-tendon unit is an equally important factor that leads to Achilles tendinitis. When the calf muscles are shortened by hard running or exercise, the tension of the muscle-tendon complex is increased. The greater the tension, the more likely you are to get tendinitis. Here are other factors which contribute to tendinitis.

- *Heel strike*. If your heel turns suddenly inward or outward when you land, you put more stress on the muscle-tendon unit. The poor positioning of your heel bone is due to its shape, something you inherit from your parents.
- *The shoes you run in*. You must have athletic shoes with a firm heel counter and a sole design which absorbs shock. A good heel counter is essential in providing stability of the heel bone and the Achilles tendon complex. You can tell if the heel counter is substantial by squeezing it together. A soft heel portion of the sole absorbs some shock during the heel strike phase of running and lessens the amount of force passed into the irritated tendon.
- *The surface you run on*. Run on a soft surface: Grass is the best, but a composition track is good. Both have a little give. Avoid running on concrete or asphalt.

Achilles tendinitis is unique in that the pain is more intense when you first rise in the morning and lessens as you use the tendon. When you start a workout your tendon will be extremely painful, but as you continue the pain abates.

I find that if the pain decreases with exercise, athletes do not stop exercising. They go back to performing the same exercise that caused the problem in the first place.

For the sake of simplicity, I divide Achilles tendinitis into acute and chronic cases. If you have had tendinitis for less than six weeks, I consider it *acute* tendinitis. If the inflammation continues beyond six weeks, I call it *chronic* tendinitis. The reason for the distinction is biologic. All true tendons—including the Achilles tendon—occupy a sheath or tube. The lining of this tube is composed of cells which produce lubricating fluid to allow for smooth gliding of the tendon within the tube. This lining, called *synovium*, is the same material that lines all joints in your body. In acute tendinitis, it becomes inflamed and produces extra fluid, which makes the tube tight, limiting the tendon's function. The inflammation and the extra fluid produce the pain of acute tendinitis.

If the process continues for many weeks, in addition to the

extra fluid, adhesions develop between the tube and the tendon. These inhibit the normal motion of the tendon. Furthermore, these adhesions bring new small blood vessels into the tendon which normally do not belong there. Once these adhesions have formed, you have permanently altered the anatomy of the tendon-tube unit.

DIAGNOSIS AND TREATMENT

Achilles tendinitis comes on in an insidious fashion. You will notice a dull, draggy sensation after running or exercising. This usually intensifies over a two-week period. At first, you can run or exercise through the pain and complete your athletic activities. But, you pay the price six to twelve hours later when the pain intensifies. Almost always, your tendon area will be stiff in the morning. This tenderness is usually about one and one half inches above the tendon's attachment into the heel bone. Finally, the pain becomes too much. At this point, most athletes and fitness buffs check with a doctor.

On physical examination, I look for some puffiness or swelling around the inflamed tendon. The tenderness is usually about one and one half inches above the tendon's attachment to the heel bone. I feel along the tendon, starting at the heel bone. As I move up the tendon, I can usually isolate a particularly tender area which is approximately one inch in length.

I ask you to move your ankle and foot up and down. In severe cases of tendinitis, I can feel a grinding with my fingers in the area of the tendinitis. We call this crepitus. There is never any tenderness in the muscle belly of the calf. The ankle joint is normal.

The best treatment is to stop running or exercising immediately. Let the tendinitis cool down for one week. During that week, ice the tendon for twenty minutes twice daily, and take six to eight aspirin per day with meals. If the aspirin is not effective, then I prescribe an oral anti-inflammatory medicine.

I do not usually inject cortisone because I am concerned that some of the medication will go into the substance of the tendon and weaken it. If it is weakened, then it could rupture. From a surgeon's point of view, it is very hard to know if the needle tip is sitting between the sheath and the tendon or is in the tendon itself.

After the seventh day, start Achilles tendon stretching exercises twice a day. These should be done for ten to fifteen minutes before any athletic activity and again before bedtime. Tendinitis leaves the muscle-tendon unit short, tight, and less flexible. The stretching exercises keep the muscle-tendon unit at its normal length and prevent adhesions from forming.

In my experience, stretching is the only successful long-term treatment for acute Achilles tendinitis. Even after the pain quiets down, I recommend that patients with tendinitis always stretch before exercise.

As the tendon responds to treatment, you can return to

CARL YASTRZEMSKI'S ACHILLES TENDINITIS

In 1979, Carl Yastrzemski became the first American League player to get both 400 home runs and 3000 hits. Halfway through his nineteenth season, his Achilles tendons started to bother him. At the time, Yaz was batting .306 with 16 home runs and 53 runs batted in. The remainder of the season, he hit only 5 more home runs and finished the season with a .270 average.

"The Achilles tendinitis started after Yaz played two consecutive games on two different surfaces," recalls Dr. Arthur Pappas, medical director of the Boston Red Sox. "One baseball field was very soft. The other was a very firm artificial surface.

"Dr. Southmayd, trainer Charlie Moss, and I all examined Yaz's Achilles tendons. We decided to rest him, prescribed anti-inflammatory medication, and supported the Achilles tendon with tape in various ways. But whatever we did, we couldn't control Yaz's discomfort. I was concerned that his heel cord might rupture. Pitcher Ferguson Jenkins had presented a similar history of Achilles pain before the tendon ruptured completely and required surgery.

"Yaz's problem stemmed from his foot mechanics. The ligaments in the foot were very loose. Therefore the foot had a tendency to pronate excessively or flatten in the heel and arch area. This flattening put excessive stretch on the Achilles tendon."

In the off season Yaz had a special baseball shoe designed to overcome the problem mechanically. In 1980, he wasn't bothered by the problem. His doctors can't say whether the rest or the new shoes was the main reason for alleviation of the problem.

"One thing I learned from this injury is that a power hitter needs his heel cords," says Dr. Pappas. "It is an important part of a batter's balance and power. Without it, Yaz lost the power below his knees. Consequently, his homer production and batting average tumbled."

WALL PUSH-UPS

Standing at least four feet away, face a wall. Place your palms on the wall, keeping your back straight. Bend your elbows so your upper body will move closer to the wall. If you keep your heels on the ground, the calves and Achilles tendons will be stretched. Hold this position to the count of ten. Release the tension on your Achilles tendon by straightening your elbows and pushing your body away from the wall. Then repeat at least ten times.

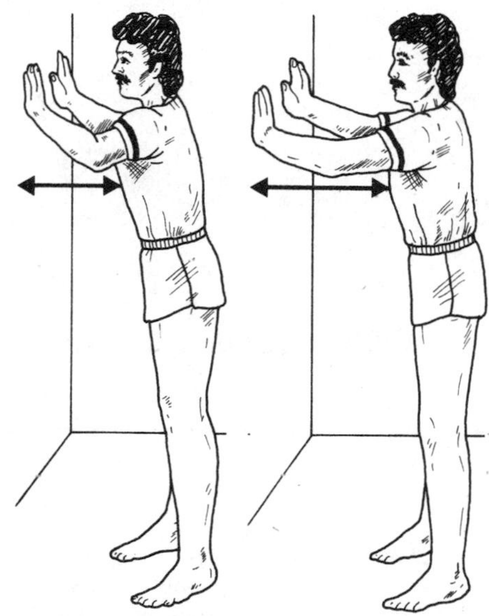

BOARD STANDS

Place a board at least one foot long by one foot wide on the floor, with one end against the wall. Put any solid material two inches thick under the part of the board that is not against the wall. With your back and heels touching the wall, stand on the board for ten to twenty minutes. To avoid boredom, read a book or watch television.

After a week or two, your tendon will be stretched out and you can raise the height to four inches off the floor so the board is inclined even further. As your tendon becomes more flexible, you will be able to put a book under the stud and progressively increase the incline.

Because you can lose your flexibility within a short time, you should stretch daily.

limited action. It is often helpful to tape the tendon and heel area for the first week of action. The tape—called a "check rein"—runs from the mid-portion of the back of the calf down under the heel itself. Its purpose is to restrict motion of the ankle so the tendon cannot be stretched. In addition, I add three-eighths of an inch of padding to elevate the heel, which takes a lot of strain off the tendon. If you do not have a trainer or doctor who can apply a tape support, a plastic heel cup often does the job. The cup seems to stabilize the heel and absorb some shock during running. It lessens the amount of force passed into the irritated tendon and therefore helps prevent reinflammation.

I recommend brands of athletic shoes which have soft heel padding and strong heel counters. I also ask you to start running or exercising on a cinder track or grass football field. These surfaces help cushion your heel strike.

If you have recurrent attacks of Achilles tendinitis, even after stretching, I prescribe an orthotic support for your foot. If your heel is not perpendicular to the ground when you are standing still, this device is very effective.

If your heel bone turns inward, then the support is designed to build up the outer aspect of the shoe. If the heel bone tilts outward, then a buildup on the inner aspect of the shoe helps. Your sports medicine specialist should be the one to determine what kind of a foot support you need.

What about chronic tendinitis? It sets in when acute tendinitis does not respond to treatment. These are the cases that worry me most. Soon, it is not a question of being limited from sports activities, but rather a question of severe pain and disability from daily activities. These people limp into my office and can hardly walk. Climbing stairs is agony.

I place the whole foot and tendon in a walking cast. I want to let the unit rest from the tendinitis. That usually takes one month. After I remove the cast, I start you on a limited range-of-motion exercises. Within two weeks, you can start doing Achilles tendon stretching exercises. This seems to be effective in stopping the tendinitis in 80 percent of cases. In the remaining 20 percent, surgery is an alternative. But only one in every one hundred athletes who has Achilles tendinitis resorts to surgery.

The operation, which I have had to do only ten times in my career, involves removing the tube and the adhesions that have formed between the tube and the tendon. This is done through a skin incision on the side of the tendon. A new tendon tube grows back without inflammation.

It gives the tendon a whole new start. I have never had to repeat the surgery. After surgery, I do not use a cast, but rather give you range-of-motion exercises. Within three months, you can return to full sports activities.

Achilles Tendon Rupture

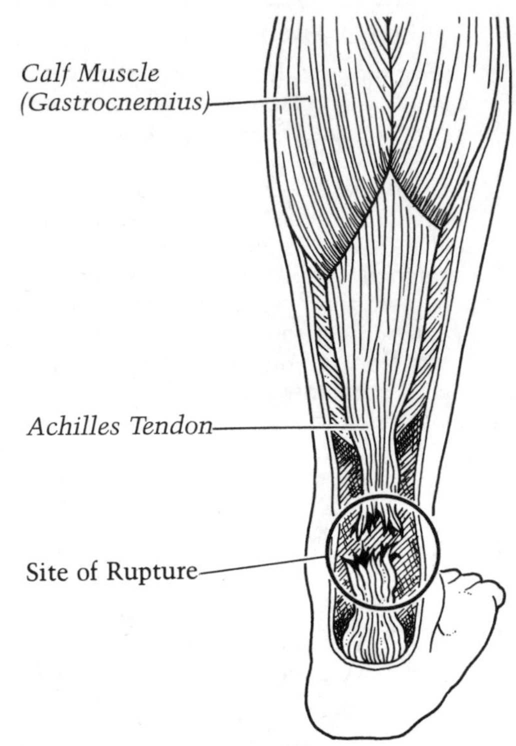

Achilles Tendon Rupture

CAUSES

A rupture of the Achilles tendon is not a common event. Tendinitis is fifty times more common. Experimentally, it has been shown that a normal Achilles tendon can withstand a force of 1000 pounds without tearing. Despite its strength, it is the most commonly ruptured tendon in the body.

The weekend athlete, usually over thirty, is the person who most often tears his Achilles tendon today. Eighty percent of those injured are male. A sudden movement such as running out for a pass in football or going for a tennis serve is the most common trigger of this injury.

In my opinion, the Achilles tendon must be in a somewhat weakened condition in order to rupture. Medical studies confirm that. In people over thirty, the blood supply to the portion of the tendon that most often rips is less than normal. Without the normal amount of blood, the tendon suffers a nutritional deficiency, resulting in weakness. The average thirty-year-old athlete or fitness enthusiast still has strong muscles attached to his tendons, and these are capable of producing enough power to cause the tendon to rip. Since most right-handed

TONY'S ACHILLES TENDON RUPTURE

Tony Streeter, a thirty-four-year-old banker, was playing in a pick-up basketball game at the local high school in Winchester, Massachusetts. Halfway through the game, he came down with a rebound and began heading up court. Suddenly, he felt a sharp pain in the back of his right heel.

"I thought that I had been shot," he explained to me. "I heard a loud pop which sounded like a pistol and I was suddenly unable to run. I limped off the court and had a sinking feeling in the pit of my stomach that my basketball playing days were over."

I examined Tony in the emergency room. I started feeling down at the heel bone level. I noticed that the contour of one Achilles tendon was far different on the injured side than the opposite side. As I followed the tendon up from the heel bone, I could feel that the continuity of the tendon had been disrupted. There was a gap between the torn ends. I could also feel a blood clot forming between the ends.

I had him kneel on a chair with his feet hanging down toward the floor. I squeezed the calf on the left leg and his foot quickly responded in a downward direction. I squeezed the right calf side to side and the foot did not move at all. The Thompson test was positive. The heel cord had definitely been ruptured.

I felt surgical repair of the tendon was in Tony's best interest. I explained that only surgery would guarantee that the tendon would be the right length in the future. I explained that if a muscle-tendon unit does not function at its normal length, it cannot develop normal power. Any lengthening of the tendon reduces the strength of the muscle unit attached to that tendon. I told him that it would be four to six months before he would return to full sports participation. He agreed, and I did the surgery that night.

Following surgery, I put Tony in a long leg cast for one month. This put his knee and ankle in a relaxed position and allowed the tendon to heal properly. One month after the surgery, I removed this long cast and put him into a short walking cast. In this cast, I positioned his ankle much closer to the right angle than it had been in the original cast. After a month in the long cast, the tendon had regained some strength and could tolerate some stretching.

Two months after surgery, I removed the short cast and put a quarter-inch elevation in the heel of his right shoe. I had him wear inserts for two months. He was also started in physical therapy to regain full motion in the ankle and strengthening of the calf muscles. In my opinion, complete rehabilitation after this injury cannot happen unless you go into physical therapy.

Three years after surgery, Tony plays all sports. He does not have any stiffness and soreness in his tendon. His calf has returned to normal circumference and he has no weakness in his legs.

people start running by pushing off the left foot, there is a higher incidence of rupture of the left tendon.

DIAGNOSIS AND TREATMENT

When your Achilles tendon ruptures, your ankle will suddenly go weak. You will feel like you've been shot. You will not be able to walk normally. Most people limp. Pain is not a prominent symptom immediately after the injury.

The one telltale sign of a ruptured tendon is the inability to rise up on your toes. Normally, I can feel the gap between the torn ends of the tendon, which is usually one and one-half inches above the heel bone.

I ask the athlete to kneel in a chair. I squeeze the calf together from the sides. If the foot moves downward, then your tendon is intact. If the foot does not move at all, your Achilles tendon is ruptured. We call this a positive Thompson test.

X-rays do not help; tendons do not show up on x-rays. Occasionally, you will have had several weeks of pain in the heel-cord area. This can be a warning signal. Pain that is not relieved with several days of rest should be reported to a physician. However, most often no pain precedes rupture of this tendon.

Once I make the diagnosis of a ruptured Achilles tendon, there are two courses of treatment. One choice involves application of a cast with the foot hanging down toward the floor. You wear the cast for six weeks, allowing the torn ends of the tendon to heal by natural scar formation. After removal of the cast, you wear a heel lift for two months, which helps to prevent stress from passing across the newly healed tendon. This method has several drawbacks. The tendon heals at a longer than normal length, which allows the foot to come up toward the shin at a much greater degree than before the injury. This reduces the power of the calf muscles, which in turn weakens push-off power. In addition, 20 percent of the tendons so treated will tear again when the patient returns to athletics. Also, the calf muscles shrink. Some people lose 20 percent of their muscle size and strength.

The second treatment involves surgery. The rupture causes severe damage. In the operation I use, I place stitches through the torn ends of the tendon. Sometimes, the plantaris tendon, lying alongside the torn heel cord, reinforces this repair. Following the surgery, a cast is applied above the knee and worn for six weeks. Then, an additional month in a short walking cast follows. This gives the patient the best chance to resume normal athletic activity. Surgery works extremely well for all patients except professional athletes. For example, restoring 95 percent of calf power to a professional football running back usually does not allow him to continue his career. However, for fitness buffs, this level assures a return to action.

How can this injury be prevented? Stretch! Stand on a slanted board for ten to fifteen minutes before exercising. An ounce of prevention is truly worth a pound of cure for this problem.

ACHILLES TENDON STRETCHING

"One of the simplest ways to stretch the Achilles tendons is to stand on a slantboard," notes Dr. Rob Roy McGregor. "But, many people do the stretch incorrectly. They don't invert their foot. Therefore, they don't get the full stretching of the board standing. The simplest way to invert the foot is to curl your arches. I do that by curling the toes downward. I prove this to an individual very simply. I have them stand in the appropriate position with their knees locked. I say, 'Let your foot go flat. Do you feel your Achilles tendon pulling?' No, they don't feel their Achilles tendon pulling at all."

Dislocating Peroneal Tendons

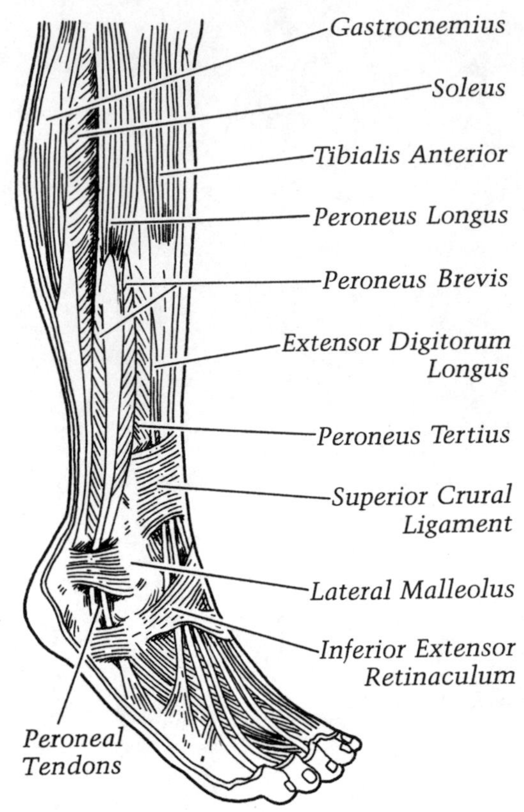

Peroneal Tendons

CAUSES

Feel your outer ankle knob. Directly behind the knob you will feel two distinct cord-like structures. These are the tendons of the *peroneal muscles*. The purpose of this muscle-tendon group is to evert or pull the foot up and out. The tendons attach to the foot bones. The tendons pass around and lie in a bony groove on the back of the outer ankle knob. The ankle knob acts as a pulley to give mechanical advantage.

The depth of this groove varies from person to person. In addition, a sheet of fascia material lies directly over the tendons to hold them on the outer ankle knob pulley.

The injury happens when the tendons slip out of the groove because of its shallowness. In medical parlance, this is called a *dislocating peroneal tendon*. Downhill snow skiiers—who use their peroneal muscles to keep their feet up and out as they dig their ski edges into the side of a mountain—are the main victims of this injury.

The injury occurs when you fall and your foot is suddenly driven under you. The tendons continue to work hard in trying to stabilize the foot, but in so doing they can actually jump the pulley. The force can be so great that the tendons rip through the covering fascia and come completely off the track. This is an acute injury and very dramatic. If not properly treated, the tendons continue to be unstable and can produce long-standing disability.

When the peroneal tendons slip off their track, you will know it. The tendons will slip around the corner of the outer ankle. You can see it. The jumping off the track is painful. It ruins your concentration because you never know when it is going to happen. They will snap back on their own. Mild swelling occurs around the outer ankle knob. The pain will diminish over the next half-hour.

The first dislocation often causes sufficient damage that the restraining fascia heals stretched out. The result: The fascia won't keep the tendons on the pulley. Repeated painful dislocations follow—and cause loss of confidence in ankle function.

DIAGNOSIS AND TREATMENT

I diagnose dislocating peroneal tendons by having you bring your toes up and out away from your ankle. This motion activates the peroneal muscles and tendons. If they are unstable, I can actually see them jump the track and swing around the corner of the outer ankle knob.

The acute injury is unmistakable. You will be unable to get back up and use your ankle and foot. When I see you immediately after the injury, the tendons are lying over the outer ankle knob. The tendons can be easily pushed back into place.

The best immediate treatment for an acute case is RICE. The swelling usually starts two to three hours after the injury.

Take six to eight aspirins a day with meals to alleviate the pain.

If it is the first dislocation, you should be placed in a short-leg walking cast for four to six weeks. This gives the best chance to achieve complete healing of the torn fascia. "Rest the injured part" is the old surgical adage. You must have rehabilitation after the cast removal to regain motion and strength in your ankle.

If the condition has existed for more than six months, it is hopeless to try conservative treatment. Surgery is the only answer. It is impossible to stabilize the tendon any other way. Some surgeons deepen the bony groove on the back of the ankle knob. I use an operation devised by Dr. Thomas B. Quigley of Boston. It is fairly straightforward in concept. The outer ankle knob is exposed through an incision directly over it. The bone covering it, called the periosteum, is elevated in a flap-like fashion. The stretched-out restraining fascia is then stitched back to its normal size. Surgically, I turn back the periosteal flap over the tendons. This assures a new strong restraining barrier. I install a short-leg walking cast for six weeks to allow the ankle and tendons to heal up.

Once you come out of the cast, you start on whirlpool baths and a range-of-motion exercises. I then cautiously start you on peroneal strengthening exercises. You can return to full sports activities three months after the injury.

MORE THAN 415,000 PEOPLE SUSTAIN ANKLE INJURIES EVERY YEAR

1979

Sport	Percent of Injuries Per Sport	Total Number of Injuries
Basketball	33.1	130,000
Baseball	13.1	60,100
Football	10.0	43,100
Bicycling	5.2	29,900
Volleyball	30.4	24,000
Motorbikes	7.3	17,400
Soccer	20.4	17,300
Kickball, stickball, tetherball, others	13.5	12,200
Roller skating	7.7	11,700
Gymnastics	13.9	7,200
Racquetball, paddleball and squash	18.7	7,000
Track and field	23.3	5,700
Snow skiing	9.8	5,000
Wrestling	7.4	4,700
Skateboarding	9.3	4,400
Playground equipment	3.8	3,800
Swimming	3.8	3,400
Hockey	7.3	3,400
Tennis	31.5	3,100
Ice skating	9.4	2,900
Sledding	5.1	2,000
Bowling	9.0	2,000

SOURCE: Consumer Product Safety Commission.

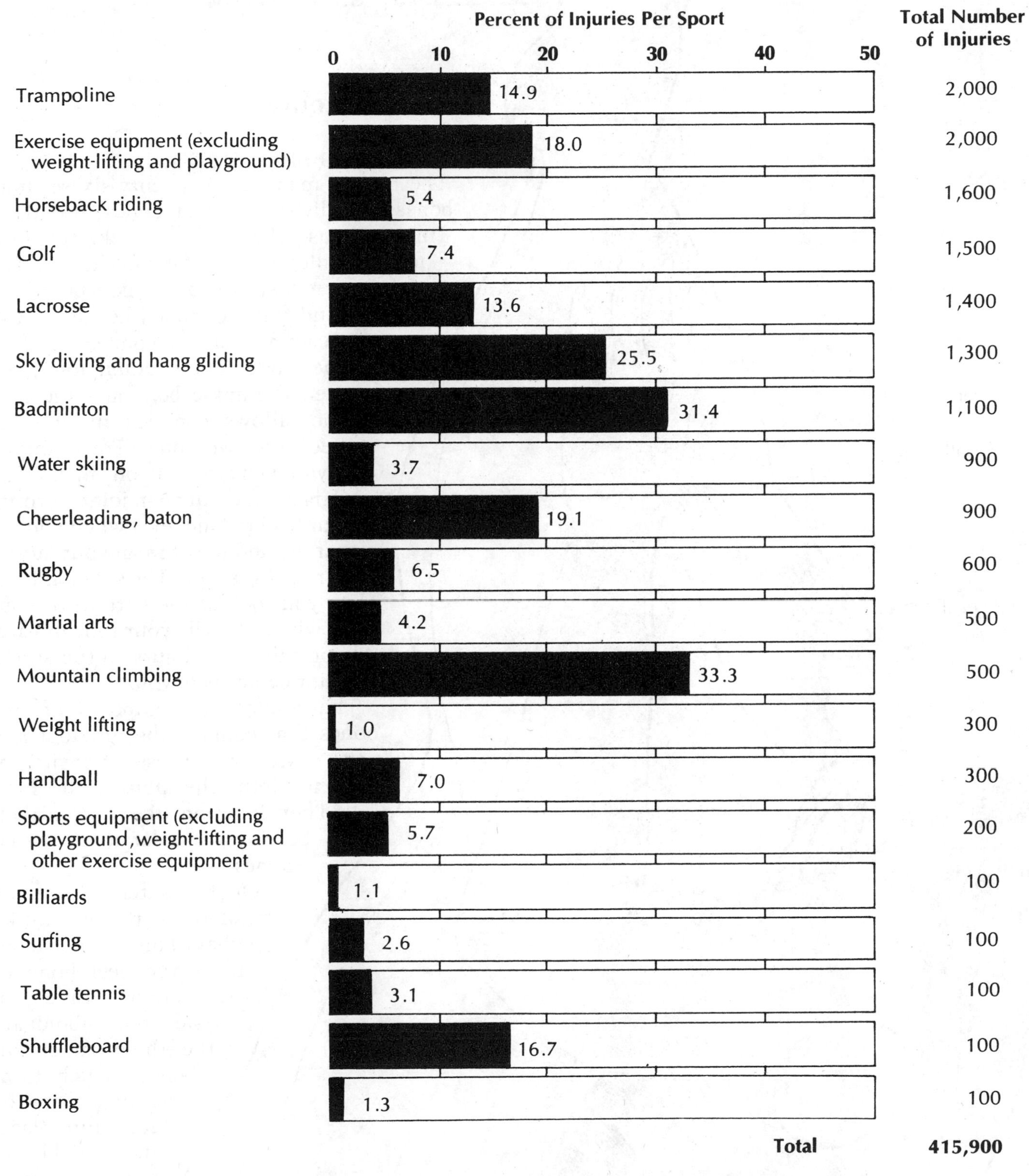

Note: These are injuries treated in emergency rooms only. At least twice as many ankle injuries are seen in doctors' offices.

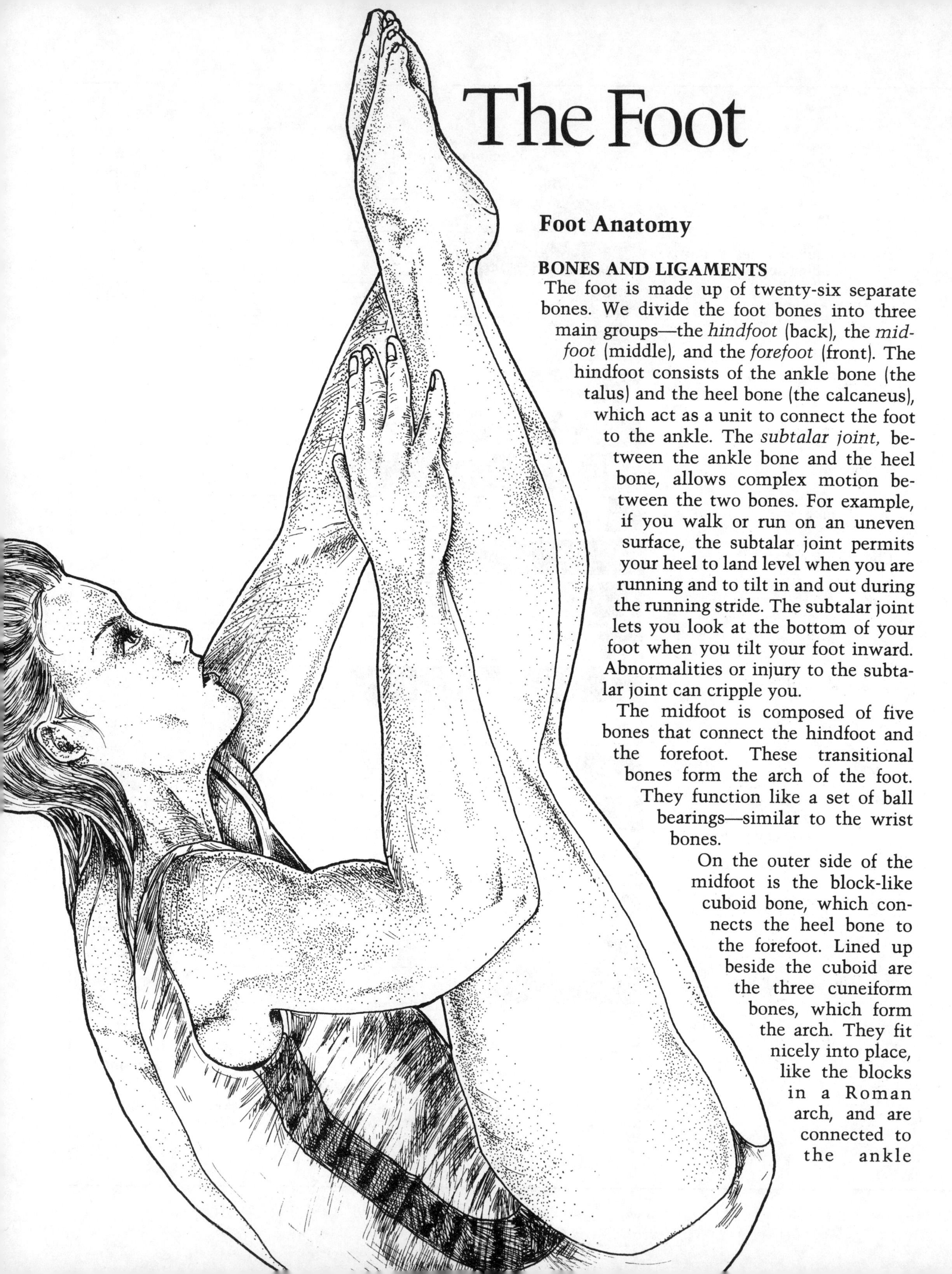

The Foot

Foot Anatomy

BONES AND LIGAMENTS

The foot is made up of twenty-six separate bones. We divide the foot bones into three main groups—the *hindfoot* (back), the *midfoot* (middle), and the *forefoot* (front). The hindfoot consists of the ankle bone (the talus) and the heel bone (the calcaneus), which act as a unit to connect the foot to the ankle. The *subtalar joint,* between the ankle bone and the heel bone, allows complex motion between the two bones. For example, if you walk or run on an uneven surface, the subtalar joint permits your heel to land level when you are running and to tilt in and out during the running stride. The subtalar joint lets you look at the bottom of your foot when you tilt your foot inward. Abnormalities or injury to the subtalar joint can cripple you.

The midfoot is composed of five bones that connect the hindfoot and the forefoot. These transitional bones form the arch of the foot. They function like a set of ball bearings—similar to the wrist bones.

On the outer side of the midfoot is the block-like cuboid bone, which connects the heel bone to the forefoot. Lined up beside the cuboid are the three cuneiform bones, which form the arch. They fit nicely into place, like the blocks in a Roman arch, and are connected to the ankle

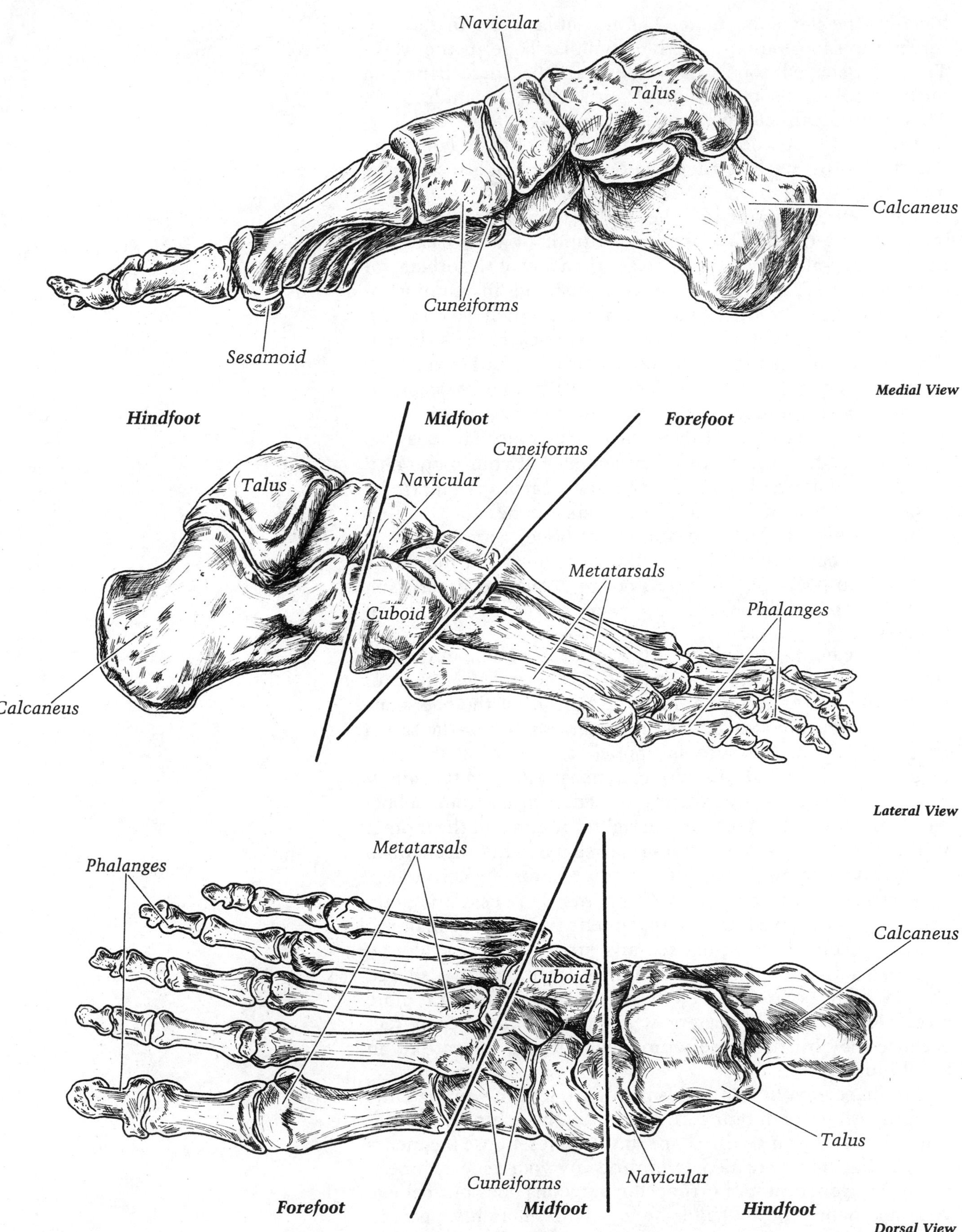

Bones of the Foot

bone by the navicular bone. The navicular bone in the foot serves the same purpose as the navicular bone in the wrist. This boat-shaped bone caps the end of the ankle bone and forms a ball-and-socket joint of sorts with the ankle bone. This anatomical structure greatly increases the foot flexibility. In the condition known as flatfoot, the midfoot bones are flattened. Most of the strong tendons of the foot attach to the midfoot.

The forefoot is formed by the five metatarsal bones and the toes. Each metatarsal attaches to the midfoot at its base. The first metatarsal is three times bigger than any of the others, for good reason. From this bone you push off in running or walking. The four other metatarsals are much less sturdy. Each becomes progressively shorter, just as each toe is shorter. The far ends of the metatarsal bones form the ball of the foot. In this area the difference in foot length primarily occurs. If you wear a size fourteen sneaker, your metatarsals are much longer than those of an athlete with a size nine. There is also great variation in the shape of these bones. If your metatarsal bones are abnormally shaped, the shoe can turn culprit by pushing on abnormally shaped metatarsal bones.

Your toes have the same number of bones as your fingers. The big toe has only two bones, like the thumb. The four other toes are composed of three little bones. Again, great variation in the length and shape of the toe bones can cause difficulty in athletes.

All these bones are held together by an incredible number of ligaments. The longest foot ligament is the *arch ligament,* or *plantar fascia,* which runs from the bottom of the heel bone along the arch and attaches to the metatarsal heads (the ball of the foot). When pressure is applied to the foot, the arch ligament flattens, but in so doing increases its own tension. It can be compared with exerting pressure on a strong rubber band. The more it is pressed, the tighter it gets and the more it wants to spring back to its normal shape. This mechanism helps provide spring to your running and jumping activities.

Because the foot has many ligaments, it is easy to sprain your foot. A sprained foot is a ligament injury; tearing of the ligament occurs because of excessive stress put on the foot by jumping activities and by running over irregularly shaped objects such as first base. I have treated a dozen or more professional baseball players who have sprained their feet because they hit first base wrong when trying to beat out an infield hit.

The shape of your foot bones is inherited from your parents. Nothing short of binding the feet, as the Chinese did for centuries, will change the shape of the bones. Loose ligaments or tight ligaments are also determined by your genetic inheritance. My generation of orthopedic surgeons does not believe you can change the shape of the foot with an arch support in childhood. I do believe, however, that with an arch support, stress can be taken off a foot that is not shaped normally. I

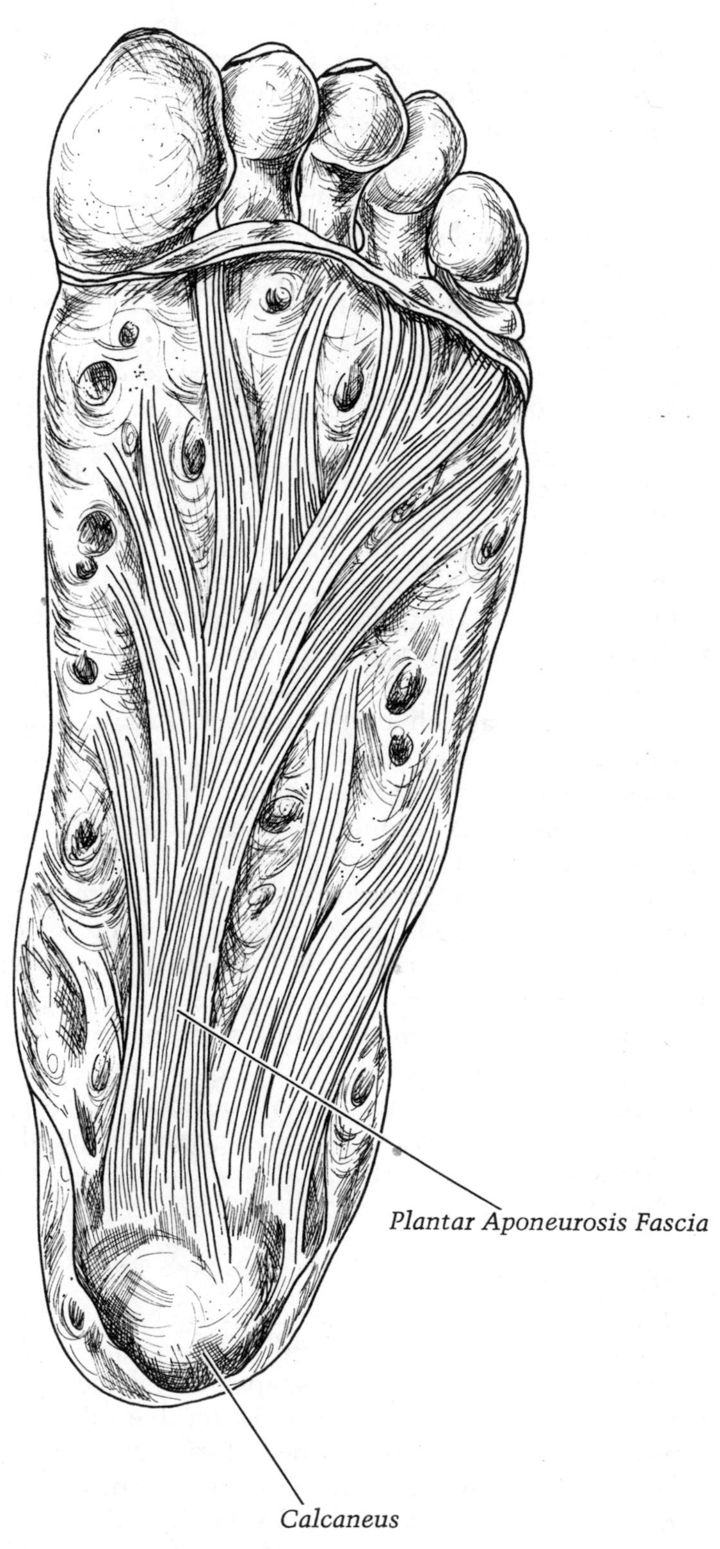

Plantar Fascia Arch Ligament

support the concept of accepting the shape and the mobility of the foot as nature has determined. This does not mean you can't be helped. If you have pain and abnormal variations in foot anatomy, sports medicine physicians and podiatrists can make you more comfortable by proper selection of your athletic shoewear and custom-made arch supports—orthotics.

MUSCLES

Picture a marionette with strings attached to the forefoot, heel, arch, and outer edge of the foot. The puppeteer can drop the marionette's forefoot by pulling on the heel string and tip the puppet's foot outward by pulling on the string on the outer border of the foot. This is exactly the way your foot works.

Many motions of the foot and toes are controlled by muscles that originate in the lower leg (extrinsic muscles). Their tendons extend to the foot area. Fine foot motions are controlled by smaller muscles that originate in the foot itself (intrinsic muscles).

For example, the string on the back of the foot is the Achilles tendon, which is the largest tendon in the body and is attached to the calf muscles. The calf muscle and Achilles tendon unit allows you to rise on your toes, climb stairs, jog, sprint, and jump. The string that attaches to the top of the foot, which allows you to bring your foot up toward your shin, is the anterior tibial tendon. Its muscle resides in the anterior compartment in the front of the lower leg. The tendon attaches directly to the top of the foot.

The peroneal muscles originate on the outer aspect of the lower leg and attach to the outside edge of the foot. The peroneal muscle and tendon unit allows you to tip your foot outward. The posterior tibial tendon, which starts on the back of the leg and goes behind the inner ankle knob, is the string that allows you to tip your foot inward.

Two small muscles and their tendons in the lower leg—the flexor hallucis longus and the flexor digitorum longus—and two muscles in the foot—the flexor hallucis brevis and the flexor digitorum brevis—give you the power to curl your toes downward. The extensor hallucis longus and the extensor digitorum longus—two muscles in the front of the lower leg—and the extensor hallucis brevis and the extensor digitorum brevis—two muscles in the foot—give you the ability to lift your toes. All in all, the foot is a complex but beautifully orchestrated structure.

NERVES

The sciatic nerve runs down into the back of the thigh. Just behind the knee, it splits into two major divisions. One branch, the *peroneal nerve*, wraps around the top part of the fibula, the outer bone of the lower leg. This nerve gives messages to the muscles in the anterior compartment, which powers the foot up (lifts the foot toward the shin) and moves the toes. The other sciatic nerve branch, the *posterior tibial nerve*, gives messages to the calf muscle and toe flexors and is responsible for powerful push-off in running. The posterior tibial nerve is deeply buried in the muscles and is seldom injured. The peroneal nerve lies just under the skin. Tom Johnson, a former hockey star with the Montreal Canadians, sustained the severing of this nerve when a skate slashed his lower leg. He was unable to control the up and the out motions of his foot, and his hockey career was ended.

SKIN

The skin is a primitive organ. When too much pressure is applied to the skin, it responds by thickening and forming calluses or blisters. The normal foot is well designed to spread the force of the body weight. When a large portion of the skin is accepting the force, an excessive callus does not develop.

The skin of the foot is like a road map. It gives clues to where the abnormal anatomy lies and where areas of excessive pressure are located. If your toes are abnormally shaped or your shoes are ill-fitting, calluses, blisters, or bunions will often develop. For example, a hammertoe deformity with the tip of the toe bent downward becomes callused on its end.

Biomechanical Factors of Running and Walking

When you walk or run, your foot strikes the ground much like an airplane landing. When the heel strikes the ground, the foot is slightly tilted inward *(supination)*. Thus, you hit the ground first with the outer part of the heel. This is why the outer corners of your running shoes become worn down first. Once the foot is on the ground, it quickly corrects this inward tilting.

As the foot moves forward, your weight is transferred to the outside part of your foot. Then your foot rolls inward, and your weight is shifted to the inside bottom part of your foot. Your arch flattens. The rolling inward *(pronation)* distributes the force of a footstrike throughout the entire foot and leg. During pronation the joints in the middle of the foot become unlocked, relieving tension in the arch ligament. This makes the foot flexible enough to adapt to uneven terrain.

The pronated (flattening of the arch) position continues until the inner part of the ball of your foot makes contact with the ground. At this time the center of gravity of your body passes directly over your foot. Then the foot rolls outward once again. As you rise on your toes, the joints in the middle of the foot lock, increasing the tension in the arch ligament and transforming the foot from a mobile adapter to a rigid lever. The foot is now ready for takeoff.

The most common result of excessive pronation is knee pain, which develops on the outside rear of the kneecap area during exercise. The excessive pronation exaggerates the normal twisting of the lower leg bone. This, in turn, causes small tears in the sleeve of the outer side of the knee. The twist motion can also cause hip pain.

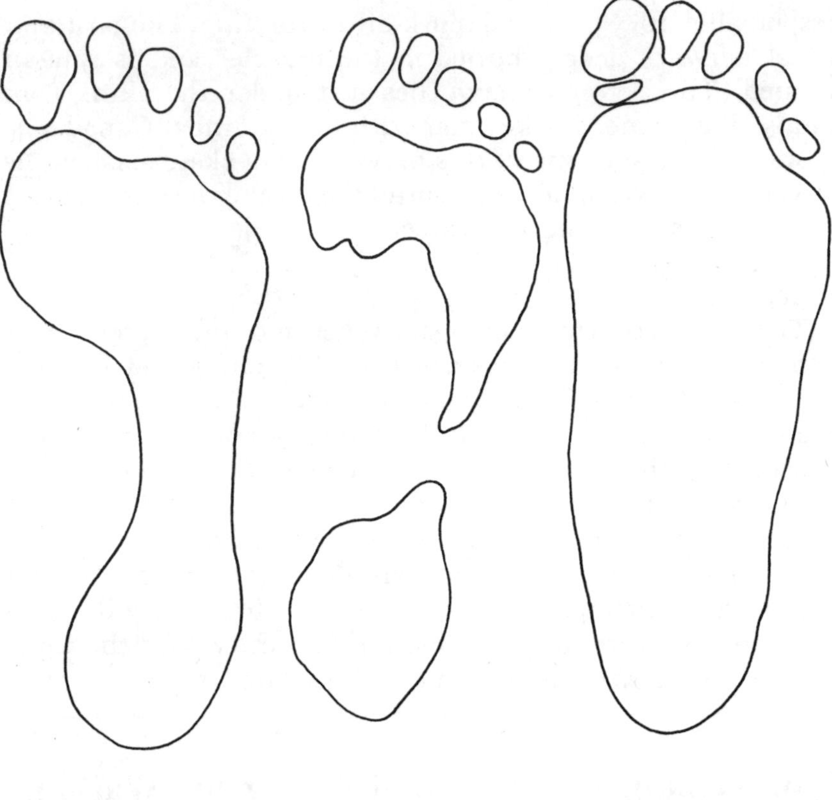
Normal High Arched Flatfooted (Pronated)

Wet Foot Test for Pronation

One method to show if you overpronate is the wet-foot test. Step in some water and then walk on dry pavement or cement. You will leave a print of the weight-bearing area of your foot. The ideal distribution shows the heel, the outer edge of the foot, and the forefoot. Your arch should not leave a print on the ground. Persons who have excessive pronation of the midfoot show a whole footprint. Because running shoes can reveal overpronating feet, I always have patients bring their shoes to my office for examination. I like to see if the outer foot area is accepting its share of the load and if the wear pattern comes across the ball of the foot to the base of the great toe (first toe). Your shoe teaches me a lot about the mechanics of your foot and how your weight is being distributed when you run.

One main reason for prolonged or excessive pronation is a forefoot supination. This means that the front of your foot is tilted upward. Thus, you must contort your foot to get your big toe on the ground. I can tell if you have a forefoot supination by checking the alignment between the heel and the ball of the foot.

I examine this relationship by having you lie flat on your stomach on the table with your feet hanging off the end. Your toes point toward the floor. I bring your heels into a "neutral position" by aligning the heel with the long axis of the lower leg. I then place a ruler on the ball of the foot. Normally the ruler would show a ninety-degree alignment with your heel.

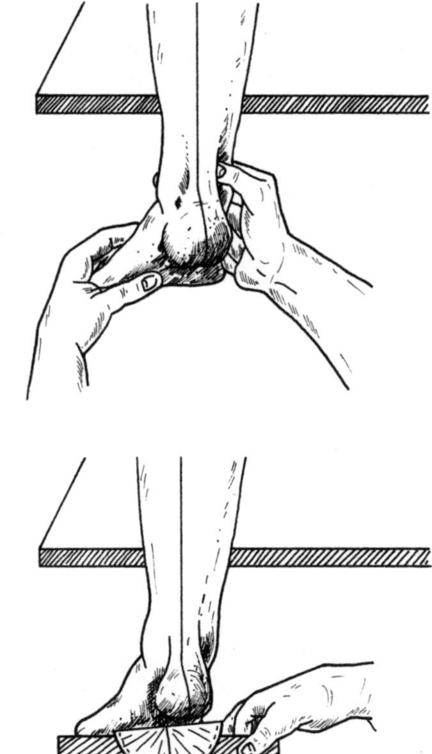
Align heel with rest of calf, "Finding the Neutral Position." Ruler on ball of foot to show the angle of forefoot supination.

Examination of Foot and Ankle

SHERPAS' FEET

Sherpas' feet are totally remarkable. They are as long as they are wide. They have a very wide heel, which provides a very steady base. The foot is covered with a "brown hide"—something a little thicker than elephant hide. Literally, they can walk across ground glass and small carpet tacks. I don't think they feel anything at all. They only complain when they have been walking across ice and snow for more than a couple of hours.

Richard A. St. Onge, M.D.

This is a well-balanced foot. For many runners who complain of foot, ankle, or leg pain, the ruler points upward away from neutral some fifteen to twenty degrees. We call this fifteen to twenty degrees of forefoot supination. In simple terms, a foot shaped in this manner must compensate for the upward tilting of the forefoot by flattening of the arch.

Stan James, M.D., of Eugene, Oregon, found that out of 180 injured and pain-afflicted runners, almost 60 percent of the group exhibited excessive pronation. Prescribed orthotic devices helped 80 percent of the overpronating athletes, and they were able to resume their preinjury training programs. In these patients the orthotic filled the space between the upward-tilted front of the foot and the ground. With orthotic devices the foot does not have to go into its prolonged or excessive pronation phase. Rather, the foot pushes on the orthotic, which then pushes on the ground; the foot does not have to contort itself to function in running. This is an important concept for all running athletes to understand. Failure to compensate for the structural abnormality leads to improper mechanics in running and to eventual injury.

The foot carries a great load as you run. What is the magnitude of these forces? When you land on the ground while running, your foot absorbs a force two to three times your body weight. In effect, you are really coming down from a small jump, which makes the landing impact greater than your body weight. With at least 800 footstrikes per foot per mile, a 150-pound athlete endures about 120 tons of force every mile. In a twenty-six-mile marathon, each foot must endure 3000 tons of force.

Hindfoot Problems

Plantar Fasciitis (Heel Spur Syndrome)

CAUSES

Plantar fasciitis is a partial or complete tear in the arch ligament on the bottom of the foot. It is the most common cause of heel pain in runners. It is usually characterized by pain just under the heel bone; sometimes it is called a *heel spur*. The pain can start immediately with a sudden tear during exercise or can develop gradually over many days. There is usually slight swelling.

The plantar fascia (which is really not a fascia, but a ligament) extends from the bottom of the heel bone to each of the five toes. It serves as an elastic guidewire to support the bottom of the foot, especially the arch. If sufficient pressure is exerted on the bottom of the foot—enough to spread the toes or to flatten the arch—the fascia tears. Here are four main causes of plantar fasciitis:

1. A sudden turn that exerts great pressure on the tissues of the bottom of the foot
   Peter Harvey, a high school basketball star, came down from a rebound and suddenly felt severe pain in the heel. By the next day he could hardly walk. The bottom of the heel was black and blue. The attachment of the arch ligament to his heel bone was tender. He had an acute ligament tear.
2. Shoes without adequate support for the arch
   The doctors at Sports Medicine Resource saw Celtic star Dave Cowens in 1977 for a second opinion. He had played all year without heel pain. He had a rigid foot—a poor shock absorber—and a high arch. No conventional athletic shoe could adequately have supported his foot. A special arch support had to be made to "bring his shoe up to meet his foot." He played the next year with much less pain. Although he had been told that he needed surgery, to this day it has been unnecessary.
3. Shoes with very stiff soles
   Every week, Rob Roy McGregor, the nationally known podiatrist at Sports Medicine Resource, sees at least five runners who have plantar fasciitis caused by track shoes with very stiff soles. Every time you run, you land on your heel and step off on your toes. If the sole is very stiff, you have to use extra force to bend it. This extra force is concentrated on the plantar fascia and can tear it. Flexible-soled shoes and arch supports are the best treatment.
4. Feet that pronate excessively
   Persons whose feet pronate excessively—that is, their feet flatten and roll inward when they walk or run—are more likely to develop plantar fasciitis. By dropping the arch and spreading the toes, pronation puts added tension on the plantar fascia. The treatment is orthotics.

For the majority of athletes, plantar fasciitis seldom comes from one specific injury. At first the pain will be low-grade and you will be able to endure the discomfort. After you finish the activity, your heel will stiffen and you will limp a bit. As the condition worsens, it will become impossible to function because of the pain. This is when athletes consult Rob Roy McGregor or me.

Because you have to walk each day—even if you quit sports—you continue to aggravate the fascia tear. The injury attempts to heal itself, but you repeatedly interrupt the process. Severe inflammation of the area follows and produces the intense pain. In this way, it is similar to tennis elbow.

DIAGNOSIS AND TREATMENT

The diagnosis of plantar fasciitis, or heel spur, is not difficult. Very little else can cause pain in that part of your foot or

JOHN DENNY'S PLANTAR FASCIA

While racing to back up first base, Cleveland Indian hurler John Denny tore his arch ligament, the plantar fascia.

"Plantar fasciitis is a tough injury for a pitcher," said trainer Jim Warfield, "because the back foot is so involved in the pushoff motion."

Initially Dr. Earl Brightman, the team orthopedist, put Denny on crutches to take the pressure off the bottom of the foot. "To reduce the swelling, I iced the foot for two days. I also taped and massaged it."

Denny was fitted with orthotic arch supports, his baseball shoes were altered, and the spikes on the foot were adjusted.

"All these treatments helped relieve some of the pain," said Warfield. "But plantar fasciitis is a longevity-type injury. To stay in shape, John worked out on a stationary bicycle. Dr. Brightman believed that John's running program was hurting him. He told John that he was beating his feet to death by trying to rush back into the pitching rotation.

"After three months of conservative treatment, he still had some pain on the bottom of his foot, but the pain was diminishing. We think that Denny will be healed for the 1981 season."

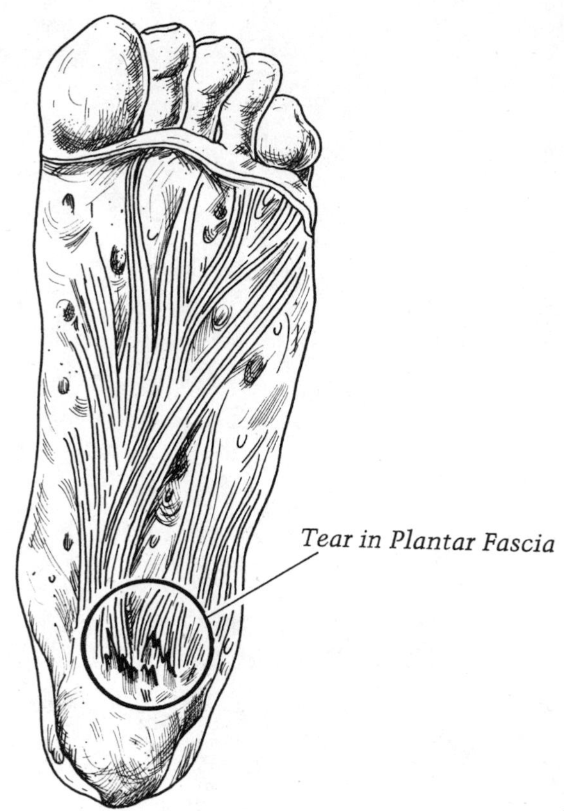

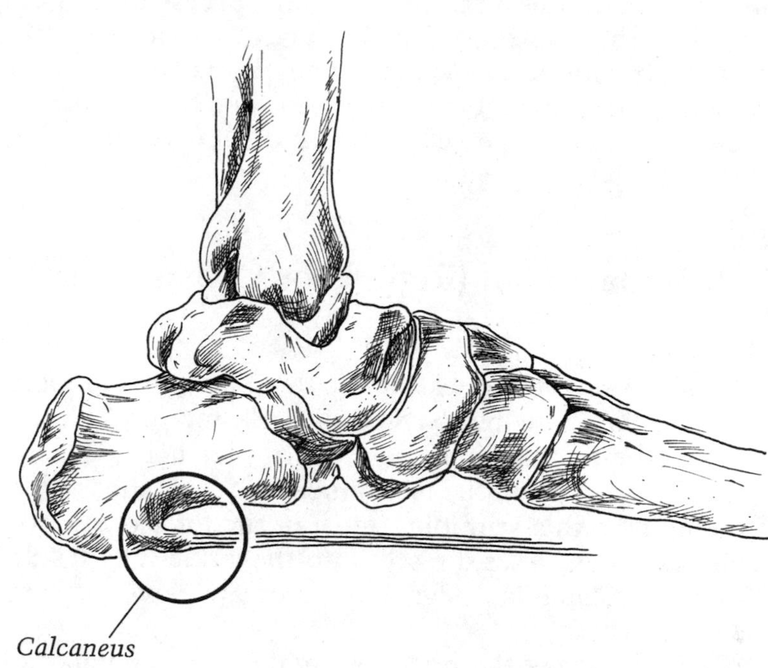

Plantar Fasciitis

heel. You can point to the bottom of the heel and poke deeply into the skin to describe the location of the discomfort. In 45 percent of the cases, it is exactly where the arch ligament attaches to the heel bone. It is on the sole of your foot, about one inch from the back edge of the heel. In the morning the heel is very stiff. You often have to limp to the bathroom.

If you have plantar fasciitis, you have already discovered that prolonged periods of rest ease the discomfort; use makes it worse.

During the physical examination, I find intense tenderness to pressure over the ligament at its heel-bone junction. I push on it through the sole of your foot. I can feel a fullness in the area due to the swelling and inflammation.

Rising on your toes causes discomfort. Walking on your heels causes discomfort. Both strain the arch ligament. The heel bone itself is not tender and the skin is normal.

I always take x-rays. In 95 percent of my cases, the x-ray reveals a normal heel bone with a spur of one-fourth to three-fourths inch sticking forward from its lowest point, which is exactly where the arch ligament attaches. This is why the condition is also called a heel spur. The x-ray also rules out any other type of bone pathology such as a tumor or a fracture.

I treat this problem in two ways.:

- I examine your shoes. Do they take the pressure off the arch ligament? Are the soles too stiff or too soft? Normally, a shoe change will help significantly.
- If the shoe change does not help, I prescribe an orthotic device, which mechanically supports the arch and keeps the arch ligament from straining.

I have all my patients put a one-fourth-inch heel pad in their shoes. I find that heel cups help some runners. I use a cortisone shot only if the shoe alterations are not successful. Sometimes a series of three injections is necessary to cure the problem. Surgery is rarely necessary; but if all else fails, a cut into the ligament at its bone attachment gives good relief of pain. This is called a Steindler procedure. It's like a Bosworth procedure for tennis elbow.

High-Impact Foot (Rigid Foot Construction)

CAUSES

The shape of your bones and the flexibility of your joints are inherited. Some people have very little motion in the joint between their ankle bone and their heel bone. This joint, called the *subtalar joint*, is the most efficient shock absorber in the foot. With a stiff joint, most of the force of a footstrike during exercise passes directly into the ankle. The result: the ankle joint, which is designed to go up and down, is asked to rotate or swivel.

"The ankle pays the price for an abnormal subtalar joint," points out Rob Roy McGregor. "In time the ankle joint stiffens and develops spurs. Loose chips can even form in the ankle." This is called a *high-impact foot*. About 5 percent of the foot patients who are examined at Sports Medicine Resource have a high-impact foot.

DIAGNOSIS AND TREATMENT

Bill Burnham, an all-American halfback at the University of New Hampshire, has a high-impact foot. At the end of his junior year Rob and I were asked to examine him. During the football season Burnham experienced ever-increasing ankle pain. His trainers taped his ankles tighter and tighter in an attempt to support them. Finally it came to the point where he was unable to run.

We examined his feet. His ankle joints had almost normal motion, but x-rays of his ankles showed spurs on the ankle bone as well as on the end of the lower leg bone (the tibia). The ankle was obviously being punished. Fortunately he had no loose chips in the ankle, but we found calcium deposits around the joint.

Rob lay Bill flat on the table, with his feet pointing toward the floor. I tried to move his heel from side to side. He had almost no motion in his subtalar joint. The joints in the midfoot were fully mobile and his toes were fine. Rob watched Bill walk and pronounced the verdict. "A great example of the high-impact foot."

Our recommendation: stop the ankle taping. What Burnham needed was mobility, not support. We told him to wear the oldest pair of football shoes he had. The extra motion he was getting out of the loose-fitting shoes helped substitute for

DAVE COWENS' HIGH IMPACT FOOT

"Dave Cowens, a former center for the Boston Celtics, is a perfect example of a person with a high-impact foot, a foot that is relatively immobile," says Dr. Rob Roy McGregor D.P.M. "Cowens was referred to me with acute heel pain. All I could do for the foot was to change the universe that the foot operates in slightly. I ultimately made a very simple soft orthotic for him.

"Dave also had problems with chronic ankle sprains. Some of this can be attributed to the high-impact foot. I also found that he had a muscle imbalance in the posterior and the anterior leg muscles. His anterior tibial muscles were weak. He worked hard to strengthen them.

"Most sports medicine physicians believe that the pain of a high-impact foot comes from jumping up and down. I tend to think that the greatest problem comes from running up and down the court, especially when a basketball player comes to a sudden stop. Breaking one's foot creates incredible forces."

the motion he lacked in the subtalar joint. His ankles were no longer being abused.

In 1978, I examined David Koza, a third baseman in the Red Sox minor league system. He, too, had had ankle pain over a two- to three-year period. He had even had surgery on his right ankle to remove chips, but new chips formed. I found that Dave had very little subtalar motion and all the force of his baseball activities was being absorbed by his ankle joint. I advised him to stop taping and supporting his ankle and got him to wear very flexible baseball shoes with short cleats. I did not want the cleats to dig solidly into the ground and reduce his ankle mobility. Once again the proof was in the result. His discomfort was relieved.

How do you know if you have a high-impact foot? Only your doctor can tell for sure. People who have high-impact feet complain about foot pain. The pain is mechanical. That means, if you do very little physically, it won't hurt; if you try to use your feet in strenuous physical activity, your ankles will complain.

Rob and I diagnose the high-impact foot by first examining the foot and testing the mobility of the subtalar joint. We x-ray both feet to evaluate the bone construction.

Once the condition is properly diagnosed, the best treatment for a high-impact foot is flexible shoewear. The sole of the shoe should bend easily. The heel counter, the back part of the shoe that cradles the heel, should be soft. The shoe tops should be made of material with plenty of give. Rob has found that an arch support, or orthotic, does not help a high-impact foot. Medically there's not much else we can do for you.

Retrocalcaneal Bursitis (Pump Bump)

CAUSES

Instead of a heel bone that tapers nicely to meet the Achilles tendon, some athletes have a box-shaped heel bone. If you are one of these people, you will find that your athletic shoewear presses hard against one of the sharp corners of the heel bone. The great pressure over the small area causes the skin to thicken. But, more important, it inflames the bursa sac that lies at the junction of the Achilles tendon and the heel bone. If there is inflammation for a long time, the bursa walls thicken and the entire sac enlarges. The enlarged bursa sac looks like a bump on the top of the heel bone, beside the Achilles tendon. It is much more common on the outer aspect of the heel. I see pump bumps mainly in skaters, skiers, and long-distance runners. Because a heel bursitis can be caused by wearing high-heel shoes, it called a *pump bump* by lay persons. In medical terms it is called *retrocalcaneal bursitis*.

If you did not wear shoes, you would never get a pump bump. The direct pressure of the footwear against this area causes the problem. Also, I have found that two extreme foot

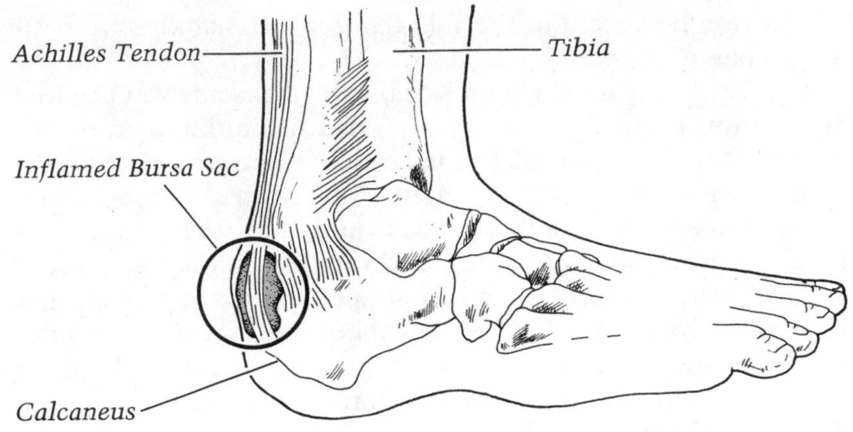

Retrocalcaneal Bursitis (Pump Bump)

shapes correlate with the pump bump—a foot with a very high arch, which generally does not pronate or flatten enough, and a foot with a very low arch.

DIAGNOSIS AND TREATMENT

The symptoms of a pump bump come on gradually. Generally it takes two to three months before the bump becomes very painful. Initially the bump is somewhat spongy, consistent with the bursitis. With continual irritiaton, the area becomes much more rigid. At this point the athletic shoewear makes this prominence very uncomfortable.

I diagnose a pump bump by touching it. When I push on the bump, which is sometimes the size of a pecan, it is tender. Some large bumps do not hurt much, and some small bumps are very painful. It depends on the degree of bursitis involved. I also check the running gait; normally the heel is rigid.

Here are the best ways to treat this problem:

- Anti-inflammation medication. Take two aspirins with each meal. If this is not effective, I prescribe a stronger anti-inflammatory medicine by mouth for at least three weeks.
- Orthotic. If you have a flatfoot, I recommend an orthotic device, a custom-made arch that controls the flattening of the foot. It is constructed by taking a mold of your foot.
- "Relieving" the shoe. This is done by stretching the leather or fabric over the bump. This should be done by a cobbler who has a special plier-like device to stretch the shoe away from the bump.
- A dispersion element. I use a one-eighth-inch moleskin with adhesive on one side and a soft, fuzzy surface on the other side. This is placed in the back of the shoe—the heel counter area. Covering all areas except over the bump, it relieves pressure on the inflammation.
- Cortisone injection. I use this as a last resort. I inject the cortisone into the bursa sac with a very small needle. After the shot, you can't participate in sports

for five to seven days. A series of three injections is often necessary to shrink the inflamed bursa.

- Surgical intervention. If all else fails, surgery is indicated. An incision is made directly over the inflamed area. The inflamed bursa sac is removed and the sharp corner of the heel bone is smoothed with a special bone instrument. By reshaping the contour of the heel bone, a pressure point is eliminated and the condition is cured. In my experience only one in fifty pump bumps needs surgery.

Inferior Calcaneal Bursitis

CAUSES

You have a bursa sac that lies between your heel bone and your heel pad. As you use your foot, the heel pad allows the skin to slide on the heel bone. The heel pad is made up of many tiny compartments. Each compartment is stuffed with fat globules, which absorb foot shock. Between heel pad and heel bone lies the inferior calcaneal bursa, which is about the size of a fifty-cent piece. It also acts as a slide.

When this bursa sac becomes inflamed, it is called *inferior calcaneal bursitis*. It is most often caused by a single traumatic event such as jumping from a height or landing awkwardly on pavement. I see this more in basketball players than in runners, more in males than in females. It is only half as frequent as a pump bump.

DIAGNOSIS AND TREATMENT

Inferior calcaneal bursitis causes a localized pain directly under the heel bone. When I push in this area with my thumb, most patients complain bitterly. The pain is a bit farther back on the heel than the attachment of the arch ligament.

If this bursa is inflamed, I feel a fullness when I touch it. Normally, because of pain, you will limp. Running is impossible in severe cases. The x-rays of the foot show no abnormality because the bones are normal.

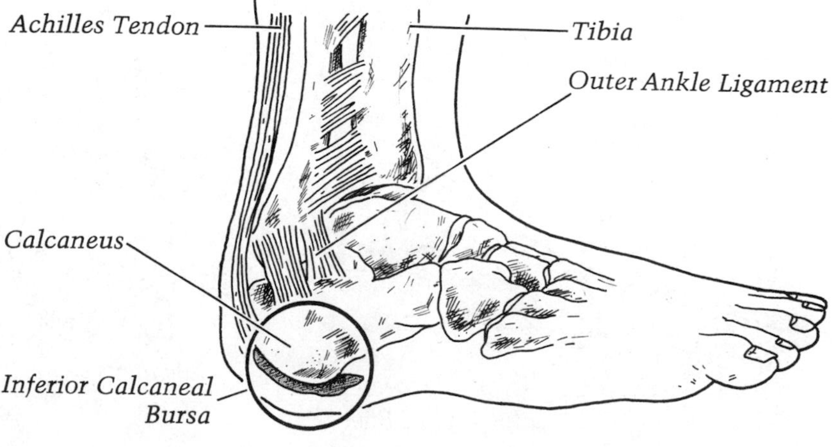

Inferior Calcaneal Bursitis

The best treatment is to stop exercising for two weeks. Take two aspirins with each meal for two weeks. I design a one-fourth-inch felt pad for the heel, which softens the effect of walking. If these treatments are not effective, I inject cortisone into the bursa sac. The cortisone usually does the trick. Occasionally a series of three injections is necessary to eliminate the bursitis. Surgery is rarely necessary to remove the inflamed bursa and smooth the heel bone.

Calcaneal Stress Fracture (Stress Fracture of the Heel Bone)

CAUSES

When I was in the military, I treated hundreds of *calcaneal stress fractures*—small cracks of the heel bone. (Stress fractures have been discussed in the fracture section.) If you get pain in the heel bone two to three weeks after you start your training season, be suspicious of a stress fracture.

Runners who suddenly increase their training schedule over two- to three-week periods are also potential victims of stress fractures.

The pain, at first low-grade, becomes steadily more severe to the point that running and sports activities are impossible.

DIAGNOSIS AND TREATMENT

I make the diagnosis of a heel-bone stress fracture by grasping the heel bone sideways. This produces great discomfort.

Initial x-rays are often negative; but as long as the squeeze test is positive (that is, it produces pain), you should not exercise for the next two weeks. I repeat the x-rays in two weeks. Usually I can see new bone production in the mid-portion of the heel bone. This looks like a hazy, white cloud.

The second x-ray usually confirms the stress fracture, which takes four to six weeks to heal. I almost never cast the foot. If you are in enough pain to limp, I give you crutches or use a gel cast—a semi-rigid support. Heel-bone stress fractures never cause permanent disability. Most athletes are back at full duty two months after the onset of pain.

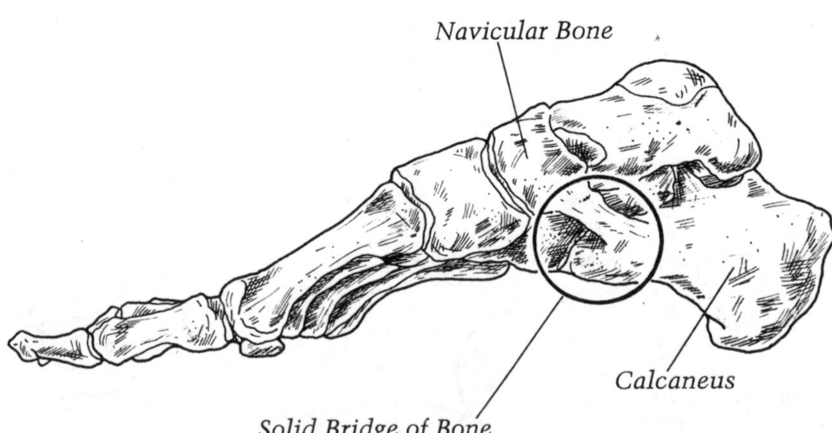

Calcaneonavicular Coalition

Midfoot Problems

Tarsal Coalition

CAUSES

In normal feet the two bones of the hindfoot and the five bones of the midfoot are all separate structures that join each other through joints. Some people, however, are born with bone bridges between some of these bones, which limit foot motion. In medical terms the bone bridging is called *tarsal coalition*. Occasionally it is not all bone, but rather a bar of cartilage that connects them.

Although a tarsal coalition starts at birth, it is not a major problem until the teenage years. The older you get, the larger the foot bones become, and the more restrictive the problem seems.

Because of the tarsal coalition, other bones of the foot have to absorb foot stress. It is very much like a back fusion in that the fusion puts stress on the other bones and discs of the back. The muscles and tendons that support the foot have to work overtime to deal with the stress. In the foot the peroneal muscles and tendons on the outer aspect of the lower leg take a particular beating. Because they overwork constantly, they go into muscle spasm. The spasm tends to flatten the arch of the foot and you end up with a condition that orthopedists and podiatrists call a *peroneal spastic flatfoot.* The spasm of the peroneal muscles goes hand in hand with the bony bridging in the hindfoot and the midfoot.

How does tarsal coalition affect your athletic performance? I compare it with a high-impact foot. Because the foot strikes the ground at lcast 800 times per mile, the restricted motion of the foot tends to create wear and tear in the remaining normal joints. The joint pain, plus the spasm of the peroneal muscles, leads to discomfort in your foot and the outer portion of your lower leg. It comes slowly over a six- to nine-month period and finally becomes so severe that athletic participation is impossible.

DIAGNOSIS AND TREATMENT

When I examine your foot, I find little motion in the subtalar joint as I attempt to move your heel from side to side. It is a stiff-feeling foot. The front of the foot moves nicely and the toes are normal. When you stand, your foot flattens. I can feel the spasm in your peroneal muscles and tendons. When you walk, you spend less time putting weight on the painful foot than on the normal foot.This is called an *antalgic gait;* you want to remove your weight from the sore foot.

I take x-rays from many angles. Because the bridges are bony, they show up on the x-rays. If I have any doubts, I do tomograms, x-ray slices through the bony area. They show the abnormal bone bridges. The most common bridges are between the heel bone and the navicular and the ankle bone and the heel bone.

I treat these patients very much like those with high-impact foot. The best treatment is to rest for at least two weeks. I recommend sloppy shoewear to give you every chance to get motion. If I tape you or give you a stiff shoe, you will suffer. I use an orthotic to take the pressure off the arch and to try to relax the peroneal muscles. If I can get you out of spasm, you are more comfortable.

Because the pain of running is often too great to stand, sometimes you must change your sport. Cycling or sculling are ideal substitutes. If you want to continue running and the bony bridge is small, I can remove it. It is a simple operation; I break the bridge with a surgical chisel. I like to place a small section of muscle between the two bones that had the bridge. This keeps the bony bridge from growing back. In cases where the bridge is quite large, it is necessary to do a triple arthrodesis of the hindfoot. The surgery permanently fuses together the ankle bone, the heel bone, the navicular bone, and the cuboid bone. This gives a stable, pain-free foot. The problem is that it does restrict the foot motion greatly, and running is quite difficult. This operation is done only for pain relief. It would not be performed with the goal of returning you to full sports activities.

Accessory Navicular

CAUSES

The navicular bone lies at the peak of the arch of your foot, just in front of the ankle bone. In some persons this is two bones instead of one. The junction between the two bones forms a false joint, or pseudarthrosis. Because the joint has mobility, it can produce pain in those who use their feet vigorously.

The extra piece of the navicular *(accessory navicular)* always lies on the inner side of the arch. This means that it comes in direct contact with shoewear, which adds to the pressure on the area and to the motion in the false joint; 20 to 30 percent of athletes with this structural abnormality complain of discomfort. Accessory navicular is a frequent problem in young female athletes.

DIAGNOSIS AND TREATMENT

The pain usually comes on gradually and is located in the extra navicular bone. When I examine the foot, the extra bone appears as a bump on the top of the arch, which is tender. Occasionally the skin over this area is thick and red.

If you have an extra navicular, it will hurt you to rise on your toes because the posterior tibial tendon pulls on the small extra navicular bone. Most people with an extra navicular have it in both feet. The accessory navicular can easily be seen

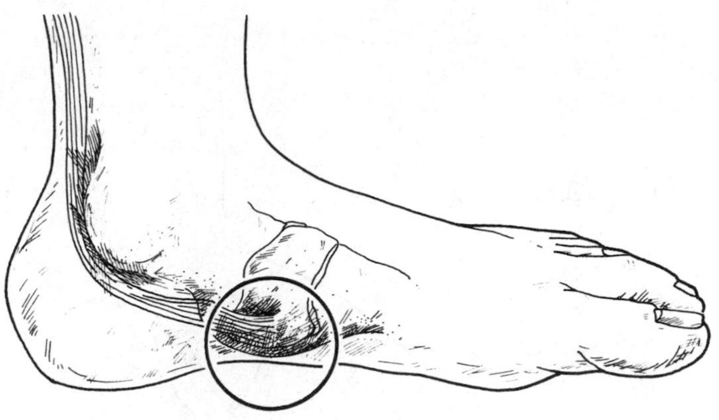

Accessory Navicular

in x-rays as separate from the main body of the bone. The line on the x-ray that separates the two portions represents the false joint. The wiggling of this false joint causes the pain.

I treat this problem with orthotics to take the stress off the extra bone. The stress comes from the posterior tibial tendon. Before the orthotic is fabricated, I strap felt padding to the foot with adhesive tape. This tells me if the orthotic will be helpful. You should wear the strapping for five days. When you shower, you must cover the strapping with a plastic bag. A cortisone injection into the tender area is sometimes helpful, but is never curative. If you are not helped by the orthotics, surgery is usually inevitable. The purpose of the operation (the Kidner procedure) is to remove the posterior tibial tendon from its attachment on the extra navicular bone. It is replaced on the regular navicular. The extra navicular bone is removed.

After the surgery you are immobilized in a walking cast for one month. You then undergo whirlpool baths and strengthening exercises. Although it takes six to nine months for the patient to recovery fully from this surgery, I have enjoyed good success with the operation. The principle behind the surgery is straightforward. It is like having your appendix out: remove the cause of the pain, and the pain goes away.

Cavus Foot

CAUSES

The *cavus foot* is nothing more than a foot that has a high arch. Visualize the arch as a cave-like structure. The root of the word *cavus* in Greek means "hollow" or "cavity." Why do some persons have high-arched feet? It is in the genes. The shape of the bone and the tightness of the ligaments in your foot are determined by your ancestors. There is no way to change the shape of your foot. You must accept it and treat it if it causes pain.

Twenty percent of the runners who see doctors for foot complaints have high-arched feet. What kinds of problems do they experience? A high-arched foot tends to be more rigid than a normal foot. Therefore, in running, a great amount of

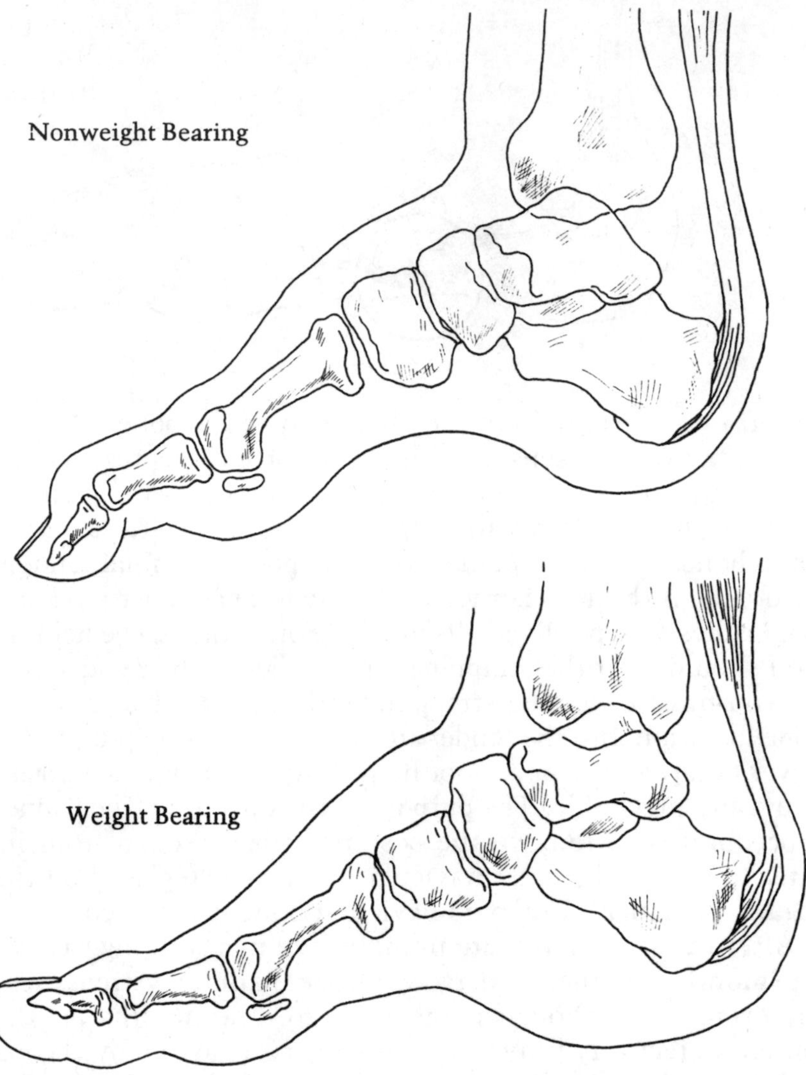

Cavus Foot

stress is not absorbed by the foot itself, but rather is passed into the lower leg, knee and hip area. Leg complaints are thus quite common. In the foot itself, a strain is placed on the arch ligament because of the high-arched architecture. Pain along the arch is not uncommon. Often the weight of walking and running are borne inordinately on the second metatarsal area. This produces a heavy callused area under the first metatarsal. This produces pain. The high-arched foot has difficulty fitting into many shoes. When the shoe is tied, pressure is put on the top of the foot (the dorsum) and pain and tendinitis may follow. Finally, persons with high-arched feet develop claw toes, which are cocked up at the first toe joint. The foot comes to the ground at a sharp angle, and the toes go away from the ground at a sharp angle. This means that the tendons on the top of the toes are not so strong as the tendons on the bottom. The bottom tendons pull the toes into the clawed position, which can cause intense pain and callus formation over the cocked-up first toe joint.

DIAGNOSIS AND TREATMENT

The diagnosis is obvious. When you examine the patient in a sitting position, the foot has a high arch and the ball of the foot is lower than the level of the heel bone. The toes are often clawed to varying degrees. When you feel the heel area, it is quite stiff and does not have much side-to-side give. I next have the patient stand on the ground to see what shape the foot has while bearing weight. In a very rigid cavus foot, the arch barely depresses at all. Some people have a semi-rigid cavus foot in which the arch partially flattens. There is always a high-arched contour to the foot in either case.

We have pointed out that 20 percent of the runners who see doctors for their foot complaints have cavus feet. This is beautifully discussed in a July 1980 article by Steven I. Subotnick, D.P.M., in *The Physician and Sportsmedicine* magazine. Dr. Subotnick described his treatment goals. We have had a similar experience at Sports Medicine Resource. A three-sixteenths-inch heel lift is often helpful to relax the plantar fascia and to place more shock absorption material in the heel area. This is necessary to counteract the rigidity of this foot. The heel lift can be built into a semi-rigid orthotic, which helps support the arch and takes away the strain on this part of the foot. We like the semi-rigid orthotic more than the hard plastic one because the plastic seems to aggravate the plantar fascia (the arch ligament). The orthotic tends to distribute the pressure of walking and weight bearing over larger areas of skin and therefore reduces the callus buildup under the second metatarsal.

The running shoe must have excellent shock absorption qualities. If pain and callus formation in the ball of the foot continue despite the orthotic, extra rubber may be added to the sole of the running shoe to take the pressure away from this painful area. "Shoe goo" is not effective in doing this. We like to elevate the shoe sole and place more rubber between it and the last of the shoe. This requires specialized equipment.

The claw toes are responsive to stretching exercises. I ask the patient to push down on the toes at least thirty times each day to keep the toes flexible. Another trick is to use a moldable podiatric compound. It comes as a powder and is mixed with water to make a moldable mass of material that easily fits in the palm of your hand. The material is placed under the toes, and the toes are flattened down into the compound. It can be comfortably worn in your street shoes or your athletic shoes and will keep your toes flat and flexible.

Careful attention should always be paid to callus formation. The calluses should be carefully shaved every two weeks. This prevents excessive callus buildup and also prevents high-force buildup over the small bit of thick callus. The treatment will diminish blistering along the edge of the callus as well. It is also important to keep the muscles of the lower legs stretched out by exercising with an inclined board. Stretches should be done alternately with the heel on the bottom of the incline and then on the top of the incline.

Dorsal Exostosis

CAUSES

Look down at the top of your foot. Normally the top of the foot is smooth and even, meaning that the tops of the joints between the ankle bone and the other parts of the foot are well aligned. In some persons, however, the bones stick up to form the joint. This is simply the way that foot is shaped. It doesn't become a problem unless you wear tight-fitting athletic shoewear such as ski boots, ice skates, or hiking boots. Any shoewear that laces tightly over the top of the foot and extends above the ankle joint can cause the bony prominences to become irritated or enlarged. The condition is called *dorsal exostosis*. *Dorsum* means "top"; *exostosis* means "bony prominence." As these bumps grow larger on the top of the foot, they put pressure against the laces of the boot, causing further growth of the exostosis.

DIAGNOSIS AND TREATMENT

It is easy to spot a dorsal exostosis; it is about as big as an almond and lies at the junction of the navicular and the cuneiform bones (the cuneiform metatarsal junction). These bones are on the top of the foot closest to the arch. The skin is often somewhat thickened and reddened over the bump.

You can walk without pain only without shoes. Your foot mobility is usually normal, and you have full and free motion of the tendon over the bump. X-rays reveal normal bone structures except for a prominent lipping of the bones as they form the joints of the midfoot.

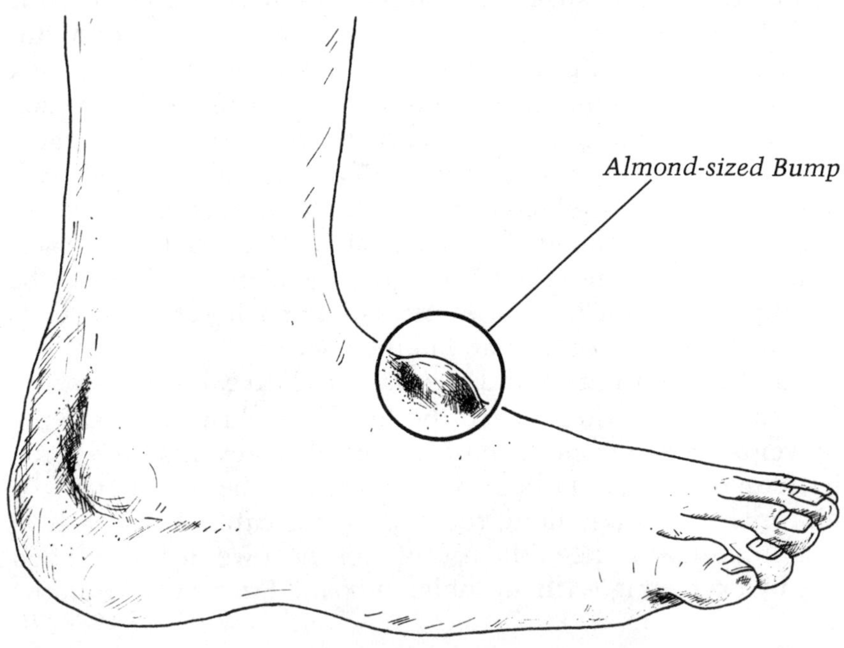

Dorsal Exostosis

Treatment is usually a modification of shoewear. The shoe tongue can be stretched to provide more room for the bump. Moleskin can be placed on the tongue to elevate it above the bump. If the dorsal exostosis continues to expand despite these conservative measures, surgery is sometimes necessary. I have performed this operation only once in twelve years. This was on an eighteen-year-old hockey player who could not wear his hockey skate because of the size of the exostosis. The surgery was successful, and he skates today without difficulty. He has a residual scar on top of his foot, however, which is sometimes sensitive. This patient traded a bump for a scar; in his case it was well worth it.

Extensor Tendinitis

CAUSES

The tendons that run down the top of your foot direct your toes. Their muscles originate in the front of your lower leg in the anterior compartment. This muscle-tendon group allows

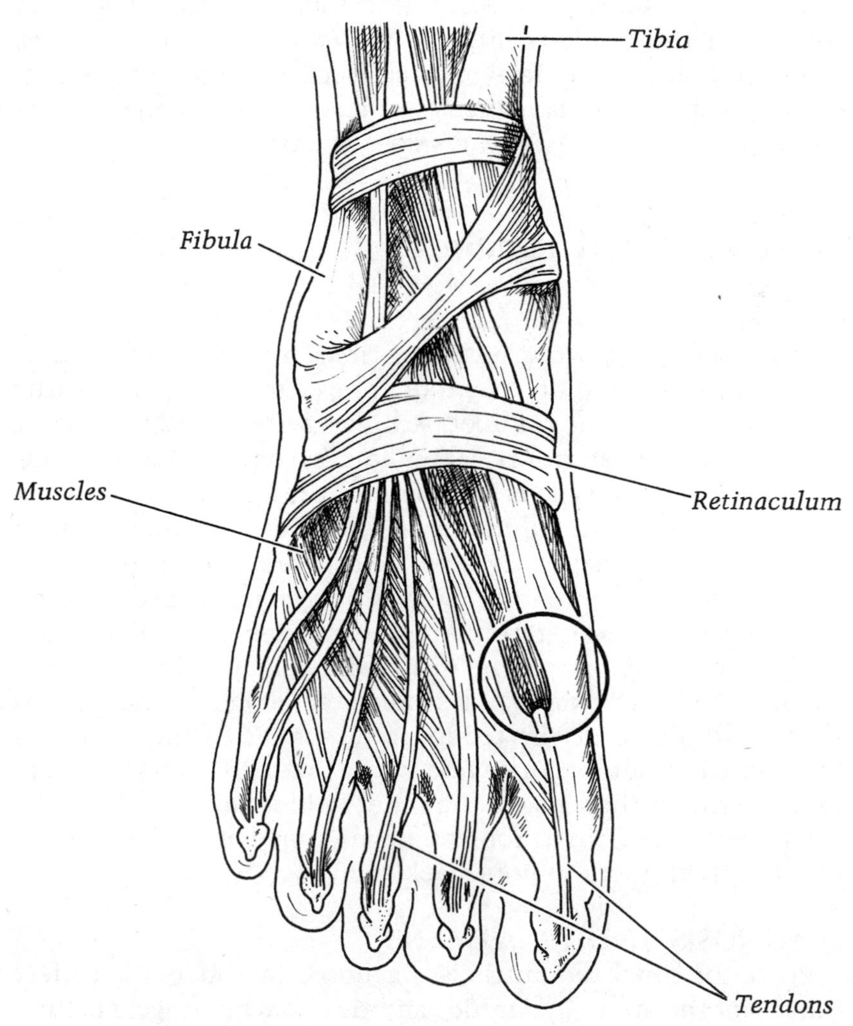

Extensor Tendinitis

your toes to straighten vigorously. The tendons inflame when they are pressured by improperly fitting athletic shoes or laces that are too tight. The inflammation, called *extensor tendinitis*, usually comes on gradually and is seldom the result of a single injury. I rarely see it in both feet.

DIAGNOSIS AND TREATMENT

Pain on the top of the foot is almost a sure sign of extensor tendinitis. It is made worse by running. Normally the pain intensifies and produces a limp.

When I look at your foot, I usually find some swelling around the tendons. The top of the foot often looks puffy. I always compare both feet. Although wiggling your toes brings discomfort, other tests on the foot are normal.

I start the tendinitis treatment immediately. You should stop exercising for one week. Ice the area for twenty minutes twice daily. Take six to eight aspirins a day with meals. I check your shoewear to see if the tongue of your athletic shoe is ill-fitting or poorly padded. This could cause the inflammation. I tell all my patients with extensor tendinitis to tie their laces in stepladder fashion rather than crisscross. I cut felt strips one-fourth inch by one-fourth inch and lay them between the tendons. This keeps the tongue of the shoe from pressuring the tendons directly. In 95 percent of the cases, these measures are curative. Occasionally the tendinitis continues despite the treatment, and a cortisone injection is necessary. This almost always does the trick.

Posterior Tibial Tendinitis

CAUSES

The major job of your posterior tibial muscle and tendon is to hold up the arch of your foot. The tendon attaches to the navicular bone in the midfoot. Athletes with flat feet put great tension on the posterior tibial tendon. In some athletes it leads to injury of the muscle at its origin in the calf. This is called a *posterior tibial shin splint*. In other athletes the tendon unit takes a beating; the result is *posterior tibial tendinitis.*

Posterior tibial tendinitis comes on gradually, usually over four to six weeks. It rarely starts with one particular injury such as stepping on a stone or missing a step, where the foot suddenly tilts outward and strains the tendon. The pain is sharply localized at the tendon attachment to the navicular bone in the midfoot. In severe cases the tenderness and pain extend around the back of the inner ankle knob into the calf.

I have treated this problem in all types of athletes, from mountain climbers to horseback riders.

DIAGNOSIS AND TREATMENT

Posterior tibial tendinitis is diagnosed by touch. First I feel the top of the arch of your foot and then rub my fingers behind

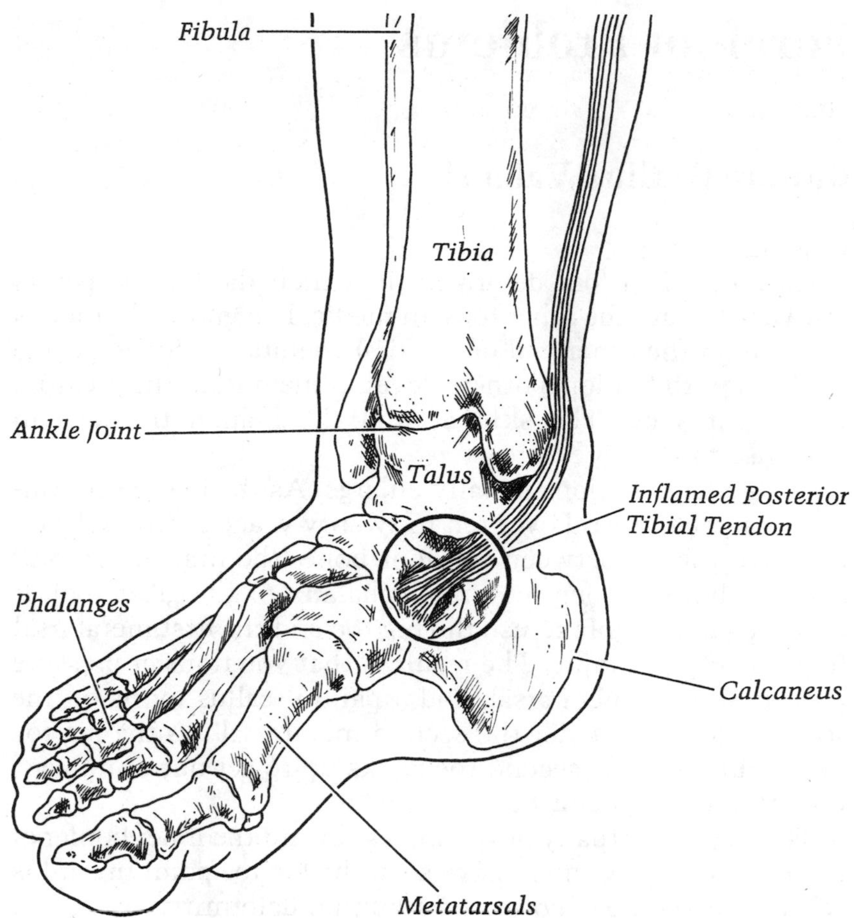

Posterior Tibial Tendinitis

the inner ankle knob. I can feel the posterior tibial tendon through the skin; pushing on it causes extreme discomfort. There is also pain if you jump up and down. Here is the treatment:

- Stop exercising for two weeks.
- Ice the tendon twice a day for twenty minutes.
- Take anti-inflammation medicine, two aspirins with each meal for two weeks.
- Place an adhesive tape strapping or a felt pad in under the arch to take some pressure off the posterior tibial tendon.
- Use a semi-rigid orthotic device. Supporting the arch takes the tension off this muscle-tendon unit and prevents recurrence.
- Consider cortisone. If the aspirins and strapping are not 100 percent effective in seven to ten days, I inject cortisone along the tendon, but never into the tendon itself. I have seldom operated on a posterior tibial tendinitis.

With the preceding treatment plan, within three weeks you should be completely free of pain.

Forefoot Problems

Bunion (Hallux Valgus)

CAUSES

A *bunion* is a foot deformity in which the big toe points outward toward the other toes. In medical language a bunion is a bump on the inner end of the first metatarsal. Some people are born with bunions; others develop them from the pressure of a tight shoe. The skin around the bunion thickens in response to shoe pressure.

With aging, bunions usually enlarge. As the big toe continues to angle outward over ten to twenty years, arthritis develops in the joint between the foot bone (the first metatarsal) and the big toe bone (proximal phalanx of the great toe). It becomes impossible to use the big toe and the first metatarsal to push off in running. The result is that you transfer pressure to your second metatarsal, and a painful callus forms on the sole of your foot under the second metatarsal. As the big toe drifts outward, the second toe cocks up to get out of its way, resulting in two deformed toes.

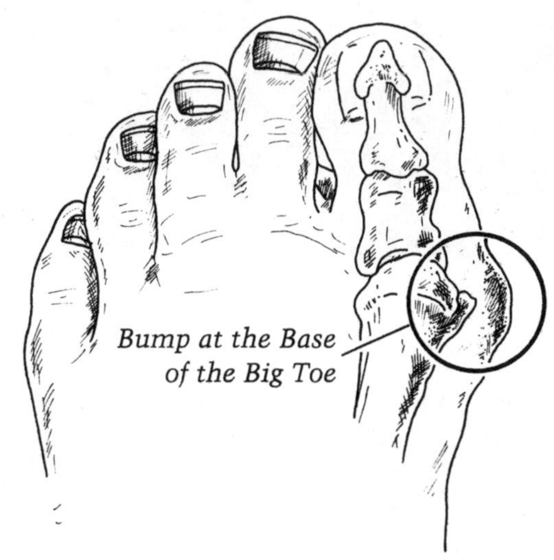

Bunion

The big toe virtually never angles toward the inner border of the foot. The only time I have seen the big toe point inward is when a surgeon overcorrected a bunion deformity.

Bunions are much more common in females and tend to run in families. But you probably would not have pain if you did not wear shoes. Although the toe would still tend to angle, there would be no pressure on the metatarsal at the bottom of the toe. Therefore, the painful bunion would never form.

DIAGNOSIS AND TREATMENT

One look at your own foot and you know you have a bunion. The big toe points toward the four other toes, and you have a prominence on the inner border of the foot. Sometimes the second toe cocks up and even overlaps the big toe. A callus, or thickened skin, in the sole of the foot under the second metatarsal is almost always present. By the time you come to see me, you have pain over the bunion and some swelling and skin thickening.

X-rays of the foot show an inward tilting of the first metatarsal; it is not parallel with the other metatarsals. Because the metatarsal inclines inward, your shoe pushes the toe outward. In medical terms the inward tilting of the first metatarsal is called *metatarsus primus varus*. This is the root of the problem.

For comfort in running, it is helpful to place a soft sponge rubber pad between the big toe and the second toe. This keeps the big toe aligned as you participate in sports and run. Commercial eliptically shaped pads, which fit between the toes, often are too large for comfort in athletic shoewear.

Rob Roy McGregor and I have found that arch supports, or

orthotics, are not helpful in treating this condition.

If your bunion becomes particularly sore and inflamed, take two aspirins with each meal for two weeks. Ice the area for twenty minutes after you run or engage in sports. Occasionally a cortisone injection is necessary to relieve an acute attack of an inflamed bunion, but nothing short of surgery can change the ultimate course. Only 10 percent of the athletes McGregor and I see need surgery.

What happens when surgery is necessary? The purpose is to straighten the first metatarsal so that it is parallel with the other metatarsal bones. The operation (McBride procedure) involves cutting all the tight ligaments on the outer aspect of

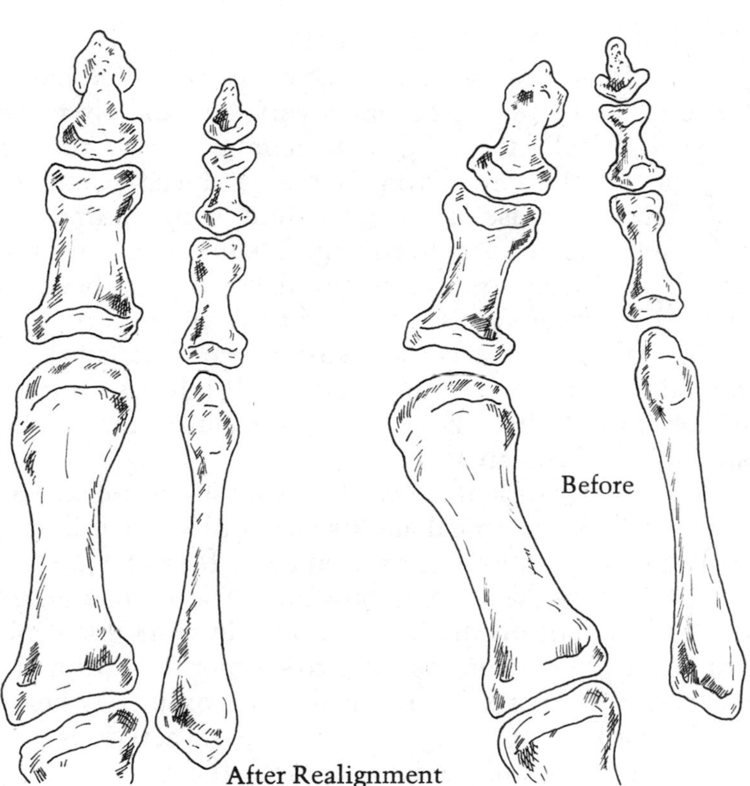

Bunion Surgery

the toe that force the toe outward. Second, I shorten the ligaments on the inner aspect of the toe to hold it straight. Once the toe alignment is surgically corrected, I protect the toe with a cast so that nothing will disturb the healing. After six weeks in a walking cast, you may begin to exercise and walk in a sandal. It takes three to six months to get back to full sports activities.

Because a serious bunion takes time to develop, it is seldom necessary to perform surgery in young people or those who are very active. I have never done a McBride procedure on a professional athlete. Most of the candidates for this operation are middle-aged recreational athletes who are having pain when they engage in tennis, jogging, or racquetball.

Turf Toe

CAUSES

Turf toe is a recently described injury that started with the advent of artificial turf. When you play on artificial turf, it grabs your athletic shoe and you stop short—more abruptly than on grass. The problem is that your foot cannot stop as fast as the shoe. This means that the foot continues to drive forward and the toe is driven hard against the end of the shoe. Your longest toe, normally the big toe, is jammed into the shoe front. This is like hitting the end of the toe with a hammer. The punishment is absorbed at the base of the toe. Physicians call this a compression injury because the surfaces of the joint are driven hard together. Athletes call it turf toe.

DIAGNOSIS AND TREATMENT

How do you know if you have turf toe? The joint at the bottom of the big toe becomes very sore and often swells. It looks and feels like a ligament sprain because the jamming together of the joint surfaces has produced swelling in the joint itself. Because of the extra fluid, you will find it hard to bend your toe. It hurts to run and it's impossible to run at full speed. The faster you run, the more you have to bend the joint at the bottom of your big toe and the more it hurts.

Normally, turf toe is the result of one specific injury, but it can be the culmination of many small injuries. In my examination I find that the whole joint hurts, rather than one sharply localized area.

The best treatment is rest for at least one week. Take two aspirins with each meal and use ice for twenty minutes twice daily to try to take the irritation away from the joint.

Although x-rays do not show any serious bone injury, they do show a dent on the joint surface. In time this dent heals, but it may take as long as three to six months. Fortunately you can return to sports before the dent is completely healed.

I ask patients to refrain from full sports activities until they have full motion of the joint. I check the motion of the same joint on the other foot. It can be as long as two to three weeks before you regain full motion. It is pointless, however, to wait longer. Even if you have mild pain, you can return to sports.

Morton's Neuroma

CAUSES

Touch the tips of your toes lightly with your fingers. The fact that your toes are being touched is relayed to your brain almost instantly by the digital nerves, which run between the metatarsal bones of your foot. Each nerve supplies sensation to the outside of one toe and the inside half of the toe next to it. In some persons the metatarsal bones have unusually large prominences. If you wear tight-fitting shoes, the prominent areas of the metatarsal bones can squeeze the digital nerves. The nerve responds by swelling, which is called a neuroma. Because this was first described by T.G. Morton, a United States surgeon, it is called *Morton's neuroma.* I have treated this ailment in many athletes; it is not more common in one sport than in another.

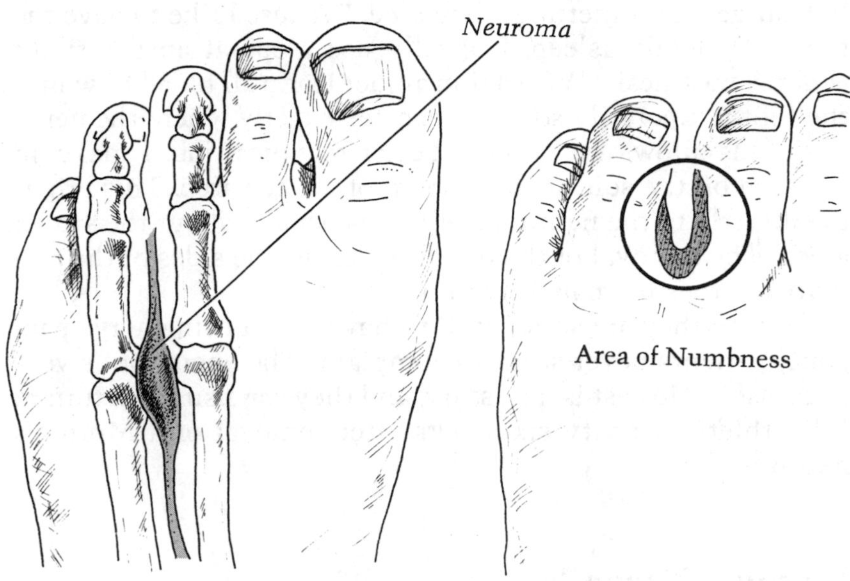

Morton's Neuroma

DIAGNOSIS AND TREATMENT

If you have a Morton's neuroma, you will complain of a fiery pain in the ball of your foot that often radiates into the toes. It is usually a singular pain that is precisely localized in one of the five nerves. If the condition is severe, your toes will get numb. Usually the pain subsides if you walk barefoot. Tight shoes definitely make the condition worse.

In the physical examination I squeeze the ball of the foot together, which drives the bones against the swollen nerves. Usually this elicits shouts and yells. If I feel between the individual metatarsal heads at the base of the toes, I am pressing directly on the neuroma. You won't like me to do this. The most common location of Morton's neuroma is between the third and fourth toes. X-rays usually show a narrowing of the area between the third and fourth metatarsal bones in the ball of the foot, which explains the pressure on the nerve.

The simplest treatment is to modify your shoes to allow plenty of room for the neuroma area; this diminishes the irritation and swelling around the nerve. It is very important to have adequate shoe width at the level of the ball of the foot to keep the bones from squeezing the neuroma. Secondly, I start patients on anti-inflammatory medication by mouth three times a day for three weeks. Sometimes this shrinks the swelling around the irritated nerve. If this is not effective, I inject cortisone directly around the neuroma. These modalities often bring months of relief. In 25 percent of my patients, the change of shoes, anti-inflammatory medication, and cortisone injection are totally curative. I often repeat the injection; three injections is my maximum. I inject the cortisone into the top of the foot, placing the needle down toward the ball of the foot. This is more comfortable for the patient.

If the cortisone injections fail to provide long-standing relief, surgery is sometimes indicated. Because I like to have the foot completely asleep, I usually use a spinal anesthetic or general anesthesia. With a tourniquet I stop all blood flow into the foot temporarily so that I can see exactly where the nerve lies. This allows me to do a better technical job. I make an incision on the sole at the bottom of the toe area. This brings me directly to the neuroma, which is usually about the size of a pea. The removal of the neuroma leaves the sides of the two involved toes permanently numb.

Because they are so relieved to have the burning nerve pain gone, patients never seem to complain. The operation is very successful. No cast is necessary, and they can usually return to full athletic activity six weeks after removal of a Morton's neuroma.

Ingrown Toenails

CAUSES

When the edge of your toenail becomes jagged and digs into the skin along the edge of the nail, you have an *ingrown toenail*. Because the big toe has broad and stiff toenails, it is most frequently affected by this problem. The two most common reasons that your toenail becomes jagged are that you get stepped on and that you wear ill-fitting athletic shoes.

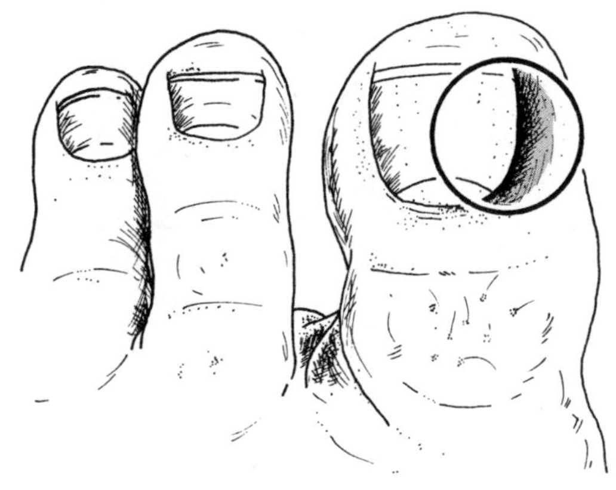

Ingrown Toenail

When the skin of the toe becomes very irritated, it "weeps"; the skin secretes liquid. The weeping softens the nail, making it still more jagged. This is the start of a vicious circle that leads to more weeping, redness, and irritation of the skin along the nail. Slowly the nail festers and can even abscess and fill with pus. At this point, ingrown toenails are extremely tender.

I had a chronically inflamed ingrown toenail of the big toe in my freshman year at Harvard. One of my most painful injuries, it cast a cloud over my entire football season. I still remember the discomfort.

My ingrown toenail started by chance. The big toe of my right foot was stepped on in three consecutive scrimmages. The nail became ragged and began to dig into the skin along the inner border of the toenail. It became swollen, red, and painful.

Joe Murphy, who was Dr. Quigley's (the Harvard team physician) right-hand man, explained my problem in detail. "You should cut a V in the middle of the toenail, which makes the nail more flexible," he advised. He placed a small pad of cotton along the nail to try to keep the area dry. "You will have some pain until the nail grows all the way out and forms a new, even edge," he said.

I didn't know at the time that it takes six to ten weeks for the toenail to grow out completely. At one point I even had to take antibiotics because of pus that formed around the toenail. By the end of each practice session, I was limping. I had no choice but to finish out the season this way. At the end of the season I visited John Rowbotham, a surgeon at the Harvard health center, who recommended surgery on the toe. Under local anesthesia Dr. Rowbotham removed the inner border of the nail and the growth portion at the bottom of the nail. Although this permanently narrowed the toenail, it completely cured the condition, and I have had no trouble since.

As a sports medicine physician, I have performed this operation (Winograd procedure) on many athletes. It is highly successful, and in my opinion it certainly beats living with the problem.

DIAGNOSIS AND TREATMENT

Everyone recognizes an ingrown toenail. The pain is sharply localized and the toe is red and irritated. Just because the conservative treatment did not work for me, it does not mean that it is ineffective. Most people are helped by the use of a small bit of cotton at the edge of the nail to keep the area dry. Cutting a V shape in the middle of the nail also helps. Filing the nail's top surface to make it thinner is an excellent idea. Try to avoid direct pressure against the edge of the nail. Wear a wide athletic shoe to take the pressure off the inflamed area.

When you cut your toenails, cut straight across. Never cut into the corner of the nail. This just invites a sharp, ragged edge that can create an ingrown toenail. If the nail festers for two or three months despite conservative treatment, surgery may be the only answer. The operation is performed under local anesthesia. The purpose is to remove the jagged toenail.

Sesamoiditis

CAUSES

To gain a strong pushoff in running, you must be able to bring your great toe down against the ground quickly and forcefully. Aiding in your pushoff are two little bones *(sesamoid bones)* embedded in the tendons that run to your big toe. The sesamoids act like a large pulley, making the tendons more efficient mechanically. I compare the sesamoid bones to the kneecap.

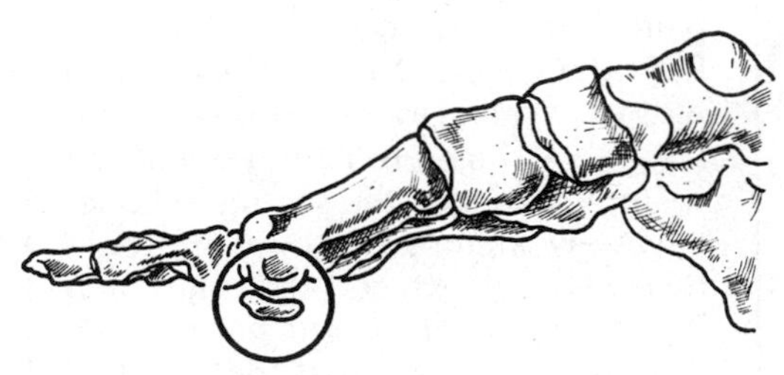

The Seasmoid Bones

Because of their location, the sesamoids activate every time you step off. They can become irritated or even fractured in sports. The irritation is called *sesamoiditis*.

I have seen sesamoiditis mainly in endurance runners and baseball pitchers. The pitcher's problem is easier to understand. When you pitch a baseball, you push off from the rubber strip on the mound with the base of your great toe—where the sesamoids lie. Thus, pitchers put the sesamoids under tremendous pressure. Experienced pitchers like Mike Torrez of the Boston Red Sox, who has pitched for fourteen seasons in the big leagues, take special precautions to avoid sesamoiditis. Mike always pads this part of the foot with one-fourth-inch foam rubber pads, giving an extra layer of padding over the sesamoids so that they cannot be directly pressured from the firm pitcher's mound.

DIAGNOSIS AND TREATMENT

Sesamoid pain is localized in the ball of your foot, directly under the end of your first metatarsal at the base of the the big toe. When I push on this area with my thumb, I find extreme tenderness. Either of the two sesamoids or both can be inflamed. The inflammation, sesamoiditis, is a smoldering problem.

I x-ray the foot. Occasionally I find that one of the sesamoids has fractured—a hard diagnosis to make because many people have two-piece sesamoids normally. Sometimes I am lucky. An outstanding young gymnast developed a broken sesamoid. Fortunately I had x-rayed her foot earlier for a different fracture. Therefore I could be sure that the sesamoid was originally one piece and that it had fractured into two pieces. I placed her in a cast for six weeks, and the fracture healed. Without the previous x-rays, I would have diagnosed the problem as sesamoiditis.

For the pitchers of the Boston Red Sox, Dr. Arthur Pappas (team medical director) and I x-ray their feet periodically to determine if arthritis is developing in the sesamoid-metatarsal joint. The first signs are spurs on the edges of the sesamoid as it rubs against the underside of the first metatarsal.

The best treatment for painful sesamoiditis is rest. Don't exercise for at least a week. Take six to eight aspirins a day with meals to reduce the inflammation. Pad the sesamoid when you return to sports. Use one-fourth-inch sponge rubber. Cut a pad approximately two inches long and one and a half inches wide. Tape it in place with half-inch adhesive tape. It is easiest to use sponge rubber with a sticky surface. Ace adhesive or benzoin sticky spray also will help the pad adhere to your foot. The extra padding will absorb shock in the sesamoid area. If these measures are not effective, cortisone injection is my next choice. A series of three injections is sometimes necessary to eliminate the symptoms. Surgery is seldom required. I have performed the operation to remove the painful sesamoid only twice in my career.

BILL LEE'S SESAMOIDS

"Bill Lee, when he was pitching for the Boston Red Sox, came to me with pain in his right foot," says Dr. Rob Roy McGregor, D.P.M. "I examined his foot. When I touched his sesamoid bone, he cringed in pain. I employed a simple pad over the sesamoid bone. It was my attempt to find out if I could help him. He pitched that night and had no pain. He came back a few days later and said that the pad had worked. I made him an orthotic and he was quoted in the papers as saying that I not only fixed his foot, but helped him with his shoulder pain."

Foot Calluses

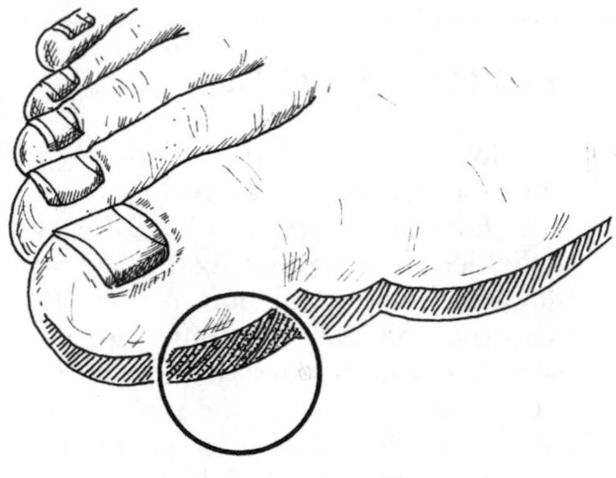

Calluses

CAUSES

When the skin is pressured, it thickens, and what we commonly call a *callus* develops. The pressure can come from many sources—tight-fitting shoes, an oddly shaped foot, a cocked-up toe. Marshall Hoffman, my coauthor, gets thick calluses on the tip of his second toe, the longest toe on his foot. The toe rubs against his shoe as he runs.

Calluses make sense from the body's point of view. If your skin did not protect itself by thickening, you would have a series of open sores. Callus formation, however, can become counterproductive. As a callus builds up, the skin becomes more prominent and pressure against the shoe increases. This, in turn, leads to additional callus formation. When heavy callus builds up, blisters form at the edge of the callus. Blisters occur because the shearing effect is at the junction of the callus and normal skin.

Calluses commonly form in certain locations and conditions such as the following:

- The heel, especially if your heel is tilted to one side or the other.
- The ball of the foot, especially if you have a bunion and use your second metatarsal for pushoff.
- The condition known as Morton's toe, in which the first metatarsal is short and the second metatarsal is quite prominent. (This is Marshall Hoffman's problem.)
- The top of the first toe joint in persons who have cocked-up, or claw, toes. This occurs most commonly in high-arched feet.
- The inner side of the big toe in those who have a bunion deformity.
- The outer aspect of the base of the little toe.

Calluses seldom form in the arch. They usually develop in areas where the foot rubs against the shoe.

DIAGNOSIS AND TREATMENT

How do we treat calluses? You have to keep fighting back. As the calluses build up, they must be trimmed down. I use a sharp scalpel blade or a special callus shaver knife. Most athletic trainers use a callus file—similar to a nail file—to smooth them. It is important to taper a callus as it joins normal skin to avoid blister formation around the edge.

What if calluses return despite your best efforts? You have to determine the root of the problem. For example:

- A prominent bone lying directly under the skin. A pad or adhesive tape can be used to cover the bone. If conservative methods fail—which happens only in a small percentage of cases—the bone can be shaved surgically.

- A crooked heel. If you land on one side of the heel, a heel cup can spread the force of the heel's landing over a large area of skin.
- A change of shoes. It has been my experience that changing shoes helps 50 percent of the time. "I had to buy three different pairs of running shoes before I could find one that did not cause calluses," testified Marshall Hoffman. "It was a $120 callus problem."

Once the root of the problem is eliminated, calluses disappear. You are left with perfectly normal skin in an area that was once thickly callused. Your skin is an organ just like your liver or your heart is. It is dynamic, is constantly changing and growing, and has the ability to correct itself if given a fair chance.

Fractured Toes

CAUSES

Breaking a toe is a common event in sports. "Imagine ten giants with size fifteen feet, jumping in a ten-foot area," points out Ray Melchiorre, Boston Celtics trainer. "It is almost impossible for anyone to land without coming down on somebody else's foot. That's the main cause of fractured toes in professional basketball."

Fracturing a toe is also common in other sports—soccer, football, lacrosse, hockey, baseball, and weight lifting.

Baseball players get broken toes from fouling balls off their toes. Fred Lynn, the all-star outfielder, fractured his big toe with a foul ball at the end of the 1980 season. The most spectacular broken toe I have ever witnessed was during my playing days at Harvard. In football practice an end, running a crisscross pattern, was looking back for the pass. In those days the goal post sat only a few yards behind the goal line. The end, concentrating on the ball, ran into the goal post. He ended up with a fractured toe and a bloody nose.

DIAGNOSIS AND TREATMENT

Immediately after the break, you feel very little pain. The toe is in shock. Within thirty minutes, however, the pain intensifies and the swelling starts. Nothing else makes a toe swell as much as a fracture does. The player is usually in agony by the time I see the toe.

The best immediate treatment is RICE—rest, ice, compression, and elevation. Even if you think that your toe isn't broken, start RICE. Do not wait. Even a severe bruise will produce swelling. Thus, RICE will benefit you. The ice will keep down the swelling and you will heal faster.

By the time I see a fractured toe, you are unable to walk. The

FRED LYNN'S BIG TOE

In late August 1980, the Red Sox were moving back into the pennant race. Fred Lynn, the Red Sox all-star outfielder, was batting against Steve McCatty of the Oakland A's. With two strikes on Lynn, McCatty threw a hard slider that broke in on Lynn.

Lynn swung to spoil the pitch. The ball hit off the lower part of his bat and ricocheted off his right foot. The umpire called it a foul ball. On the next pitch, Lynn doubled to right field. As he rounded first base, it was clear that he was not running properly. When he pulled up into second base, he was limping. At the end of the inning, he came off the field.

"I think my big toe is busted, Doc," Lynn said.

I drove him immediately to the hospital. His x-rays showed a fracture of the big toe. It was broken in five different pieces. Fortunately for Lynn, the toe was straight and the pieces were well aligned.

"This is going to take four to six weeks to heal," I told Lynn as I made him a plaster splint.

Fred was heartbroken. "We were just getting back into the pennant race, and this had to happen. I think I knew my toe was broken because it began to swell immediately. The numbness wore off as I rounded first base. I guess you just can't run if your big toe isn't working right."

"At least you won't need a cast," I said.

"How can I keep my legs strong while my toe is healing?"

I explained that this was no reason for him not to use the exercise machines in the Red Sox clubhouse. "You can do everything except toe raises," I said.

The next day Arthur Pappas, a consulting physician for the team, fitted Lynn with a cast boot. He taped the big toe to the second toe, a human splint. Five days after the injury, Lynn started his workouts. If the Red Sox had made the playoffs in 1980, Lynn could have played.

toe is twice the normal size and often black and blue. You will not be able to wiggle your toe because of the intense swelling and pain.

I have no way to be sure that your toe is broken without x-rays. In 99 percent of the cases, the break occurs in the middle of the bone, not in the portion of the bone that forms the joint.

If the toe is crooked at the fracture site, it must be straightened. I numb the toe with Novocain and then push on it until it straightens. Often I place a pencil or a pen between the toes to serve as a fulcrum to straighten the toe. If the top of the toe is broken but remains straight, I do not have to manipulate it.

I tape the broken toe to the one next to it, using one-fourth-inch tape because it contours to the shape of the toes. I seldom cast a fractured toe. If it is summer, cut the front out of an old sneaker. Most patients do fine without crutches.

Most toe fractures heal in three to six weeks, but it might be twelve weeks before the toe is completely pain-free. Longer disability from a toe fracture is rare. The toe bone usually heals without any residual stiffness of the joints around the fracture. Surgery to set the bones in toe fractures is rare.

Soft Corns

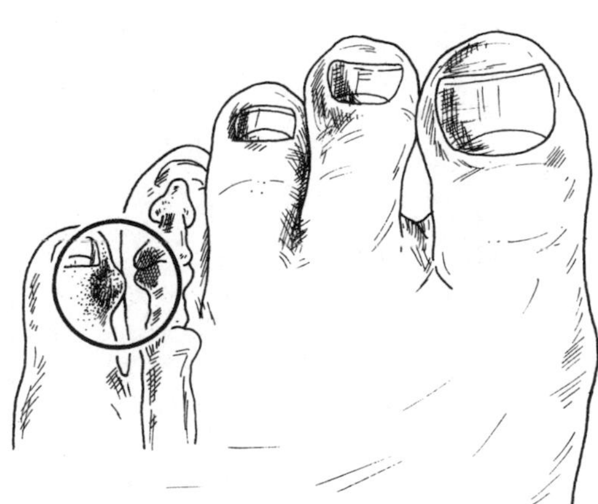

Soft Corns

CAUSES

A *corn* is a type of callus that forms between your toes. It looks like a small, wet volcano crater that has been scooped out of the skin. It forms because of the pressure of tight-fitting shoes or a bony spur on one toe. Instead of thickening and forming a regular callus, the skin keeps peeling and floating away. This, in part, happens because the foot perspires between the toes. Eventually a crater forms. The skin at the bottom of the crater becomes very thin. The thinning-out process produces pain.

Corns commonly form between the fourth and fifth toes. In some people the fourth toe has an odd shape. It has a little extra bony prominence on the outer side, and this prominence pushes into the inner side of the fifth toe. The constant pressure of this bony spike leads to the formation of a corn.

DIAGNOSIS AND TREATMENT

I see corns when they are fully developed and very painful. I always x-ray them to determine if bone pressure is causing the problem. The treatment for corns:

- Get wider shoes. This eliminates forcing the toes together.
- Place a small bit of cotton or lambs wool between the toes. This soaks up the moisture and distributes the pressure over a large patch of skin. Commercialy available corn pads serve the same purpose.

With the pressure off, the corn usually heals itself. Surgery is rarely necessary. But when it is required, it is simple, similar to surgery for a bunionette. I simply remove the bony prominences causing the trouble. The operation is performed under local anesthesia on an outpatient basis. You can return to full sports activities in three to four weeks.

Blisters

CAUSES

Blisters are localized collections of fluid in the outer part of the skin. Whether the blister forms on your hand or on your foot, all are produced the same way. They are usually the result of repeated rubbing against the skin in a relatively small area. The rubbing causes a shearing stress on the surface, which means that the top surface of the skin is moving back and forth while the bottom skin layers are stationary. In time the stress separates the two layers, and the space fills with tissue juices.

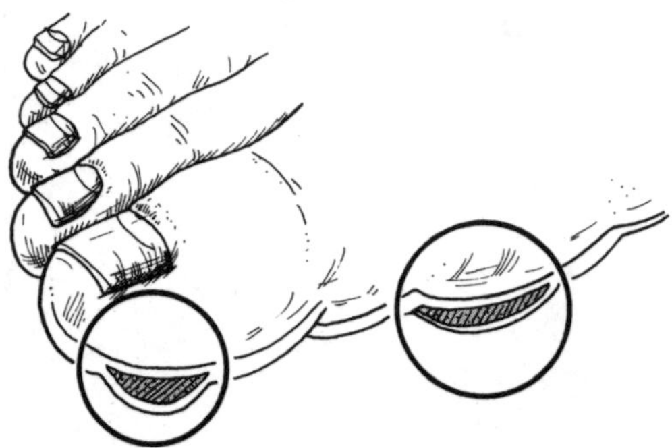

Blisters

The outer layer of the skin forms the covering of the blister. Next is the fluid. The bottom is the red, raw crater.

A blister aches because the skin is damaged. It is very much like a burn; in fact, burns produce blisters.

Blisters occur early in the sports season. Pitchers get them on their pitching hands from gripping the baseball. Bicycle racers get them on their behinds from rubbing against the seat. Running athletes get them on their toes and the soles of their feet.

The problem is that the skin is not ready for the wear and tear of athletics. Second, in running you usually start the season with a new pair of shoes. The combination almost always produces foot blisters.

In running athletes most blisters are on the big toe and the ball of the foot, the stress areas in running. However, I have treated blisters on the outer border of the foot and on the heel.

DIAGNOSIS AND TREATMENT

When a blister forms, you feel a warm, uncomfortable sensaton. The area becomes red. This is the time to limit the blister's development. Here are my recommendations:

- Apply benzoin sticky spray or tough skin to the reddened area. Both wash off with alcohol.
- Apply adhesive tape to the reddened area; do not use gauze. The shoe will rub against the tape instead of the skin.
- Apply petrolatum. Ray Melchiorre and some marathon runners I know prevent foot blisters by covering their feet with petrolatum. If you are prone to blisters, try rubbing petrolatum into the shoe at the site where it rubs against your skin.

- Do not wear socks. Athletic shoes are designed to fit like a glove. Thick socks prevent a glove-like fit. About half the athletes examined at Sports Medicine Resource shun socks. In my opinion the main reason to wear socks is to absorb sweat and prevent odor.
- Stretch the shoe. If one area of your shoe causes blisters, stretch the shoe at this point or even put a small cut in it to relieve the pressure.

If you get fluid production in the blister, you have already permanently damaged the top layer of skin. Medical studies show that blisters heal faster when the fluid is drained. First, cleanse the blister site with alcohol. Second, sterilize a needle by heating it over a flame until it turns bright red. When the needle is cool, use it to puncture the edge of the blister.

Drain the fluid by pressing the top of the blister. Do not remove the top of the skin over the blister because you will be creating an open wound, which is susceptible to infection.

Then tape a small doughnut-shaped pad over the blister. Many times you can resume exercising immediately. Some people place adhesive tape (with no gauze) tightly over the blister. Both methods work. After you open the blister, wash your feet at least once a day with surgical soap—Phisohex or Hibiclens—to prevent infection. New skin will form in ten to fourteen days. Wear protective covering until then.

Black Toenail

CAUSES

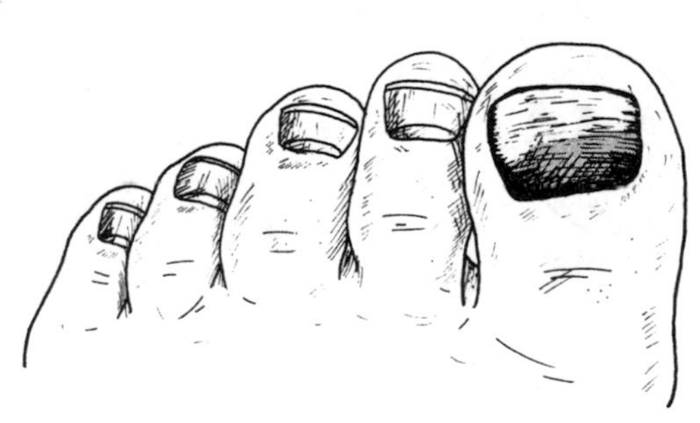

Black Toenail

Before 35 million Americans took up jogging, black toenail was a rare ailment. Now I see it almost daily.

In medical terms, *black toenail* is a subungual hematoma. *Subungual* means "under the toenail." *Hematoma* means "collection of blood." The black under the toenail is old blood.

In running athletes, the injury occurs when the toe jams into the end of the shoe. It happens often when you buy shoes that are too tight or you run down too many hills.

"When you run down hills, you have a natural tendency to brake yourself," says Marshall Hoffman. "Your toes are always slamming into the front of the shoes. If you like to run down hills or you train on hills, you are a candidate for black toes."

At first the injury is painful. The blood collected under the nail creates tremendous pressure. Besides giving you pain, the pooled blood disturbs the normal nutritional pattern of the nail, restricting the blood supply. Within two to three weeks the nail usually falls off, exposing the sensitive skin under the nail. More often than not, I see black toenails just before they are ready to drop off.

If I can see this injury when it first happens, I can save the nail. The trick is to make a small hole in the center of the toenail to release the accumulation of blood. You can do this yourself. Heat a paper clip over a flame. When the end of the paper clip becomes red, push it down into the center of the nail. The heat produces the hole and the pressure is relieved.

Most people shy away from fiery-hot paper clips. They prefer to have the doctor do the job. If the puncturing procedure is not done within the first forty-eight hours, it is pointless to do it at all. You can save yourself many hours of misery by relieving the pressure as soon as the blood collects.

If you are unable to produce a hole in the nail and black toenail forms, it is good to use a small Band-Aid or a piece of adhesive tape to keep the nail attached to the toe. This should be done for three weeks to allow the skin under the toenail to toughen and be ready to face the outside world when the nail falls off. There is no reason to stop running except during the first two days of extreme discomfort. A new toenail always grows back within three to six months.

It is important to think about why the injury happened and to evaluate your shoewear. If the shoes are too small, they should be changed during the healing phase of black toenail.

BASKETBALL TENNIS TOE

"Black toes are very common in basketball players," says Ray Melchiorre, trainer for the Boston Celtics. "Gerald Henderson lost two big toenails. Kevin McHale, rookie Don Newman and Dave Cowens lost one each.

"Because basketball players are always stepping on one another. It's very common.

"Secondly, many players mash their toes just stopping quickly."

Bunionette

CAUSES

A *bunionette* is a painful enlargement over the outer part of the ball of your foot, at the end of the fifth metatarsal bone at the base of the fifth toe. The difference between a bunionette and a bunion is location. A bunion is on the inner edge of the foot.

Some persons have foot anatomy that makes them prime candidates for bunionettes. The fifth metatarsal bone has a prominent peak just beneath the skin. In tight shoewear the skin is forced against the bursa sac between this bony prominence and the skin. The bursa sac inflames and enlarges, leaving less room in the shoe for the foot. The greater the swelling, the tighter the shoe; the tighter the shoe, the greater the swelling. The pain becomes so severe that wearing shoes is impossible.

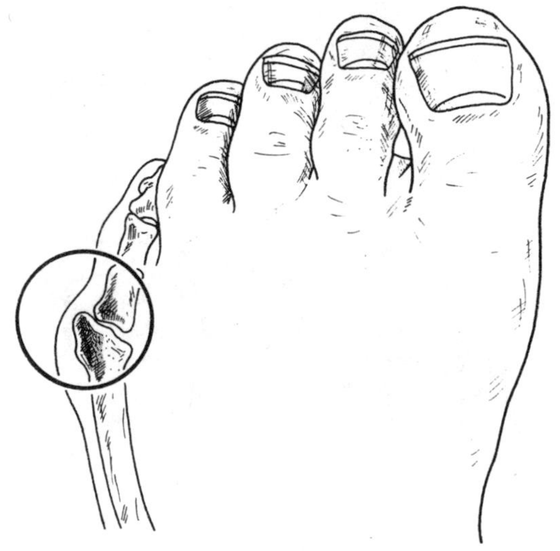

Bunionette

DIAGNOSIS AND TREATMENT

"It aches right here," and the patient points to the outer edge of the foot. "I have had to buy wider shoes in order to walk comfortably." These are two statements that bunionette sufferers always make.

The bunionette—sometimes red and swollen—is tender to the touch. X-rays reveal a prominence on the outer aspect of the far end of the fifth metatarsal. This is the culprit. Normally I do not find other foot abnormalities.

The best treatment is wider shoes to get the pressure off the bunionette. Sometimes I have you cut a hole in your shoes. As with all bursitis conditions, I prescribe anti-inflammatory medicine. Start with six to eight aspirins daily with food, and ice the area for twenty minutes twice a day. In a week the bursitis should quiet down. If the anti-inflammatory medicine and ice routine does not help, I inject cortisone into the bursa sac. This shrinks the inflamed bursa sac during the next five to seven days. If necessary, two additional cortisone injections can be made into the local area.

If all the treatments fail and pain persists, simple surgery is the only answer. I remove the bony prominence on the outer aspect of the end of the fifth metatarsal. The inflamed bursa sac is removed at the same time. It can be done under local anesthesia and requires a two-day hospitalization.

Metatarsal Stress Fractures

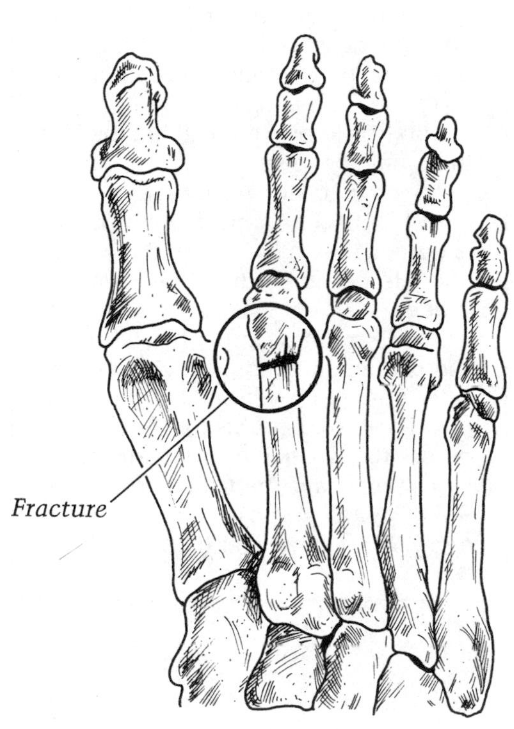

Metatarsal Stress Fracture

The metatarsal stress fracture is the most common stress fracture seen in the human body. It was so commonly seen in the military, that it was termed "the march fracture." This came about during wartime, when military recruits who had relatively sedentary life-styles were brought into service. They were then marched on many long jaunts, and the pounding that their feet took produced these stress fractures.

The fracture occurs as a cumulative result of the force applied to the foot in running or marching. No one foot strike is sufficient to break the bone, but the millions of times that the foot pounds onto the ground is enough to cause a stress fracture. More details on the mechanism of stress fracture production are given in the fracture chapter.

Your foot has five metatarsal bones. The first metatarsal bone, which leads to the bottom of your big toe, is so stout and strong that it never gets a stress fracture. The other four metatarsals are much skinnier and are candidates for stress fracture production. However, the most likely site of a stress fracture is the fourth metatarsal bone, which leads to the bottom of the fourth toe. The site of the stress fracture is always at the far end of this bone—in the area of the ball of your foot.

DIAGNOSIS AND TREATMENT

I listen to the athlete's story very carefully. Usually, a stress fracture in this area comes on two to three weeks into the season. It takes this long for the stress on the bone to take its toll. It can also occur two to three weeks after a radical increase in the training routine. This occurs when people are building up for major events such as marathons. As far as the metatarsal bone is concerned, it cares about any increased level of activity beyond that to which it is accustomed. So, starting a season or drastically increasing your training routine are both events which can produce a metatarsal stress fracture.

The pain is often vague at first. It will become sore after you run, but you will be able to run through the pain. However, when the bone crack becomes well developed, it is much too painful to run at all. Often, in fact, the athlete limps when trying to walk. By this time, the pain is steady and like a toothache. Examining the foot, you can make this diagnosis by dead reckoning on the anatomy. By this I mean that you can apply pressure directly over the suspected stress fracture and there will be sharply localized tenderness in this area. I start in the arch of the foot area and work my fingers down toward the toes. You can do this on each of the metatarsal bones separately. When you get to the ball of the foot area on the fourth metatarsal, the patient usually protests violently. Often, there will be a sensation of swelling around the fracture itself. If the fracture is four to six weeks old, there will also be a pecan-size mass of healing new bone around the crack.

I always take x-rays. The initial x-ray may show no evidence of a crack. You must have a high index of suspicion that this is the problem. I let two more weeks go by and take them out of running activities. I then take another x-ray, and usually you can see the crack at this point. You will also begin to see healing bone present one month after the start of symptoms. This fracture requires two to three months to completely heal and allow you to return to full activity. Application of a cast is seldom necessary. I do this if the patients have to limp when they walk and have steady, disabling pain. I would rather put them on crutches for five to seven days to allow the tenderness to subside. However, a walking cast is very little nuisance and can be applied easily for three to four weeks until healing has progressed to a final stage.

I have seen a few patients who have actually had a stress fracture return. To prevent this, gradually increase training activity once the fracture is healed. You should never increase your mileage more than two to three miles a week in preparation for a running event. If you increase faster, you run the risk of a stress fracture.

ROB ROY MCGREGOR, D.P.M.

Rob Roy McGregor, D.P.M., is one of the founders of Sports Medicine Resource, Inc. He was the first member of Sports Medicine Resource to bring his message to the people. For two years he did segments on sports medicine for the television show, "Evening Magazine" in Boston. Later he made appearances with model Cheryl Tiegs on "Good Morning America." Jim Fixx, author of the *Complete Book of Running,* calls Rob "one of the the nation's half-dozen best-known podiatrists." Peter R. Cavanaugh, author of *The Running Shoe Book,* says Rob "fits under the umbrella of a 'philosophical podiatrist.' "

Rob Roy knows more about the foot than anyone else I know. In thirty years of practice, he has seen more than 180,000 patients. He has put his experience with the foot into practical use: He designed the Etonic running shoes.

Rob Roy has an enthusiasm for sports medicine. Some people equate it to a religious fervor. He tours the country to explain his "global theory of sports medicine." Hundreds of thousands have benefited from Rob Roy's high purpose and scientific mind. He is a pioneer in the field.

McGregor's Eight Imperatives For Proper Shoe Fit

Rob Roy McGregor, D.P.M.

Over the years—as a result of being asked the question "What should I look for in the proper fit of a running shoe?"—I've come up with a set of working principles that I give to patients and offer at lectures. I call them the *eight imperatives* because each aspect is necessary—indeed, imperative—for a proper fit.

1. *Heel Height*
   Running in heels that are too low causes excessive pull on the calf muscles and the Achilles tendon. This can lead to chronic pain in these two structures. Remember, forces up to three times the body weight are activated every time the foot strikes the ground. The legion of limpers are often helped simply by raising the heel height, either with heel lifts or new shoes.

MCGREGOR'S PHILOSOPHY

A single injury is easy to treat. The problem is that all the body parts are connected—the ankles, the feet, the knees, and the hips. The single injury can turn into multiple injuries if you have a structural abnormality. For example, you may have pronating feet. Until the day of your injury, you may have been able to do very well. Now, your body is deficient. Suddenly, a whole new symptom complex develops. The pronating feet aggravate your Achilles tendon. The tendinitis makes you hobble, which puts pressure on your knee and calves.

Patients say to me, "I've always pronated."

"That's right," I say, "but, now you have a weakened member and the weakened member does not give your body a chance to heal. Pronation now becomes a problem."

An unlikely chain of events? No! I see it every day in my office.

Rob Roy McGregor, D.P.M.
Chief of Podiatry
Sports Medicine Resource, Inc.

2. *Heel Cushion*
   If the heel cushion is too hard, the heel becomes bruised. I call this "jogger's heel," the first socially acceptable foot problem. Conversely, if the heel has too much cushioning, you sink into your shoes and lose some of the rebound energy that accompanies each footstrike. A soft heel cushion leads also to fatigue.
3. *Heel Stability*
   The heel counter, at the back of the shoe, encircles and holds the heel in place. It should be stiff to control heel motion. The more it prevents excessive rolling in or out, the better.
4. *Wedge Support*
   The foot needs support against the rolling in or rolling out of the arch that takes place in running. The best support is achieved by adding a wedge from the heel to the ball area of the shoe.
5. *Forefoot Cushion*
   Biomechanists have proved that the greatest amount of vertical force the foot must absorb is just behind the ball of the foot. To protect the foot, cushioning must be built into the shoe sole. The reason a person who runs in tennis sneakers feels "burning" in the ball of the foot is that the sneakers do not have much cushioning in that area.
6. *Forefoot Flexibility*
   The shoe should bend where the foot bends—at the ball. If the shoe is too stiff, it can cause shin splints, Achilles tendinitis, or lower leg pain. The stiff sole causes the muscles in the foot and the leg to work excessively.
7. *Toe Clearance*
   To function properly, toes should have clearance both above and straight ahead. You should be able to wiggle your toes easily up and down. If there is extra pressure

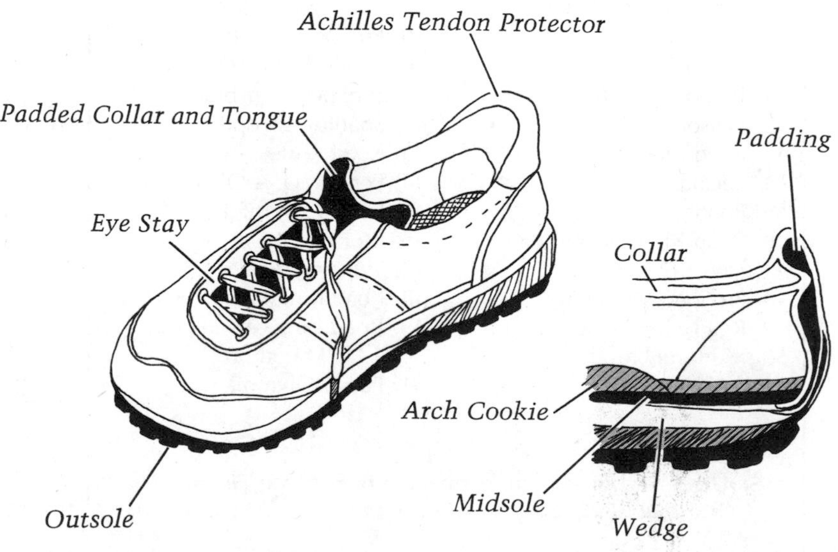

Elements of a Good Running Shoe

RATINGS FOR 100 MODELS OF RUNNING SHOES

Men's Training Shoes

SHOCK RATING (Gs)	MOTION CONTROL (Rank)
1. Brooks Nighthawk 450	1. Brooks Nighthawk 450
2. New Balance 420	2. Nike Daybreak
3. Nike Tailwind	3. New Balance 420
4. Saucony Jazz	4. Saucony Jazz
5. Tiger X-Caliber	5. Brooks Vantage
6. New Balance 620	6. Nike Tailwind
7. Adidas TRX Trainer	7. Tiger X-Caliber
8. Tiger Ultra-T	8. Adidas TRX Trainer
9. Nike Daybreak	9. Brooks Vantage Supreme
10. Brooks Vantage	10. Pony Shadow
11. Adidas Marathon Trainer	11. Adidas Marathon Trainer
12. Brooks Hugger GT	12. Etonic Roadworker
13. Etonic Roadworker	13. Brooks Hugger GT
14. Brooks Vantage Supreme	14. Tiger Ultra
15. Pony Shadow	15. Adidas SL80
16. Adidas L.A. Trainer	16. New Balance 620
17. Nike Bermuda	17. Reebok Aztec
18. Brooks Super Villanova	18. Saucony TC-84
19. Reebok Aztec	19. Osaga KT26
20. Saucony TC-84	20. Etonic Eclipse Trainer
21. Saucony 1980 Trainer	21. Brooks Super Villanova
22. Etonic Stabilizer (KM520)	22. Ambi Trailblazer 2001
23. Osaga KT26	23. Nike Bermuda
24. Adidas SL80	24. Adidas L.A. Trainer
25. Jox Road Handler	25. Ambi Warrior 9001
26. Etonic Streetfighter	26. Nike Waffle II
27. Saucony Hornet 84	27. Saucony Hornet 84
28. Penney USA Olympic Gold	28. Saucony Hornet
29. Nike Waffle II	29. Brooks Silver Streak
30. Etonic Eclipse Trainer	30. Etonic Streetfighter
31. ProSpecs Munich II*	31. Penney USA Olympic Gold
32. Puma Elite Roadrunner	32. Brooks Villanova
33. Ambi Trailblazer 2001	33. LeCoq Sportif-V
34. Brooks Villanova	34. ProSpecs Munich II
35. New Balance 455	35. New Balance 455
36. Brooks Silver Streak	36. Saucony 1980 Trainer
37. Saucony Hornet	37. Jox Roadhandler
38. Ambi Warrior 9001	38. Puma Elite Roadrunner
39. LeCoq Sportif-V*	39. ProSpecs MV 80
40. Spaulding Topflight	40. Etonic Stabilizer KM520
41. ProSpecs MV80*	41. ProSpecs Innsbruck
42. Wilson Force 5	42. Spaulding Topflight
43. Road King	43. Road King
44. LeCoq Sportif-T*	44. Penney USA Olympics
45. Penney USA Olympic	45. LeCoq Sportif-T
46. Road King Mission Bay	46. Morton Flites
47. Fayva Olympian	47. Wilson Force 5
48. ProSpecs Innsbruck*	48. Road King Palos Verdes
49. Road King Palos Verdes	49. Road King Mission Bay
50. Morton Flites	50. AAU Marathon
51. AAU Marathon	51. Fayva Olympian

Women's Training Shoes

SHOCK RATING (Gs)	MOTION CONTROL (Rank)
1. New Balance 420	1. Nike Liberator
2. Nike Tempest	2. Brooks Lady Nighthawk
3. Brooks Lady Nighthawk	3. New Balance 420
4. Saucony Jazz	4. Nike Tempest
5. Brooks Lady Vantage	5. Adidas TRX Trainer
6. Adidas TRX Trainer	6. Saucony Jazz
7. Etonic Roadworker	7. New Balance 620
8. Pony Shadow	8. Brooks Lady Hugger GT
9. Brooks Lady Vantage Supreme	9. Pony Shadow
10. New Balance 620	10. Brooks Vantage
11. Brooks Lady Hugger GT	11. Etonic Roadworker
12. Reebok Aztec Princess	12. Adidas Marathon Trainer
13. Tiger Tigress	13. Tiger Tigress
14. Nike Liberator	14. Brooks Lady Super Villanova
15. Adidas Marathon Trainer	15. Etonic Eclipse Trainer
16. Brooks Lady Super Villanova	16. Brooks Lady Super Villanova
17. Etonic Eclipse Trainer	17. Etonic Stabilizer
18. Saucony MS Trainer	18. Ambi Princess
19. Ambi Princess	19. Saucony TC 84
20. Etonic Streetfighter	20. Saucony Ms Hornet 84
21. Saucony TC 84	21. Reebok Aztec Princess
22. Ambi Trailmate	22. Saucony MS Trainer
23. Etonic Stabilizer	23. Brooks Lady Villanova
24. Saucony MS Hornet 84	24. Brooks Lady Silver Streak
25. Brooks Lady Silver Streak	25. Etonic Streetfighter
26. New Balance 455	26. Wilson Force 5
27. Brooks Lady Villanova	27. New Balance 455
28. ProSpecs Innsbruck	28. Ambi Trailmate
29. Wilson Force 5	29. ProSpecs Innsbruck

Men's Racing Shoes

SHOCK ABSORPTION	MOTION CONTROL
1. Adidas TRX Super Comp	1. Tiger Ultimate
2. Nike Eagle	2. Adidas Marathon 80
3. Etonic Eclipse	3. Adidas L.A. Comp
4. Adidas Marathon 80	4. Nike Eagle
5. Tiger Ultimate	5. Saucony Silver Streak
6. Nike Elite	6. Adidas TRX Super Comp
7. Adidas L.A. Comp	7. New Balance Comp 100
8. Brooks John Walker RT-1	8. Nike Elite
9. Reebok 10K	9. Etonic Eclipse
10. New Balance Comp 100	10. Brooks John Walker RT-1
11. Saucony Silver Streak	11. Reebok 10K

Women's Racing Shoes

SHOCK ABSORPTION	MOTION CONTROL
1. Adidas TRX Supercomp	1. Adidas Marathon 80
2. Etonic Eclipse	2. Adidas TRX Comp
3. Adidas Marathon 80	3. Saucony Silver Streak
4. Adidas TRX Comp	4. Adidas TRX Super Comp
5. Brooks Lady John Walker	5. New Balance Comp 100
6. Saucony Silver Streak	6. Etonic Eclipse
7. New Balance Comp 100	7. Brooks Lady John Walker

SOURCE: Reprinted with permission of *Running Times*.

(Asterisks denote models which were not eligible for the Gold or Silver Shoe Rating because either they were obtained directly from manufacturers rather than "off the shelf" from stores, or because at the time of publication they were not for sale in stores in sufficient quantity.)

THE CELTICS' ORTHOTICS

"I've made temporary orthotics for half the players on the team," says Ray Melchiorre, trainer of the Boston Celtics.

on the toes from the shoe, irritations like blisters, calluses, corns, or runner's toe (a blood blister under the nail) can form.

8. *Comfort*
   Do not buy a shoe that is not comfortable. It should have the proper configuration at the outset. That does not mean, however, that the shoe cannot become more comfortable. Two old sayings that apply are: "If the shoe fits, wear it" and "comfortable as an old shoe."

Orthotics

Rob Roy McGregor, D.P.M.

An *orthotic* is a special shoe insert made from a cast of the patient's foot or from the foot itself. It is a custom-made foot support. The purpose of an orthotic is to prevent the foot from pronating excessively or the arch from flattening.

Because pronation distributes the force of the footstrike (sometimes three times the body weight) over the entire foot, the turning-in is good. The problem arises when the foot twists excessively. The overpronation causes the lower leg to twist inward, which, in turn, sometimes causes the kneecap to rub against the long bone of the thigh—that is, runner's knee. The twisting can also extend to the hip (runner's hip). Overpronation can cause certain back, arch, ankle, or foot conditions.

Orthotics are rigid or soft. Rigid orthotics offer better foot control, but soft orthotics are more comfortable to wear. Only a small percentage of athletes and fitness buffs wear rigid orthotics in their athletic shoes.

Orthotics are like eyeglasses. If you need a pair, you need your own. You cannot wear somebody else's and have them work. My personal bias is that orthotics do not correct bodies; they correct functions of the body. An orthotic might even permit the body to function so well that it may not need constant or ongoing mechanical help. It can fail, too.

The use of orthotics is still controversial. After thirty years of practice I have finally figured out why. Foot scientists and podiatrists are not sure what must be done to control foot motion. This does not mean that foot science is not moving ahead. It is just a statement of fact. There are many variables of the footstrike that cannot be measured with scientific accuracy to create a valid control system.

Only recently have the scientific tools become available to collect and analyze data on how the foot works in motion. The studies combine information gained from the use of a force plate, high-speed photography, and a computer. The force plate identifies activity in the vertical, fore and aft, and side-to-side axes. The high-speed camera takes 400 frames per second. When information from these two sources is combined and fed into a computer, a graphic representation of the forces can be displayed by the computer. Such data, however, relate only to *gross* motion—or combined forces. The state of the art is not sufficiently advanced to tell us the *distribution* of force—that is, where the force is at any moment in the foot and shoe relationship. Scientists really need a force plate in the shoe, not under it. Dr. Peter Cavanaugh, a biomechanist at Penn State University, is currently working on this problem.

I do not have a well-defined historical perspective regarding the use of mechanical aids for foot function or comfort. The history, however, seems to be longer than some might have us think. I can remember in my student days approaching a professor with a slight tinge of superiority, almost flaunting my newfound knowledge about some concept or aid for foot control, only to be told he was taught the same things. I now find myself in that position with today's students.

In the last analysis, the key is to recognize when the body needs help and to supply it.

The Manufacturing and Fitting of Orthotics

Alan R. Nowick, of Sports Medicine Resource, makes our soft orthotics. He uses a technique pioneered by Rob Roy McGregor in 1973.

"I start the process by heating three pieces of Styrofoam-like plastic at 300 degrees in a convection oven," Nowick explains. "Next, I have the patient stand in the heated plastic for thirty seconds. I provide cotton socks so that the feet do not burn. (In a manner of speaking, I do not want to give patients a hotfoot.) I place the patient's foot in the neutral position.

"I trim the orthotic with a pair of scissors. Next, I grind the orthotic so that it fits the shoe wall-to-wall. I check and readjust the orthotic so that it corrects the biomechanical abnormality. The soft orthotic should be comfortable to wear. In the first week the orthotic compresses 10 to 15 percent."

The soft orthotics have these advantages:

- They are comfortable.
- They are easy to get used to.
- They are easy to adjust.
- They provide extra cushioning.
- They are built slightly longer than rigid orthotics; thus, they give somewhat better forefoot correction.

Blisters may develop on the bottom of your foot along the inside border. They may be considered part of the breaking-in process and should go away almost as quickly as they came.

The soft orthotics are washable, but not in the washing machine; use soap, water, and sponge. Let them air-dry.

You should not have to change your regular shoe size to accommodate the orthotic. The question "Should I wear them

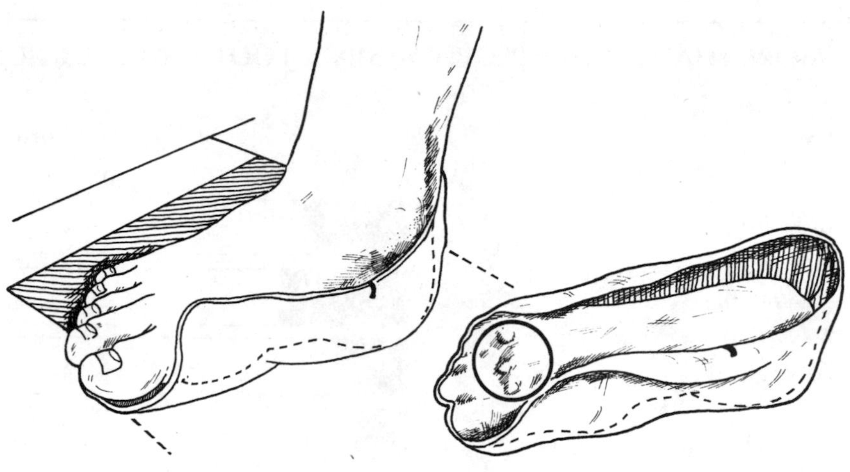

Fitting of Orthotics

all the time?" is answered simply—try with and then without. *You* judge.

Orthotics are made of lightweight materials that eventually wear out. The average pair lasts eighteen months. Hard orthotics—made of hard plastic—are also designed to correct biochemical abnormalities of the foot. They are more durable than soft orthotics. Many athletes and fitness buffs wear them in walking shoes, but only a small percentage use them in their athletic shoes because they are so rigid that they can be uncomfortable.

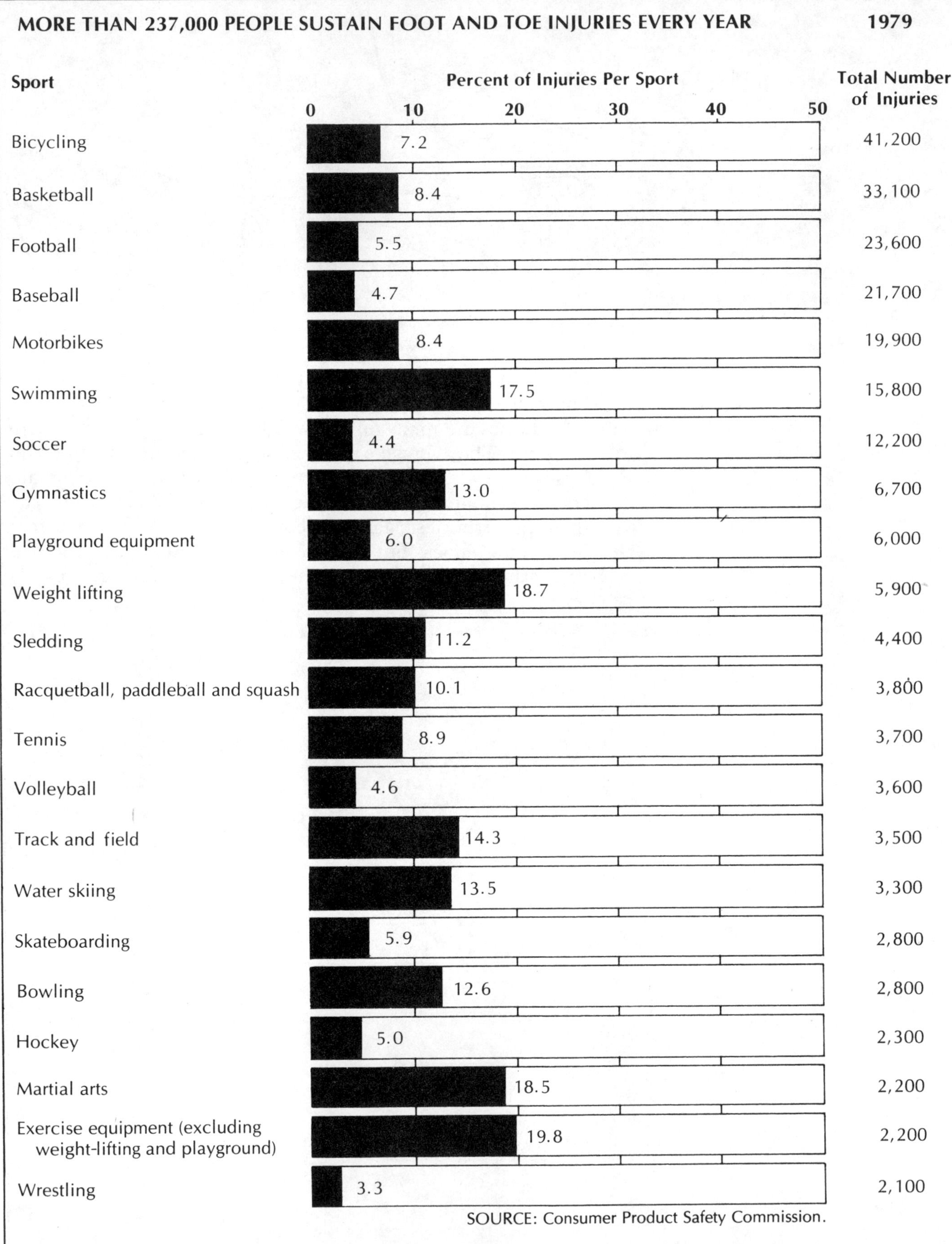
MORE THAN 237,000 PEOPLE SUSTAIN FOOT AND TOE INJURIES EVERY YEAR
1979
Sport
Percent of Injuries Per Sport
Total Number of Injuries
0
10
20
30
40
50
Bicycling
7.2
41,200
Basketball
8.4
33,100
Football
5.5
23,600
Baseball
4.7
21,700
Motorbikes
8.4
19,900
Swimming
17.5
15,800
Soccer
4.4
12,200
Gymnastics
13.0
6,700
Playground equipment
6.0
6,000
Weight lifting
18.7
5,900
Sledding
11.2
4,400
Racquetball, paddleball and squash
10.1
3,800
Tennis
8.9
3,700
Volleyball
4.6
3,600
Track and field
14.3
3,500
Water skiing
13.5
3,300
Skateboarding
5.9
2,800
Bowling
12.6
2,800
Hockey
5.0
2,300
Martial arts
18.5
2,200
Exercise equipment (excluding weight-lifting and playground)
19.8
2,200
Wrestling
3.3
2,100
SOURCE: Consumer Product Safety Commission.

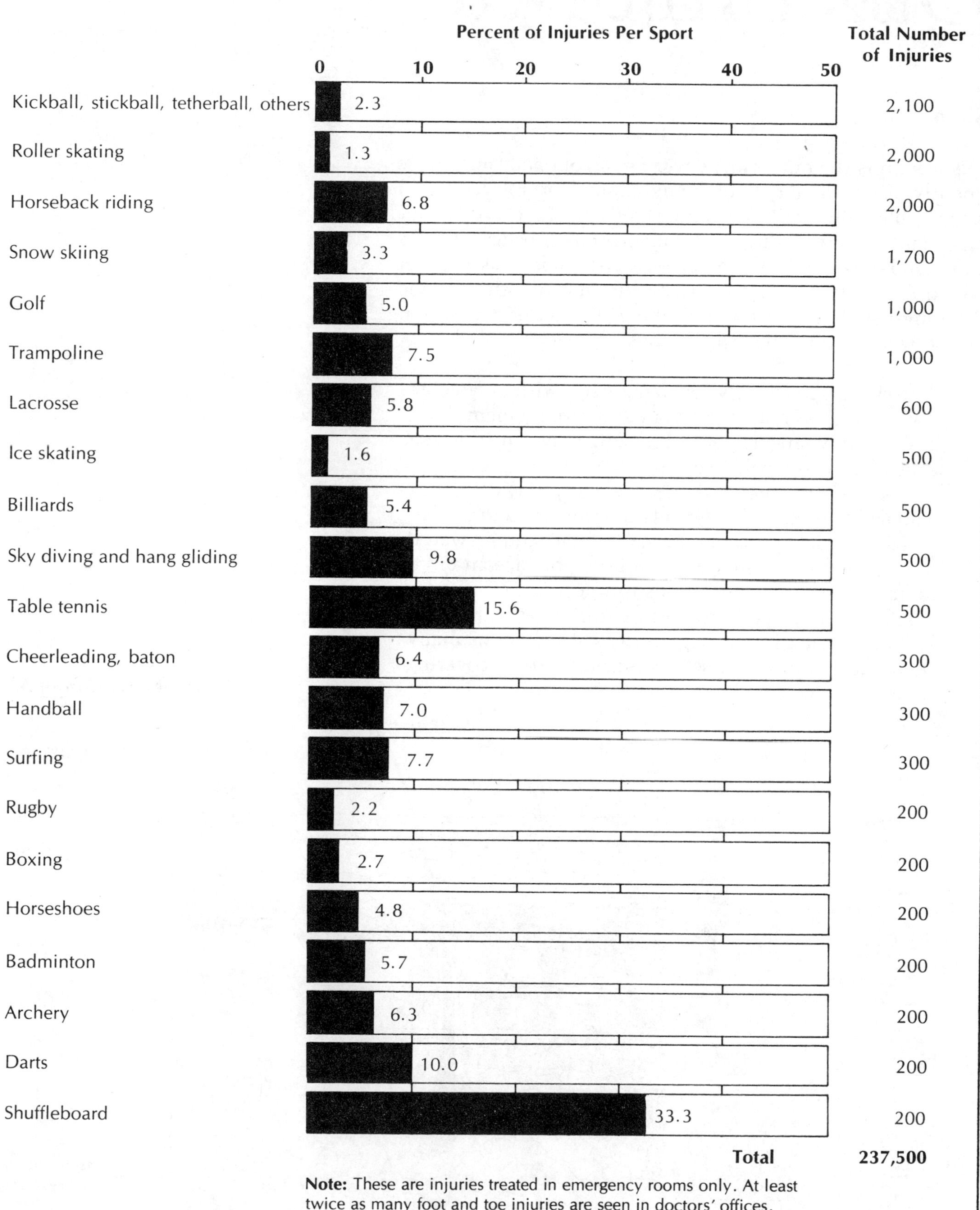
Percent of Injuries Per Sport
Total Number of Injuries
0
10
20
30
40
50
Kickball, stickball, tetherball, others 2.3 2,100
Roller skating 1.3 2,000
Horseback riding 6.8 2,000
Snow skiing 3.3 1,700
Golf 5.0 1,000
Trampoline 7.5 1,000
Lacrosse 5.8 600
Ice skating 1.6 500
Billiards 5.4 500
Sky diving and hang gliding 9.8 500
Table tennis 15.6 500
Cheerleading, baton 6.4 300
Handball 7.0 300
Surfing 7.7 300
Rugby 2.2 200
Boxing 2.7 200
Horseshoes 4.8 200
Badminton 5.7 200
Archery 6.3 200
Darts 10.0 200
Shuffleboard 33.3 200
Total 237,500
Note: These are injuries treated in emergency rooms only. At least twice as many foot and toe injuries are seen in doctors' offices.

Other Injuries

Skin

The skin is the outermost covering of your body and consequently is at high risk of being injured or infected. It is remarkable how well the skin stands up to the continual onslaught of germs, dirt, pollutants, scrapes, bangs, friction, sun, wind, heat, and cold. Your skin lasts as long as you live, which is many times the usual lifetime of other surface coverings, such as paint and clothing. You skin lasts so long and so well for the following reasons:

- The skin is constantly replacing itself with new cells. You grow a new surface layer every twenty-eight days. If you cut yourself, your skin will grow up to seven times faster to repair itself.
- The skin is specially adapted in many ways to resist damage. For example, it tends to maintain a dry outer surface, which prevents germs (which require water) from surviving or growing. Wetness provides an environment that frequently leads to infection.
- The skin can respond to protect itself. For example, if skin is frequently and vigorously rubbed, as in many sports, it will protect itself by forming a thick covering called a callus.

Cross Section of Skin

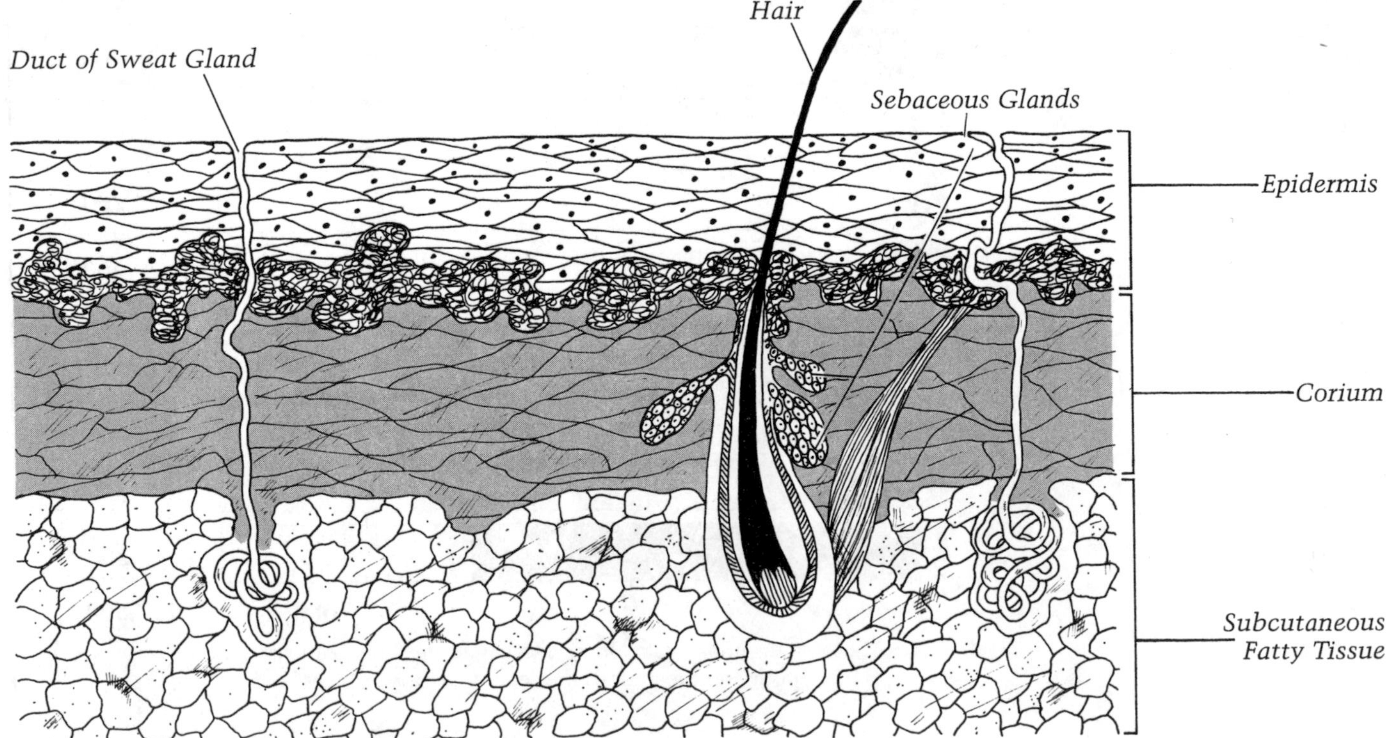

Skin injuries are often the result of predisposing functional or structural abnormalities. For example, an athlete who does not tan easily and cannot form a protective layer of pigment is likely to sunburn when exposed to the noonday sun. An exerciser whose fourth and fifth toes are particularly close together, a condition that prevents perspiration from evaporating, is predisposed to infection such as athlete's foot. Many skin injuries, like injuries to other tissues, can be prevented by correcting or compensating for the predisposing factors.

Abrasions*

DILONE'S STRAWBERRIES

In 1980 Mike Dilone, the Cleveland Indian outfielder, led the American League in stolen bases and abrasions.

"Dilone's abrasions look like strawberries," said Cleveland trainer Jim Warfield. "They are bright red. Sometimes they bleed a little. He gets them on his elbows and buttocks from sliding and diving.

"My main concern is that the strawberries get infected. I clean them the minute he comes off the field. I put zinc oxide ointment on the abrasions and cover them with moleskin. To help protect his lower body, I ordered sliding pants for him. They are made primarily for football players, but they seem to help."

CAUSES

An *abrasion* is an injury in which some of the skin is scraped off through some unusual or abnormal mechanical process. It is frequently the result of falling onto a hard, rough, or jagged surface. Cyclists get abrasions from falling on the pavement. Football players get them from falling on artificial turf. Boxers can get them when they are hit. Rarely are abrasions deep enough to cause any serious problems. You usually don't develop a scar unless the wound cuts through to the deeper layers of skin.

DIAGNOSIS AND TREATMENT

What do you do when you suffer an abrasion? The most important part of the treatment is to remove all foreign material from the wound, such as dirt or glass. Run hydrogen peroxide or water over the area to wash it thoroughly.

A wound may be too painful or too deep in the skin to be washed thoroughly. In that case, see a physician. The physician will inject an anesthetic near the edges of the wound so it can be cleaned without discomfort. You may also be given an antibiotic to kill germs, and for a deep wound, a tetanus shot.

Don't apply alcohol to an abrasion, because it will irritate your skin. Instead, apply an antibiotic ointment, which you can buy in a drugstore without a prescription.

If possible, don't cover a wound. A tight covering keeps the area moist and encourages bacterial growth. To prevent your clothes from rubbing on the wound and dirt from getting into it, you may need to apply a light gauze pad loosely over the antibiotic ointment. Remove the pad each night, however, to let the wound dry.

Healing time depends on several factors that include your age and the location of the wound. Children heal much faster than older persons. Also, a small facial wound will heal in three days, while the same size wound on the leg may take more than a week. The reason is that the face has a better blood supply, which carries nutrients to the wound and promotes healing.

* Ronald Shore, M.D., from Gabe Mirkin, M.D., and Marshall Hoffman, *The Sportsmedicine Book*, Little, Brown & Co., 1978.

There are several things you can do to prevent abrasions. Always wear protective equipment when participating in sports in which falls or body contact are common. When possible, take measures to improve the condition of the playing field.

Lacerations

CAUSES

A *laceration* is a cut that extends through both layers of the skin, the epidermis and the dermis. A deep laceration can also cut into the structures that lie beneath the skin—muscles, tendons, nerves, or blood vessels. Lacerations usually result from an encounter with a sharp object such as the edge of a skate, a baseball cleat, or a ski pole. Some lacerations occur because the skin is tightly stretched over a relatively sharp bone. For example, this happens when your facial skin is stretched over the bones around the eyes. Boxers get this type of laceration from punches. The laceration is actually a "burst" injury of the skin. I am an expert on lacerations. Every time I operate, I create a laceration.

DIAGNOSIS AND TREATMENT

Why are lacerations bad? Because they open the body to infection. Before the discovery of antibiotics, infection in a cut could lead to blood poisoning and death. If the laceration is caused by a bacteria-covered hockey puck, the chances of infection are fifteen in 100. In an operating room the chances of infection are one in 200. Lacerations can damage the underlying structures. Complete cutting of major tendons, arteries, or nerves is a serious surgical problem.

What first-aid measures can you take when you are lacerat-

SCAR FACES

"The average National Hockey League player gets fifty to sixty stitches in his face during his career," said Bert Zarins, the team physician of the Boston Bruins.

"The worst laceration I ever treated was sustained by Bobby Schmatz," recalled Dr. Zarins. "He got hit in the face with a hockey stick that cut through his lip into his mouth. The laceration bled profusely. Dr. Wilkins, a team physician, and I took him to the hospital to suture the lip back together. The swelling was enormous. He was out for ten days."

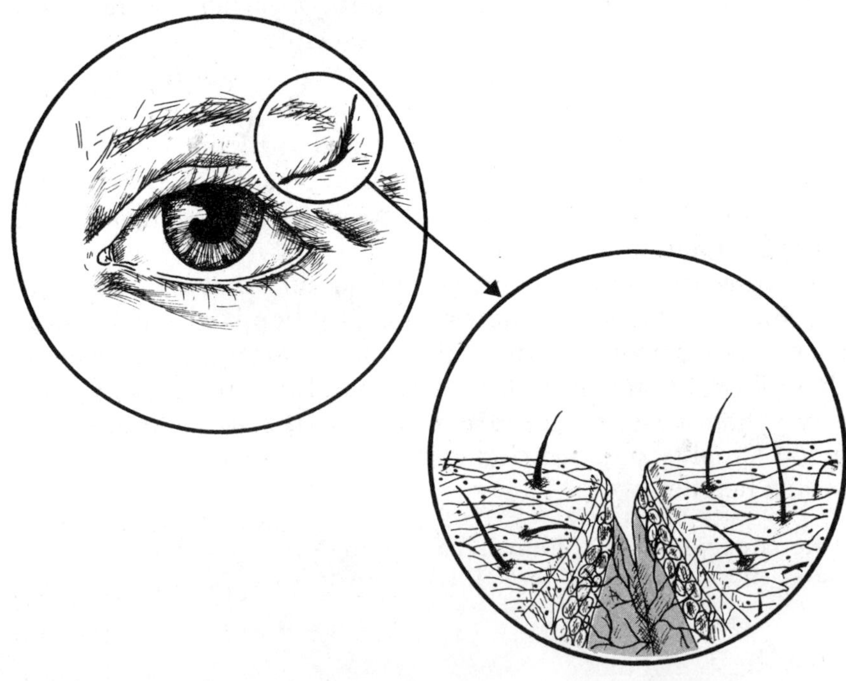

Skin Laceration

ed? First, apply a sterile gauze pad over the cut to reduce or eliminate further contamination. If the edges of the cut bleed vigorously, apply pressure with your hand; an Ace bandage wrap can be used as a pressure bandage, but never wrap it so tightly that it cuts off the circulation.

Only a doctor should judge the severity of a laceration. I examine the cut to determine its depth. I test for nerve, tendon, or blood vessel damage. Sometimes I have to inject Novocain into the cut so that I can thoroughly probe and examine the damage.

Before a laceration is stitched or taped shut, it should be cleaned. I wash the wound with two to three pints of sterile salt solution to remove the debris, dirt, and bacterial organisms. This is called lavage of the wound. First, I numb the edges of the skin with Novocain.

Many lacerations require stitches to hold the edges together while healing takes place. If the cut is less than one-half-inch long, I use special adhesive tapes called steri-strips or butterflies to pull the wound together. The tapes are changed every two to three days.

Usually I use stitches in cuts around joints, because joint motion tends to stretch the overlying skin and thus pulls the laceration apart. Nylon is the material most often used for stitches. It is nonreactive—that is, the body does not object to its presence in the skin. Nylon is strong and is easy to remove when the wound is healed. Laceration healing varies; it depends on the blood supply of the area. I remove stitches after healing:

- Face and scalp—three to five days
- Shoulders, arms, and chest—seven days
- Abdomen—ten days
- Hips and legs—two weeks

Stitches should always be put in within twelve hours of the injury. If they are put in later, there is an increased risk of infection. If I see a laceration late, I put a gauze pack, soaked with antibiotic solution, into the cut. I leave it in the wound for five days. Then I stitch the laceration closed. We call this delayed primary closure.

How do lacerations heal? The same way any tissue heals—with scar tissue. Normally, a laceration leaves a permanent mark. In children, however, this scar tissue does not grow with the rest of the body. The scar becomes relatively small.

How can lacerations be prevented? You have to pad and protect likely areas. Sally, my wife, spent $125 to outfit my son, David, for ice hockey. His helmet, face mask, elbow pads, gloves, shin pads, and skates are all designed to prevent lacerations.

I received several facial lacerations playing football. They were stitched so soon after I received them that the doctor did not use Novocain. My skin was in shock; all I could feel was the stitch being pulled through the skin. Taking stitches out does not hurt at all.

O'REILLY'S LACERATIONS

Terry O'Reilly, the Boston Bruin forward, is the most aggressive player on the team. He is also the most battered Bruin.

"Some of these guys get stitched so frequently, it's hard to tell which wound came from which game," said Jim Kausek, sports therapist for the Bruins. "In the playoff series with the Long Island Islanders, O'Reilly and Clark got into a fight in almost every game. In one game his lip was opened up. We stitched him. His eyebrow was cut. We stitched him. But all he wanted to do was get back at Clark. A third laceration required another eight stitches to close. All in all, he had twenty stitches in that one game.

"His teammates called him ten-time O'Reilly—ten stitching sessions a season."

The worst sports laceration I have ever worked on was for a local semi-pro hockey player who was hit with a stick. He was not wearing a helmet. It took seventy-three stitches to close the cut on his scalp.

What about a tetanus shot? All children need tetanus shots about every three years to make them immune to the dangers of tetanus bacteria. The immunization material is called tetanus toxoid.

Tetanus bugs grow in dead tissue in lacerations that have been stitched closed. If they do, you get lockjaw, a disease in which all the muscles in your body go into spasm. The jaw muscles lock, and you can't eat. The disease can be fatal. Treatment involves giving premade antibody against the toxin by vein. This is called tetanus antitoxin. The disease is totally avoided if all devitalized tissue is removed from the cut before it is stitched closed.

Concussions

CAUSES

When I played football at Harvard in the early sixties, we called a hard hit to the head "having your bell rung." Medically it is called a *concussion*, a mild bruising of brain tissue. I have treated concussions from a punch, a hockey stick, and a hard tackle.

I sustained a concussion in my sophomore year in a game against Princeton. It was an important game in the Ivy League championship race. I tackled the ball carrier from one side. A teammate tackled him from the other side. My teammate and I collided head-on. I saw stars. I released my grip of the ball carrier for an instant and he scooted away. I can remember watching him go down the sidelines and into the end zone for a touchdown.

I felt a little fuzzy for the next two or three minutes, but continued playing. After the halftime my thought processes were cleared up. Fortunately, we went on to victory that day. That night I had a mild headache, but by Sunday morning I was completely normal once again. I remember asking the doctors what had happened to me. They said, "Don't worry, son, you just had a concussion."

DIAGNOSIS AND TREATMENT

With most concussions you do not lose consciousness. You just feel a little woozy. You do not lose your orientation; you know where you are, who you are, and what day it is. We call this being oriented to place, person, and time.

A concussion should be checked by a physician. It is important to determine that a concussion is a concussion, not a skull fracture or an epidural hematoma. That is why Dr. Pappas and I send every player who is hit on the head with a baseball to the hospital for x-rays.

MY BATTERING RAM

"The only injuries I had in high school football were a series of five concussions," explained Richard A. St. Onge, an orthopedic and hand surgeon. "We played a brand of football that required you to use your head as a battering ram."

WALT MICHAELS' TEST

In a game against the San Francisco Forty-Niners, a New York Jet lineman sustained a concussion, but he wanted to stay in the game. Coach Walt Michaels, a savvy veteran, asked the lineman, "Where are we?"

"In San Francisco," said the lineman.

"What field?"

"Shea Stadium."

"What year is it?"

"1980."

"Son, it's 1981, and this isn't Shea Stadium. You've got to go to the hospital and get checked out."

FRANCO'S CONCUSSION

One reason the Pittsburgh Steelers did so poorly in the 1980 season was the concussion sustained by Franco Harris, one of the best running backs in football history. Franco, who is rarely out, missed three games in midseason, and the Steelers fell out of the race.

> **HIT AT 100 MILES AN HOUR**
>
> Almost all goalies in the National Hockey League wear masks. Gil Gilbert, the goalie for the Boston Bruins, was saved from a serious injury in the 1979-80 season because he was wearing a mask. In a game against the Toronto Maple Leafs, Gilbert was hit in the forehead, just above the eyebrow, by a 100-mile-an-hour slapshot.
>
> "It knocked him out for five or six seconds," remembered Jim Kausek, sports therapist for the Bruins. "By the time I got out on the ice, he was conscious again. He had a big, big knot right over his forehead. But he finished the game. It took two weeks for the swelling to go down."

After a concussion you may have a mild headache. Like any other body part, the brain swells when it is hit hard. The treatment is to ice the point of the concussion and take aspirin. By morning you should feel fine. There should be no residual problem. If there is, see a neurologist or a neurosurgeon.

Epidural Hematoma

CAUSES

Your brain is your most important organ. Controlling all your physical and mental functions, it is the best and most efficient mini-computer known. It correlates, coordinates, and interprets every movement, feeling, sensation, and thought of your body. It is the chief planner of the body organization.

The brain is beautifully protected by the skull and the dura, a sac that contains the same fluid found around the spinal cord. The thick bone covering of the skull is the first line of defense against injuries to your brain. In adults the skull is made of

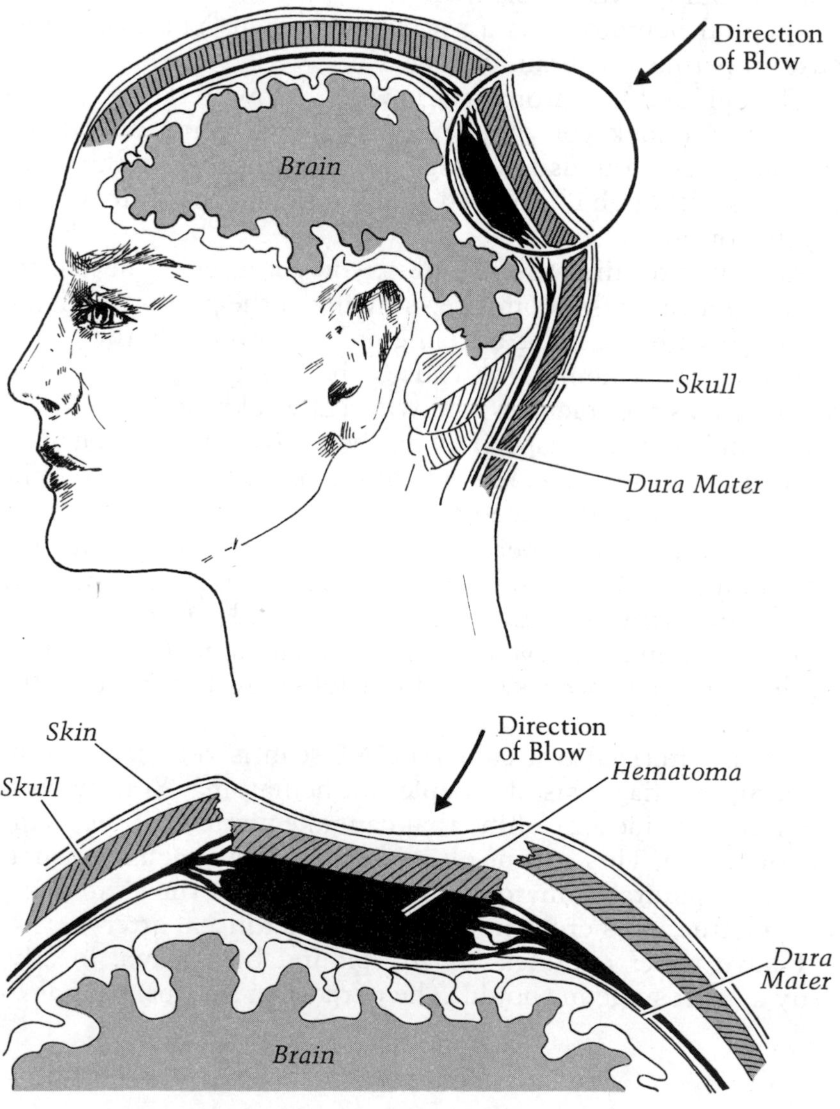

Epidural Hematoma

rigid bone. In children it has some give, because a child's brain is growing and its covering must grow with it. The growth areas of a child's skull, which are called sutures, make the skull flexible. They are like the edges of a patch in a quilt.

What is an epidural hematoma? First, it is a skull fracture—a crack in the skull. It happens when you do not wear head gear on motorcycles, bicycles, and horses. It occurs in baseball, hockey, rugby, and lacrosse. The crack can be felt in extensive fractures. The fractured part of the skull can sink below the level of the rest of the skull. In medical terms this is called a "depressed" skull fracture.

Second, the skull fracture "tears" the middle meningeal artery, which lies between the skull and the dural sac. The ruptured artery pours forth blood, which collects in a large pool. The pooling of blood is called a *hematoma*. The hematoma pressures the outside of the dura and affects the brain's function. Because it is outside the dura, it is called an *epidural hematoma.*

DIAGNOSIS AND TREATMENT

Epidural hematoma is a serious condition. You can die or sustain permanent brain damage.

The epidural hematoma starts with a blow to the head. The blow may knock you out, or it may not. If you are rendered unconscious, you usually regain consciousness within five minutes, although I have read about athletes who were unconscious for hours.

After you regain consciousness, you get a severe headache. You lose awareness. Vomiting, slowing of the pulse, and rising blood pressure are signs of rising pressure on the brain.

The epidural hematoma causes the pupil to dilate on the same side as the fracture. You will feel weakness in the arms, legs, hand, and face, especially on the opposite side of the body from the fracture. (The left side of the brain controls the right side of the body, and vice-versa.)

According to the American College of Surgeons Committee on Trauma, if you sustain an epidural hematoma, you will have some symptoms. Even during the "lucid period," you will be confused and may even fall asleep if left alone. I never let an athlete who has been knocked out return to competition the same day.

A computerized test called a CAT scan is very accurate in making the diagnosis of an epidural hematoma. You lie with your head inside a machine that can take pictures of the shape of your brain. This is done electronically. The machine visualizes the epidural hematoma. Regular x-rays of the skull show the fracture lines crossing the middle meningeal artery.

I have never operated on an epidural hematoma. Ideally, only a neurosurgeon should. The purpose of the operation is to

BEANING OF DWIGHT EVANS

In late August 1978, Dwight Evans, the Boston Red Sox right fielder, was struck on the head by a pitched ball and knocked unconscious.

"I examined him on the field," recalled Dr. Arthur Pappas. "A head injury can be very serious. I immediately checked to see if fluids—either clear or bloody—were coming from his ears or nose. Bloody fluid indicates internal bleeding. Clear fluid indicates spinal fluid, which could mean a fracture through the skull into the brain area.

"When I brought a light to his eyes, he reacted, which indicated that his level of unconsciousness was not deep. He was just stunned. His teammates and I carried him off the field on a stretcher. He was hospitalized immediately and stayed for four days. He was examined by a neurologist, a neurosurgeon, and an opthalmologist. He was carefully observed, tested, and x-rayed.

"It appeared that Dwight's head injury was resolving favorably. Obviously he had a headache for a few days, but after the headaches disappeared, he started to have persistent dizziness. He would lose eye contact

MOUNTAIN-CLIMBING DOCTOR

"In 1976, I went to Scotland to study hand surgery," recalled Richard A. St. Onge, a world-class mountain climber and orthopedic surgeon. "It was there I started to climb mountains. I wasn't very good in those days, but I learned quickly.

"My first difficult climb came about because I was a doctor. I was asked to be a member of a rescue team. I was told that three climbers had reached the top of Coryschniter, a famous Scottish mountain. On their way down they were pulled off the mountain by a falling cornice. The cornice caught the lead climber and knocked him off his ice stance. He was hanging from the end of the rope about 300 feet in the air. His partners, dewith the object he was watching. He would have to stop to regain balance.

"My consultants and I were unsure whether the problem stemmed from the brain, the nerves of the eyes, or the balancing mechanism of the ears. He also had a 20 percent loss in hearing. The hearing and the balance mechanisms are just millimeters away from each other.

"Ultimately we determined that the injury was to the balancing mechanism in the ear. In time, the injury healed itself. Some players, however, who had this same problem never returned to baseball.

"I gave him medication to decrease the dizziness. He came back during the season. Every time I thought he was improving, he would get into a critical situation. He'd try to look up for a ball, and he'd feel himself getting dizzy. For the remaining six weeks of the season, he did not play in top form. During the winter he was examined many times. By spring training the dizziness had disappeared.

"One postscript. If Dwight hadn't been wearing a batting helmet, the injury might have been much more serious!"

relieve the pressure on the brain. The neurosurgeon drills a one-inch hole at the site of the fracture and evacuates the blood. The hole is covered with a plastic plate. After surgery you have to wait at least three months to return to sports.

layed by a piece of rock, could not get to him. They were shell-shocked by the sight of their companion dangling helplessly.

"I started up the mountain with an experienced New Zealander who had been climbing mountains for years. This climb was technically beyond my capabilities. I was frightened half out of my mind, but the New Zealand climber kept saying, 'Aw, don't worry about a thing. Just do what I do and you'll be all set, Doc.'

"I watched him carefully. He was making some very delicate moves on fine, brittle ice. I was terribly scared and wasn't sure my crampons would hold. My calves were aching and the pain in my head was worse.

" 'Well, you are doing just fine, Doc,' said the New Zealander. 'Come on now, a little to your right. Yes, Doc, now a little to your left. Easy does it.'

"Slowly we ascended the vertical ice face. We were only a couple of hundred yards from the victim when an RAF helicopter, its blades thundering, approached us.

"The New Zealander let out a string of swear words that only another New Zealander could have understood. He finally got the RAF pilot's attention by waving his ice ax and sputtering into his radio communicator, 'Bug off, you crazy lunatic.'

"The noise from the helicopter was making the ice vibrate. This made the situation far more tense.

"We finally reached the climber, who had been hanging upside down for two hours in the cold. I estimated that the windchill factor was easily fifty degrees below zero that day. He was bleeding from his ears and nose and probably had a basal skull fracture.

"Despite our best efforts, we could not retrieve him. The closest we could get was fifteen feet. It was a situation that I will never forget; I felt completely helpless. We just couldn't reach him. We backed off the mountain and called the helicopter back. The helicopter maneuvered and removed him from the mountain with a sling. But it was too late. He was dead."

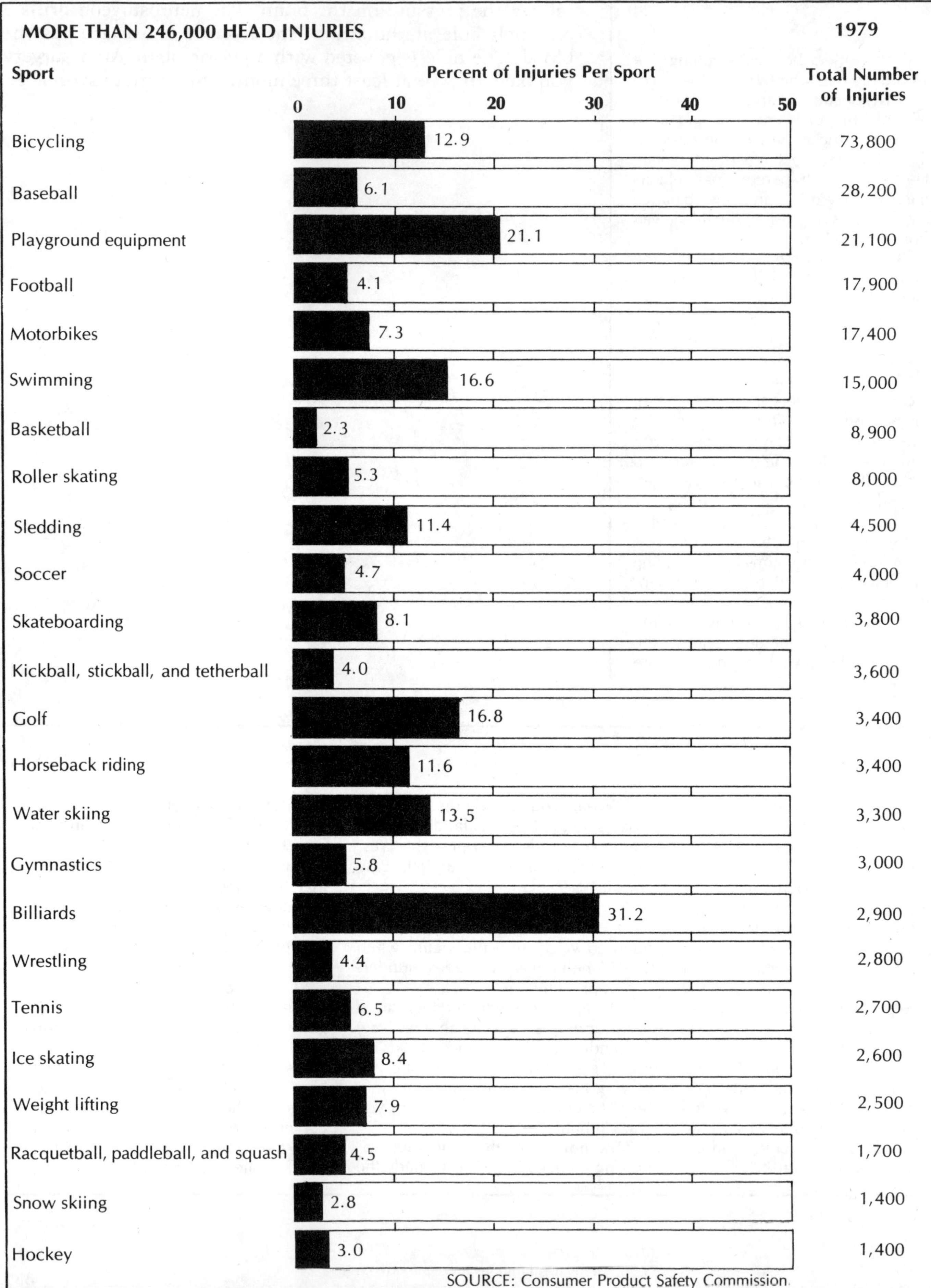
MORE THAN 246,000 HEAD INJURIES
1979
Sport
Percent of Injuries Per Sport
Total Number of Injuries
0
10
20
30
40
50
Bicycling 12.9 73,800
Baseball 6.1 28,200
Playground equipment 21.1 21,100
Football 4.1 17,900
Motorbikes 7.3 17,400
Swimming 16.6 15,000
Basketball 2.3 8,900
Roller skating 5.3 8,000
Sledding 11.4 4,500
Soccer 4.7 4,000
Skateboarding 8.1 3,800
Kickball, stickball, and tetherball 4.0 3,600
Golf 16.8 3,400
Horseback riding 11.6 3,400
Water skiing 13.5 3,300
Gymnastics 5.8 3,000
Billiards 31.2 2,900
Wrestling 4.4 2,800
Tennis 6.5 2,700
Ice skating 8.4 2,600
Weight lifting 7.9 2,500
Racquetball, paddleball, and squash 4.5 1,700
Snow skiing 2.8 1,400
Hockey 3.0 1,400
SOURCE: Consumer Product Safety Commission

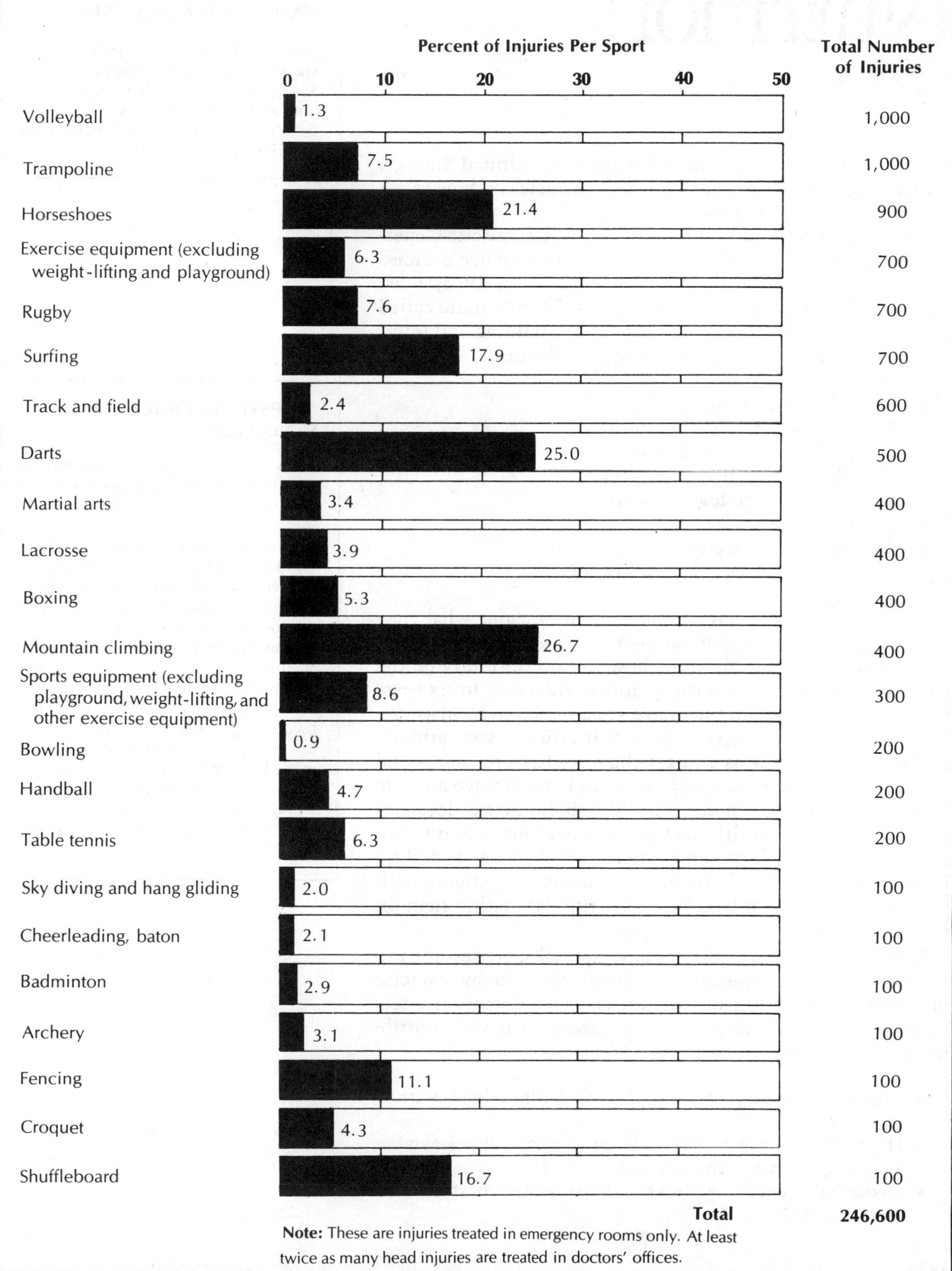

Note: These are injuries treated in emergency rooms only. At least twice as many head injuries are treated in doctors' offices.

Nutrition

Nancy Clark

> **NANCY CLARK, M.S., R.D.**
>
> Nancy Clark, M.S., R.D., is the sports nutritionist for Boston's Sports Medicine Resource, Inc. She is the author of *The Athlete's Kitchen*, published March 1981 by CBI Publishing Co., Boston, MA 02210.

As one of the few sports nutritionists in the United States, I counsel both professional and weekend athletes on how to get the most out of their diets.

A registered dietitian, I trained at the Massachusetts General Hospital and Boston University. I have also studied exercise physiology and consider myself an athlete. In 1978, I led fifteen cyclists across the country on a 4273-mile route called the "Bikecentennial Trail." Because of my training and interest in sports, I am often asked questions like these:

- Should I take vitamin pills?
- Will junk food hurt my performance?
- Does it matter if I skip breakfast?
- What should I eat before competition?
- Does carbohydrate loading work?
- Do I need to eat meat?
- What is my ideal weight?
- How many calories a day do I need?

The majority of athletes I counsel worry about what they eat. "I don't have time to eat right" is their most common complaint. They want to know how to eat healthfully on the run and are insecure about the quality of their diet. In six years of practice I have found they have good reason to be insecure.

Food is important for your general health and your athletic performance. As long as you get the forty basic nutrients and enough calories, you will perform well. I do not have a magic diet for my patients or for you. Metabolic magic does not exist. Promises by health food promoters about vitamin, protein, and quick-energy supplements, crash diets, and fluid replacements are mainly hokum. I educate my patients with nutrition information based on scientific fact rather than on hope.

Because competitive athletes have special sport-specific nutritional needs, I translate the latest research by exercise physiologists and nutritionists into useful information. I recommend specific foods and eating patterns that will contribute to optimal performance.

- Marathoners ask, "Should I carbohydrate-load with fruits or pasta?"
- High school football players want to know, "How can I gain weight that is muscle but not fat?"
- Basketball players wonder, "What is the best fluid replacement?"

> **MY PSYCHOLOGICAL SANDWICH**
>
> My friends and patients are curious about what I eat. They assume that I possess the magic keys to good eating.
>
> Like most busy professionals, I eat on the run. I try to keep my refrigerator well stocked with handy foods. My secret is that when I cook, I double the recipes. Tonight's dinner can be quickly heated for tomorrow's instant meal.
>
> When I am planning my athletic day—hiking or cycling—I make my favorite high-energy food: peanut butter and honey sandwich on homemade banana bread. This special treat has psychological as well as nutritional value. It gives me endless energy, and my friends have trouble keeping my pace.

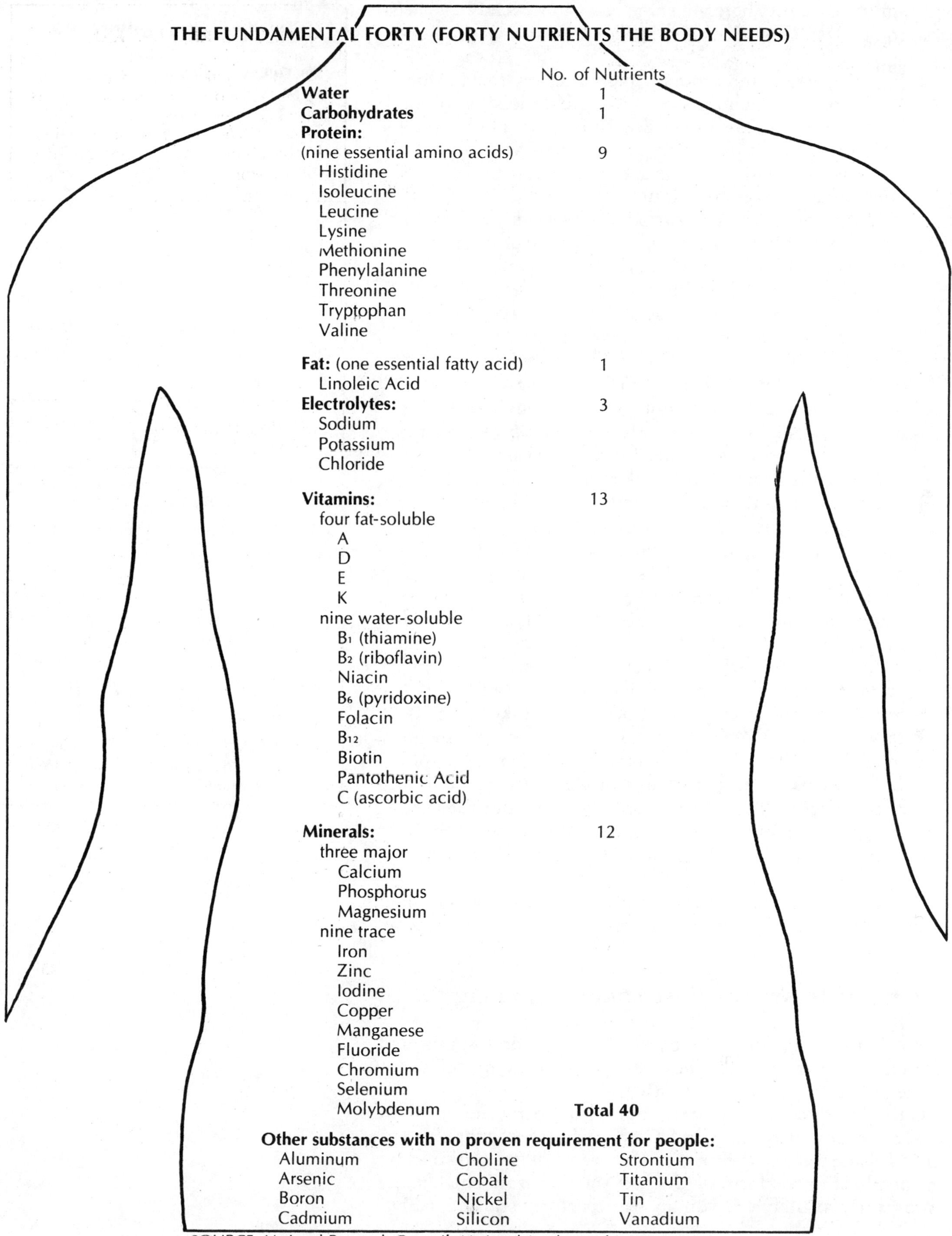

THE FUNDAMENTAL FORTY (FORTY NUTRIENTS THE BODY NEEDS)

	No. of Nutrients
Water	1
Carbohydrates	1
Protein:	
(nine essential amino acids)	9
Histidine	
Isoleucine	
Leucine	
Lysine	
Methionine	
Phenylalanine	
Threonine	
Tryptophan	
Valine	
Fat: (one essential fatty acid)	1
Linoleic Acid	
Electrolytes:	3
Sodium	
Potassium	
Chloride	
Vitamins:	13
four fat-soluble	
A	
D	
E	
K	
nine water-soluble	
B_1 (thiamine)	
B_2 (riboflavin)	
Niacin	
B_6 (pyridoxine)	
Folacin	
B_{12}	
Biotin	
Pantothenic Acid	
C (ascorbic acid)	
Minerals:	12
three major	
Calcium	
Phosphorus	
Magnesium	
nine trace	
Iron	
Zinc	
Iodine	
Copper	
Manganese	
Fluoride	
Chromium	
Selenium	
Molybdenum	**Total 40**

Other substances with no proven requirement for people:

Aluminum	Choline	Strontium
Arsenic	Cobalt	Titanium
Boron	Nickel	Tin
Cadmium	Silicon	Vanadium

SOURCE: National Research Council, National Academy of Sciences.

From my case file, here are some ways in which I have been able to help athletes with their specific nutritional problems.

- In 1979, a top marathoner, preparing for the Boston Marathon, was planning to carbohydrate-load on pizza and cheesecake, both loaded with fat instead of carbohydrates. I advised him to eat spaghetti with tomato sauce as well as sherbet. He set a personal record of two hours and twenty-one minutes.
- A high school wrestler, who was always struggling to lose weight by crash dieting and by dehydrating himself, came to me for advice. His mother said, "My son is eating practically nothing. Tom's completely neurotic about his weight. His grades in school have fallen." I asked him to write down everything he ate for four days. I discovered that Tom's diet was loaded with high-calorie fats—butter, salad dressing, and fatty meats. Second, he was not getting enough carbohydrates to fuel his muscle. He was eating almost no fruits. I designed a new diet for him, and now he is in college with a wrestling scholarship.
- A nationally ranked cross country skier had a problem maintaining his weight during the ski season. This athlete was a vegetarian who filled up on bulky salads and fruit. His problem was that he was not eating enough high-calorie foods like beans, dried fruits, peanut butter, and nuts. He made the dietary change and maintained his weight during the ski season.
- A professional ballet dancer was beset by muscle cramps. After a comprehensive interview with her, I found that she was limiting her fluid intake to maintain her weight. She was drinking only two glasses of juice a day. Thus, she was dehydrating herself. I told her to drink at least six cups of water a day. "Won't that make me heavier?" she asked. I told her that she might weigh more, but that she would not be fatter. In a few days the muscle cramps disappeared.

The discussion that follows is based on some of the most popular questions I get in my practice.

GALLUP YOUTH SURVEY

Do You Want to Gain or Lose Weight?

	Gain	Lose	Stay Same
Nationwide	24%	50%	26%
Boys	35	33	32
13-15 years old	33	37	30
16-18 years old	37	30	33
Girls	14	65	21
13-15 years old	15	62	23
16-18 years old	13	70	17

Note: The survey results are based on a representative sample of 1012 teenagers from across the nation, interviewed by telephone during October 1979.

SOURCE: E. Darden, "Gaining Bodyweight," *Nautilus,* Summer 1980.

AMERICANS EAT TOO MUCH FAT AND SUGAR

	Average American Recommended Diet Percent of Calories
Carbohydrates	40 (mainly from refined sugar)
Fat	40
Protein	20

SOURCE: U.S. Department of Agriculture and U.S. Senate Select Committee on Nutrition and Human Needs.

What Foods Are the Best Sources of Energy?

Foods are divided into three groups: carbohydrates, fats, and proteins. All three groups provide energy for the muscles, but some are more efficient than others.

Carbohydrates are the best sources of energy for muscles that are strenuously exercised. Carbohydrates are stored in the muscles and liver in the form of glycogen—muscle sugar—and in the blood in the form of glucose—blood sugar. This energy is instantly available. If you are unexpectedly startled, carbo-

HOW MUCH PROTEIN DO YOU NEED?

A 125-pound female needs 44 grams of protein. For a 150-pound male, the recommended daily allowance is 56 grams, which can be obtained from the following foods:

Two glasses of milk	16 grams
Four ounces of meat, fish, poultry, or cheese	28 grams
Additional food (bread, cereal, vegetables)	12 grams
Total	**56 grams**

Diet Percent of Calories
58 (mainly from refined sugar) fruits, vegetables, and starches
30
12

CHOLESTEROL CONTENT

Recommended limit = 300 mg per day

Food	Cholesterol
Liver, 3 oz. cooked	370 mg
Egg, 1	250
Shrimp, 3 oz.	130
Beef, 3 oz.	75
Chicken, 3 oz.	65
Fish, 3 oz.	55
Milk, 1 cup	35
Cheese, 1 oz.	30

SOURCE: J. Pennington and H. Church, *Bowes and Church's Food Values of Portions Commonly Used*, Harper & Row, New York, 1980.

hydrates give you the energy to escape from the scare.

Carbohydrates include both sugar and starch. Sugars bond together to form long chains of starch and are readily absorbed into the system, whereas starches first need to be digested. Some examples of foods that are rich in sugar are juice, fruit, sugar, candy, soda, jelly, and honey. Foods that are rich in starch are bread, muffins, crackers, pasta, rice, potatoes, and vegetables.

Fats are the most concentrated form of energy. Fat supplies more than twice as much energy (calories) as do carbohydrates. A little bit of fat supplies enough energy to go a long way. Not even a marathoner burns a pound of fat during the twenty-six-mile run.

Fat is the primary source of energy during rest or light activity. While you are reading this line, your body is burning mainly fat. If you start to exercise strenuously, you will switch to burning carbohydrate. During a marathon a runner burns about 70 percent carbohydrate (glycogen) and 30 percent fat. Excessive calories from any source, as well as fat itself, can be converted into fat. Some examples of fat are butter, margarine, salad dressing, mayonnaise, gravy, frying oil, cream, and lard.

Proteins build tissue. The typical American diet, however, provides two to three times the recommended fifty-six grams (for males) of protein. The extra protein is not stored in the body, but is converted to energy and to a waste product called urea, which is carried by the blood to the kidney and discharged in the urine.

Eat small portions of meats, enough to satisfy the requirement for building tissue, but not excessive amounts that are converted into sugar for energy. Four ounces of protein—one-fourth pound of hamburger or a medium piece of chicken—plus two servings of milk, yogurt, or cheese supply all the protein you need for the entire day.

Meat is a very expensive energy source and is high in saturated fat and cholesterol content, both linked to heart disease. The health-conscious athlete or exerciser should choose carbohydrates as the dietary mainstay; meat should be a secondary source of calories.

Some examples of protein from animal sources are beef, lamb, pork, fish, chicken, turkey, eggs, cheese, and milk. Examples of protein from vegetables are peanuts, nuts, dried beans, dried peas (split peas), and grains (cornmeal, barley).

When you choose foods for your active life-style, focus first on breads, cereals, pastas, fruits, vegetables, and other carbohydrates. They are the best fuel for your muscles. Eat fatty foods—beef, pork, sausage, ice cream, butter—in moderation. They are high in saturated fats and cholesterol. Second, fatty foods are harder to digest than carbohydrates.

How Many Meals Should You Eat in a Day?

What you eat throughout the day is more important than the number of times you eat. Eating three meals a day is not a prerequisite for a well-balanced diet.

Energy requirements vary among individuals. Grandpa feels content with two meals per day. The high school football player eats breakfast, lunch, and dinner plus continual snacks in between to obtain the 4000 plus calories he requires for growth and exercise.

Three to five small meals are ideal for maintaining a high energy level through the day. Studies of students and factory workers indicate that both productivity and concentration improve when snack breaks are scheduled during the day. (The rest and socializing, in addition to the food, are undoubtedly connected with the improvements.) Meal skippers not only miss out on nutrients, but also the relaxation and pleasure of an enjoyable meal.

If you sleep through breakfast, grab a candy bar for lunch, and gorge yourself at suppertime, you are overloading your stomach. Feed your system regularly, and you will feel better throughout the day. Meal skippers are nutrient skippers. Munching on cookies after dinner does replace calories, but not the vitamins, minerals, and protein you missed at breakfast and lunch.

Assuming that you make wise choices, snacking can be good for you. If you plan mini-meals as a part of your daily diet, you will more likely eat better than if you haphazardly grab snacks that lack the forty nutrients your body requires. Keep on hand foods that do not spoil: dry-roasted nuts, raisins, dates, other dried fruits, whole wheat crackers, peanut butter, breadsticks, or granola bars. When mid-afternoon hunger hits, you will at least have the chance to eat healthful pick-me-ups.

Wise mini-meals help you maintain high energy and stamina throughout the day. They contribute the vitamins, minerals, and protein you need for optimal athletic performance. High school athletes, cross country runners, bike racers, and others who burn 4000 to 5000 calories a day frequently need an additional meal before bedtime.

Can Junk Food Hurt Your Performance?

Junk food will not *help* your performance. Candy, soda, chips, and other sugary or greasy foods supply lots of high-calorie sugar and fat for fuel. But they lack the vitamins, minerals, and protein the muscles need to function optimally. You fill your car with gasoline, but it still needs spark plugs to run.

Limit your intake of foods enrolled in the "junk food hall of shame": sugar, salt, butter, shortening, saturated fats, candy without nuts, soda, alcohol, beer, and wine.

Eating sweets thirty to forty-five minutes before exercise frequently contributes to hypoglycemia (low blood sugar). The sugary snacks are quickly absorbed from the stomach into the blood. This stimulates abnormally large insulin secretion to transport the sugar into the muscles.

The high insulin level, when combined with exercise (which increases the rate of sugar into the working muscles), drops the blood sugar to an abnormally low level. The athlete may feel light-headed, shaky, uncoordinated, and hungry. Sugar in any form, such as candy, soda, raisins, or juice, can contribute to hypoglycemia.

A COLA DRINK (8 OUNCES)

Calories	96
Protein	0
Vitamin A	0
Vitamin C	0
Thiamine (B_1)	0
Riboflavin (B_2)	0
Niacin	0
Calcium	0
Iron	0

SOURCE: U.S. Department of Agriculture.

Is There a Good Way to Get All the Nutrients I Need?

Yes! The National Academy of Sciences established the recommended daily allowance (RDA) for many of the body's main nutrients. This is the amount of each nutrient you need to remain healthy.

It is not necessary to know the RDA for each of the nutrients. To ensure that you eat the right foods in the right amounts, follow the Four-Food Plan, a simple diet guide published by the United States Department of Agriculture. Each food has been assigned to one of the four groups of foods similar in nutritional value. The system is based on the principle that it doesn't matter whether your vitamin C comes from broccoli, an apple, or an orange, or your protein from meat, corn, beans, chicken, or fish. Your body breaks down all food, regardless of its source, into its basic elements before using it.

The beauty of the Four-Food Plan is its flexibility and simplicity. You don't have to have a Ph.D. in nutrition to choose a balanced diet. If you follow the Four-Food Plan, all your nutritional needs will be fulfilled:

- two servings of dairy products (four for teenagers)

THE FOUR-FOOD PLAN					
Food Group	**Recommended Number of Servings**				
	Child	**Teenager**	**Adult**	**Pregnant Women**	**Lactating Woman**
Milk 1 cup milk, yogurt, OR 1½ slices (1½ oz.) cheddar cheese 1 cup pudding 1¾ cups ice cream 2 cups cottage cheese	**3**	**4**	**2**	**4**	**4**
Meat 2 ozs. cooked, lean meat, fish, poultry 2 eggs 2 slices (2 oz.) cheddar cheese ½ cup cottage cheese 1 cup dried beans, peas 4 Tbsp. peanut butter	**2**	**2**	**2**	**3**	**2**
Fruit-Vegetable ½ cup cooked or juice 1 cup raw Portion commonly served such as a medium-size apple or banana	**4**	**4**	**4**	**4**	**4**
Grain, whole grain fortified, enriched 1 slice bread 1 cup ready-to-eat cereal ½ cup cooked cereal, pasta, grits	**4**	**4**	**4**	**4**	**4**

By eating a variety of foods, you get the variety of nutrients. For example:

Milk group—Calcium, riboflavin, protein
Meat group—Protein, niacin, iron
Fruit/vegetable group—Vitamins A and C
Grain group—Carbohydrate, thiamine, iron, niacin

SOURCE: National Dairy Council.

- two servings of protein foods (total of four ounces)
- four servings of fruit, juice, and vegetables
- four servings of bread, cereals, grain products

With these servings you get approximately 1500 calories that contain the forty nutrients the body needs. I call these the "fundamental 1500."

People who exercise rigorously—at least an hour a day—may require twice this amount of calories. For extra calories, eat additional servings of the four groups. You may eat extra meat and dairy products for the additional calories, but I have some reservations about these foods. They are high in fat and cholesterol and low in carbohydrates. They do not replace muscle glycogen.

Are Training Tables Necessary?

A special menu for sports teams is unnecessary. Eating special foods nourishes the mind more than the muscles. Athletes have two particular nutritional needs: (1) They require extra calories for energy, and (2) they require extra fluids to replace sweat losses. Both needs are easily met in any dining hall.

Many football and soccer players avoid training tables that offer roast beef, steak, eggs, and the other traditionally high-protein, high-fat foods. They prefer to feed their muscles with foods rich in carbohydrates: spaghetti, potatoes, casseroles, bread, and fruits. Active muscles do not need extra protein, but they do need extra carbohydrates.

Training tables have limited selections. Freedom to choose foods that are tasty, enjoyable, and easily digested is important. Each athlete has individual tastes and preferences. Pizza and beer may be the magic meal for one player, steak and eggs the choice for another. The psychological value of food is frequently more important than the nutritional value.

Training tables may limit not only food choices but also companionship. A football player may want to eat with his classmates rather than his teammates. Socializing at mealtime is important. It relaxes you, which aids digestion. A well-balanced life, in addition to a well-balanced diet, nourishes the athlete best.

Should a Weekend Athlete Eat Like a Professional?

No! People are people. No matter what your level of performance is, your body requires certain amounts of vitamins, minerals, proteins, and calories to exercise well and maintain good health. Strenuous exercise does not significantly change nutrient requirements other than for calories.

Weekend athletes need to adjust their caloric intake. For example, an intense tennis match may require 900 more calories than a stressful business meeting. Eating larger weekend meals will easily meet this increased requirement. Comes Monday, however, when the activity subsides, cutting back on portions will be important to practice.

Should Children in Sports Eat Differently?

Children and adults alike need the same vitamins and minerals for health and well-being.

Children have slightly higher needs for certain nutrients for growth. For example, each day an adolescent should eat four servings of dairy products (that is, cereal with milk, a cheese sandwich, yogurt, or pizza) for calcium. The calcium is important for the formation of strong bones. An adult requires only two servings of dairy products to get sufficient calcium for maintaining strong bones. (Bones are living tissue with a lifelong requirement for calcium. Adults do not outgrow their need for calcium.)

Iron is a second example of a nutrient needed by active children. Iron is fundamental for making red blood cells. Growing children have a growing blood volume. They require proportionately more iron than the adult who needs to maintain only a certain iron level.

Active children and adults can both meet their nutrient requirements by eating a wide variety of wholesome foods. Children should learn to eat wisely and make healthful choices. They are establishing lifelong eating habits that will affect not only their present performance, but also their future health and well-being. The child who snacks on candy, cookies, and Coke may easily become the overweight adult who continually fights a sweet tooth and the battle of the bulge.

What Is the Best Pregame Meal? Does It Vary for Different Sports?

The magic meal that assures success is still a mystery. Certain precompetition meal plans, however, will help you perform at your best. At a time when you are tense and nervous, the precompetition meal should focus on high-carbohydrate, low-fat foods. Carbohydrates are best for the following reasons:

1. They digest faster and more easily than protein and fat. Fats remain in the stomach longer and may cause discomfort during competition.
2. They are stored in the muscles as glycogen and are readily available energy.
3. They maintain the normal blood glucose level and prevent hypoglycemia with its symptoms of weakness and light-headedness.

Fruit, juice, cereal and skim milk, muffins, whole wheat bread, low-fat yogurt, crackers, and soup broth are wise low-fat choices.

Sugary-sweet carbohydrates (candy, soda, honey) are poor choices. The athlete may experience a "sugar high," but then will crash with a "sugar low" (hypoglycemia).

Protein foods contain hidden fats and take longer to digest than carbohydrates. Protein digests into urea, a waste product that is excreted via the kidneys. The athlete will need to urinate more frequently and may be inconvenienced during competition.

IRON-RICH FOODS

RDA for iron = 10 mg—men
18 mg—women

Food	Iron
*Cereal, 100% fortified (MOST, TOTAL) ¾ cup	18 mg
*Cream of Wheat, ½ cup	9
Liver, 3 oz.	7
Dried apricots, 12	6
Turkey, 3 oz.	5
Pork, 3 oz.	5
Prune juice, ½ cup	5
Dried dates, 9	5
*Raisin Bran, ¾ cup	5
Beef, 3 oz.	4
Shrimp, 3 oz.	3
Baked beans, ½ cup	3
Kidney beans, ½ cup	3
Molasses, 1 Tbsp., blackstrap	3
Raisins, ½ cup	2
Spinach, ½ cup	2
Bean curd (tofu), 4 oz.	2
Brewer's yeast, 1 Tbsp.	2
Chicken, 3 oz.	1
Fish, 3 oz.	1
Egg, 1 large	1
Peas, ½ cup	1
*Bread, 1 slice	1
*Spaghetti, ½ cup	1
Molasses, 1 Tbsp. regular	1
Wheat germ, 1 Tbsp.	1

*Commercially enriched with iron

Note: Iron from meat is absorbed more readily than iron from fruit and vegetables. Vitamin C helps iron absorption. Eat vitamin C-rich foods (broccoli, coleslaw, green peppers, oranges) along with the iron-rich foods.

SOURCE: Nancy Clark, Sports Medicine Resource, Inc.

Low-fat proteins that are an acceptable accompaniment to the precompetition meal include small portions of chicken, turkey, eggs, low-fat cottage cheese, skim milk, and low-fat cheese. Some pregame meal suggestions are:

- Cereal with skim milk and banana
- Poached egg on dry toast
- Low-fat cottage cheese with peaches
- Yogurt and applesauce
- Sliced chicken sandwich without mayonnaise
- Vegetable soup with crackers

If it is a small meal (oatmeal with raisins), the precompetition meal should be eaten two to three hours before the event. A larger meal (a big steak) should be eaten four to five hours before. This allows sufficient time for the food to digest and leave the stomach, but not long enough for the athlete to feel hungry during competition.

The precompetition meal should include two to three glasses of fluids. Water, juice, and skim milk are good choices; coffee and tea may overstimulate the already-nervous athlete. Some long-distance athletes, however, use coffee successfully to prolong endurance. (This is discussed in a later section.)

Dehydration reduces the ability to perform at your best. A 3 percent drop in body weight (five pounds for a 150-pound person) may result in a 20 to 30 percent drop in performance. The athlete should enter competition fully hydrated. Drinking lots of beverages one to two days before the event is equally as important as one to two hours before.

The precompetition meal should be appealing and enjoyable. Some athletes break all the "rules" and enjoy steak, pizza, or an ice-cream sundae before a game. They feel fine and play well. Favorite foods offer special psychological benefits that should not be overlooked.

The precompetition meal should include familiar foods. To experiment with a new diet before an important event may lead to disappointment. The super-duper high-energy formula may give you diarrhea. Some people reject new foods. By pretesting the new foods, you will know how your body and mind react to the special meal.

Carbohydrate loading—exercising your muscles empty of glycogen and reloading them by eating foods high in carbohydrates—is unnecessary for the average athlete, football player, and jogger. Eating a carbohydrate-rich meal, however, such as spaghetti or tuna-noodle casserole the night before a football game or a ten-kilometer race will beneficially saturate the glycogen stores. This substantial meal is more important than breakfast the next morning because the food will be digested and ready to use for energy. Marathon runners, cross country skiers, and other endurance athletes may benefit from eating mostly carbohydrates for two to three days before their race.

NUTRIENTS FOR HEALTH

Nutrient	Important Sources of Nutrient	Some major physiological functions	
		Provide energy	Build and maintain body cells
Protein	Meat, Poultry, Fish Dried Beans and Peas Egg Cheese Milk	Supplies 4 Calories per gram.	Constitutes part of the structure of every cell, such as muscle, blood, and bone; supports growth and maintains healthy body cells.
Carbohydrate	Cereal Potatoes Dried Beans Corn Bread Sugar	Supplies 4 Calories per gram. Major source of energy for central nervous system.	Supplies energy so protein can be used for growth and maintenance of body cells.
Fat	Shortening, Oil Butter, Margarine Salad Dressing Sausages	Supplies 9 Calories per gram.	Constitutes part of the structure of every cell. Supplies essential fatty acids.
Vitamin A (Retinol)	Liver Carrots Sweet Potatoes Greens Butter, Margarine		Assists formation and maintenance of skin and mucous membranes that line body cavities and tracts, such as nasal passages and intestinal tract, thus increasing resistance to infection.
Vitamin C (Ascorbic Acid)	Broccoli Orange Grapefruit Papaya Strawberries		Forms cementing substances, such as collagen, that hold body cells together, thus strengthening blood vessels, hastening healing of wounds and bones, and increasing resistance to infection.
Thiamine (B_1)	Lean Pork Nuts Fortified Cereal Products	Aids in utilization of energy.	
Riboflavin (B_2)	Liver Milk Yogurt Cottage Cheese	Aids in utilization of energy.	
Niacin	Liver Meat, Poultry, Fish Peanuts Fortified Cereal Products	Aids in utilization of energy.	
Calcium	Milk, Yogurt Cheese Sardines and Salmon with Bones Collard, Kale, Mustard, and Turnip Greens		Combines with other minerals within a protein framework to give structure and strength to bones and teeth.
Iron	Enriched Farina Prune Juice Liver Dried Beans and Peas Red Meat	Aids in utilization of energy.	Combines with protein to form hemoglobin the red substance in blood that carries oxygen to and carbon dioxide from the cells. Prevents nutritional anemia and its accompanying fatigue. Increases resistance to infection.

SOURCE: National Dairy Council.

Regulate body processes
Constitutes part of enzymes, some hormones and body fluids, and antibodies that increase resistance to infection
Unrefined products supply fiber—complex carbohydrates in fruits, vegetables, and whole grains—for regular elimination. Assists in fat utilization.
Provides and carries fat-soluble vitamins (A, D, E, and K).
Functions in visual processes and forms visual purple, thus promoting healthy eye tissues and eye adaptation in dim light.
Aids utilization of iron.
Functions as part of a coenzyme to promote the utilization of carbohydrate. Promotes normal appetite. Contributes to normal functioning of nervous system.
Functions as part of a coenzyme in the production of energy within body cells. Promotes healthy skin, eyes, and clear vision.
Functions as part of a coenzyme in fat synthesis, tissue respiration, and utilization of carbohydrate. Promotes healthy skin, nerves, and digestive tract. Aids digestion and fosters normal appetite.
Assists in blood clotting. Functions in normal muscle contraction and relaxation, and normal nerve transmission.
Functions as part of enzymes involved in tissue respiration.

Can or Should You Eat During Exercise?

Whether or not you can eat during exercise depends on your gastrointestinal comfort and your physiological need for energy. Intensely active athletes such as runners, swimmers, and basketball players do not usually eat while exercising.

The reason is that exercising with a belly full of food and digestive juice is uncomfortable. Runners, in particular, dislike having food bounce along in the stomach. Cyclists, on the other hand, can eat because their stomachs are stationary.

Second, even carbohydrate snacks may take two to three hours to digest. Therefore, if you exercise continuously for two hours or more, snacks can contribute to your energy level. They can maintain a normal blood sugar and prevent hunger pangs.

Moderate exercise does not interrupt digestion. You digest food during a tennis match as you do during sleep. The process may be slightly slower, however, depending on the muscles' demand for blood. With intense activity the stomach receives less blood, and the working muscles get more.

The size and type of the intra-exercise snack is important. A hefty peanut butter sandwich loaded with hard-to-digest fat and protein will probably cause lead-in-the-belly blues. A little bit of low-fat fruit, yogurt, sherbet, fruit, juice, or other carbohydrates is a better choice. Small, frequent feedings are easier to digest. Liquids, such as juices, are best during long-term moderate activity—cycling, hiking, cross country skiing. They replace not only calories, but also fluids.

How Good Are Liquid Meals?

For some athletes a liquid meal is convenient, easily digested, and soothing to a nervous stomach. It may offer a special psychological lift. For other athletes the liquid formula is a poor substitute for a favorite pregame magic meal. Some liquid meals are nutritious, high in carbohydrates, low in fats, and readily digested in one and a half to three hours. In my opinion, however, they are expensive and offer no significant benefit over regular juice or low-fat milk in a pregame situation. You should experiment with a liquid meal during training rather than pick one before an important event. You will then learn whether the meal will cause diarrhea, nausea, or gastric discomfort.

As a general rule, substituting a formula for a normal meal is undesirable. Natural, wholesome foods offer a greater variety of vitamins and minerals, and they develop better eating habits.

Do Some Foods Burn Muscle Sugar Faster?

No! The intensity of exercise, rather than diet, influences the rate that muscle sugar (glycogen) is burned. Glycogen is carbohydrate energy that is stored in the muscles. When your muscles work hard, they burn glycogen because it is most readily available. With less strenuous exercise, they burn fats from the blood. Diet plays an insignificant role.

Do Some Foods Produce More Lactic Acid?

No! Lactic acid results from lack of oxygen. Oxygen is fundamental for converting glucose into energy. With insufficient oxygen, glucose is only partially metabolized. It is converted into lactic acid instead of completely metabolizing into energy.

Diet does not affect lactic acid production. Rather, the intensity of exercise does. With intense, high-power exercise, as in a quick sprint, your muscles accumulate lactic acid and become quickly fatigued as a result.

Does a Candy Bar Give Quick Energy?

Eating a candy bar for quicky energy thirty to forty-five minutes before exercise can actually hinder more than help your performance. As noted in the section on "junk food," the sugar quickly enters the bloodstream, stimulating the release of a high level of insulin to clear the blood of sugar. The extra insulin, when combined with exercise, may cause the blood sugar to drop to an abnormally low level (hypoglycemia). Symptoms include light-headedness, shakiness, lack of coordination, and hunger.

A candy bar eaten during exercise will merely sit in the stomach. It will draw body fluids from the tissues into the stomach to dilute the strong sugar concentration. This dehydrates the body and bloats the stomach. The sugar is slowly absorbed, but offers little immediate value to the strenuously exercising muscles.

Quick energy already exists within the muscles in the form of glycogen. Glycogen is muscle sugar that is readily available for immediate action.

Does Carbohydrate Loading Work?

Carbohydrate loading is a dietary means to increase the muscle glycogen. When the supply of muscle glycogen depletes, the muscles become exhausted; the athlete "hits the

BLOOD: HOW IT CIRCULATES

	Rest	Hard Exercise
Muscle	20%	80%
Gut	25	5
Kidneys	20	4
Skin	5	6*
Other (brain, lungs, external surface of heart, fat and bone)	30	5

NOTE: In hot weather more blood may circulate to the skin as a way to cool your body.

HYPOGLYCEMIA

Exercise physiologist David Costill studied the effect on athletic performance of 300 calories of sugar taken forty-five minutes before hard running. The sugar caused elevated blood sugar and insulin levels. When the subjects began to exercise, the combination of high insulin and muscular activity caused their blood glucose to be removed too quickly, and it dropped abnormally low. The subjects felt tired and found the running difficult.

SOURCE: David Costill, Ph. D., Ball State University.

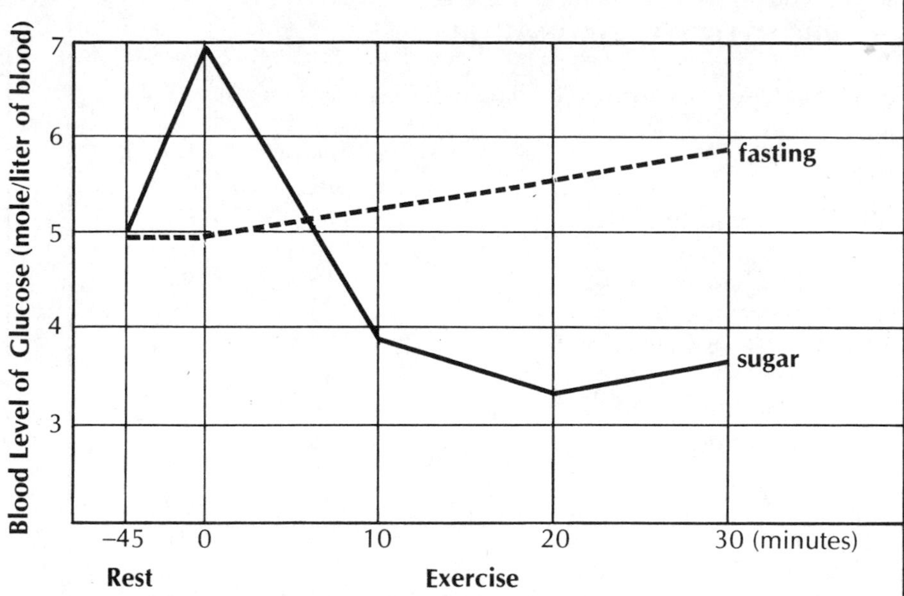

wall." The exercised muscles lose their strength and begin to hurt.

Carbohydrate loading is beneficial primarily for the endurance athlete who will be exercising for more than two hours. In one study this allowed the muscle to work as much as 30 percent longer. The 2700 calories needed to run a marathon without hitting the wall can be supplied if the muscles are super-saturated with glycogen.

The traditional means to carbohydrate-load is to follow a week-long regimen before the endurance event. This involves an exhaustive training session to deplete the muscle glycogen, followed by a three-day depletion phase with a low-carbohydrate diet. Three days before the event, the athlete switches to a high-carbohydrate diet. The starved muscles soak up excessive carbohydrates; they become super-saturated with glycogen. This extra stock pile of energy will aid the marathoner, long-distance bike racer, cross country skier, or other endurance athlete.

Most highly trained athletes do not follow this regimen.They realize that training, rather than special diets, is the more important key for increasing muscle glycogen. With training you repeatedly deplete the glycogen store and then super-load them when you next eat. Depletion is the key to super-loading. Eating quantities of sherbet or noodles without a previous hard workout has little value. For the competitive athlete, I recommend eating high-carbohydrate meals two to three days before the endurance event. An untrained person may have thirteen grams of glycogen per kilogram of muscle, whereas the trained person may have thirty-two grams—more than twice as much. The "loaded" muscle may have thirty-five to forty grams.

CARBOHYDRATE LOADING DIET

3800-calorie carbohydrate loading diet (one day's meals)

	Number of calories	Percent of carbohydrates
8 oz. orange juice	110	93
½ cup Grapenuts	210	88
1 medium banana	140	94
8 oz. low-fat milk	110	48
1 whole wheat English muffin	130	83
1 Tbsp. jelly	50	100
2 oz. turkey	155	—
2 slices bran bread	175	74
lettuce/tomato	25	82
8 oz. apple juice	120	100
1 cup lemon sherbet	275	91
3 cups spaghetti (6 oz. uncooked)	625	82
½ cup tomato sauce with mushrooms	75	64
2 Tbsp. Parmesan cheese	55	—
¼ loaf French bread	300	75
1 cup apple crisp with ¼ cup raisins	400	82
1 cup orange juice	110	93
2 cups cranberry juice	290	100
6 fig cookies	330	80
Total calories	**3815**	
Percent of carbohydrates	**80**	

SOURCE: Nancy Clark, Sports Medicine Resource, Inc.

What Are the Dangers of Carbohydrate Loading?

The most common danger in carbohydrate loading is psychological. The athlete may see the week-long regimen as an unhealthful switch from the usual meal patterns. During the first low-carbohydrate phase, the athlete gives up rice, pasta, bread, and fruits. Instead, lots of chicken, fish, peanut butter, and cheese—foods rich in protein—are eaten. This is a major change.

The athlete who is discontented with the diet will be at a psychological disadvantage. Many endurance athletes try the regimen once and dislike it. They stick to their tried-and-true training diet the next time around.

Physiological dangers are also inherent in the regimen. For example, the severe carbohydrate restrictions of the depletion phase may cause ketosis. This results from an accumulation of the end products of fat metabolism—ketones. Ketosis leads to personality changes such as irritability, and it can cause kidney damage.

Athletes with any history of cardiac problems should not carbohydrate-load. The depletion phase has a high-fat, high-cholesterol content and will temporarily increase blood lipids.

The diet may also upset the normal electrolyte-fluid balance, resulting in irregular heartbeats.

Carbohydrate loading is accompanied by water loading. Three grams of water are stored with each gram of glycogen. The well-loaded athlete may gain four pounds of weight. The muscles will feel stiff, tight, and heavy. The athlete may feel uncomfortable.

Can Vegetarians Perform as Well as Meat Eaters?

Certainly! Vegetarians can obtain sufficient protein from grains and vegetables to maintain muscular strength and stamina.

Vegetarians who eat eggs and dairy products, in addition to plant foods, easily consume sufficient protein. These foods are life-sustaining and highly nutritious. For example, an egg yolk

TOO MUCH RED MEAT

"Most of the Cleveland Indians are meat-and-potato men," says Cleveland trainer Jim Warfield. "Most players eat before they come to the park. In my opinion, most players eat too much red meat, but that is a tradition with baseball players."

"I DON'T EAT CANDIES"

"I don't eat a lot of junk," says all-star catcher Carlton Fisk of the Boston Red Sox. "Basically, I like all kinds of meat. I only eat red meats two or three times a week. I know that carbohydrates are important for me, so I eat spaghetti and pasta. I go for fish and chicken for lunch."

STEAK-EATING HOCKEY PLAYERS

"Most of the Boston Bruins are steak-and-potato men," says Jim Kausek, sports therapist for the Boston Bruins. "It's an old habit, but it's hard to break. I know they are eating too much protein and tell them that their muscles will only burn carbohydrates. It's a constant education program. I know that they need 4000 calories a day, but they are getting it the wrong way. They need more carbohydrates."

ZINC-RICH FOODS

RDA for zinc = 15 mg

Amount	Food	Zinc
Meats:		
10 average	oysters	80.00 mg
3 oz.	liver	5.0
3 oz.	ham	4.0
3 oz.	hamburger	3.8
3 oz.	chicken leg	2.4
7	shrimp	2.0
1	egg	0.5
Dairy:		
1 oz.	cheese	1.0
1 cup	milk	0.9
Beans:		
1 cup	lentils	2.0
2 Tbsp.	peanut butter	1.0
Grains:		
1 oz.	bran flakes	3.7*
¼ cup	wheat germ	3.2
1 cup	oatmeal	1.2
1 slice	wheat bread	0.5

*Enriched breads and cereals are good sources of zinc.

Note: The best sources of zinc are foods of animal origin. It is better absorbed than zinc from plants.

SOURCE: J. Pennington and H. Church, *Bowes and Church's Food Values of Portions Commonly Used*, Harper & Row, New York, 1980.

nourishes an embryo; milk nourishes babies. Eggs and milk enhance the nutritional valuc of bread, cereal, nuts, and other vegetable sources of protein.

The vegetarian athlete can easily consume the needed protein as well as vitamins and minerals. Sometimes vegetarians do not get enough iron and zinc. Meats are the richest sources of these minerals. Iron is important for transporting oxygen from the lungs to the working muscles. Zinc is a part of many enzymes that help the body function properly. Although vegetables and grains provide fair amounts of these minerals, they do have phytates, which interfere with mineral absorption. To alleviate this problem, vegetarians should include in each meal a rich source of vitamin C as orange juice, cantaloupe, broccoli, or green peppers.

Are There Foods That Build Muscles?

Training, not diet, increases muscular strength. Muscles require slightly more protein during development. This additional need, however, is more than abundantly met by the average American diet, which contains two to three times the RDA for protein.

Special high-protein supplements are expensive and unnecessary items in the athlete's diet. Exercise increases the need for more carbohydrate, but not for more protein.

The key to developing muscle size is repeated hard exercise, which stresses the muscles. For example, weight lifting, a resistance exercise, develops bulky muscles.

Do You Need Extra Vitamins?

Vitamins are chemical substances the body cannot manufacture. You must obtain them in the food you eat. Vitamins are metabolic catalysts; they regulate the chemical reactions within the body and are called partial enzymes.

At present, scientists have discovered thirteen vitamins. Each has a specific function. For example:

- Thiamine (B_1) helps convert glucose into energy.
- Vitamin D controls the way the body uses calcium in bones and cartilage.
- Vitamin A is part of an eye pigment that helps us see in diminished light.

By eating a variety of wholesome foods, you can obtain all the vitamins you need to maintain a healthy, active body. Vitamin supplements are an unnecessary part of an athlete's diet.

Extra vitamins will not give you a winning edge. Extra vitamins do *not*:

- Increase performance.
- Increase strength or endurance.
- Prevent injuries or illness.
- Provide energy.
- Build muscles.

Supplements offer no benefits for either the average person or the athlete. Nevertheless, many athletes are vitamin pill poppers. They believe, "If a little is good, a lot will be better." Not true! Vitamins are similar to spark plugs in a car. A few extras won't make the engine run stronger.

Excessive amounts of vitamins may be dangerous. The body functions best when the systems are in proper balance. A large dose of vitamins may upset that balance. Money spent on them could better be used to buy additional fruits and vegetables.

THIAMINE (B_1)

RDA = 0.5 mg/1000 cal = 1.5 mg

Pork chop	3 oz.	1.2 mg
Brewers yeast	1 Tbsp.	1.2
Wheat germ	¼ up	0.4
*Dry cereal	1 oz.	0.3
Granola	1 oz.	0.03
*Bread	2 slices	0.15
Broccoli	1 cup	0.2
Green peas	½ cup	0.2

*Enriched with thiamine

SOURCE: J. Pennington and H. Church, *Bowes and Church's Food Values of Portions Commonly Used,* Harper & Row, New York, 1980.

VITAMIN A

RDA = 5000 IU

The following portions supply 100 percent of the recommended daily allowance.

Dark-colored vegetables

½ medium carrot
⅓ cup spinach
⅔ cup winter squash
3 large tomatoes

Liver

½ oz. beef liver
½ medium chicken liver

Vitamin A is stored in fat and liver.

Toxic dose = 5 x RDA for months
= 25,000 IU per day for several months

Toxic Symptoms

headache	blurred vision
nausea	hair loss
insomnia	scaly skin

SOURCE: Nancy Clark, Sports Medicine Resource, Inc.

What Vitamin Deficiencies Are Common in Athletes?

Only in unusual circumstances would an athlete become vitamin-deficient. Strenuous exercise does not significantly increase vitamin requirements.

The fundamental 1500 calories from a variety of dairy products, grains, fruits, vegetables, and protein foods provide your vitamin requirements. In your daily 4000 to 6000 calories, you may eat three to four times your vitamin requirements.

Vitamin C, for example, is important for strong, healthy tissues and for fighting infection. One small (six-ounce) glass of orange juice provides sixty milligrams of vitamin C—100 percent of the day's requirement. More thirsty athletes drink a whole quart of orange juice. They get five times the RDA for vitamin C. Deficiency is not a cause for concern.

VITAMIN C

RDA = 60 mg

The following portions supply 100 percent of the recommended daily allowance.

¾ cup orange juice
½ large grapefruit
½ medium cantaloupe
½ medium green pepper
1 cup cabbage, raw
2 medium tomatoes
½ cup broccoli
1 large potato, baked

Vitamin C is destroyed by air, heat, water, and light.

Vitamin C Loss	Percent
Orange Juice—eight days in refrigerator	15
Broccoli—frozen one year	50
Potato—stored eight months	40
Tomato—peeled	75
Kale—two days at room temperature	40

Vitamin C destruction by cooking

	Loss Percent	Recoverable in Water Percent
Spinach		
underdone	40	31
optimum	55	24
overdone	60	12

SOURCE: R. Harris and E. Karmas, *Nutritional Evaluation of Food Processing,* AVI Publishing Co., Westport, Conn., 1975.

CALCIUM

RDA = 800 mg
The following portions supply 100 per cent of the recommended daily intake.

2½ cups milk
2½ cups yogurt
4 oz. cheese
2½ cups collards
6 cups broccoli
20 oz. bean curd

MINERALS THE BODY NEEDS

Required in large amounts: calcium
chloride
magnesium
phosphorus
potassium

Required in trace amounts: chromium
copper
fluoride
iodine
iron
manganese
selenium
zinc

Found but not required: arsenic
cadium
cobalt
nickel
silicone
tin
vanadium

Note: Why are they called *trace elements?* Because you need only tiny amounts. For example, the RDA for iodine is 150 micrograms. One level teaspoon of iodine is 5,000,000 micrograms. You need only a trace of iodine to satisfy your body requirement.

SOURCE: National Research Council, National Academy of Sciences.

Do Vitamin Supplements Help Heal Fractures?

No. A well-balanced diet is more important than vitamin pills, not only for recovering from a fracture but also possibly for preventing one. The following nutrients are essential for proper healing:

- *Calcium and phosphorus* are fundamental for developing and maintaining strong bones. To obtain the recommended intake, you need two servings of dairy products per day if you are an adult, four servings if you are a growing adolescent. Excellent choices include cereal with yogurt, macaroni and cheese, cottage cheese with fruits, and pizza—as well as a glass of milk.
- *Vitamin C* is important for fighting infection and maintaining strong tissue. Have at least one serving of a vitamin C food each day: orange, cantaloupe, broccoli, baked potato, grapefruit juice, strawberries, green pepper, tomato.
- *Zinc* is also involved in the healing process. Animal sources of zinc are best: beef, lamb, pork, liverwurst, eggs, seafood. Vegetarians should buy breads and cereals that are enriched with zinc. Read the food label; look for zinc in the list of ingredients.

What Kinds of Minerals Does the Body Require?

Minerals are basic elements found in the soil. They are picked up by plants. Humans eat the plants, or meat from the animals that ate the plants, and obtain the needed minerals.

A 150-pound man contains about seven pounds of minerals. He has enough salt to fill a small saltshaker, enough iron for three hair pins, and enough calcium phosphate to sculpt a figurine.

Fifteen minerals that we now know are required for maintaining body functions. Each mineral has a specific job. For example, calcium and phosphorus form sturdy crystals that give the bones strength and durability. Sodium (in salt) regulates water distribution. Iron in the red blood cells carries oxygen to the working muscles.

Should You Drink More Water When You Exercise?

Drink as much as you comfortably can. Drinking too much water is almost impossible. Your kidneys will excrete the extra. Drinking too little water, however, is very common.

Before exercise you should drink two to three cups of water, juice, or other fluids. This saturates your tissues so they will initially be well hydrated.

During exercise a cup (eight ounces) of water every twenty minutes is ideal. This is the amount you can absorb. Keep in mind that you may be sweating three times that amount, so you will still have a deficit.

After exercise drink to quench your thirst, then drink even more. Thirst does not sufficiently indicate your need for fluids. You may feel fully hydrated, but still need water.

If you urinate infrequently after exercising or if your urine is a dark color, you are not drinking enough. You may have a headache and feel irritable.

PRECOMPETITION HYDRATION

Drink at least eight glasses of fluids the day before.

Drink up to two hours before competition.

Drink one to two glasses five minutes before competition.

FLUID LOSS

You lose more fluid than you can replace during strenuous exercise in the heat.

Sweat Loss: 33 oz./30 minutes
Absorption: 9 oz./30 minutes
Deficit: 24 oz.

Does Ice Water Cause Cramps?

Studies conducted with cold fluids have confirmed that drinking ice water does not cause cramps. Large volumes of water hurriedly gulped may, however, cause stomach distress.

Cold water (forty degrees F) is preferable to warm water for the following reasons:

- It is absorbed faster. Cold water increases gastric activity; the water enters the intestines more rapidly, where it is absorbed into the system.
- It cools the body better. Cold water can reduce the stomach temperature by twelve to thirty degrees. This in turn cools the body core and reduces heat stress during strenuous exercise.

Does Coffee Improve Your Performance?

Marathon racers, bicycle racers, cross country skiers, and other endurance athletes may benefit from the caffeine in coffee. The studies by David Costill, an exercise physiologist at Ball State University, indicated that trained cyclists worked 7 percent harder in two hours of cycling when they were given caffeine. In a second test they had 19 percent more endurance.

Caffeine's stimulant effect is not responsible for increasing your endurance. Rather, caffeine lessens the amount of glycogen burned by the muscles. It stimulates the release of fats from the tissues into the bloodstream. When more fats are available, the muscles burn them in preference to the glycogen. This spares the glycogen stores, and the muscles are able to work longer.

Caffeine overdose is common with athletes who drink three to four cups of strong coffee. Two cups does the trick. The suggested dose is about one and a half to two milligrams per pound. A 150-pound runner may benefit from 300 milligrams of caffeine, the amount in two average cups of coffee.

CAFFEINE PROLONGS ENDURANCE

Sixty minutes before exercise, the subjects drank 330 mg caffeine (equivalent to two cups of coffee). They exercised fifteen minutes longer.

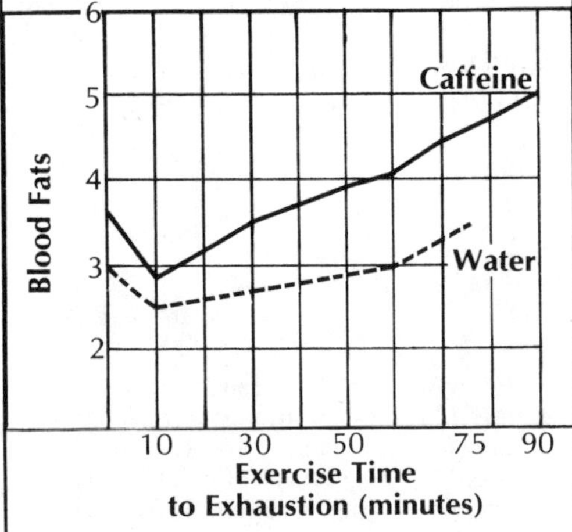

SOURCE: David Costill, Ph.D., Ball State University

CAFFEINE CONTENT OF COMMON BEVERAGES

COFFEE:		**CARBONATED SODA**	
Drip	145 mg/standard five-oz. cup	COKE	65 mg/twelve-oz. can
Percolated	110	TAB	50
Instant	65	PEPSI	45
Decaffeinated	5		
TEA:		**COCOA**	15
five-minute brew	45 mg/standard five-oz. cup		
one-minute brew	30		

SOURCE: M. L. Bunker and M. McWilliams, "Caffeine Content of Common Beverages," *J. Am. Diet. Assoc.*, 74:28, 1979.

What Is the Best Postgame Drink?

Water! Your main concern after strenuous exercise is to replace the water lost through sweating. Replacing the electrolytes (sodium, potassium, magnesium, calcium) lost along with the sweat is of secondary importance.

Water is best. It is absorbed more quickly than special athletic drinks, beverages, and juices that contain sugar, glucose, sodium, potassium, and other "ingredients." The higher the concentration of ingredients, the slower the rate of absorption.

After you exercise in very hot, dehydrating weather, first drink two to three glasses of cold water. Then enjoy some juice. The water will replace your lost fluids more quickly. Juice or food taken twenty to thirty minutes later will replace the lost electrolytes.

Should Athletes Drink Beer?

Drinking beer before or during an athletic event is bad. It depresses your nervous system so that you cannot function quickly and skillfully. Drinking beer immediately after an event is also unwise. It inhibits the release of the antidiuretic hormone (ADH), which retains water in the body. Instead of replacing fluids, you will urinate more frequently and lose body fluids. After games or exercise, first drink two to three glasses of water to replace sweat losses. Then enjoy a few beers.

Many athletes drink beer as a means to carbohydrate-load. That's a poor choice again. A twelve-ounce can of beer has only sixteen grams of carbohydrate. The same volume of orange juice has thirty-nine grams. The calories in beer come mostly from the alcohol. These calories are neither stored as glycogen nor used for muscular energy.

A can of beer has less nutritional value than a slice of bread. Although beer has significant amounts of riboflavin (B_2), you would need to drink eleven cans to meet the recommended daily allowance for this one nutrient. Thiamine (B_1) is used by the yeast during fermentation, so only a little is left in the

WATER LOSS

On a hot day, a starting pitcher can lose two to three pounds. But some lose far more than others. Veteran pitcher Gaylord Perry loses as much as eight pounds. Marshall Hoffman, my coauthor loses about three pounds on his 10-mile run in the summer. In a forty-five-minute racquetball game, I lose about two pounds.

finished brew. Small amounts of the B vitamins (niacin, biotin, pyridoxine, and pantothenic acid) are in beer, but a slice of whole wheat bread will offer you more of these vitamins.

CALORIC QUENCHERS

"I exercise everyday . . . and *still* can't eat very much without gaining weight!"

More than likely, the primary reason is that you drown your thirst in calories. Check the following comparisons and you may find you have been eating more than you thought.

- One quart of orange juice is equal in calories to a peanut butter sandwich.
- Two cans of Coke could be traded in for a large ice-cream cone.
- One beer is the caloric equivalent of six cups of popcorn (unbuttered).

If you drink large quantities of fluids, acknowledge them as being food and count their calories.

Calories/8-ounce glass		Calories/12-ounce can	
Lemonade	100	Lipton iced tea	130
Grapefruit juice	100	Sprite	140
Orange juice	110	Budweiser	150
Apple juice	120	Pepsi	160

The best low-calorie fluid is water. Sound boring? Alternatives are:

- Club soda with lemon or lime
- Iced tea, herbal or regular
- Iced coffee, decaffeinated or regular

Other suggestions are:

- Drink V-8 or tomato juice instead of higher-calorie fruit juices.
- Put lots of ice in your juice to dilute it.
- Flavor ice water with a little milk and almond or vanilla extract.
- Stretch red wine with club soda to make a larger wine spritzer.
- Reduce the amount of sugar and cream you put in your coffee. A large cup with two creamers (forty calories) and one packet of sugar (sixteen calories) costs you fifty-six calories. Your four cups per day add up to 224 calories, the caloric equivalent of a big bran muffin.

By counting the liquid calories that sneak past your lips, you may discover the answer to your dietary dilemma.

SOURCE: Nancy Clark, Sports Medicine Resource, Inc.

AMMONIA WATER

"Because it is hot in Cleveland in the summer, I worry about the players becoming dehydrated," says Cleveland trainer Jim Warfield. "I make up something which the players call ammonia water. I take a prepared electrolyte drink and dilute it with ice and water. I also keep ice bags, ice buckets and wet towels in the dugout."

AMOUNTS OF CARBOHYDRATE, ALCOHOL, AND CALORIES IN ALCOHOLIC DRINKS					
Alcoholic Drinks	**Household Measure**	**Total Grams**	**Grams of Carbohydrate***	**Grams of Alcohol†**	**Calories (Approx.)**
Whiskey–Bourbon, Irish Rye, and Scotch	one brandy glass (1 oz.)	30	none	10½–12	75–85
Brandy, Gin, and Rum	one brandy glass (1 oz.)	30	none	10½–13	75–90
Liqueurs and Cordials	one cordial glass (⅔ oz.)	20	4–10	4–7	50–80
Malt liquors–ale beer, porter, and stout	one glass (8 oz.)	240	7–14	7–14	80–150
Wines— Sweet domestic	one wine glass (3½ oz.)	100	8–14	13–15	140–165
Sweet imported	one wine glass (3½ oz.)	100	3–20	10½–18	110–175
Dry domestic	one wine glass (3½ oz.)	100	½–4	10–11	75–90
Dry imported	one wine glass (3½ oz.)	100	½–3	8–14	60–110

*Every gram of carbohydrate supplies four calories.
†Every gram of alcohol supplies seven calories.

SOURCE: *Joslin Diabetes Teaching Guide,* Joslin Diabetes Foundation, 1977.

SODIUM LOST IN SWEAT

On a twenty-mile run (60°F), 1.8 to 5.5 grams of sodium may be lost. Most fluid replacements contain little sodium:

Orange juice	0.003 gram/8 oz.
Coke	0.001
Gatorade	0.125
Beer	0.010
Tomato juice	0.480
Bouillon	0.425/cube

Note: You replace your sodium losses mainly in the foods you eat.

SOURCE: Nancy Clark, Sports Medicine Resource, Inc.

Should You Take Salt Tablets?

No! Salt tablets are unnecessary and potentially dangerous. They draw water from the tissues into the stomach to dilute the abnormally high sodium concentration from the tablet. This gastric reaction dehydrates you, and you perform suboptimally.

Although you lose some salt when you sweat, salt losses are minimal compared with total body sodium and total dietary intake. Here are the figures:

- One liter of sweat contains 0.9 to 1.4 grams of sodium.
- The average male body contains ninety grams of sodium.
- The average American diet contains five to twelve grams of sodium per day.

You do not need to replace sodium losses while you exercise; the amount of sodium in your blood increases during prolonged exercise. You lose proportionately more water than salt in sweat.

You can easily replace the sodium at your next meal or snack. For example, three ounces of cheese contain six-tenths of a gram of sodium. A large, fast-food hamburger contains a gram of salt. A can of chicken soup contains two and three-tenths grams.

If you pop salt tablets regularly, you increase your chance of developing high blood pressure, which can lead to heart attack, stroke, kidney failure, blindness, and even death. If your body requires extra salt, you will crave it. Consequently, you will prefer salty foods.

SALT CONTENT OF FOOD

Foods High in Salt	grams sodium
1 oz. bologna	0.360
1 slice bacon	0.075
1 oz. ham	0.280
1 (2-oz.) hotdog	0.550
1 oz. Cheddar cheese	0.200
½ cup cottage cheese	0.250
1 oz. American cheese	0.320
½ cup canned vegetables	0.200
½ cup tomato juice	0.200
1 bouillon cube	0.425
1 turkey pot pie	1.100
Chicken TV dinner	1.100
1 cup canned spaghetti	1.220
½ cup vegetable beef soup	1.250
1 Tsp. mustard	0.065
1 Tbsp. catsup	0.175
1 Tbsp. salad dressing (bottled)	0.200
1 packet salt	0.500
1 Tbsp. soysauce	1.350
¼ cup salted peanuts	0.150
10 potato chips	0.200
1 slice chocolate cake	0.450
2 large pickle spears	0.900

Foods Low in Salt	grams sodium
1 oz. plain chicken	0.020
2 Tbsp. peanut butter	0.050
1 egg	0.060
8 oz. milk, yogurt	0.125
½ cup ice cream	0.050
½ cup fresh fruit	0.005
½ cup fresh vegetables	0.010
1 cup pasta (unsalted)	0.004
1 slice bread	0.150
1 oz. dry cereal	0.250
1 pat margarine	0.050

SOURCE: J. Pennington and H. Church, *Bowes and Church's Food Values of Portions Commonly Used*, Harper & Row, New York, 1980.

How Many Calories Do You Need For Exercise?

Exercise increases your need for calories. A calorie is a unit of energy, measured in the form of heat. One calorie increases the temperature of a liter of water by one degree centigrade. Food provides potential energy (calories) for the working muscle. When your muscles exercise hard, you need more calories. Either food or body fat supplies this need.

In training, you burn two or three times the calories required by a sedentary student. The actual requirements vary, depending on the following factors:

- Big people need more calories than small people do. A tall, 170-pound jogger burns fifteen calories per minutes, whereas a shorter, 115-pound female companion burns only twelve calories per minute. Women, on the average, need fewer calories than men.
- Even when they are the same height, men need more calories than women because men have more muscle tissue, which metabolizes more calories than fatty tissue. The body of an average male is 45 percent muscle; a female, 36 percent.
- Younger people need more calories for growth than mature adults do. Caloric needs slowly decline with age. At twenty years you may burn 3000 calories; at thirty years your requirements drops to 2700 calories; at fifty years, 2400 calories. A woman's caloric needs are about 25 percent less than a man's.

Although exercise increases caloric needs, many people overcompensate. You may expend 200 calories walking to the store, and consume 300 calories rewarding yourself with ice cream.

What Sports Burn the Most Calories?

Active sports that involve speed, motion, and vigor burn the most calories. Competition running, scull rowing, cycling, squash, handball, and cross country skiing burn the most calories of all the sports rated by Robert E. Johnson, M.D., professor of biology at Knox College. All these sports use both the arms and the legs; they are total body sports.

Also, they are aerobic sports; they require a lot of oxygen. They require you to breathe deeply, which saturates the blood with oxygen. They make your heart pump hard, which circulates blood to the exercising muscles. Lying down or sitting, a person burns about one-tenth as many calories as in hard exercise.

MARSHALL HOFFMAN'S CALORIC NEEDS

(175-pound male, 5′ 10″, 38 years old)

Resting metabolic rate (calories needed with no activity)	1775
Light activity (energy required for day activity, normally one-half the resting metabolic rate)	800
Exercise (5 miles daily run at 8-minute-per-mile pace)	750
Total Calories Needed	**3325**

WILLIAM SOUTHMAYD'S CALORIC NEEDS

(175-pound male, 5′9″, 38 years old)

Resting metabolic rate	1725
Light activity	875
Total calories needed	**2600**

SOURCE: Nancy Clark, Sports Medicine Resource, Inc.

CALORIC REQUIREMENTS VARY

Age	Size	Climate
Sex	Health	Pregnancy
Basal metabolic rate	Activity	Lactation

SOURCE: Nancy Clark, Sports Medicine Resource, Inc.

How Many Calories Must You Burn to Lose One Pound of Fat?

A pound of fat is the equivalent of 3500 calories in stored energy. Every time you burn 3500 more calories than you consume, you use one pound of fat. By burning 500 more calories per day than you eat, you will lose one pound of fat in a week.

Keep in mind that one pound of fat is equal to one and two-tenths pounds of fatty tissue. The tissue includes blood vessels, membranes, and a small amount of water.

What Is Too Fat or Too Thin?

The best and most convenient way to measure body fat is with a skinfold caliper, which looks something like a giant pair of pliers with a gauge.

It works on the principle that fat is stored under the skin. The more fat you have in your body, the thicker the layer of fat under your skin. The caliper is used to measure the thickness of the fat layer under your skin when it is pulled away from the muscle. From tables your percentage of body fat can then be read.

Because they provide little information about the composition of an individual's body weight, popular height and weight tables do not work for athletes. For example, muscle is 22 percent denser than fat. A muscular, 160-pound athlete may be "overweight" according to the standards, but he will not be "overfat."

Body-fat measurement via underwater weighing in an immersion tank, although most accurate, is used primarily for research purposes. The method is complex, time-consuming, and inconvenient. It works on the principle that muscle is denser than fat. A runner's 150-pound muscular body is smaller than a bus-driver's 150-pound flabby body. The volume of water displaced by your body is used to compute your percentage of body fat.

A nutritionist or trainer can best measure your body fat and determine your ideal weight. A second, less professional method is the quick and easy pinch test. Can you pinch an inch of fat between your waist and lower rib cage? If so, you have excess fat.

What Is the Optimum Percentage of Body Fat?

A certain amount of fat is essential to maintain a healthy body. Men have 3 percent essential fat, women 13 percent. Women have more fat so that they can nurture a child. The average male in college is 15 to 18 percent fat; the average female is 25 to 28 percent fat.

Fat does not contribute to athletic strength, but rather consumes energy that could better be used to fuel the muscles. Your sports nutritionist or your trainer should determine the ideal percentage of fat for you, based on your athletic needs, age, and sex.

What Is the Best Way to Lose Weight for Sports?

Eating less and exercising more are the two keys to successful weight loss. The following guidelines will help you reach and maintain your goal.

1. **Determine your ideal weight.** Using a skinfold caliper, a nutritionist or a trainer should calculate what percentage of your weight is fat. Fat, rather than total weight, is your main concern. Based on your percentage of body fat, the nutritionist determines your ideal weight for your sport. For example, long-distance runners perform best when they are light. Elite male runners may be 6 to 10 percent fat, as compared with the average male, who is 15 to 18 percent fat. Swimmers perform better with a little more fat for insulation and buoyancy.

2. **Set a realistic goal.** I recommend a loss of two pounds a week for women, three pounds for men. More than four pounds a week is too much. Is your goal to lose ten pounds?

ENERGY EXPENDITURE BY A 150-POUND PERSON IN VARIOUS ACTIVITIES*

Activity	Calories Per Hour
A. Rest and Light Activity	**50–200**
Lying down or sleeping	80
Sitting	100
Driving an automobile	120
Standing	140
Domestic work	180
B. Moderate Activity	**200–300**
Bicycling (5½ mph)	210
Walking (2½ mph)	210
Gardening	220
Canoeing (2½ mph)	230
Golf	250
Lawn mowing (power mower)	250
Bowling	270
Lawn mowing (hand mower)	270
Fencing	300
Rowboating (2½ mph)	300
Swimming (¼ mph)	300
Walking (3¾ mph)	300
Badminton	350
Horseback riding (trotting)	350
Square dancing	350
Volleyball	350
Roller skating	350

MARSHALL HOFFMAN'S ACCOUNT OF HIS WEIGHT

"When I came to Boston on April 23, 1980, to write this book, I was heavy. Nancy Clark, the nutritionist at Sports Medicine Resource, found that I weighed 183 pounds with body fat of 20 percent. That was the heaviest I had been since college. The next day I started a program of jogging twenty miles a week. After three months of running, I increased to thirty miles a week. In October Nancy measured me again. I was 174 pounds with 15 percent body fat. Most of the change was in my measurements. During a five-month period I lost three inches from my waist."

Activity	Calories Per Hour
C. Vigorous Activity	**over 350**
Table tennis	360
Ditch digging (hand shovel)	400
Ice skating (10 mph)	400
Wood chopping or sawing	400
Tennis	420
Water skiing	480
Hill climbing (100 ft. per hr.)	490
Skiing (10 mph)	600
Squash and handball	600
Bicycling (13 mph)	660
Scull rowing (race)	840
Running (10 mph)	900

*The standards represent a compromise between those proposed by the British Medical Association (1950), Christensen (1953), and Wells, Balke, and Van Fossan (1956). Where available, actual measured values have been used; for other values a "best guess" was made.

SOURCE: Robert E. Johnson, M.D., Ph.D., Professor of Biology, Knox College, Coordinator, Knox Rush Medical Program, Scientific Consultant, Department of Medicine, Presbyterian St. Luke's Hospital, Chicago.

Allow four to five weeks. Gaining the weight took time; so will losing the weight.

Wrestlers, dancers, gymnasts, and other lightweight athletes should plan in advance of the season. Trying to shed pounds at the last minute results in weakness and poor performance. If you lose the match, "making weight" for wrestling is worthless.

3. **Stay away from fad diets.** Claims for "eat all you want," "no willpower," and "spot reduction" do not work. There is no quick and easy way to lose weight. Fad diets are unbalanced, unrealistic, and unhealthy because:

- They generally eliminate food groups such as carbohydrates—an important part of an athlete's diet.
- They drastically reduce your caloric intake to unrealistic levels, which sometimes accounts for personality changes along with the weight loss.
- They promote muscle deterioration and water loss more than fat loss. When you diet without exercising, you lose 25 percent muscle. If you regain the weight, you put on 100 percent fat. Athletes who try to lose weight by starvation and dehydration compromise their health and performance.

4. **Eat smaller portions.** Plan with your nutritionist a personalized meal pattern that includes the foods that you normally eat and are readily available. Do not drastically change your eating habits to include "diet" foods. The total number of daily calories, not the source of the calories, determines the amount of weight loss. Two hundred calories from cottage cheese is just as fattening as 200 calories from bread.

Watch portion size: 1/2 cup of rice = 70 calories
3/4 cup of rice = 115 calories
1/2 cup of cottage cheese = 110 calories
2/3 cup of cottage cheese = 135 calories

By eating foods you normally enjoy, you will feel psychologically satisfied rather than denied. You must learn to eat smaller portions instead of "pig-out plates." Cut your normal intake by one-third: two slices of pizza instead of three, a small ice-cream cone instead of a sundae.

5. **Choose more fruits, vetegables, and low-fat foods.** Eat crackers, bread, skim milk. Creamy, greasy, and fatty foods are calorie disasters. Fat in any form is concentrated calories. Limit your intake of:

butter	cream cheese	fried foods
margarine	peanut butter	fatty meat
oil	sour cream	gravy
salad dressing		mayonnaise
sauces		

Beware of hidden fats in cheese, meat, nuts, and bakery items.

6. **Eat at regular times.** Meal skippers frequently overcompensate at the end of the day when they feel uncontrollably ravenous. The pizza at 10 P.M., which you claim you "deserve," is more fattening than the breakfast and lunch which you omitted.

7. **Exercise more.** The more active you are, the more calories you burn. In addition to your daily training, incorporate new ways to step up your energy output. Walk to the post office, park your car at the far end of the lot, take an activity break instead of a coffee break, use stairs instead of the elevator.

8. **Quench your thirst with water.** Sweaty athletes guzzle lots of liquids, lots of calories. Convert these calories into solid food, and you can easily see them as fattening additions to your diet:

- One quart of orange juice is equal in calories to a peanut butter sandwich.
- Two cans of Coke can be traded in for a large ice-cream cone.
- One beer is the equivalent of six cups of unbuttered popcorn. Water is a low-cost, low-calorie alternative. It is best for replacing sweat losses.

9. **Learn how to eat for a lifetime.** Trade in your fattening eating problems for healthful habits. Your nutritionist can teach you how to eat wisely, lose weight, and maintain that loss. Good nutrition is an investment in your present performance as well as your future health.

FATS ARE FATTENING

Added fats, as well as hidden fats in meat and cheese, give you fattening calories.

	calories
½ English muffin	70
with 1 pat margarine	130
1 medium potato	80
with 2 Tbsp. sour cream	140
3 oz. hamburger	200
1 hamburger roll	90
1 oz. cheese	110
4 saltines	50

BODY FAT

	Average (College-age)	Athlete	Elite Athlete
Male	15	10 to 15	5 to 8
Female	28	16 to 20	11+

	Essential Fat	Storage	Total
Male	3	12	15
Female	13	15	28

All figures expressed as a percent of total body weight.

Why Do People Lose Inches Instead of Pounds?

Inches represent fat loss. By eating less and exercising more, you burn your fat for energy. Your body takes the fat from where you store it—the stomach, hips, thighs, arms, and breast. As you draw on the excess fat, you lose inches from your body. Exercise builds muscles. Muscle tissue weighs one and two-tenths times more than fat tissue. Therefore you may gain muscle from your exercise program, lose fat from your dietary program, and still maintain your weight.

A nutritionist can help you monitor these changes in your physique by taking skinfold measurements before and during your shape-up campaign. This will help you determine your ideal weight, based on your fat weight rather than total body weight.

What Is the Best Way to Gain Weight for Sport?

Gaining weight, or maintaining one's ideal weight throughout strenuous sports seasons, is a problem for many athletes. The solution is clear-cut and mathematical. To gain one pound in one week, each day you must consume 500 calories more than you burn. You can do this by:

- Eating more often or snacking.
- Eating larger portions or second helpings at meal times.
- Choosing higher-calorie foods.

The following guidelines will help you reach your goal.

1. Determine your daily caloric needs. A nutritionist will calculate this information based on your age, height, weight, sex, and activity. You may require 4000 to 7000 calories per day. That's a lot of food!
2. Record everything you eat and drink for two to three days. Your nutritionist will assess how many calories you have been eating and make suggestions for higher-calorie choices. For example, grape juice has more calories than orange juice, granola more calories than raisin bran. A twelve-ounce glass of milk has eighty more calories than the same amount of beer.
3. Focus on polyunsaturated fats. They are a more concentrated source of calories than carbohydrates. Eat toast with corn oil margarine, salads with dressing based on safflower oil, thick sandwiches made with old-fashioned peanut butter. Limit your intake of saturated animal fats (butter, ice cream, cheese, and fatty meats). A diet high in saturated fat and cholesterol will contribute to weight gain and also to heart disease. Your overall health is equally important to your sports career.
4. Plan your heaviest meals or snacks after you have exercised. I eat breakfast after my morning run. The Boston Red Sox have a heavy meal after a ball game. To gain weight, you should eat until you feel full, then eat a little bit more. Food, for you, may be like medicine.

Why Are Runners and Cyclists So Thin?

Runners and cyclists expend tremendous amounts of energy when they train and race. A 150-pound male runner burns 1000 calories on a ten-mile training jaunt. Competitive cy-

BOOST YOUR CALORIES

Small changes in your normal eating habits can make big changes in your weight. Compare these two diets:

Usual choices	Calories	High-calorie choices	Calories
8 oz. orange juice	110	8 oz. cranberry juice	170
1½ cups bran flakes	200	1½ cups granola	780
8 oz. milk	160	8 oz. milk/¼ cup dried milk	220
1 slice toast	80	1 slice toast	80
1 pat margarine	50	1½ pats margarine	75
½ Tbsp. jelly	30	1 Tbsp. peanut butter	95
1 cup coffee	0	1 cup hot cocoa	225
1 tsp. sugar	15		
Quarter-pounder hamburg	420	Big Mac	550
Large fries	215	Large fries	215
Large cola	160	Chocolate shake	360
Large apple	130	Large banana	170
1 cup vegetable soup	80	1 cup cream mushroom soup	135
7 oz. baked chicken	330	7 oz. baked chicken	330
1 cup mashed potato	150	1 cup noodles	200
1 pat margarine	50	2 pats margarine	100
1 cup green beans	40	2 Tbsp. Parmesan Cheese	110
½ pat margarine	25	1 cup peas	105
½ cup jello	70	1 pat margarine	50
8 oz. milk	160	1 cup pudding	200
8 oz. tomato juice	40	8 oz. milk/¼ cup dried milk	220
1 English muffin	130		
1 pat margarine	50	8 oz. pineapple juice	150
1 Tbsp. honey	60	1 bagel	180
		2 pats margarine	100
		2 Tbsp. cream cheese	200
Total calories	**2755**		**5020**

Note: The small changes contributed 2265 extra calories! One pound of fat has 3500 calories.

SOURCE: Nancy Clark, Sports Medicine Resource, Inc.

clists burn at least that much in an hour on the roads.

Replacing all the expended calories can be a problem for very active athletes. They frequently lose weight and maintain a low percentage of body fat.

Runners strive to be thin and light. Each extra pound is unnecessary weight to carry along mile after mile. Cyclists are not so weight-fanatical; the bicycle supports their weight. They realize, however, that excess fat consumes oxygen that the exercising muscles could better use. Because they prefer to carry less excess fat baggage, they design a diet to maintain their thinness.

Why Do Professional Athletes Get Fat?

Fatness indicates overeating. When you eat more calories than you expend, you store the extra. If you store 3500 extra calories, you gain one pound of fat.

Professional athletes are accustomed to eating quantities of food to meet their requirements of 4000 to 6000 calories per day. When they quit sports, they frequently continue to eat in the same fashion. They may need only half that amount. Hence, they gain weight.

Athletes who have always been active, thin, and carefree about calories must learn to curb their appetites when they stop participating in sports either for the off season or for retirement. To change this carefree pattern is difficult and certainly not so much fun as eating large meals. Creeping obesity transforms Mr. Fit into Mr. Fat.

For those who want to read more on sports nutrition, we recommend the following books.

Textbooks:

Astrand, P., and K. Rodahl, *Textbook of Work Physiology,* McGraw-Hill, New York, 1970.

Mathews, D. K., and E. L. Fox., *The Physiological Basis of Physical Education and Athletics,* Saunders, Philadelphia, 1976.

General:

Clark, Nancy, *The Athlete's Kitchen,* CBI Publishing, Boston, 1981.

Costill, D., "The Scientific Approach to Running," *Track and Field News,* 1979.

Darden, E., *Nutrition and Athletic Performance,* Athletic Press, San Marino, Calif., 1976.

Katch, F. I., and W. D. McArdle, *Nutrition, Weight Control and Exercise,* Houghton-Mifflin, Boston, 1977.

Nutrition for Athletes, A Handbook for Coaches. Am. Assoc. for Health, Physical Education, and Recreation (with cooperation from the Am. Dietet. Assoc. and the Nutr. Foundation), Washington, D.C. 1971.

Smith, N. F., *Food for Sport,* Bull Publishing, Palo Alto, Calif., 1976.

Williams, M. H., *Nutritional Aspects of Human Physical and Athletic Performance,* Thomas Publishing, Springfield, Ill., 1976.

Zohman, L. R., *Beyond Diet . . . Exercise Your Way to Fitness and Heart Health,* Prentice-Hall, Englewood Cliffs, N.J., 1974.

Children and Sports

Arthur M. Pappas, M.D.

I am one of the few physicians in the United States who specializes in sports medicine and orthopedics for children and young adults. In my twenty years as an orthopedist I have treated thousands of children with every conceivable kind of sports ailment, among them Osgood-Schlatter's syndrome, Little League elbow, osteochondritis dissecans, and Köhler's syndrome.

About 40 percent of all sports injuries happen to children and young adults. Nationally, nearly 4 million individuals less than fifteen years of age are injured each year while engaged in sports. Yet there are very few specialists dealing with the pediatric aspects of sports medicine. And, even among the general public, few people recognize that the injuries sustained by a favorite celebrity differ markedly from those suffered by a growing youngster.

The traditional role of the sports physician is to handle individual injuries and general team health problems. This role is changing as physicians extend their involvement to include advice on conditioning techniques and general fitness. In addition, we must counsel the young athlete, his or her parents, and the coach on matters of health, injury prevention, and the mutual responsibilities within the athlete/parent/coach triangle.

As the chart on page 446 indicates, six activities—bicycling, football, baseball, basketball, roller skating, and playground equipment recreation—account for 75 percent of all the sports injuries to children. By far the largest number of children are injured in bicycling accidents—a whopping 1.2 million injuries.

Sixty percent of all sports injuries to children fall into these categories: sprains, strains, contusions, abrasions, and lacerations. Less than 15 percent are fractures.

Despite the large number of injuries to children and young adults, the rate of injury for any given number of children participating in sports is actually the lowest of any age group. That rate, according to a special study prepared for this book by the Consumer Product Safety Commission, is less than 10 percent. Of those injured, less than 2 percent require hospitalization.

The injury rate changes dramatically as players and athletes become older, the competition keener, and the physical involvement more intense. Older participants are larger, heavier, and stronger. When I was the team physician for the Harvard football team, three-quarters of the players were injured every year, an injury rate of 75 percent. Almost every player on the Boston Red Sox is injured at least once during the season.

Why are the injuries of children less severe than those of adults? There are two reasons. First, because children and young adults are physically smaller than adults, less force is involved. For example, when a sixty-pound child falls on his knee, he does so with one-third the force of a 185-pound man. Less force usually results in less severe injury.

Second, children's bones are more resilient and flexible; the ligaments are more elastic and the joint surfaces, because they have a small blood supply, are capable of repairing themselves.

Injuries of the ends of major bones in children and young adults can cause major deformities, however. The ends of bones are growth centers where growth takes place. Such injuries can cause growth to stop partially or totally. In such cases, the result may be that one limb remains shorter than another or it may grow at an angle from the point at which only part of the growth center is injured.

Why are so many children injured in sports? One reason is that sports are a normal part of childhood. Studies show that virtually all children up to the age of ten participate in some form of organized play or sports. This is largely due to the fact that children have the time and a need to participate in physical activity. Behavioral studies reveal that in the summer children can spend as much as forty hours a week playing in both street and organized sports. Sports and games are an important form of social interaction.

As you grow up, you have less time for sports. Yet, the older you get, the more you need sports to develop and maintain good health, and generally, the more competitive the sports become. However, because of the competition, fewer people play. When I was in grade school, most of the boys played neighborhood football or baseball. When I played football in high school, the number engaged in the varsity sports program dwindled to a handful. The process became even more selective in college, when I played on the Harvard football team. A small percentage of the student body played varsity sports, and only a handful of Harvard undergraduates ever became professional athletes. Therefore, participation may be thought of as being shaped like a pyramid. Less than 25 percent of teenagers who are seventeen, eighteen, and nineteen years of age remain involved in sports or athletics. One reason is the sports pyramid.

I've asked teenagers why they are not playing sports. The answers I've gotten clearly demonstrate the progressive disengagement from sports that begins in junior high school:

- Sports are boring when I cannot participate.
- I was criticized by my coaches, friends, and family, so I quit.
- I was cut from the team because I wasn't good enough.
- I am afraid to lose.
- With my playing ability, I couldn't find a team to play on.

These answers reflect how youngsters respond to the team competitive sports we encourage. If youngsters were as encouraged to develop skills in individual sports, more might grow up with enthusiasm for sports participation and a concern for lifelong physical fitness.

My personal experience follows the national trend. For example, more than 5 million secondary school students play varsity sports. In college, the number drops to 450,000 students. The pro ranks are about 6600. Only one of 100,000 secondary school athletes achieves professional status. Hardly an encouragement for children to emulate the professional model.

Children are children; they are not miniature adults. Doctors recognize this. That is why there are specialties in pediatrics and pediatric orthopedics.

There are good reasons why you should seek out a specialist in sports medicine for young children's sports injuries:

- Children are physically and emotionally different from adults.
- Even outstanding physicians who regularly don't treat children may be unfamiliar with some of the diagnoses and treatment of children's problems.

Up to the age of ten, boys and girls are physically similar. The size of their bones and muscles are the same. They have the same athletic skills.

Between the ages of ten and thirteen, girls start to mature—come into the age of adolescence. With the start of menses the girls experience hormonal changes in their bodies. This causes sex-organ changes—the growth of breasts and pubic hair. It also brings on the growth spurt when a girl will grow as much as four to five inches a year. The growth spurt usually ends by age fourteen. Three-quarters of the boys will start their growth spurt between the ages of twelve to fifteen. The maturational growth period will usually terminate between fifteen and eighteen.

Once the growth spurt has slowed down, there is an increase in muscle mass. This usually occurs six to twelve months after the growth spurt has ended. As the muscle tissue develops, the adolescent increases dramatically in strength and weight. As one may expect, studies show that adolescents who mature early are more likely to succeed in junior high and high school athletics. I recall that most of the boys on my high school football team shaved regularly by the age of fifteen.

How do children grow? The chemical changes that trigger growth are complex. But the mechanism of growth is simple. The growth occurs at the upper and lower end of each bone (called the *epiphysis).* The growth occurs mainly in the physis—the growth plates or growth centers. The growth plates are cartilage-bone cells that are not fully formed. On x-rays they look like a cloudy mist about one-half inch to an inch from the bone end. The growth cells multiply and eventually

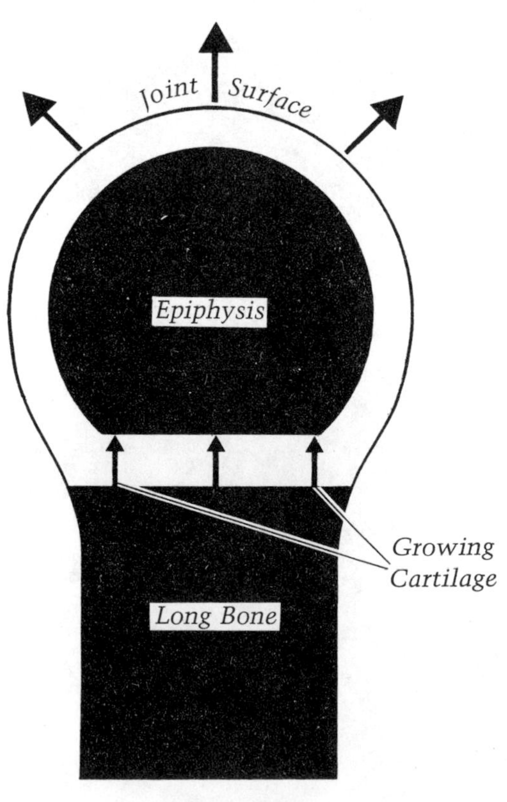

Bone Growth

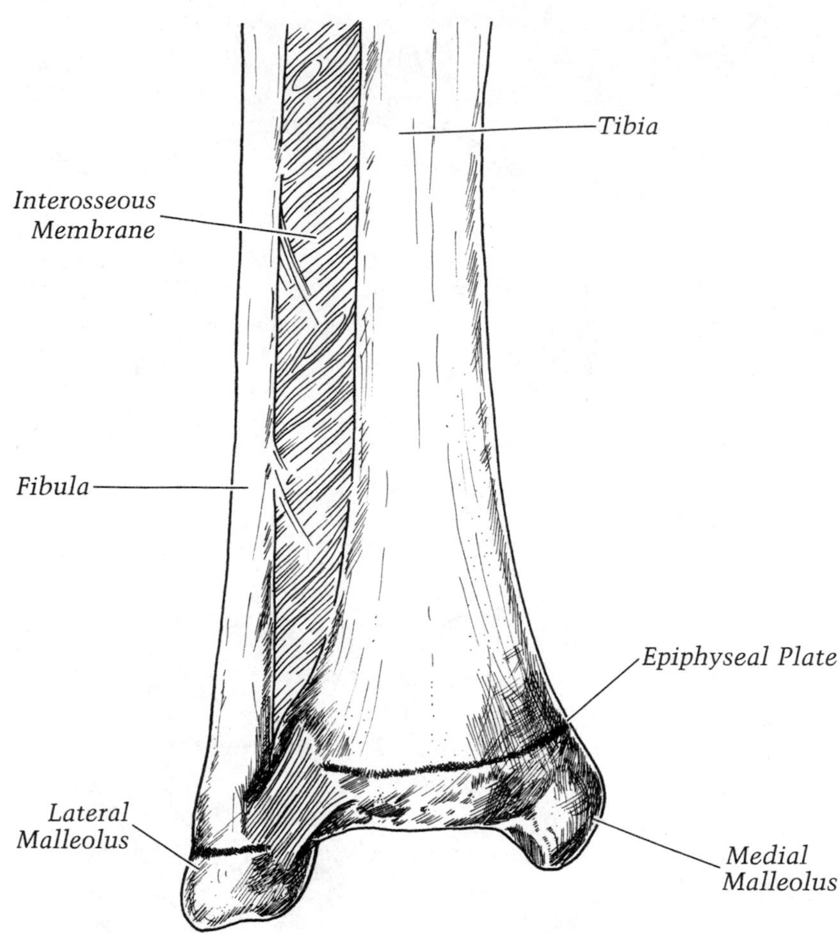

Growth Plates of Fibula and Tibia

harden and form bone. Because growth plates are softer than fully formed bones, they are far more likely to be injured than bone. They are the weakest links in the skeletal chain.

Growth rates of different bones vary. For example, the femur—the thighbone—may grow three-quarters of an inch to an inch per year. The radius, a forearm bone, may grow one-quarter to one-half an inch per year.

The same bone can grow at a different rate on each end. For example, 80 percent of the growth of the upper arm bone comes from the shoulder end of the humerus. The lower end forms the upper end of the elbow joint.

The growth rates increase tremendously with the onset of maturation. Most boys can grow between four and eight inches per year during their growth spurts. In this period, the growth plates are working overtime.

There are other types of bones called flat bones, such as the bones of the pelvis and of the scapula (shoulder blade), that have an epiphysis around their periphery. They gradually grow away from themselves in a pattern very similar to long bones. Major muscles and tendons either originate or attach to these flat bones. Thus, a sudden pulling away of a muscle from one of the flat bones may also pull away the growth center and result in a type of athletic injury we see only in childhood and adolescence. This is known as an *avulsion fracture injury.*

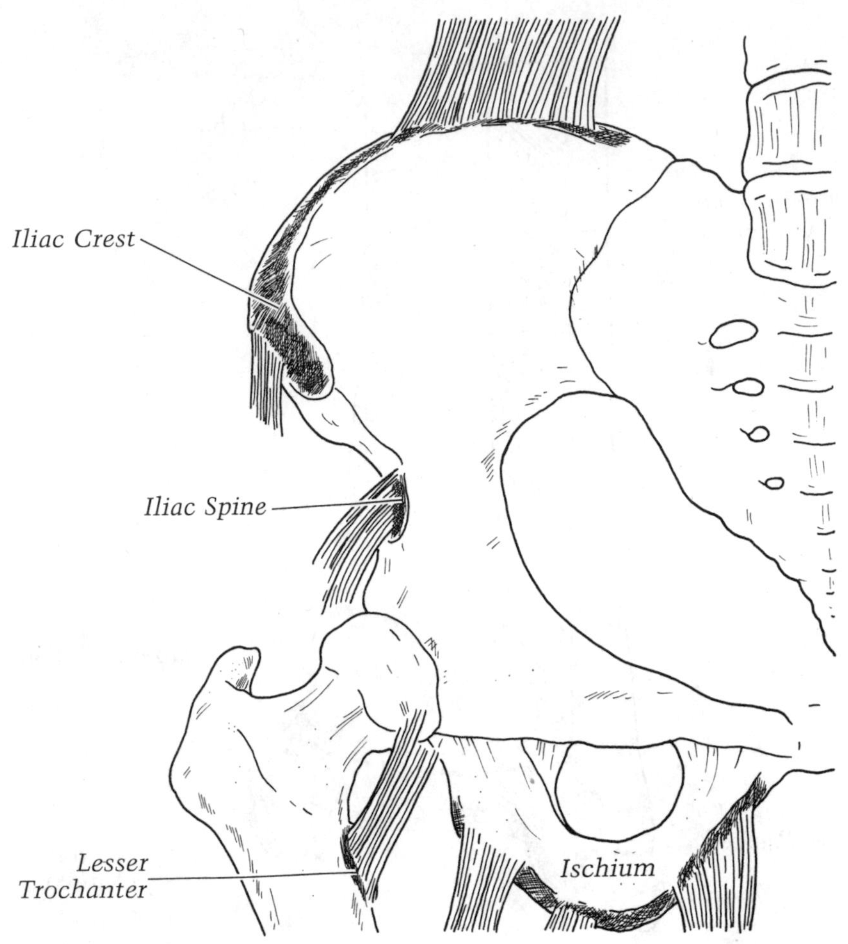

Areas of Pelvic Avulsion Injury

The development of muscle mass in children is closely related to their bone growth and genetic traits. In general, it is impossible to encourage the rapid growth and strengthening of muscle during childhood because the hormonal factors that relate to strength development are not present until adolescence and later. The fat content of the growing child is, in part, genetically determined but it is influenced to a much greater degree by nutrition. Childhood is the time when athletic events may serve as a ground for long-term health and nutritional understanding.

Parents often ask me if their children should play sports. The questions are usually phrased this way: "Is it safe for my son or daughter to play sports?" and "Is my son or daughter old enough to play sports?"

Among physical and mental health professionals there is no question that an appropriately balanced athletic program is a positive factor for general physical and emotional development. We should endorse such programs within our schools and communities.

All athletic events expose the participants to some degree of physical and emotional risks, but our attention should be directed to adequate management for maximum safety, while permitting active participation. This requires attention to rules, their modification and enforcement, i.e., face guards are

PELVIC AVULSION INJURY

John Gray, a fourteen-year-old ninth-grade football player, was running wind sprints at the end of practice. Two-thirds of the way through the sprints, he felt a pull in his buttock area. The pain was immediate and agonizing. John limped off the field and was examined by his family physician, who diagnosed his condition as a hamstring pull. If he had been a full-grown adult, this would have been the correct diagnosis.

The family doctor recommended rest and heat treatments. Two and one-half weeks later, John was still limping badly and could not play football. His mother became worried and brought him to me.

In an athlete his age, I suspected a bone avulsion injury of the ischium. Examining him, I found tenderness and fullness in the depths of the buttocks over the ischium. I put him flat on his back on my examining table. I raised the injured leg straight off the table at the hip joint. This pulled on the hamstring muscles and he felt pain.

Next, I had x-rays taken of his pelvis and hips which showed an avulsion injury of the prebone of his right ischium. It was pulled away from the pelvis about an inch. Some new bone had begun to form between the main pelvis and the detached bone.

I told John's mother this injury is often confusing to doctors who do not treat athletes regularly. No harm had been done in delaying the correct diagnosis. The healing had gone on just the same. He was walking and the x-rays showed enough healing that I did not recommend crutches. Initially, he would have been more comfortable with crutches. Two weeks after I examined him, he had very little tenderness. I began him on gentle jogging exercises. Weight lifting and rehabilitation exercises were started six weeks after the injury, with gradual return to sports. Ten weeks after the injury, he returned to full sports activity and enjoyed a successful basketball season.

In my experience, growing athletes more commonly will pull the mechanically weak prebone away from the pelvis than pull their hamstring muscles.

necessary in youth hockey. The careful selection of and education of coaches and support of competent officials and a planned medical/health care program before, during, and after the season are essential to a good sports program. Sports programs present an ideal format for children to learn about dealing with stress, and offer a valuable long-term resource. Each athlete's ability should be assessed and realistic expectations defined. Unrealistic expectations of less talented athletes are a major reason youngsters withdraw from physical education and sports. Some children excel in sports because of positive feedback from their peers, as well as their coaches and parents. Others seem to be motivated by a fear of failure. The former group handles stress better than the latter. A third group, the handicapped, are frequently excused from sports participation when, on the contrary, these children benefit more from and require greater physical attention than do their able-bodied counterparts.

At the other extreme, the young athlete who presents extraordinary potential is a challenge for the physician, coach, and family. Too often, our society permits talented young athletes to develop a distorted view of themselves. Sometimes this results in a depreciation of academic performance and personal growth. Maintaining a balance between a youngster's academic and athletic achievement is a goal which should never be overlooked.

Another positive aspect of the concern for sports health and safety is reflected in more deliberate concern whether and when a youngster should become a single-sport athlete. Once this decision is made, it requires a different life-educational pattern for the young athlete. Travel and practice time may demand up to six to eight hours each day and the physical demands of repetitive drills and performance frequently result in "overuse syndromes" of stress-related anatomical areas. These children are necessarily exempted from the opportunity to participate in many community and social events during the childhood and adolescent developmental years.

If the athlete is fortunate enough to possess the talent to become a national or international competitor, the rewards are perhaps significant enough to compensate. However, only a few make it to the national and international level. There are many others who lack the skill or have been injured and are unable to continue. Unless parents and coaches offer sensitive guidance and support, these children frequently suffer a major physical and psychological loss. At times the injury that forces a youngster to withdraw from participation may be minor; however, in an intensely competitive situation, a loss of 1 or 2 percent in ability can remove one from the highest ranks. The emotional fall is frequently much more severe than the physical injury.

Another problem that effects the highly committed, excelling athlete is his acceptance by his siblings and other members of the family. I have had the opportunity to observe a number of families where the successful athlete was treated as so special by the parents that other siblings become progressively resentful. Again, the psychological disruption far exceeded the physical damage of any sport injury.

These comments are not made to discourage the development of superior skill or the development of a world-class athlete if the potential is present. Rather they should encourage awareness of the need to provide and sustain a family and community structure appropriate to the young athlete's successes and achievements.

"Is my son or daughter old enough to play sports?" In my opinion the question is not age but size and maturity. Boys in junior high range in size from four feet six inches to six feet and weigh from 75 to 200 pounds. The main point at issue is whether your child will be physically mismatched. The bigger the differences in height and weight in contact sports, the higher the chances that your child will get injured both physically and emotionally. Some organized sports take these factors into account and require participants to be both of a certain age and weight.

Too often children and parents accept what team sports are available or elect not to participate. But there are numerous alternatives available to all children, in other team sports and in many individual sports. The important considerations are to combine the child's ability and physical stature with a realistic and safe sport-achievement goal.

I recently examined a large, immature fifteen-year-old football player. Because of genetic predisposition he is six feet tall and 210 pounds. He is as large as most of the athletes on his high school football team. Therefore, he is encouraged to participate in football. The problem is that many of his growth centers are continuing to contribute to his growth. I explained to the boy's father that these growth centers are considerably weaker than other parts of the body. "He is strong and will be competing with people of equal size and strength," I said. "His actively growing growth centers are much more prone to injury and secondary growth disturbances because of the injury. Maybe your son would consider another sport until his skeleton matures." His father withdrew the boy from football for two years.

Physical mismatches usually result when indiscriminate scheduling causes children to be clustered by age rather than by physical characteristics and skeletal development. These physical mismatches boost the injury rate, sometimes as much as 50 percent.

Frequently individuals who are opposed to contact sports cite the injury rate as the reason to drop these sports from the junior high and high school athletic curricula. I have been invited to some high schools to explain the high injury rate to school administrators, coaches, and parents. In general, those against contact sports and those favoring contact sports seem to be aligned more along socioeconomic lines rather than sound sports rationale. In neighborhoods of a higher socioeconomic level, physical assertiveness and contact sports sometimes have negative connotations. In these areas, competitive contact sports are generally discouraged, whereas, in lower socioeconomic neighborhoods, physical assertion, competition, and body contact sports are associated with an athletic image, increased social esteem, and the possibility of greater success in athletics at a higher level of competition. In the eyes of some, this is taking unfair advantage of the athletic system; for others, it presents an opportunity for some to achieve an education and a better life.

In any event, contact sports are frequently an issue of personal philosophy. The real issue is how to minimize injuries if contact sports are a part of the interscholastic or community athletic environment. A good sports program requires careful assessment of personality traits, physical aptitude, and maturational factors to determine who should participate. Then, a careful review of equipment and rules can minimize the likelihood of injury. Besides poor equipment and training techniques, it is mainly physical mismatches that push up the injury rate.

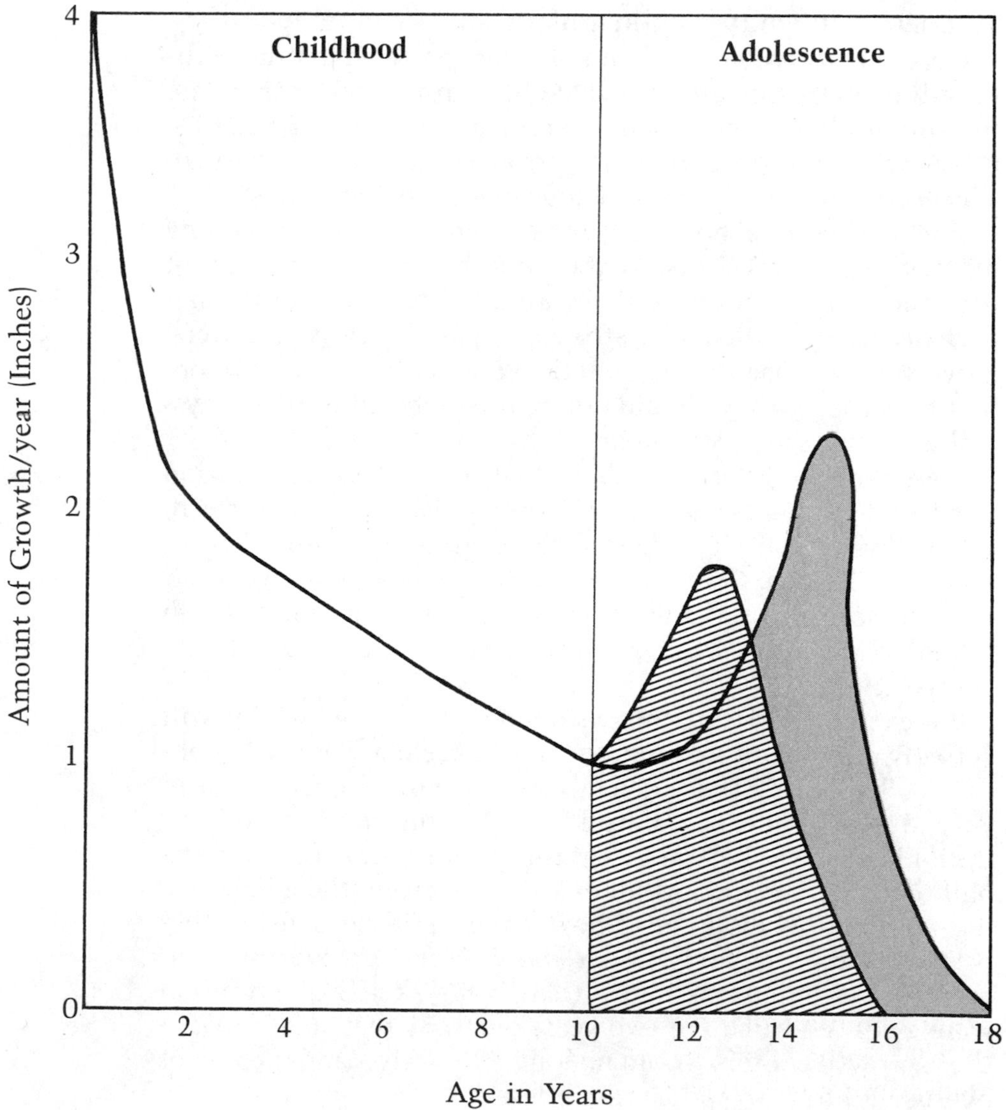

One way I tell the relative physical age and maturity of a child or a young adult is by a physical exam. There are characteristic physical changes that occur during the various phases of adolescence. A second way is to x-ray the wrist and hand, which have twenty-nine individual growth centers. The rate of development of these growth centers determines the relative skeletal age of an individual.

Because girls mature earlier than boys, many are larger, stronger, and more agile than boys of the same age. During this period—usually between the ages of ten and thirteen—many girls are better athletes than boys. This fact has resulted in a series of requests by parents and coaches for increased participation by girls in coed contact sports—football, hockey, boxing, and the like.

Coed sports are important for psychosocial development as well as for the development of physical skills. Sports are as beneficial to girls as to boys. But as a physician I think that girls competing with boys in contact sports like football are only avoiding the reality of physical inequality between boys and girls. It will only be a matter of time before the boys surpass girls in size and strength. When this occurs, many girls experience difficulty both in athletic and social terms.

Children whose skeletal and physical patterns mature earlier will be more powerful and able to compete better than late maturers. This is evident in all studies of adolescent athletes; those who mature earlier are stronger and more likely to succeed in junior high school and high school athletics.

The problem is how to counsel late maturers regarding athletic and sports skills. By the time they are twenty years of age, the late maturers will, in all likelihood, be taller and perhaps more skilled and stronger than the early maturers. However, because of the genetic variance in their developmental pattern, they should not be mismatched to their physical and emotional detriment.

Boys do not experience their peak of skeletal maturation until approximately age fifteen. Some will start at age thirteen, others do not complete their skeletal maturation until seventeen or eighteen. Thus, the majority of boys undergo their growth acceleration and physical maturation during high school years. This difference becomes significant in athletic competition.

For example, if a tall, rangy fifteen-year-old boy, who is still actively growing, were found to be an excellent baseball pitcher, yet had a relative skeletal maturation of only fourteen years, it would be wise counsel for this young man not to pitch until his relative skeletal maturation were at least fifteen and a half to sixteen. With a skeletal age of fourteen, the epiphyses, growth centers about his elbow would still be considerably weaker and, therefore, much more prone to overuse syndromes and long-term injury than if his skeletal maturation were a year and a half or two years older. By that stage most of these growth centers would no longer be active, and the risk of overuse injuries would be much less.

Injuries to Growth Centers

Any injury to a growth center or cartilage is a serious occurrence. It is the growth centers that are the mechanism for growth of the long and the flat bones of the body. Injuries to growth centers can stunt growth or result in permanent joint deformity.

An injury can happen in many ways—a fall, a twist, a turn. The growth center can be fractured, compressed, or torn. Some or all of the germination cells may die.

The injury may heal without any long-term deformity, growth may slow temporarily until the fracture or compression of the growth center heals. Normal growth then continues. However, the growth center can stop growing altogether.

The potential for deformity resulting from injury is tremendous. Take, for example, an injury to the growth center of the

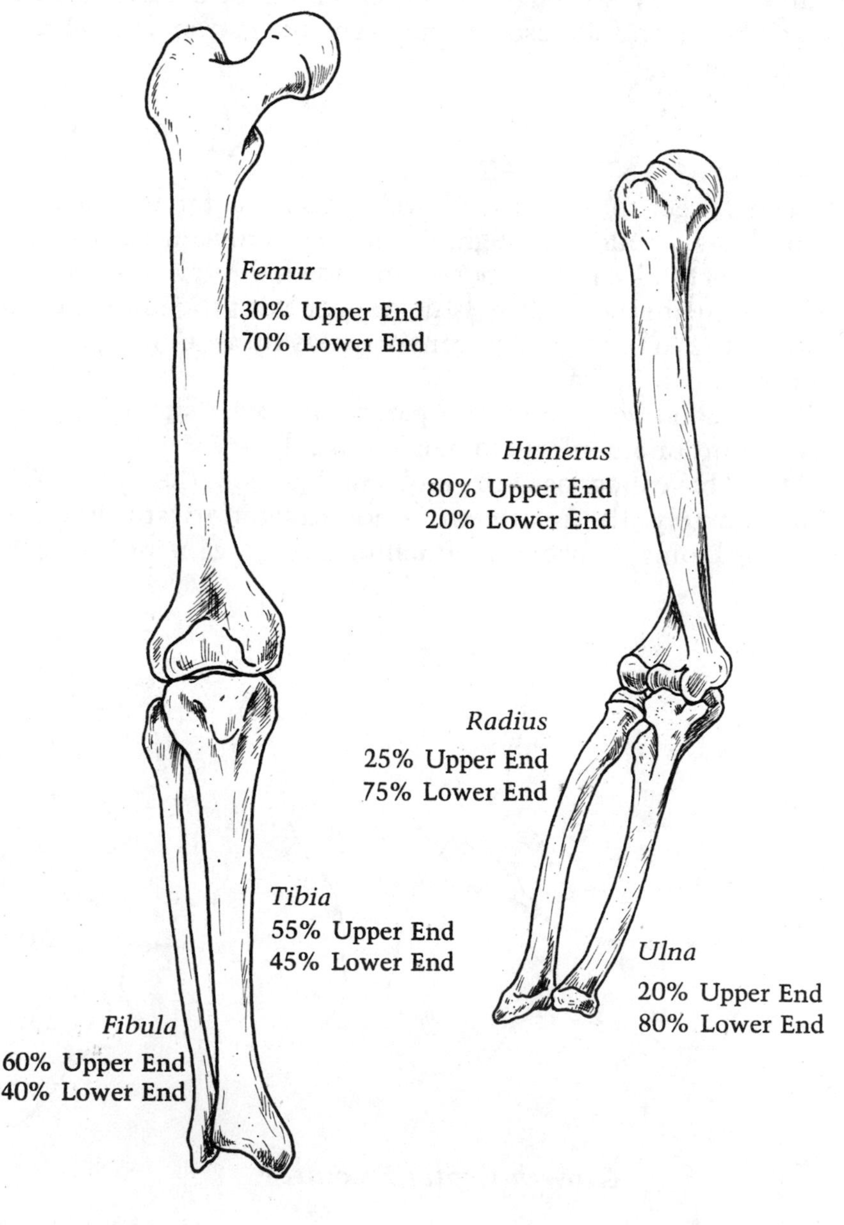

Growth Contribution

lower femur (thighbone). While the injured center is healing, the other leg is growing. That can mean a difference of one-half inch per year. For the average twelve-year-old boy that may account for one leg becoming two inches shorter at the end of his growth period.

Sometimes growth centers are only partially injured. This is an even more difficult medical situation because the growth centers cause only part of the bone to grow. In this case, if the child's bones grow out of line, they are angular or crooked.

Where a growth center injury occurs in the lower leg about the ankle or in the upper arm about the wrist, there are two bones growing in that portion of the extremity. If only one is injured, then the effect of the growth discrepancy and the normal growth of the noninjured bone must be looked at carefully for their effect on the function of the joint.

In an adolescent who is growing rapidly, an injury to the knee from a side tackle is much more likely to be a growth center (epiphyseal) fracture than it is a ligament injury. A ligament injury would be more common in a fully grown person but in the adolescent, the ligament is stronger than the growth center.

DIAGNOSIS AND TREATMENT

I usually get to see these injuries under two circumstances. Sometimes a child is brought to me with a routine fracture or joint problem. I x-ray the bones, including the growth centers. I also compare the healthy bone or joint with the injured one. In time the injured growth center shows up as an irregularity on the x-ray.

The second way is when a parent notices a small limp or that a joint or a bone is growing crookedly.

After I have diagnosed the problem, I propose the treatment. Unfortunately, the treatment is not straightforward, like removing a joint mouse or reattaching a torn ligament.

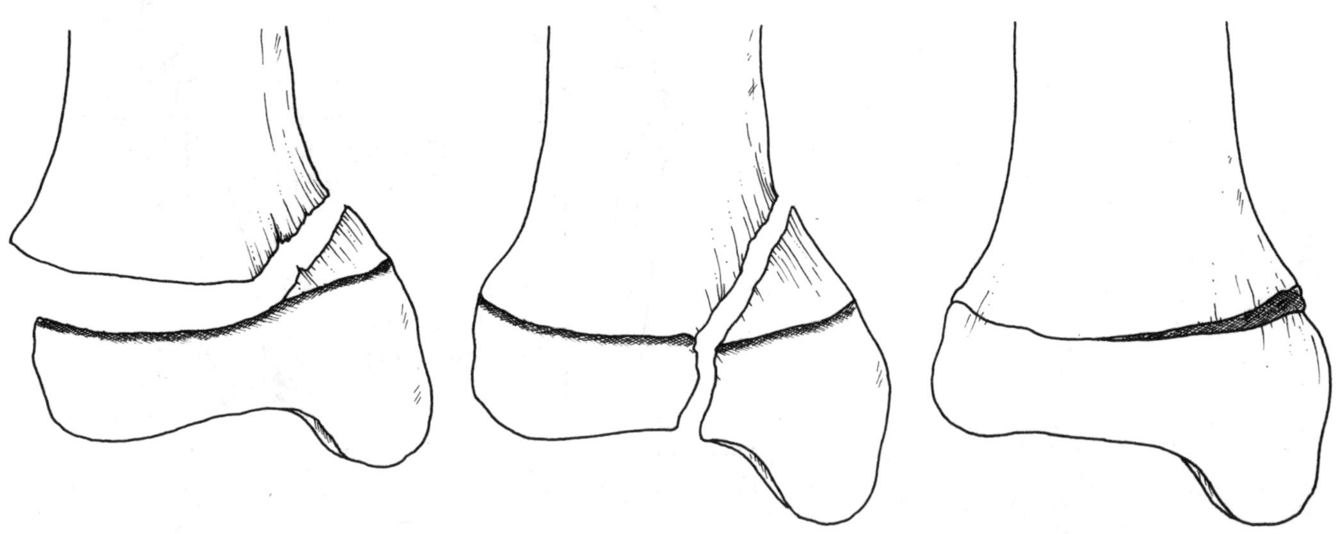

Growth Center Fractures

Deformity Due to Partial Closure

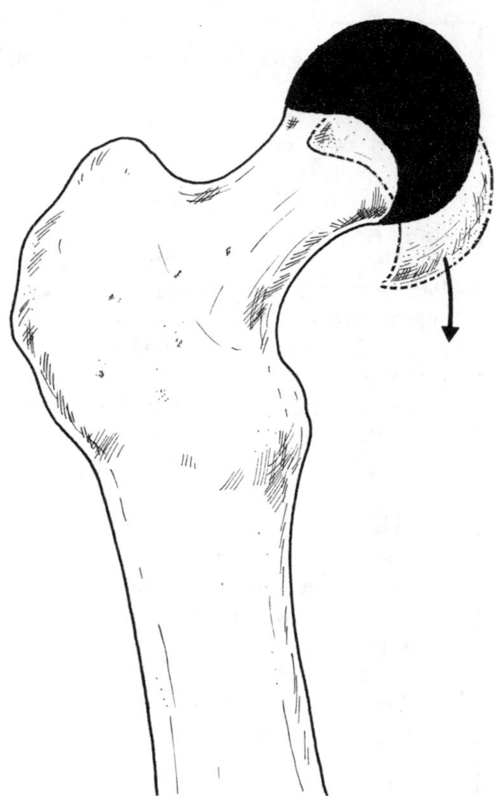

Slipped Femoral Epiphysis

There are three procedures to consider. They are all surgical. First, if a limb is crooked, I straighten it. Secondly, I can operate on the uninjured limb to slow its growth, or I can shorten the uninjured limb. Lastly, I can surgically lengthen the injured limb.

Depending on the joint or the bone involved, your child can be in the hospital for only a few days or for as long as a month. Sometimes one operation is not enough. Children in their growth spurts have radically changing bodies.

Injuries to growth centers in joints present very different problems for the adolescent age group. An injury may result in a fracture of the joint surface cartilage and the maturing bone beneath the joint surface. This is called a *intra-articular fracture.* It is identified by careful x-ray examination of the joint. Usually I can see small fragments of bone attached to the cartilage. Invariably such a fracture will require an operative procedure for proper reestablishment of a smooth joint surface.

At the time of early adolescence, I frequently see children with a unique growth plate injury, a slipped epiphysis. The most frequent location is a slipped capital femoral epiphysis—the ball of the hip joint becomes diplaced. This may happen gradually (chronic slip) or suddenly (acute slip) during an athletic event. The young person is usually of chubby body type. Surgery is required to prevent further slipping of the epiphysis. There is a 25 percent chance of this happening to the other hip therefore, close observation is important.

An *avulsion fracture* is the name that physicians give to fractures of the growth centers of the flat bones. The most

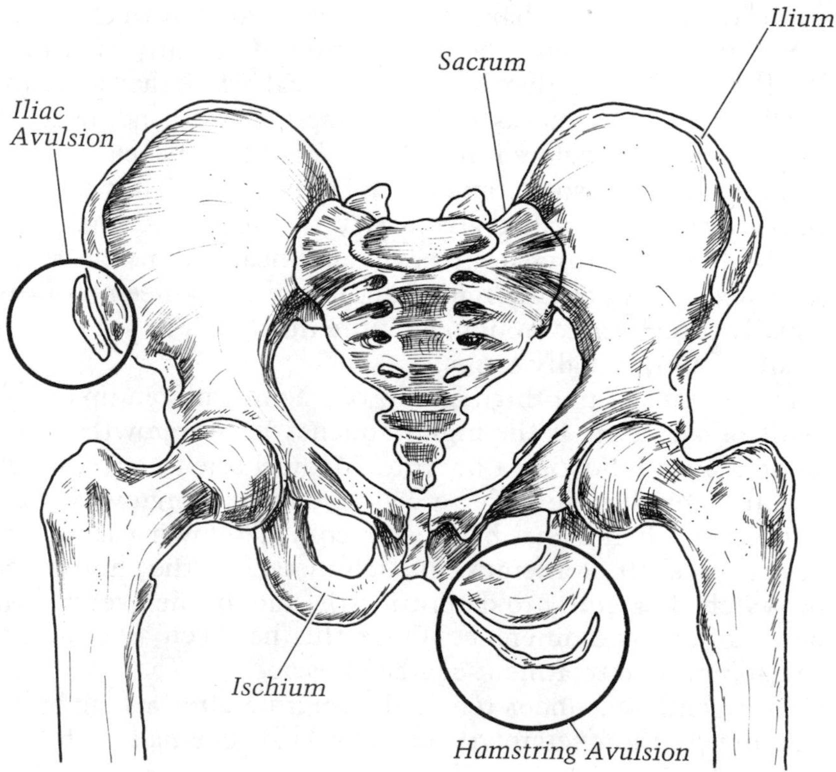

Pelvic Avulsion Injuries

frequent sites of avulsions are the growth centers of the pelvic bones. The large muscles of the back and abdomen extend into the top of the pelvis bone at the growth centers. Children and adolescents who participate in sports that demand sudden twisting maneuvers frequently have avulsion injuries.

Avulsions involve much more than a muscle pull or strain. They are true fractures. The significance of diagnosing them correctly is that the healing time is significantly longer than that of a pull or strain. An avulsion can take two to six months to heal. In general it is not necessary to operate on an avulsion injury; the best treatment is RICE, rest and protection, and incremental rehabilitation and return to sports.

Long-Bone Fractures

The bones of children and young adults are different from adult bones in three major ways: they are softer, blood flows to them at a higher rate than in adult bones.

Every orthopedist takes these factors into consideration when treating fractures. For example, I rarely use a pin or a plate on fractured long bones in children. Rather, I reduce the fracture and align the broken bones. In about 25 percent of fractures of long bones in adults, it is considered necessary to pin, screw, or plate the bones to maintain alignment.

One consequence of the different healing properties of children's bones and adults is that children need remain in casts for less time than adults. Johnny White, the seven-year-old son of Jane White, the business manager of Sports Medicine Resource, Inc., fell out of a tree house and fractured his femur. The boy was in traction for four weeks and in a body cast for another four weeks. If the same fracture had happened to an adult, the cast would have had to remain for ten weeks.

Secondly, children's bones fracture differently. Because their bones are soft, they bend a great deal before they actually break. It is the same as taking a sapling and bending it. It bends a long way before it finally cracks. When it does crack, it splinters but doesn't come completely apart. That is why these fractures are called green-stick fractures. (See page 92 .)

Children's fractures unite easily and heal better than adult fractures. X-rays of children's fractures after eighteen months usually reveal perfect union. Fracture lines can't be seen. In an adult, a bump usually forms.

In fractures of the thigh and lower leg in children up to ten years of age, there is the high frequency of overgrowth of the injured bone following a fracture. Medical science cannot yet explain why the injured leg grows faster. An overgrowth of up to one and a half inches has been recorded following a fracture of the midshaft of a femur. In such instances, the child must be watched carefully to determine the rate of the overgrowth and the ultimate difference. Once this has been ascertained, the appropriate treatment can be chosen.

Most children under ten will overgrow after a long-break fracture. If the overgrowth is more than one-half inch, the

THE DANGERS OF A TRIPLE JUMP

Priscilla Hill, an internationally ranked, free-style skater, was plagued by an avulsion injury to the iliac crest. In 1977, the injury sidelined her for three months.

"The onset of pain came during the last revolution of a triple jump," explains Arthur M. Pappas, her physician. "The muscles that permit you to rotate attach at the iliac crest. The gap where the muscles had pulled away from the bone showed up on x-rays. The gap was at least a quarter of an inch.

"There wasn't much that I could do for her. The avulsed bone was very tender to the touch. To alleviate the pain, she took aspirin and had heat treatments.

"After the sensitivity abated, I started her on stroking exercises—skating slowly by pushing off one leg at a time. This is a time injury. Only time will heal it.

"Soon after Priscilla Hill's injury, I reviewed twenty-two injuries of a similar nature. I found that twenty out of the twenty-two were either skaters or sprinters. Although they all had injuries that showed up on x-rays, they all recovered without surgery and returned to their preinjury level of competition."

We would call this injury a hip pointer if it had happened in an adult.

child should wear an elevation in his shoe on the opposite foot to equalize the limb lengths. This prevents abnormal stress on the lower back. If the overgrowth does not correct itself before the end of the growth period and remains greater than one-half inch, an operation may be done to slow growth in the longer limb so that at the end of the growth period the two limbs are equal in length.

Overuse Syndromes

Overuse syndromes happen when a child or teenager exercises or trains too much. Frequently the tissue cannot tolerate the strain and becomes injured.

Overuse syndromes in children and young adults are very common. I see teenage swimmers who swim six hours a day, and gymnasts who live with their coaches so that they have more time to work out. I have treated many world-class skaters who spend more than forty hours a week on ice. There is an upper limit to the amount of work that even a highly conditioned body can perform.

In some individuals, there is a genetic predisposition to overuse syndromes. In others, it may be hastened by loose ligaments or loose-jointedness.

There are several levels or grades of overuse syndromes:

Grade I: The child complains of a mild ache during sports or exercise. Sometimes the pain continues for a short time after exercise ends. X-rays generally do not show tissue changes. Sometimes I can produce pain upon examination.

Grade II: The pain occurs during the performance, shortens participation time, and continues after the performance. X-rays show early changes of bone and growth centers, especially when they are compared to the normal tissue. I usually can produce pain at the site of the injury.

Grade III: The pain is significant during performance as well as afterwards. The pain may linger for many hours. The injury site swells. I notice a difference in motion between an injured and uninjured limb or joint. X-ray changes are identifiable, usually in the form of excessive bone development or bone fragment development. I can produce pain by moving the joint or limb. At this stage, I advise the child to refrain from athletics for six to eight weeks, and sometimes longer. On occasion, I apply a cast to the limb or a joint to facilitate rest.

Grade IV: The child is unable to perform for an extended period because of persistent pain. After exercise or sports end, the pain lasts for

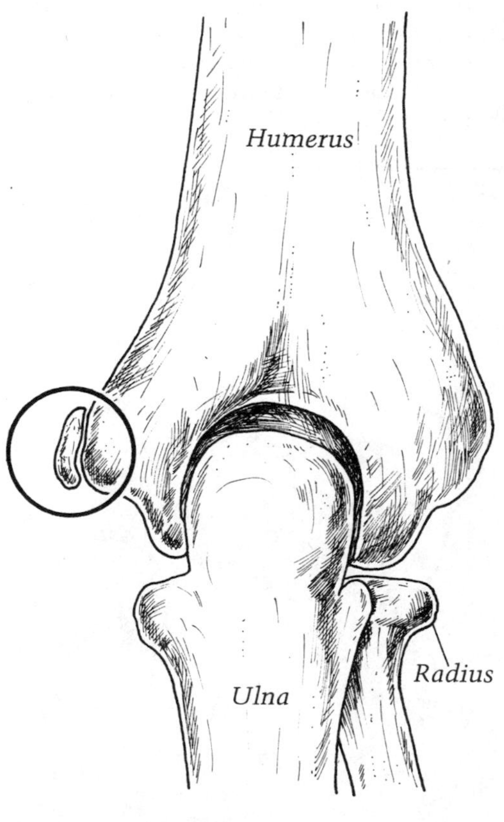

Little League Elbow

> more than twenty-four hours. X-rays show definite changes. The child or young adult has lost the full range of motion. When I examine the involved area, I find excessive pain and swelling. In this final stage, the child is in trouble. I advise rest for months—frequently a change in sports selection—and occasionally an operation.

Listen to children and encourage them to speak up when they are hurt. Too often, parents, coaches, and teammates encourage children to play despite pain. This attitude leads to grade III and IV overuse syndromes.

Pain is nature's way of telling your child and you that something is wrong. When it speaks, you should heed its voice.

In the professional ranks, every player cannot be pain-free for every game. That may be all right for a professional; they make their living from sports. I am strongly against this philosophy for children and young adults. Playing when hurt can only complicate or prolong an injury.

The classic example of an overuse or overstress syndrome in children and young adults is that of the Little League pitcher. Because of the potential for long-term injury, the National Little League Administration has limited the number of innings a child may pitch per week. It is unusual to have any child who follows these guidelines develop an overuse syndrome. A child who is encouraged to pitch more often than the guidelines recommend, because of the opportunity to participate on multiple teams or because a parent or coach feels he possesses extraordinary talent that must be developed at an early age, is a sure bet to develop overuse syndrome.

In some situations, other conditions such as an osteochondroses or stress fracture (gymnast back) may mimic an overuse syndrome. Undoubtedly there is a relationship between the repetitive forces of certain athletic events and anatomical consequences.

The prime treatment of overuse syndromes is prevention or early recognition. The tragedy is when a grade IV problem is diagnosed, and the sad news must be shared with child and parents.

Osteochondroses

Osteochondroses are a group of disorders that occur in the growing areas of bones. They usually begin with some single or repetitive episode of trauma to a bone or joint. Frequently this occurs with overuse syndromes. Episodes of this sort affect the blood supply to the developing bone. Osteochondroses (aseptic avascular necrosis) is simply a noninfectious interruption of blood supply which results in injured tissue.

FROG KICK KNEE

"About three years ago, a twelve-year-old swimmer from Pennsylvania came to me with knee pain," recalls Dr. Arthur Pappas.

"His mother had taken him to other physicians who couldn't explain the pain and even suggested that the pain was psychological.

"On examination, I discovered that the boy—a breast stroker—swam six hours a day. In breast stroking, a swimmer snaps his knee, placing the major force on the inner aspect of his knee. I call it a 'frog kick.' In an hour of swimming, the knee was being stressed more than 5000 times.

"I felt his knee. The discomfort was at the junction of the femur and the growth plate, almost at the top of the kneecap. You would have thought that he had a knee sprain.

"X-rays revealed that the bone area (growth plate) was wider than normal, which indicated to me that the area was under tremendous stress. In effect, I diagnosed it as an undisplaced, incomplete fracture.

"The recommendation was to cut down the workouts. Also, I advised that he work a different stroke for two months. This gave his knee a rest.

"It was just like a fracture. It took time to heal."

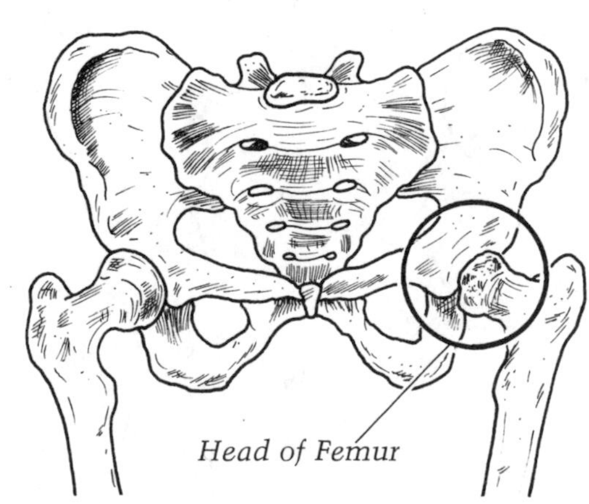

Legg-Perthes Syndrome

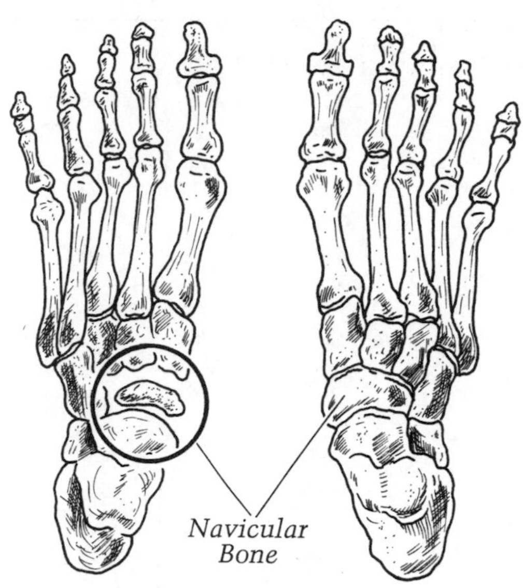

Köhler's Syndrome

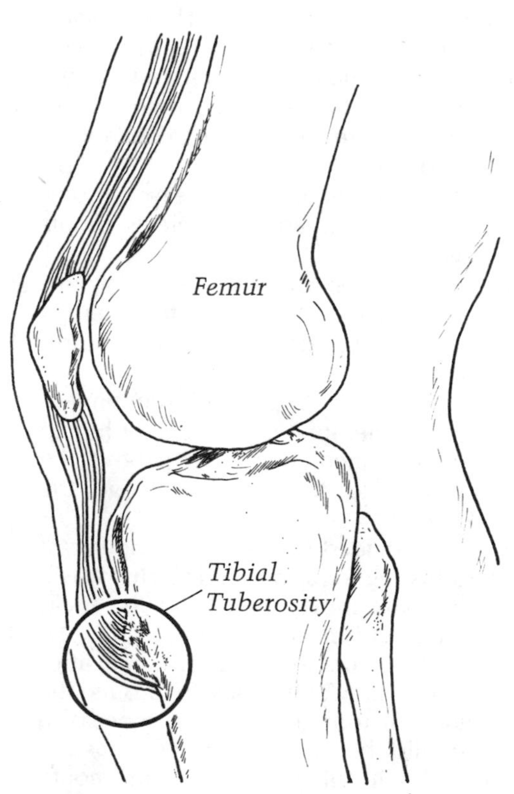

Osgood-Schlatter's Syndrome

The symptoms usually follow excessive physical demands of sports and are most frequently identified during the early stages of growth development. Any unusual compressive or shearing force may produce some change in the configuration of the bone-cartilage unit or interfere with the blood supply of the area. This changes its development and, secondarily, its x-ray appearance. If an osteochondrosis is diagnosed at an early stage, it usually follows a self-limited natural history and heals without major deformity or disability. If it is not identified at an early stage, it is possible the symptoms will continue and result in a joint deformity, and possibly long-term impairment of function and disability. Pediatric orthopedists usually grade the severity of osteochondroses from I to IV. In general, the older the child and the higher the grade, the more concern we have about the outcome.

Two kinds of osteochondroses of childhood are Legg-Perthes syndrome, an osteochondrosis of the femoral head, and Köhler's syndrome, an osteochondrosis of the navicular bone of the foot. Legg-Perthes usually occurs in children between four and eight years of age. The specific cause is unknown, but it is associated with a restriction of blood supply to the head of the femur and becomes significantly altered with continued repetitive running and jumping. The complaint of the child is usually one of the following: pain about the hip and thigh, a painless limp, or a mild limp with pain localizing on the inner side of the knee. The complaint of knee pain when the problem is in the hip occurs all too often in children. One of my basic tenets to medical students is that when a child is complaining of knee pain, automatically question and x-ray the hip. Nonsurgical and surgical treatment are both considerations for the treatment of Legg-Perthes syndrome. In early diagnosis, nonsurgical treatment with a brace will usually result in a satisfactory end result. In the child who has a long-standing history or advanced deformity and progressive x-ray changes, a surgical procedure might be necessary. Any child with a diagnosis of Legg-Perthes syndrome will probably be kept from participating in running and jumping sports for twelve to twenty-four months.

Köhler's syndrome is a similar process that occurs in the tarsal navicular. The tarsal navicular is in the inner portion of the middle of the foot. The initial treatment is rest. A limitation of activities or immobilization in a splint or cast for six to eight weeks is usually sufficient to resolve this problem.

The most common form of osteochondrosis is Osgood-Schlatter's syndrome which occurs during the early adolescent period. Osgood-Schlatter's syndrome is located at the front portion of the upper tibia. You can feel this area as a bump, two to three inches, below your kneecap. When this cartilage converts to bone and there is sudden or repetitive stress on the attachment of the patella tendon from the quadriceps (thigh muscles), a series of micro-fractures are created. This interrupts the normal blood supply and the normal conversion of cartilage to bone, resulting in a tender bump. Localized pain and pain when pressure is applied to the bump are the major symptoms. The young athlete will complain of pain after activity such as running, jumping, and kneeling. This is a condition that will heal when growth is completed. Until that time it is important to advise the athlete to stay away from certain sports activities that increase the pain. In some instances, it is necessary to apply a cast for six to eight weeks to aid the healing.

I have not found the use of cortisone injections, radiation, diet changes, or surgery to be necessary in this condition.

Other osteochondroses occur in the vertebral bodies of the spine during adolescence and account for postural deformities such as roundback (kyphosis). This is known as Scheuermann's disease. Once again the key to success is early diagnosis, an appropriate change of activity, and prescribed exercises or the use of a cast or external bracing support.

Osteochondritis Dissecans

The bone surfaces that form joints are covered with a glistening, white cartilage that is very slippery. The cartilage, nourished and lubricated by joint fluid, has considerable elasticity and recovers from denting remarkably well. Joint surface cartilage can be damaged. It can happen with a fall, turn, twist, or tackle.

Osteochondritis dissecans is a defect in the joint-bone cartilage of any synovial joint—knee, wrist, elbow, ankle. An x-ray of the area looks like a divot out of the joint surface. In time, the piece may loosen and cause pain. The *dissecans* refers to a dissection of a loose piece from the surface of the joint.

In children, the dissecans usually remain seated in the divot. However, in athletes between ages twelve and sixteen, the divot sometimes falls free. When the piece does fall free into the joint, it is called a "joint mouse," because it may wander around the joint.

In all osteochondritis dissecans, the pain starts when the cartilage piece loosens from its bed. The wiggling of the piece causes the pain. If the child doesn't move the joint, no pain is created.

I always order plain x-rays, to try to find the loose piece. Frequently, I order special views. The piece of bone that is attached to the fragment is more dense than the normal bone because it has lost its blood supply and is attempting repair.

I cannot see the surface of the joint on regular x-rays, only bone. In order to see the surface of the joint, I put dye in the joint, which outlines the fragment and the bone surface. This test is called an *arthrogram.*

The status of a fragment can also be determined with the *arthroscope,* a body telescope that can peer inside a narrow

OSGOOD-SCHLATTER'S DISEASE

Mike Botticelli and Cheryl Franks are pairs figure skaters. They finished in the top ten in the 1980 Olympics.

"When he was thirteen, Mike Botticelli came to me with a painful patella tendon," recalls Arthur M. Pappas, his doctor. "I diagnosed the problem as Osgood-Schlatter's disease, a common ailment among athletes of his age.

"The best treatment for Osgood-Schlatter's disease is no treatment. I prescribed, rest, aspirin, and warm baths.

"But, Mike's pain still continued. His problem: it was difficult for him to rest the knee. Being competitive, he wanted to test and challenge the knee.

"I decided to apply a cylinder cast to his leg for two months. This was a dramatic move, but it did the trick. The pain was alleviated.

"Two months in a cast had atrophied many of the muscles around the knee. It took another two months to rebuild the smaller muscles.

"All in all, this was a four-month episode."

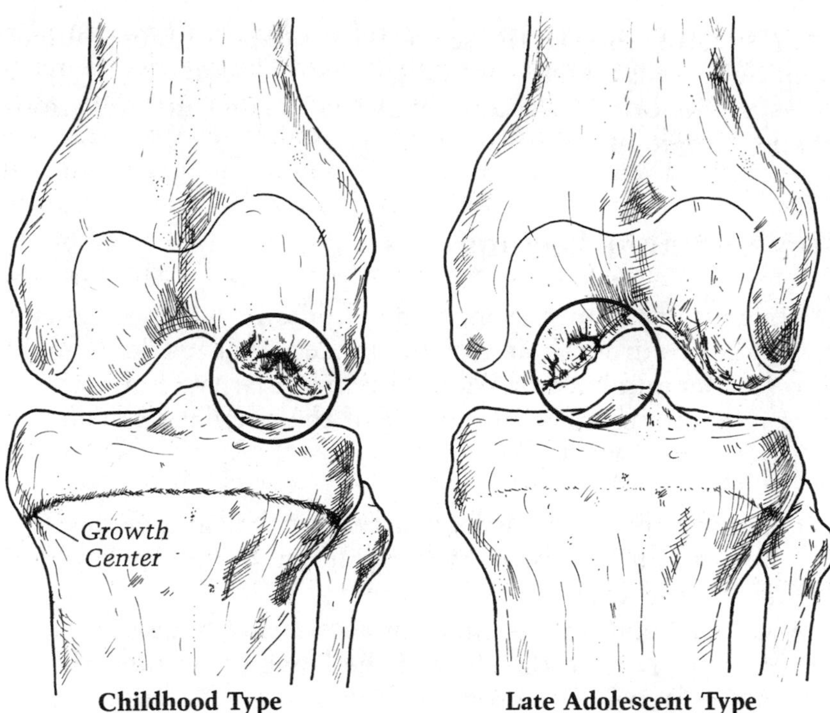

Osteochondritis Dissecans

Childhood Type **Late Adolescent Type**

opening in the joint. (See illustration, page 265.)

In young children, the number one treatment is rest. Sometimes this means a splint or a cast. Rest works in 99 percent of the cases. The reason is that the cartilage is still growing and usually covers the divot. There is usually no residual deformity. Osteochondritis dissecans in young children is mainly a problem of diagnosis—not treatment.

In the twelve- to sixteen-year-old age group, I sometimes recommend a splint or a cast, but I prefer not to. If the fragment is present on a plain x-ray and the surface is intact, I only restrict sports activities. Three to six months of no sports is usually necessary. Sometimes I will splint the joint to restrict motion. I prefer not to, because the muscle around the joint will atrophy in a cast. I x-ray the joint every six to eight weeks to follow the course of healing.

I do not use cortisone injections, radiation, or changes in diet. I have found they do not help.

If after six months the patient still experiences pain and the x-rays are unchanged, surgery must be considered.

In young people whose joints are still growing, it makes sense to try to stimulate healing of the fragment. The operation stimulates the blood supply from the bone to the fragment.

If the athlete is skeletally mature, and has completed bone growth, I have found that it is biologically almost impossible for the cartilage to heal itself. In that case, I remove the loose fragments.

Next, I drill deeply into the bone at the base of the crater. Blood is allowed to flow into the crater and form a clot. Eventually the blood clot converts to solid scar tissue. The body substitutes scar tissue in place of the missing surface material.

Osteochondritis dissecans is also discussed in the various chapters on joints.

Development Lesions

Without apparent reason, some children develop irregularities of bone during their growing process that result in minor developmental abnormalities. This irregularity of growth occurs in approximately 20 percent of the population. The two most common are:

- Benign fibrous cystic lesions that grow near the growth center. In the majority of children, these appear just above the knee joint.
- Osteochondromas—small projections of bone that extend away from the shaft of the bone. Sometimes they touch nerves or interfere with muscle function and cause pain.

In most instances, they are identified on an x-ray that is taken because of some other problem. In most instances, these developmental lesions disappear or do not cause structural interference.

DIAGNOSIS AND TREATMENT

Parents bring their children for examination after minor injury or the complaint of recurrent pain. In a small percentage of children, the osteochondromas will cause pain. In most situations, a developmental lesion is found coincidental to examination for another complaint. The cystic lesion in the bone cannot be felt during the physical examination. On rare occasions, the cystic lesion may be near the edge of the bone and an athletic injury may cause a small fracture that will result in local discomfort.

The next step is to x-ray. These developmental lesions show up on x-rays clearly.

The best treatment is instruction about the lesion. The child is usually permitted to continue with sports. I rarely apply a splint or cast. Left alone, these lesions resolve themselves.

In a rare instance, an osteochondroma will cause local discomfort or may interfere with a nerve or muscle function. In those cases, I operate to remove the abnormal bone growth.

The operation is usually a simple one, depending on the location of the lesion. The child stays in the hospital for a few days. A cast may be applied to the limb to ensure rest. The cast remains in place for two to three weeks and then is removed. I then begin the rehabilitation program to rebuild the muscles around the injury.

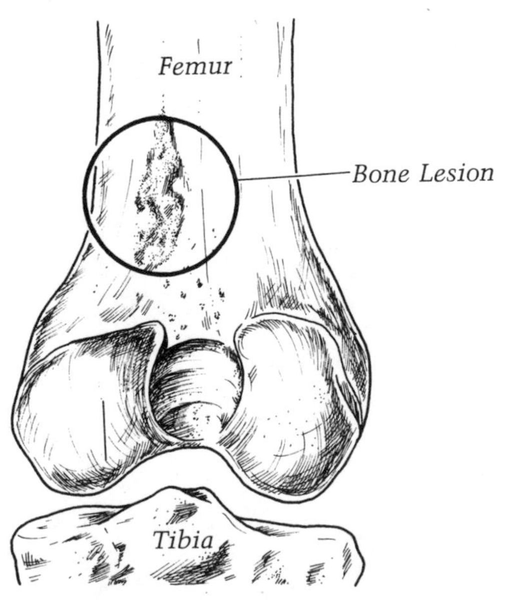

Benign Fibro-Cortical Lesion

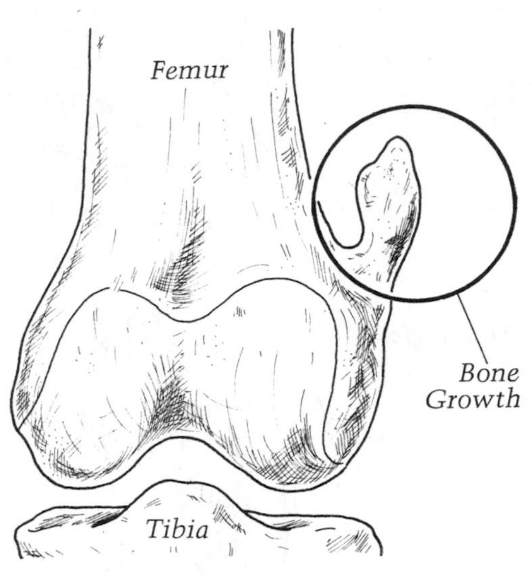

Osteochondroma

Other developmental lesions that may be of greater significance are tumors of the extremities. Not all tumors are malignant. However, even benign tumors can increase in size and cause weakness and pain in an extremity.

Tumors and tumorous conditions of bone are most frequently brought to the attention of parents and physicians by a recent injury. In many instances, it is implied that the injury is the cause of the tumor or tumorous condition. In reality, the bone has been weakened by the underlying process to a point where a small stress or trauma results in a mild fracture. This results in pain. When investigating the cause of the pain, x-rays reveal the abnormality of bone consistent with some tumor or tumorous condition. In some instances, the tumor may be a benign lesion such as a bone cyst. This cyst has increased in size to a point where just throwing or kicking a ball creates enough stress to result in a fracture. Such fractures are known as *pathological fractures.*

Other times the lesion may be caused by a bone tumor growing within the bone or one that has spread to an area of bone from another site. Again, the bone is generally weakened to an extent that mild trauma results in a pathological fracture. This causes the pain and brings the child to the attention of the physician.

Thus, whenever a child complains of pain or appears to have sustained an insignificant injury, the child should be listened to and, if questions remain, brought to the school trainer, nurse, or doctor.

Congenital Abnormalities

Some children are born with skeletal abnormalities that remain unidentified until late in childhood or adolescence. Some of these abnormalities are so minor that they show up only when a child or adolescent attempts to develop athletic skills. In other instances, the abnormality does not interfere with function until a certain stage of development of the skeleton has been achieved.

A physician brought his six-year-old son to me for a consultation. My colleague had observed that his son threw a ball awkwardly. I examined the boy's arm, and discovered that the bones of his forearm didn't rotate. The boy was unable to hold his elbow at his side and rotate his palm upward and then downward; x-rays revealed that the elbow joint was not normally formed. He had a *congenital synostosis* (bone fusion) that did not permit normal forearm rotation. He was directed to the athletic activities that did not require forearm rotation.

In 1970, I treated a nine-year-old boy who complained of neck pain whenever he had to wrestle in a physical education class at school. At first his parents and teachers thought he was faking the pain to avoid participation. However, his parents observed him at home following wrestling competition. He was brought to me for evaluation.

I uncovered a congenital abnormality of his upper neck region that resulted in instability of the neck. Whenever I applied stress to his neck, the boy screamed. His muscle went into a spasm. I advised the parents to withdraw him from sports that place stress on his neck. His discomfort subsided, but the deformity persisted and the instability increased.

In subsequent examinations, I uncovered pressure on his spinal cord. The symptoms were pain and weakness in his arms and legs. I operated on the boy. The procedure is called fusion. (See page 217.)

After surgery, his neurological symptoms subsided. I advised his parents to choose a sport that would not place stress on his neck. He has grown up normally and has developed his athletic skills in other areas without major emotional deficits and without any physical problems.

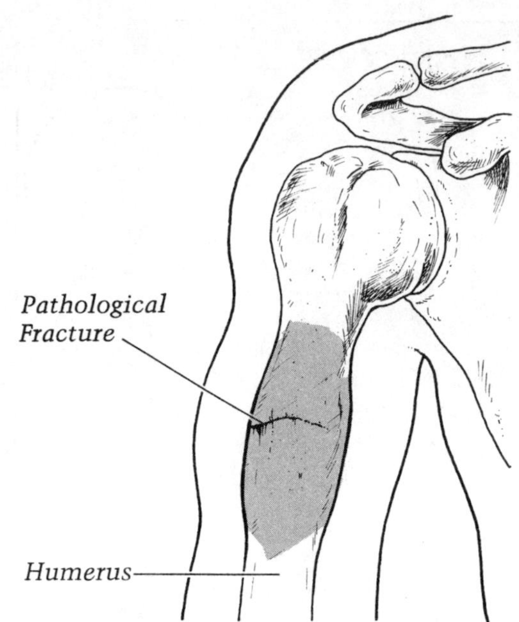

Humeral Bone Cyst

Discoid meniscus of the knee is a congenital abnormality that usually remains undetected until adolescence. Instead of the knee washer, called the meniscus, forming the traditional quarter-moon configuration, it grows thicker than a normal meniscus and in a different configuration, more like a thick disc.

Individuals with a discoid meniscus usually experience knee instability, loud clicking, recurrent pain, and swelling. The only effective treatment is surgical removal of the discoid meniscus. (See page 265.)

There can be congenital deviations or anomalies that remain undetected and asymptomatic into adolescence and, at times, into young adulthood. These may be present in various parts of the body, and the usual complaint is local discomfort or a mild interference of function. This results in the person not being able to participate at his usual level of performance.

There are certain congenital abnormalities of the lower spine that will limit the degree of normal motion or rotation. A fusion between the lower lumbar spine and the sacrum, *hemisacralization,* exemplifies this. I recently treated a nationally ranked figure skater with low back discomfort. After careful x-ray examination, the hemisacralization abormality was evident. He had recently completed the transition from a singles figure skater to a pairs figure skater. The extra demands created by maneuvering and lifting another skater brought this congenital anomaly to the level of a functional impairment. The treatment consisted in recognizing the problem and then advising the selection of other athletic activities.

There is a peculiar abnormality of the foot known as a *tarsal coalition.* In many instances, an individual will have participated in all athletic events into high school prior to the onset of symptoms from a tarsal coalition. The most frequent of tarsal coalitions is one between the calcaneus and navicular; the calcaneus is the heel bone and the navicular is the bone at the apex of the arch. There is a relationship between the outer

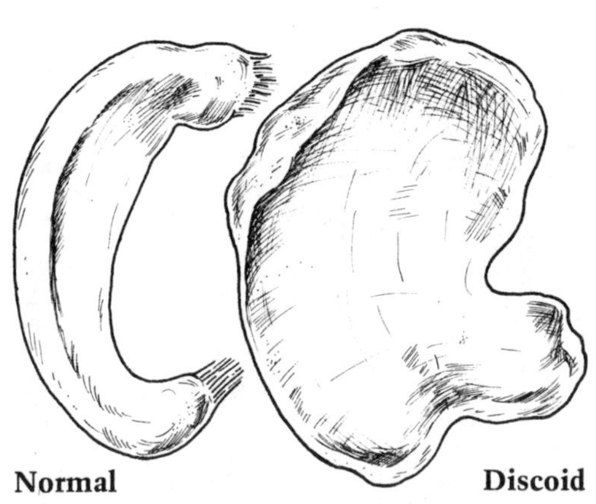

Congenital Discoid Meniscus

FUSED FOOT

Jane Burnett was a world-class hurdler in the early seventies. In a pre-Olympic meet she first experienced the onset of foot pain. She was seventeen at the time. The pain centered in the middle of her feet. As the discomfort increased during the year, her speed decreased.

When I examined her, I found spasms in the peroneal tendon and an incomplete calcaneal navicular coalition. It showed clearly on the x-rays. The two foot bones were almost fused.

I advised her that as long as the condition remained, her running days were numbered.

"What can be done?" she asked.

"The bones can be separated surgically," I said. "The operation is called a resection. You will at least be more comfortable during daily activity."

"Do you think that I will be able to run again?"

"Returning to track will be a bonus," I said. "Operating on a world-class athlete is tricky business."

Immediately after the operation I applied a cast to her foot. The foot remained in the cast for about three weeks. Then I used a removable cast. Every day she did range-of-motion exercises for the foot. I was concerned that Jane's foot might reject this stress.

In three months she had regained enough strength in her ankle and foot to start jogging. In six months she was back to running. The following summer she returned to international competition, and she was winning.

Arthur M. Pappas, M.D.

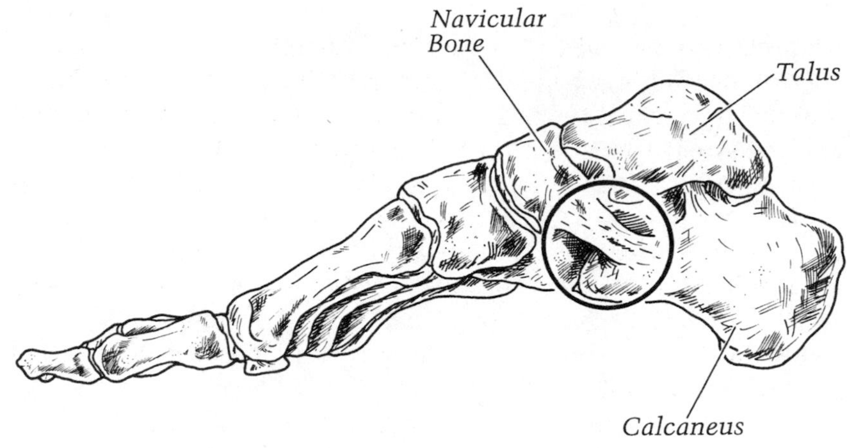

Calcaneonavicular Coalition

end of the navicular and the inner end of the heel bone where there is usually a joint. In some individuals, this joint does not develop properly, and an area of incomplete or complete bone formation across the normal joint space occurs. This will result in the onset of local pain, and usually a deformity of the foot secondary to spasm. This is called a *peroneal spastic flatfoot.*

The treatment, once the problem has been identified, is a period of rest in a cast or splint and the use of anti-inflammatory medications. If this does not result in improvement, then a decision must be made to either restrict athletic activity or undergo surgery.

The only surgical procedure that may allow a return to competition is a resection of the coalition, a removal of the abnormal bone, and the replacement of muscle between the bones. This is known as an *interpositional arthroplasty.* I have had the opportunity to perform this procedure on a number of individuals who wanted to return to athletic events.

In approximately one-third of the cases, the procedure was successful and the young person could return to unrestricted competition. In another one-third, the discomfort reoccurred, requiring a change of athletic interests. In the last group, the procedure was not successful; the pain persisted and a more extensive surgical procedure was necessary.

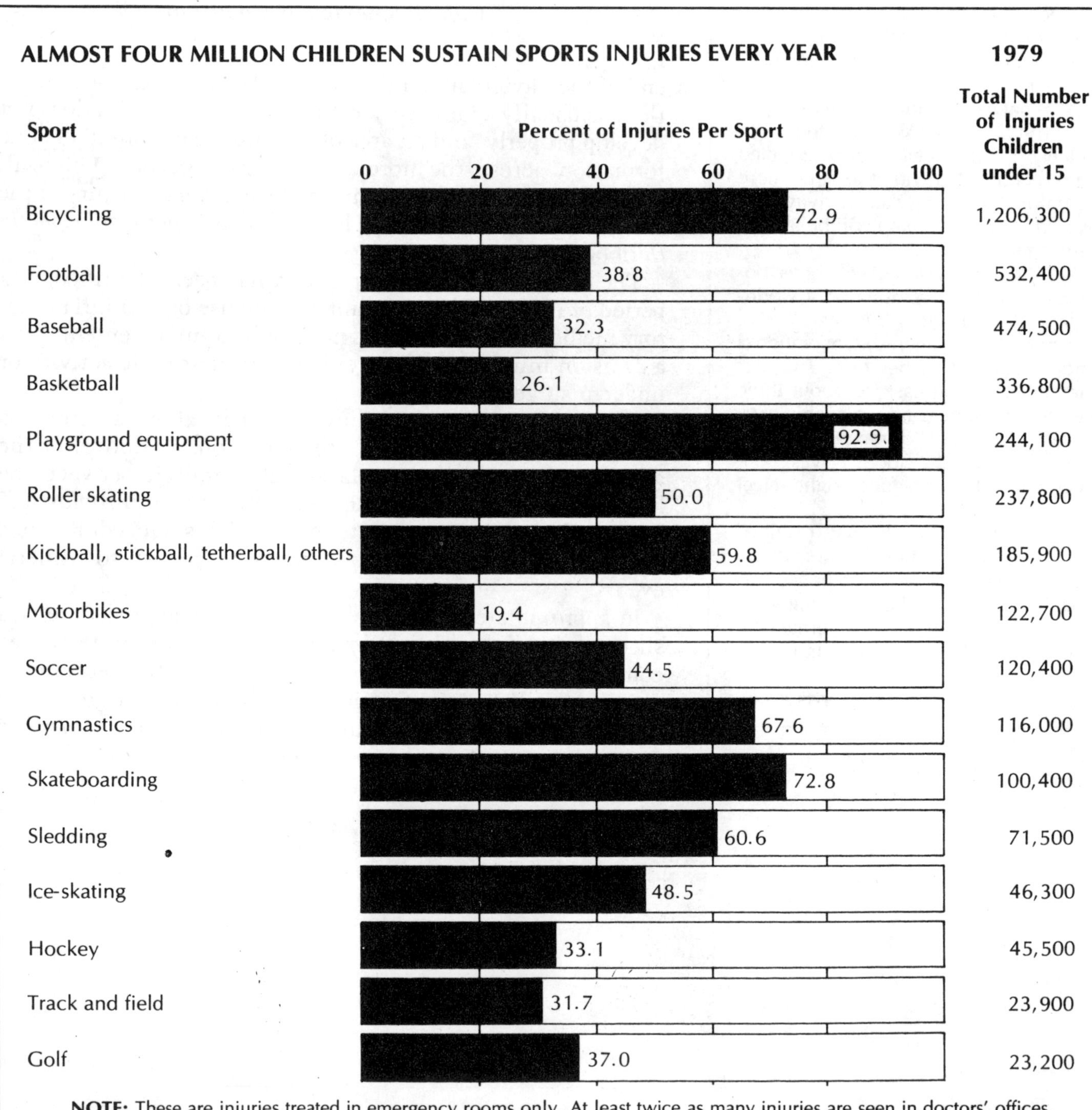

NOTE: These are injuries treated in emergency rooms only. At least twice as many injuries are seen in doctors' offices.

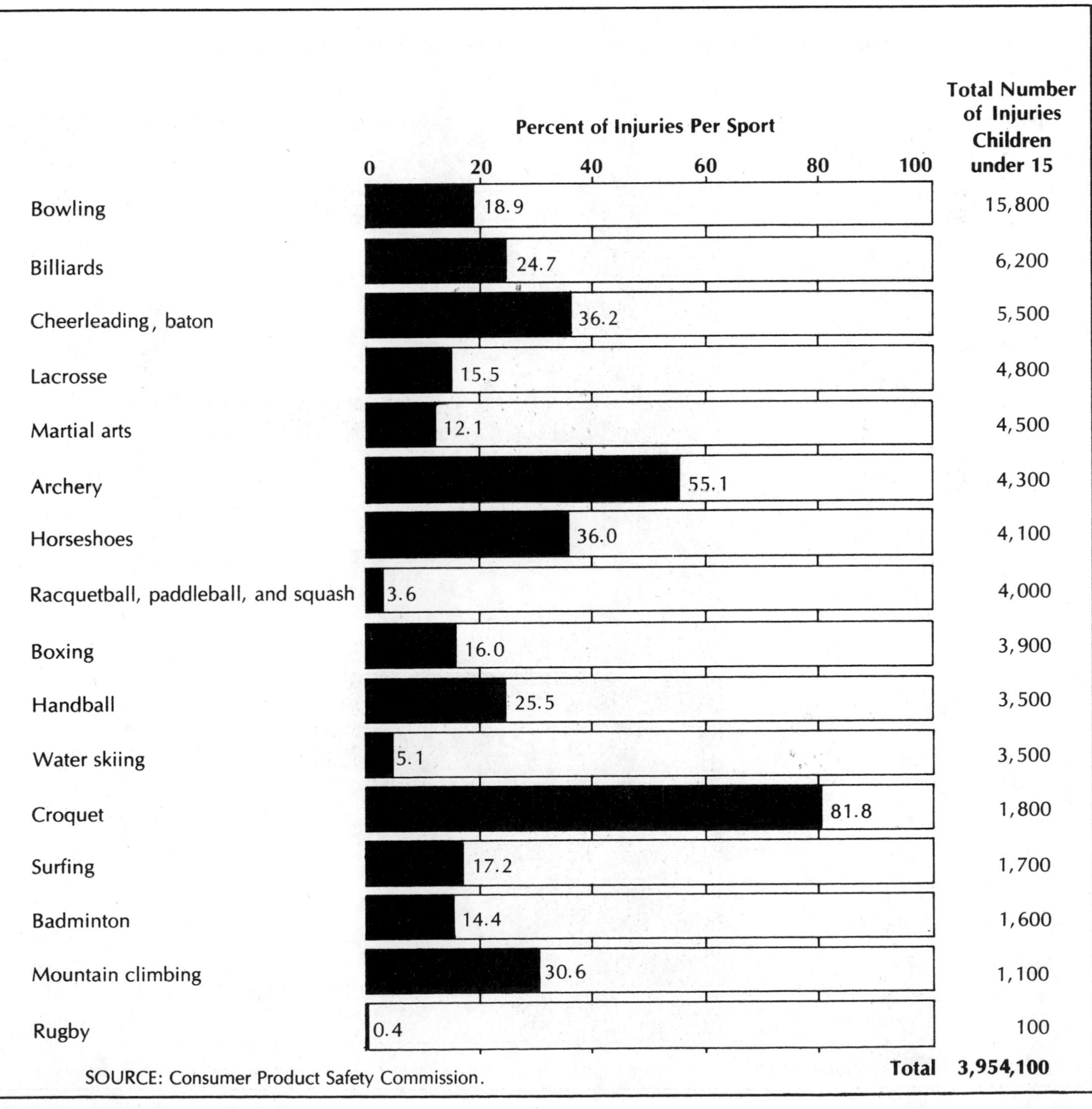

SOURCE: Consumer Product Safety Commission.

Biographical Sketches

William W. Southmayd, M.D.
William W. Southmayd studied medicine at Harvard Medical School where, as an undergraduate, he was an All-Ivy-League guard and captain of the football team. He was chief resident of orthopedic surgery at the Massachusetts General Hospital and, since 1975, has been the medical director of Sports Medicine Resource, Inc., the largest sports medicine clinic in New England. He has treated such professional athletes as Jim Rice, Jerry Remy, Brad Park, Fred Lynn, and David Cowens. In addition to his practice in Chestnut Hill, Massachusetts, Dr. Southmayd is a consulting physician to the Boston Red Sox. He is medical editor of *The Professional Athlete* and writes a column for The New York Times Syndicate. His weekly TV series, "Sports and Fitness," produced by The New York Times Syndicate and Patrick Productions, is aired in major cities across the United States.

Marshall Hoffman
Marshall Hoffman, a professional journalist who writes a column on sports injuries with Dr. Southmayd for The New York Times Syndicate, is also the executive producer of Patrick Productions, which produces "Sports and Fitness." He is the coauthor of THE SPORTSMEDICINE BOOK, a national best seller. Before his writing and TV production demanded his full-time attention, he was a writer for *U.S. News and World Report.*

Arthur M. Pappas, M.D.
Arthur M. Pappas is one of America's most distinguished doctors. His contributions to medicine in twenty-five years include more than fifty published papers and more than 150 others prepared for presentation at national and international conventions, seminars, and panels. Dr. Pappas is professor and chairman of orthopedics and professor of pediatrics at the University of Massachusetts Medical Center. He is the former chairman of the orthopedic section of the American Academy of Pediatrics and is a member of the American Academy of Pediatrics committee on physical fitness, recreation and sports. He is president-elect of the Association of Professional Baseball Physicians, and team physician and medical director for the Boston Red Sox. He served as team physician at Harvard College for more than ten years.

Index

Accessory navicular, 352–353
Ace bandage, 319, 322
Achilles tendon, 62, 64, *71*
  anatomy of 300, 301
  exercises for, 329
  rupture of, *330*–331
  surgery on, 329
  tendinitis of, *326*–329
Acidosis, 74
Actin, 72
Adenosine triphosphase (ATP), 72
Adrenal glands, 36
Aerobic exercise, 73
Agganis, Harry, 302
Albright, Tenley E., 15
Ali, Muhammad, 17
Analgesics, 53, 209
Ankle, 18, 33
  Achilles tendon in. *See* Achilles tendon
  anatomy of, 310, *311*, 312
  balance test for, 320
  dislocated peroneal tendon in, *332*–333
  exercises for, 316, *323*
  fracture of, 320–324
  injuries to according to sport, 335(*t*)
  lateral ligament instability of, 318–320
  osteochondritis dissecans, *325*–326
  sprain: grades of, 43, 315; healing time for, 42; inversion, 312, 313, 316; prevention, 317; single ligament, 315–316; two-ligament, *317*, 319
  x-ray of, 319
Ankle balance test, 320
Annulus fibrosus, 186
Antalgic gait, 351
Anterior compartment syndrone, *306*–307
Anterior cruciate ligament, *253*–258
Antidiuretic hormone (ADH), 413
Archery
  children injured playing in U.S., 447
  foot and toe injury in U.S., 381
  hand injury in U.S., 185
  head injury in U.S., 391
  popularity in U.S., 14
  wrist injury in U.S., 167
Archibald, Nate, 233
Arm. *See* Elbow; Fingers; Hand; Wrist
Arthritis, traumatic, 97–98
Arthrogram, 116, 142, 263
Arthroscope, *265*
Aspirin, 37
  allergy to, 47
  description of, 46
  history of, 46
  for injuries, 46
  precautions for, 46, 47
Athletes
  See also specific entries
  bill of rights for school and college, 31
  earnings of, 17
  playing into shape, 39
  professional, and injuries, 30, 38
  women vs. men and injury, 32–33

Bablonia, Tai, 22
Back
  See also Spine, cervical; Spine, lumbar; Spine, thoracic.
  corsets for, 34
  care for, 209
Badminton
  ankle injury in U.S., 335
  children injured playing in U.S., 447
  foot and toe injury in U.S., 381
  hand injury in U.S., 185
  head injury in U.S., 391
  wrist injury in U.S., 167
Baker's cyst, *286*–288
Bandages. *See* Casts; Splints; Taping
Bankart procedure, 112
Barry, Rick, 276
Baseball
  ankle injury in U.S., 334
  calf strain, 302
  catcher's hand, 170
  children injured playing in U.S., 446
  demands on knee, 297
  elbow injury in U.S., 152
  finger, *181*
  foot and toe injury in, 380
  hamstring injury in, 242
  hand injury in U.S., 184
  head injury in U.S., 390
  pitching injury, 79
  popularity in U.S., 14
  protective equipment for, 64
  shoulder injury in U.S., 126
  wrist injury in U.S., 166
Basketball
  ankle injury in U.S., 334
  calf strain, 302
  children injured playing in U.S., 446
  court condition for, 65
  demands on knee, 297
  elbow injury in U.S., 152
  foot and toe injury in U.S., 380
  hand injury in U.S., 184
  head injury in U.S., 390
  injuries in, 23(*t*)
  popularity in U.S., 14
  protective equipment for, 64
  shoulder injury in U.S., 126
  wrist injury in U.S., 166
Beer, 413–414
Bendix, Shana L., 15
Bennett's fracture, *178*
Benzoin stick spray, 366, 370
Biceps muscle, 70, 122–124, 131
Bicepital groove, 122, *123*
Bicycling
  ankle injury in U.S., 334
  "bonking," 72
  children injured in U.S., 446
  elbow injury in U.S., 152
  foot and toe injury in U.S., 334
  hand injury in U.S., 184
  head injury in U.S., 390
  injuries in, 19(*t*)
  muscles used in, 71, 72
  popularity in U.S., 14
  and recovery from injury, 44
  shoulder injury in U.S., 126
  wrist injury in U.S., 166
Bierbaum, Benjamin E., 15
"Bikecentennial Trail," 392
Billiards
  ankle injury in U.S., 335
  children injured playing in U.S., 447
  elbow injury in U.S., 153
  foot and toe injury in U.S., 381
  hand injury in U.S., 185
  head injury in U.S., 390
  shoulder injury in U.S., 127
  wrist injury in U.S., 168
Bird, Larry, 237
Black toenail, *371*–372
Blisters, *370*
Blood supply and healing of injury, 42–43
Body, human
  See also specific parts
  body fat, 418
  fractures. *See* Fractures
  level of performance, 38
  most injured parts in sports, 18
  muscles in. *See* Muscles
  and overtraining, 55–56
  responses during exercises, 17
  seven structures of, 45
  shape of and ability, 34–35
  skeletal structure. *See specific anatomical sites*
  weight loss in, 417, 418
Bones
  See also specific bones
  and blood supply, 43
  broken. *See* Fractures

composition of, 86
definition of, 86
purpose of, 86
types of, 86–87
Borg, Bjorn, 17
Bosworth procedure, 140
Boutonnière deformity, *182*, 183
Bowling
ankle injury in U.S., 334
children injured playing in U.S., 447
elbow injury in U.S., 153
foot and toe injury in U.S., 380
hand injury in U.S., 185
head injury in U.S., 391
popularity of, 14
shoulder injury in U.S., 127
wrist injury in U.S., 166
Boxer's fracture, *177*
Boxing
ankle injury in U.S., 335
arm weariness in, causes of, 72
children injured playing in U.S., 447
foot and toe injury in U.S., 381
head injury in U.S., 391
shoulder injury in U.S., 127
wrist injury in U.S., 167
Brachial plexus, 102
Braden, Vic, 134
Bras, 63
Brett, George, 20
Bristow procedures, 113
Buchholtz, Butch, 36
Bunions, *360*–361
Bunionette, *372*–373
Bursa sac, *120*, 133, 283, *284*, 285
Bursitis
Baker's cyst, *286*–288
elbow, *133*, 134
goose-foot, 285–286
iliopsoas (pelvis), 232–*233*
inferior calcaneal, *349*–350
knee, 283–*284*
prepatella, 284–286
retrocalcaneal (pump bump), 347–*348*
shoulder, 120–122
trochanteric (hip), *231*–232
and ultrasound, 49
Bush, George, 13
Butkus, Dick, 22, 36
Butterfly fragment, 91

Caffeine, 412, 413
Cahall, Bernard A., 58
Calcaneonavicular coalition, *445*
Calcium, 120, 402
Calf muscle, 70
anatomy of, 300, *301*, 302
strain, 302–303
Calluses, *367*–368
Calories
in alcoholic drinks, 415(*t*)
burning of according to sport, 416–418
list of according to food, 422(*t*)
need for in exercise, 416
in nonalcoholic drinks, 414
requirements for, 117
Candy bars, 404
Carbohydrates, 394–395, 402
loading, 401, 404–407
Cardiovascular fitness, 44
See also Heart
Carter, Jimmy, 13
Cartilage, knee, *259*–266
Cashman, Wayne, 251
Castino, John, 318
Casts, 32
for fractured wrist, 164
gel, 316
for sprained wrist, 158
thumb spica, 164
Catalano, Patti Lyons, 58, 232
CAT scan, 389
Cervical burners (cervical nerve stretch), 192–193
Cervical disc disease, acute, 193–196
Cervical spine. *See* Spine, cervical
Cervical traction, description of, *51*
Charley horse. *See* Contusions
Cheerleading, baton
ankle injury in U.S., 335
children injured playing in U.S., 447
elbow injury in U.S., 153
foot and toe injury in U.S., 381
hand injury in U.S., 185
head injury in U.S., 391
shoulder injury in U.S., 127
wrist injury in U.S., 167
Chenier, Phil, 217
Children
avulsion fracture in, 427, 429, *435*
Baker's cyst in, *286*–288
bone lesions in, *442*–443
bone size, *426–428*
and family acceptance of sport, 429–430
and growth centers, 430, *432*, *433*, 434
healing rate in vs. adults, 36–37
injury rate of according to sport, 424, 446–447(*t*)
Little League elbow in, 148–*149*, 437
long-bone fracture in, 436
nutrition for, 399–400
and physical mismatches in sports, 430
Osgood-Schlatter's syndrome in, 439
osteochondritis dissecans in, 440–441
osteochondroses in, 438–439
overuse syndromes in, 437–438
reasons for injuries to, 425
Clark, Nancy, 15
Clavicle. *See* Collarbone
Closed fracture, 90
Coaching, 31
importance of, 60–61, 430–32
Codman, Ernest Emery, 120
Coffee, 412
Collagen, 43, 86
Collagenase, 276
Collarbone, 100, *101*
fracture of, *108*–110
healing time for fracture, 42
Competition level and injury, 30
Compression, 37, 41
See also RICE
Concussions, 386–387
Congenital synostosis, 443
Consumer Product Safety Commission, 17, 424
Contusions, 42, *237*–239
Coracoacromial ligament, 117
Corns, *369*
Corsets, 34
Cortisone, 36, 140
Coughlin, Chris, 71
Cowens, Dave, 15, 22, 26, 344, 346
Crackback blocking, 61
Croquet
children injured playing in U.S., 447
hand injury in U.S., 185
head injury in U.S., 391
Cybex isokinetic machine, 22, 58
for ankle fracture, 324
description of, 74–75
isokinetic exercise on, 76
for rotator cuff, 116

Darts
foot and toe injury in U.S., 381
hand injury in U.S., 185
head injury in U.S., 391
wrist injury in U.S., 167
Dart thrower's elbow, 133, 134
Dean, Dizzy, 24
Decker, Mary, 307
Dehydration, 401
DeLorme, Thomas, 76
Deltoid muscle, 102
exercise for, 105
Denny, John, 344
Depomedrol, 125
DeQuervain's tendinitis, 160, *161*, 162
Diathermy, 50, 204
Discectomy, 211
Disc, slipped
See also Cervical disc disease, acute; Spine, cervical; Spine, lumbar
Dorsiflexion, 310, 316
Duran, Roberto, 17
Dutoy staple procedure, 113

Ears, 18
See also Head
Ecchymosis, 43
Eckersley, Dennis, 22, 219
Elbow, 33
anatomy of, 128, *129*
bones of, *128*
bursitis, *133*–134
chips in, 141

exercises for, *133*
injury to according to sport, 152–153(*t*)
Little League, 148, *149*
loose bodies in (joint mouse), 141–143
Panner's disease, 141
pitcher's, 148, *149*
purpose of, 128
radial head fracture in, 143, *144*, 145
sprain, 131–*132*
tennis, *134*–141
ulnar neuritis, *146*–148
valgus stress in, 141
Electrical stimulation, 50–*51*
Electromyogram, 195
Elevation, 37, 41
See also RICE
Ellison procedure, 258
Endurance
definition of, 73
and fat reserves, 74
and hyperthermia, 74
and hypoxia, 74
muscle, 73–74
and sugar, 74
and training, 74
Environment and injury
eyeguards, 67
football, 64
ice hockey, 65
running, 64
skiing, 64–65
soccer, 64
uneven surfaces, 64
Equipment, protective
bras, 63
face masks, 63
for females, 63
head gear, 63
industry for, 63
and injury, 62–63
list of according to sport, 64
mouthpiece, *63*
in schools, 63
tape, 64
Ethyl chloride, 80
Etonic stabilizers, 54
Evans, Dwight, 75, 388
Exercise
Achilles tendon, 329
aerobic, 73
ankle, 316, *323*
eating during, 402
elbow, *133*
Gallup poll on, 12
health studies in, 16
hip, *230*
isokinetic, 76
isometric, 76
isotonic, 76
knee, 262–*263*
Kocher maneuver (dislocated shoulder), 111
lower leg, 302
and muscles. *See* Muscles
neck, *192*
pendulum (shoulder), *104*
progressive resistance, 76
pulley (shoulder), *125*
stretching, 56–67
for tennis elbow, *138*
thigh, *239*
wall walk (frozen shoulder), *124*
warming up, 57
Williams' flexion (back), 208–209
wrist, *157*
Eye guard manufacturers, list of, 67
Eyes, 18
See also Face

Face, 18
See also Head
lacerations, 384
Face masks, 63
Facet joint arthropathy, 221
Facilities, sports, 31
Fadden, Jack, 44
Falmouth Road Race, 12, 62
Fast twitch fibers, 71
Fats, 74, 394, 395, 402
Femur, (thighbone), 227, 234, 244
Fencing
head injury in U.S., 391
protective equipment for, 64
shoulder injury in U.S., 127
wrist injury in U.S., 167
Fibers, 70–71
Fingers, 8
baseball, *181*
Bennett's fracture (thumb), 178
dislocation, *174*–175
flexor muscle in, 71
fracture, 176, *177*, 178
fracture, healing time, 42
jersey, 182
mallet, 182
sprain, 173–176
Fisk, Carlton, 22, 40, 52, 170, 213
Flanagan, Mike, 81
Flat feet, 304
Flexibility, 75
Fluoroscope, 116
Foot
See also Shoes; Toes
accessory navicular, 352–353
anatomy of, 336, *337*, 338, 339
arch, 341
black toenail, *371*–372
blisters, *370*–371
bunion, *360*–361
bunionette, *372*–373
calluses, *367*–368
cavus, 353–355
corns, 369
dorsal exostosis, *356*–357
examination of, 342
extensor tendinitis, 357–358
forefoot, 336, 338, 360
heel fracture, *350*
high-impact, 346–347
hindfoot, 336
inferior calcaneal bursitis, *349*–350
ingrown toenail, *364*–365
injury to according to sport, 380–381(*t*)
ligaments of, 338–*339*
metatarsal fractures, *373*–374
midfoot, 336, 351–353
Morton's neuroma, 362–353
muscles of, 340
nerves of, 340
plantar fasciitis, 62, 343–*345*
retrocalcaneal bursitis, 347–*348*
sesamoiditis, *365*–*366*
turf toe, 362
Football
ankle injury in U.S., 334
calf strain, 302
cervical sprain in, 302
chain gang in, 65
children injured playing in U.S., 446
demands on the knee in, 297
elbow injury in U.S., 152
field condition in, 65
first down markers in, 65
foot and toe injury in U.S., 380
goal posts in, 65
hamstring injury, 242
hand injury in U.S., 184
head injury in U.S., 390
injuries in, 20, 25(*t*)
neck, 198
popularity in U.S., 14
protective equipment for, 64–65
shoulder injury in U.S., 126
and strength training, 57–59
Foramina, 189
Ford, Gerald, 13
Fracture calluses, 92
Fracture hematoma, *92*
Fractures, 32, 38
aligning of, 95, *96*
ankle, 320–324
carpal navicular (wrist), *162*–165
causes of, 87
cervical, 198–199
in children, 427, 429, 436
of clavicle (collarbone), *108*–110
comminuted, *91*
complete, 89–90
detection of, 93–94
greenstick, *91*–*92*
heel, *350*
hematoma, *92*
immobilization of, 94, 97
mending of, 92, *93*, 98
metatarsal, *373*–374
oblique, *91*
open, *90*
pins for, *99*
plates for, 99
radial head (elbow), 143–145
screws for, *99*
spiral, *91*, 92
splints for, 94
stress, *88*–89, *308*–309
surgery for, 97–100
time required for healing by location, 98(*t*)
traction for, *96*

transverse, *91*
treatment of, 93–97
Fredrickson, Tucker, 266
Frozen shoulder, 124–125
Fused foot, 445

Galen, 27
Gamekeeper's thumb, 174
Gardner, Randy, 22
Garland, Wayne, 119
Gastrocnemius muscle. *See* Calf muscle
Gerberlich, Susan, 134
Gluteus maximus muscle, 228
Goldthwaite procedure, 270
Golf
ankle injury in U.S., 335
children injured playing in U.S., 446
elbow injuries in U.S., 153
foot and toe injury in U.S., 381
hand injury in U.S., 185
head injury in U.S., 390
popularity of, 14
shoulder injury in U.S., 127
wrist injury in U.S., 167
Giustina, Victor Della, 63
Gray, Susan, 112
Griffith, Edward, 58
Grimditch, Wayne, 15
Gymnastics
ankle injury in U.S., 334
children injured playing in U.S., 446
elbow injury in U.S., 152
foot and toe injury in U.S., 380
hand injury in U.S., 152
head injury in U.S., 390
shoulder injury in U.S., 126
wrist injury in U.S., 166

Hamstring muscle
anatomy of, *235, 240*
exercise for, *242, 243*
healing time for pull, 42
strain (pull), 239, 241–243
Handball
ankle injury in U.S., 335
children injured playing in U.S., 447
elbow injury in U.S., 153
foot and toe injury in U.S., 381
hand injury in U.S., 185
head injury in U.S., 391
shoulder injury in U.S., 127
wrist injury in U.S., 167
Hands
anatomy of, 168, *169*, 170
Bennett's fracture, *178*
bones of, 168, 169, 170
Boutonnière deformity, *182*, 183
contusions, 176
finger dislocation, *174–175*
finger fracture, 176, *177, 178*
finger sprain, 173–176
flexor, tendinitis of, 180
function of, *170*
injury to according to sports, 184–185(*t*)
muscles in, 171
nerves in, *172*
in sports, 168
tendons in *171, 179*–180
tendon laceration in, 181
treatment of injury to, 172–173
Hargrove, Mike, 156
Harper, Tommy, 60
Haycock, Christine, 32
Head
concussions, 386–387
epidural hematoma, *387*–389
face lacerations, *384*–386
injury to according to sport, 390–391(*t*)
Healing. *See* Rehabilitation
Health supervision, 31
Heart
and aerobic exercise, 73
strengthening of, 59
Heat, 37–38
and hydrocollators, 49
and short-wave diathermy (SWD), 49–50
and ultrasound, 49
Hematoma, 237
epidural, 387–389
Heridocus, 27
Heryer, John, 20
Hibiclens, 371
"High sticking," 62
Hill, Priscilla, 75
Hip
anatomy of, 227, *228*
bursitis of, *231–233*
exercises for, *230*
pointers, *224*–225
muscles in, 228, *229*
Hippocrates, 27
Hip pointers, *224*–225
Hisle, Larry, 116
Hobson, Butch, 22, 142
Hoffman, Glen, 20, 22, 204
Holmes, Larry, 17
Horseback riding
ankle injury in U.S., 334
elbow injury in U.S., 152
foot and toe injury in U.S., 380
hand injury in U.S., 184
head injury in U.S., 390
shoulder injury in U.S., 126
wrist injury in U.S., 184
Horseshoes
children injured playing in U.S., 447
foot and toe injury in U.S., 381
hand injury in U.S., 185
head injury in U.S., 391
Hriniak, Walter, 103
Humeral bone cyst, *444*
Hunter, Harlan C., 34
Hydrocollators
description of, *49*
uses of, 49
Hyperplasia, 71
Hypertrophy, 71
Hypoglycemia, 405(*t*)

Ice, 37
See also RICE
for bursitis, 121, 133
for common injuries, 40
Ice hockey
ankle injury in U.S., 334
children injured playing in U.S., 446
demands on knee, 297
elbow injury in U.S., 152
foot and toe injury in U.S., 380
hamstring injury in, 240
hand in U.S., 184
head injury in U.S., 390
high sticking in, 62
lumbar spine strain in, 202
popularity in U.S., 14
shoulder injury in U.S., 126
wrist injury in U.S., 166
Ice skating. *See* Skating, ice
Idiosyncrasy, 47
Iliopsoas muscle, 228
Inflammation, 38
Interpositional arthroplasty, 445
Intervertebral discs, 186
Ions, 51
Iontophoresis, 51
Iron, 400, 402
Ischium, 223
Isokinetic exercise, 76
Isometric exercise, 76
Isotonic exercise, 76

James, Stan, 343
Jessup, George, 120
Joints
deterioration from sports, 33
evaluating injury to, 41
facet, *188*
fibrous, 81
healing of, 84
immovable, 81
knee, 82, *83*
subtalar, 336
synovial, 81, *82*, 83, 84
types of, 81
Jumper's knee, *281*–283
Junk food, 397

Kickball, stickball, tetherball
ankle injury in U.S., 334
children injured playing in U.S., 446
elbow injury in U.S., 152
foot and toe injury in U.S., 381
hand injury in U.S., 184
head injury in U.S., 390
shoulder injury in U.S., 126
wrist injury in U.S., 166
Kneecap (patella)
dislocation, 268–271
instability, 274

patella chondromalica, *271*–280
subluxation, *268*–271
surgery, 275
Knees, 33
anatomy of, 244, *245*, *246*
anterior cruciate ligament injury, *253*–258
Baker's cyst, *286*–288
bursitis of, 283–*284*
cartilage injury, *259*–260
causes of knee pain, 247
demands on according to sport, 297(*t*)
exercise for, 262–263
extensor mechanism of, *267*–268
goose-foot bursitis, 285–286
joint cartilage and blood supply, 43
joint sleeve, *82*
jumper's (upper patella tendinitis), *281*–283
kneecap. *See* Kneecap
ligaments of, 246–247
ligament sprain of, *248*–249
ligament tears, *84*
locked, *261*
meniscus injury of, *83*, 259–266
Osgood-Schlatter's disease, 292, *293*, 294. *See also* Children
osteochondritis dissecans, 288, *289*–292
posterior cruciate injury, *258*–259
prepatella bursitis, 284–285
protection against injury to, 249
purpose of, 244
runner's, 294, *295*, 296, 297
sprain: 1st degree, 250; 2nd degree, 250–251; 3rd degree, 252
supports for, 270
surgery for, 265, 270
symptoms of knee problems, 247
treatment for knee pain, 147
Kocher maneuver, 111
Köhler's disease, *438*, 439
Korbut, Olga, 213, 214
Koufax, Sandy, 36
Kupchak, Mitch, 43, 221

Labrum, 110
Lacerations, *384*–386
Lacrosse
ankle injury in U.S., 335
children injured playing in U.S., 447
elbow injury in U.S., 153
foot and toe injury in U.S., 381
hand injury in U.S., 185
head injury in U.S., 391
protective equipment for, 64
shoulder injury in U.S., 127
wrist injury in U.S., 167
Lactic acid, 73, 74, 404
Lally, John, 56
Lamina, 210
Laminectomy, 210–211
Lateral humeral epicondyle, 131
Laurie, Suzanne, 122
Layne, Bobby, 44
Leadbetter, Wayne, 62
Lee, Bill, 116, 360
Legg-Perthes syndrome, *438*, 439
Leg, lower, 88, 90, 244
Achilles tendon. *See* Achilles tendon
calf muscle. *See* Calf muscle
exercises for, 302
shin splints, 303, *304*, *305*, 306, 307
stress fractures, *308*–309
Lenox-Hill derotation brace, *256*
Leonard, Sugar Ray, 17
Le Roux, Buddy, 15
Ligaments
anterior, cruciate, *253*–258
anterior talofibular, *313*
and blood supply, 43
calcaneal fiber (ankle), 312
coracoacromial, 117
deltoid, 312
foot, 338
healing process of, 84
injury to, 84
intertubercular, 122
knee, 246–247
and local anesthetics, 85
posterior cruciate, *258*–259
stout, 188
treatment of injury to, 84–85
Ligamentum flavum, 188
Little League elbow, 148, *149*
Lumbar myelogram, 210
Lumbosacral junction, 223
Lynn, Fred, 20, 44, 241, 250, 368

McBride procedure, 361
McGregor, Rob Roy, 13, 15, 344, 345–347, 360–361, 374–378
McMaster, J. H., 213
McMurray test, 263
Magnusen-Stack procedure, 113
Marr, David, 17
Martial arts
ankle injury in U.S., 335
children injured playing in U.S., 447
elbow injury in U.S., 153
foot and toe injury in U.S., 380
hand injury in U.S., 185
head injury in U.S., 391
shoulder injury in U.S., 127
wrist injury in U.S., 167
Melchiorre, Ray, 64
Membrane permeability, 49
Metacarpals, 168, 170
Micro-trauma, 327
Minerals, 411
Mintz, Ida, 35
Moore, Lenny, 317
Mortice, 310
Morton's Neuroma, 363, 365
Moss, Charlie, 40
Motorbiking
ankle injury in U.S., 334
children injured with in U.S., 446
elbow injury in U.S., 152
foot and toe injury in U.S., 380
hand injury in U.S., 184
head injury in U.S., 390
popularity in U.S., 14
shoulder injury in U.S., 126
wrist injury in U.S., 166
Mountain climbing
ankle injury in U.S., 335
children injured playing in U.S., 447
hand injury in U.S., 185
head injury in U.S., 391
shoulder injury in U.S., 127
Mouth, 18
See also Face; Head
Mouthpiece, protective, 63
Multipennate muscles, 71, *72*
Muscles, 45
adenosine triphosphase in, 72
arm, 131
in athletic events, 70, 71
biceps, 70, 131
biopsy of, 71
and blood flow, 72
and blood supply, 43
calf, *71*
and coordination, 74
definition of, 70
depletion of, 72
fast twitch fibers, 71
and fibers, 70–71
fibrils in, 72
flexor, 71
foot, 340
function of, 70–73
hand, 171–172
hip, 228–229
and hyperplasia, 71
and hypertrophy, 71
injury of, 77–79
involuntary, 70
micro-tears in, 75
multipennate, 71, *72*
and muscle sugar, 72, 73
penniform, 71, *72*
rehabilitation of, 79–81
shoulder, 102–104
size of, 71
slow twitch fibers, 71
smooth, 70
and speed, 74
spinal, 189–190
and strains, *78*, 89
striated, 70
tears in, 77
training of, 74, 75–77
unipennate, 71, *72*
voluntary, 70
wrist, 156
Myosin, 72
Myositis ossificans, 238

Nautilus equipment, 58
for neck, *200*
for shoulder, 105
Navratilova, Martina, 17
Nayorga, William, 93

Neck
broken, 198
cervical disc surgery, 199
and cervical spondylosis, 196–197
football, 198
Nautilus machine for, *200*
Nerve
arm, 131
brachial plexus, 102
cervical, 102
foot, 340–341
hip, 228–229
median, 102, 131, 156, 172
pinched, 196
radial, 102, 131, 156, 172
spinal, *189*–190
ulna, 102, 131, 156, 172
wrist, 156
Neuritis and aspirin, 46
Niacin, 402
Nicholas, James, 27
Novocain, 80, 85, 140
Nucleus pulposus, 186
Nutrients, 42
Nutrition
in alcoholic drinks, 415(*t*)
beer, 413–414
calcium, 402
calories. *See* Calories
carbohydrates, 394–395, 402, 404–407
for children, 399–400
cholesterol content in food, 395
coffee, 412
cola drinks, 397
eating during exercise, 402
energy sources, 394–396
evaluation of nutrients, 402–403(*t*)
fats, 395, 402
four-food plan, 398(*t*)
fundamental nutrients, 393
importance of, 392
iron-rich foods, 400(*t*)
junk food, 397
lactic acid in food, 402
liquid meals, 403
minerals, 411
number of meals, 396–397
pregame meals, 400–401
protein, 395–396, 402
recommended daily allowance (RDA), 397–398
riboflavin, 402
salt content in food, 416(*t*)
salt tablets, 415, 416
thirst quenchers, 414
and vegetarians, 407–408
vitamins. *See* Vitamins
water consumption, 411, 412
for weekend athletes, 399
weight gain, 421–422
weight loss, 417–420
zinc, 407

Officiating, 31
Open fracture, 90
O'Reilly, Terry, 385
Orthotics, 377–379
Orthotron machine, 76
Osgood-Schlatter's disease, 292, *293*, 294, *439*
Osteoblasts, 237
Ostoechondritis dissecans, 288–292, *325*–326, 440–441
Oxygen, 42, 74

Pain
description of, 51–52
evaluating, 41
guidelines for, 53
Palmer, Jim, 24
Panner's disease, 141
Pappas, Arthur M., 15, 48, 70, 424
Park, Brad, 15, 265
Patella. *See* Kneecap
Patella chondromalica
causes, 272
due to instability, 274
idiopathic, 279–280
synovial plica, *277*–278
traumatic, 276
treatment, 273–274
types of, 274–280
Pelvis
anatomy of, 222–*223*
bone avulsion injury in, *225*–227
purpose of, 222
Penniform muscles, 71, *72*
Periosteum, 237
Periostitis, tibial, 305–306
Peroneal muscles, 316, 319
Peroneal spastic flatfoot, 351
Perthes disease, *438*, 439
Petrocelli, Rico, 135
Phisohex, 371
Phlebitis, 46
Physicians in sports, responsibility, of, 65–66
Ping pong. *See* Tennis, table
Pitcher's elbow, 148, *149*
Plantar fasciitis, 62, 343–*345*
Plantar flexion, 310
Playground equipment
ankle injury in U.S., 334
children injured playing in U.S., 446
elbow injury in U.S., 152
foot and toe injury in U.S., 380
hand injury in U.S., 184
head injury in U.S., 390
shoulder injury in U.S., 126
wrist injury in U.S., 166
Plica, 278, 279
Polio, 87
Poquette, Tom, 114
Positive ankle drawer sign, 319
Priest, James D., 134
Progressive Resistance Exercise (PRE), 76
Pronation, 341
Proprioception, 52
Protein, 72, 395–396, 401, 402
Pubic symphisis, 223
Pubis, 223
Puttyplat procedure, 113
Pyruvate, 73

Q-angle test, 269
Quadriceps muscle, 234, *235*, 236, 244, 268
Quigley, Thomas B., 44

Racquetball, paddleball, squash
ankle injury in U.S., 334
children injured playing in U.S., 447
elbow injury in U.S., 152
foot and toe injury in, 380
hand injury in U.S., 166
head injury in U.S., 390
hip injury in, 231
popularity of, 14
shoulder injury in U.S., 152
wrist injury in U.S., 166
Rainy, Chuck, 22, 151
Ratelle, Jean, 210
Reagan, Ronald, 13
Rehabilitation
for Bennett's fracture (thumb), *178*–179
and blood supply, 42–43
and bone avulsion in pelvis, 226–227
from broken bones, 38, 94–97
and cervical traction, 51
for collarbone fractures, 109
for contusions, 237–238
of DeQuervain's tendinitis, 161–162
for dislocated shoulder, 111–113
and electrical stimulation, 50–51
guidelines for resuming activity after, 45–46
healing rate (children vs. adult), 36–37
and hydrocollators, 49
for muscle injury, 79–81
and physical therapists, 47
of plantar fasciitis, 344–*345*
and RICE, 37
of rotator cuff tendinitis, 116, 117
and short-wave diathermy, 49–50
for shoulder strains, 103–16
and surgery. *See* Surgery
techniques for, 48
time need for according to injury, 42,43
of torn cartilage, 261–266
and transcutaneous electrical nerve stimulation, 50
and ultrasound, 48–49
when to start it, 47
Remy, Jerry, 15, 20
Rest, 37, 40
See also RICE
Reuthven, Dick, 15
Riboflavin (B_2), 402
Rice, Jim, 20, 76, 77
RICE treatment (*R*est, *I*ce, *C*ompression, *E*levation), 37

ankle injury, 316
calf strain, 302
common injuries, 40–41
contusions, 237
fractured toe, 368
hand injury, 173
knee injury, 249
ligament injury, 84
muscle injury, 79
thigh injury, 236, 241
Road Runners Club, 47
Rodgers, Bill, 15
Roosevelt, Theodore, 61
Rotator cuff, 79
Cybex machine for, 116
diagnosis of tendinitis in, 115–116
exercises for, 105, *118*
muscles of, 114, *115*
tendinitis, 114, 115, 116, 117
tendons of, 115
Rote, Kyle, Jr., 75
Rugby
ankle injury in U.S., 335
children injured playing in U.S., 447
elbow injury in U.S., 153
foot and toe injury in U.S., 381
hand injury in U.S., 185
head injury in U.S., 391
shoulder injury in U.S., 127
wrist injury in U.S., 167
Rules and injury
changes in, 61–62
in football, 61
in hockey, 62
in running, 62
Running, 34
and Achilles tendinitis, 326–327
biomechanical factors of, 341–343
and hip pointers, *224*–225
"hitting the wall," 73
muscles used in, 71, 72
popularity in U.S., 14
runner's knee, 294–297
showers during, 62
Ryan, Allan J., 27

Sacralization, 219
Sacrum, 223
St. Onge, Richard A., 15, 71
Salicylate poisoning, 47
Sayers, Gayle, 22
Salt tablets, 415–416
Scapula. *See* Shoulder blade
Sciatica, 210
Scott, George, 38, 213, 215
Semon, R. L., 213
Separated shoulder. *See* Shoulder
Sesamoiditis, 365–366
Shea, William D., 15
Shin pads, 63
Shin splints
anterior compartment syndrome, *306*–307
causes of, 303
definition of, 303
posterior tibial, *304*–305, 358
tibial periostitis, *305*–306
Shoes
criteria for selection of, 374–*375*
orthotics for, 377–379
running, rating of by manufacturer, 376
Short-wave diathermy (SWD), 49–50
Shoulder, 33
anatomy of, *101*
blade, 100, *101*
bones of, 100
bursa sac, *120*
bursitis, 120–122
collarbone, 100, *101*, 108
diagnosis of separation, 106
exercises for, *104*, *105*
frozen (adhesive capsulitis), 124–125
impingement syndrome, 120
injury to according to sport, 126–127(*t*)
joint, 100, 101
muscles in, 102–104
partial dislocation of, 113–114
separation, 42, 104, *106*–108
strains, 102–104
Shoulder blade, 100, *101*
Shoulder joint, 100, *101*
Shuffleboard
ankle injury in U.S., 335
foot and toe injury in U.S., 381
hand injury in U.S., 185
head injury in U.S., 391
Skateboarding
ankle injury in U.S., 334
children injured playing in U.S., 446
elbow injury in U.S., 152
foot and toe injury in U.S., 380
hand injury in U.S., 184
head injury in U.S., 390
shoulder injury in U.S., 126
wrist injury in U.S., 166
Skating, ice
ankle injury in U.S., 334
children injured playing in U.S., 446
elbow injury in U.S., 152
foot and toe injury in U.S., 381
hand injury in U.S., 185
head injury in U.S., 390
popularity of, 14
shoulder injury in U.S., 127
wrist injury in U.S., 166
Skating, roller
ankle injury in U.S., 334
children injured playing in U.S., 446
elbow injury in U.S., 152
foot and toe injury in U.S., 380
hand injury in U.S., 184
head injury in U.S., 390
popularity in U.S., 14
shoulder injury in U.S., 126
wrist injury in U.S., 166
Skiing, snow
ankle injury in U.S., 334
demands on knee in, 297
elbow injury in U.S., 152
environment for and injury, 64–65
foot and toe injury in U.S., 380
hand injury in U.S., 184
head injury in U.S., 390
muscles used in, 71
popularity in U.S., 14
shoulder injury in U.S., 126
wrist injury in U.S., 166
Skiing, water
ankle injury in U.S., 184
children injured playing in U.S., 447
elbow injury in U.S., 153
foot and toe injury in U.S., 380
hand injury in U.S., 184
head injury in U.S., 390
popularity in U.S., 14
shoulder injury in U.S., 127
wrist injury in U.S., 167
Skin, 45, *382*
abrasions, 383
injuries to, 382–383
lacerations, *384*–386
Sky diving and hang gliding
ankle injury in U.S., 335
elbow injury in U.S., 153
foot and toe injury in U.S., 381
hand injury in U.S., 185
head injury in U.S., 391
shoulder injury in U.S., 127
wrist injury in U.S., 167
Sledding
anklc injury in U.S., 334
children injured playing in U.S., 446
elbow injury in U.S., 126
foot and toe injury in U.S., 380
hand injury in U.S., 184
head injury in U.S., 390
shoulder injury in U.S., 126
wrist injury in U.S., 166
Slow twitch fibers, 71
Smillie pins, 291
Smith, Elmore, 163
Snowmobiling, 14
Soccer
ankle injury in U.S., 334
calf strain, 302
children injured playing in U.S., 446
demands on knee in, 297
elbow injury in U.S., 152
foot and toe injury in U.S., 380
hamstring injury, 242
hand injury in U.S., 184
head injury in U.S., 390
popularity in U.S., 14
protective equipment for, 64
shoulder injury in U.S., 126
soccer neck, 191
wrist injury in U.S., 166
Sodium loss, 415
Softball, 14
Spearing, 61
Spengler, D. M., 213
Spine, cervical
See also Spine, lumbar

acute cervical disc disease, 193–196
anatomy of, 186, *187*, *188*
cervical burners, 192–193
and cervical spondylosis, 196
definition of, 188
disc surgery for, *199*
exercises for neck, *192*
facet joint, 188
fracture, 198, 201
nerves of, *189*–190
sprain, 190–192
traction for, *195*
Spine, lumbar
See also Spine, cervical
anatomy of, 189
exercises for, *208*
facet joint of, *188*
fusion of, *216*, 219, *220*, 221
and low back pain, 217–218
myelogram for, 210
rules for care of, 209
slipped disc, *205*, *206*, 207
and spondylolisthesis, 212, *213*–217
and spondylolysis, *212*–217
strain of, 202, *203*, 204, 205
Spine, thoracic, 188
Spinous process, 186
Splints
air, *94*
hand, 174
pillow, *94*
wood, *94*
Spondylolysis
cervical, 196–197
lumbar, *212*–217
Spondylolisthesis, 212, *213*–217
Sponge rubber donut, Harvard, *315*
Sports
See also specific sport
ankle injury according to, 334–335(*t*)
caloric consumption according to, 416–417(*t*)
demands on knee according to, 297
elbow injury according to, 152–153(*t*)
head injury according to, 390–391(*t*)
health studies on, 16
injuries. *See* Sports injuries
and joint deterioration, 33
most popular in U.S., 14
Nielsen survey on (1979), 12
selection of, 54–55
shoulder injury according to, 126–127(*t*)
thumb injury according to, 184–185(*t*)
wrist injury according to, 166–167(*t*)
Sports injuries, 18
according to age, 18, 20
according to level of competition, 30
amateur vs. professional, 30
ankle, according to sport, 334–335(*t*)
in baseball, 21(*t*)
in basketball, 23(*t*)
in bicycling, 18, 19(*t*)
and cortisone, 36
declining rate of, 35
elbow, according to sport, 152–153(*t*)
and environment. *See* Environment and injury
foot, according to sport, 380–381(*t*)
in football, 25(*t*)
guidelines for determining need for a doctor, 41–42
guidelines for resuming activity, 45–46
hand, according to sport, 184–185(*t*)
head, according to sport, 390–391(*t*)
heat for, 37–38
listed by sport, 28–29(*t*)
most injured parts of the body, 18
to muscles. *See* Muscles
need for immediate treatment, 37
and playing into shape, 39
prevention of, 53–59
proneness to early in season, 34
protection for school and college athletes, 31
rehabilitation after, 30, 32
and RICE. *See* RICE
shoulder, according to sport, 126–127(*t*)
and surgery. *See* Surgery
treating common injuries, 40–41
women vs. men, 32–33
wrist, according to sport, 166–167(*t*)
Sports medicine
in ancient times, 27
growth of, 27
Sports Medicine Resource (SMR), Inc., 15, 27
Sports safety, 18
bill of rights for, 31
Sprains
ankle, 43, 312–317
cervical spine, 190-192
elbow, 131–*132*
knee ligament, *248*–249
as ligament injuries, 84
and ultrasound, 49
wrist, 156, 159
Stapleton, Dave, 171
Stevens, Greer, 255
Strains, 77, *78*
hamstring, 239, 241–243
of the lumbar spine, 202–205
predisposing factors for, 78–79
of the shoulder muscles, 102–103
thigh, 234–236
Straight leg raising, 207
Strength training, 57–59
Stretching
four important exercises for, *57*, *58*
how should it feel, 56
importance of, 56
and prevention of injury, 54
rules for, 57
and warming up, 57
Structural scoliosis, 219
Subtalar joint, 336, 346
Sugar
consumption in U.S., 394
muscle, 72–74, 404
Supination, 341
Surfing
ankle injury in U.S., 335
children injured playing in U.S., 447
elbow injury in U.S., 153
foot and toe injury in U.S., 381
hand injury in U.S., 185
head injury in U.S., 391
shoulder injury in U.S.,127
Surgery, 39
Achilles tendon, 329
ankle, 319–320
anterior compartment syndrome, 307
anterior cruciate ligament, 307
bunion, 361
cartilage removal, 265
for cervical disc, 199
for chondromalicia, 275
for fractures, 97–99
lumbar disc, 210–211
rotator cuff, 117
for shoulder separation, 106–108
spinal fusion, 217
for tennis elbow, 140
Swimming
ankle injury in U.S., 334
elbow injury in U.S., 152
foot and toe injury in U.S., 380
hand injury in U.S., 184
head injury in U.S., 390
popularity in U.S., 12, 14
and recovery from injury, 44
shoulder injury in U.S., 126
and swimmer's shoulder, 117
wrist injury in U.S., 166

Taping
of Achilles tendon, *64*
ankle, *317*, 319
of fingers, 174
of hamstring, 242
importance of, 64
of wrist, *159*–160
Tapper, Edward, 35
Tarsal coalition, 351, 444
Tears
micro, 75
muscle, types of, 77
Tendinitis
of Achilles tendon, 326–329
and aspirin, 46
of biceps, 122–124
DeQuervain's (wrist), 160–162
extensor, 357–358
flexor (hand), 180
jumper's knee (upper patella), *281*–283

Osgood-Schlatter's disease, 292–299
posterior tibial, 358–359
of the shoulder, 120–122
and ultrasound, 49
Tendon, 45
abductor pollicis longus, 160
Achilles, 62, 64
avulsion injury, 181
description of, 70
extensor pollicis brevis, 160
injury to, 79
laceration of, 181
sliding, 160
Tennis
ankle injury in U.S., 334
elbow injury in U.S., 152
foot and toe injury in U.S., 380
hand injury in U.S., 184
head injury in U.S., 390
leading money winners in, 17(*t*)
popularity in U.S., 14
shoulder injury in U.S., 127
and tennis elbow, *135*–141
wrist injury in U.S., 166
Tennis elbow (lateral humeral epicondylitis)
anatomy of, *135*
Aufranc method for, 140
causes, 135, 137
surgery for, 140
treatment for, 136(*t*), *137*, 138–141
Tennis, paddle, 14
Tennis, platform, 14
Tennis, table
ankle injury in U.S., 335
elbow injury in U.S., 153
foot and toe injury in U.S., 381
hand injury in U.S., 185
head injury in U.S., 391
shoulder injury in U.S., 127
Testosterone, 71
Thiamine (B_1), 402
Thigh
anatomy of, *235*
contusions, 42, *237*–238
excercise for, *239*
hamstring, 42, 235, 239–243
quadriceps muscle, 234–235
strain, 234, 235, *236*
Thighbone. *See* Femur
Thirst quenches by calorie, 414(*t*)
Thornton, Andre, 260
Thrombophlebitis, 302
Tiant, Luis, 53, 175
Time reaction variable, 70
Toe, 388
black toenail, *371*–372
broken, healing time for, 42
fracture, 368–369
ingrown toe nails, *364*–365
Morton's neuroma, 362–365
turf, 362
Track and field
ankle injury in U.S., 334
children injured playing in U.S., 446
elbow injury in U.S., 153
foot and toe injury in U.S., 380
hand injury in U.S., 185
head injury in U.S., 391
protective equipment for, 64
shoulder injury in U.S., 126
wrist injury in U.S., 167
Training
and endurance, 73, 74
flexibility, 75
muscle, 74, 75–77
risks of overtraining, 55–56
strength, 57–59, 76
stretching, 56–57
warming up, 57
Trampoline
ankle injury in U.S., 355
elbow injury in U.S., 152
foot and toe injury in U.S., 381
hand injury in U.S., 185
head injury in U.S., 391
shoulder injury in U.S., 127
wrist injury in U.S., 167
Transcutaneous electrical nerve stimulation (TENS)
advantages of, 50
description of, *50*
Triceps muscle, 131
Trigger points, 197
Tubercle, 281
Turf toe, 362

Ulna, 154
Ular neuritis, *146*–148
Ultrasound
description of, 48–49
and heat, 49
Unipennate muscles, 71, *72*
United States
elbow injury in according to sport, 152–153(*t*)
most popular sports in, 14(*t*)
shoulder injury in according to sport, 126–127(*t*)
sports activities in, 12–13
sports injury in, 20, 35
Universal gym equipment, 58
for shoulder rehabilitation, 105

Valium, 194
Vegetarians, 407–408
Vertebral elements, 186
Visual neural pathway, 70
Vitamins
A, 402, 409
B_1, 402, 408, 409
B_2, 402
C, 402, 408, 410, 411
D, 408
deficiencies, 410
need for, 408
supplements, 411
Volleyball
ankle injury in U.S., 334
elbow injury in U.S., 152
foot and toe injury in U.S., 380
hand injury in U.S., 334
head injury in U.S., 390
shoulder injury in U.S., 126
wrist injury in U.S., 184

Waddy, Billy, 24
Walking, biomechanical factors of, 341–343
Warming up, 57
Water consumption, 411, 412
Water polo
hand injury in U.S., 185
Watson-Jones procedure, 319–320
Weight gain, 421, 423
Weight lifting
ankle injury in U.S., 335
elbow injury in U.S., 153
foot and toe injury in U.S., 380
hand injury in U.S., 184
head injury in U.S., 390
shoulder injury in U.S., 126
wrist injury in U.S., 166
Weight loss, 417, 418, 419, 420
Weights, dead, 58
Whiplash, 190
Williams' flexion exercise, *208*–209
Williams, Ted, 32, 109, 145
Wrestling
ankle injury in U.S., 334
demands on knee in U.S., 297
elbow injury in U.S., 126
foot and toe injury in U.S., 380
hand injury in U.S., 184
head injury in U.S., 390
protective equipment for, 64
shoulder injury in U.S., 126
wrist injury in U.S., 166
Wrist
anatomy of, 154, *155*, 156
bones of, 154
exercises for, *157*, 159
fracture of, *162*–165
in hockey, 62
injury to according to sport, 166–167(*t*)
muscles in, 156
sprain, 156–*158*
taping, 159
tendinitis (DeQuervain's), 160, *161*, 162

X-rays
ankle, 319, 322, 325
elbow, 142, 144, 145
finger, 174, 176
foot, 351, 356, 363
for fractures, 89, 91, 92
neck, 201
shoulder, 103, 106, 107, 116
skull, 389
spine, 191, 194, 212
Xylocaine, 80, 85

Yastrzemski, Carl, 15, 97, 157, 328

Zimmer, Don, 54
Zinc, 407, 408

Notes

Notes

Notes

Notes